MICROBIOLOGY OF THE AGRICULTURAL SKY

MICROBIOLOGY OF THE AGRICULTURAL SKY

by

K. R. Krishna, PhD

AAP APPLE ACADEMIC PRESS

First edition published 2025

Apple Academic Press Inc.
1265 Goldenrod Circle, NE,
Palm Bay, FL 32905 USA

760 Laurentian Drive, Unit 19,
Burlington, ON L7N 0A4, CANADA

CRC Press
2385 NW Executive Center Drive,
Suite 320, Boca Raton FL 33431

4 Park Square, Milton Park,
Abingdon, Oxon, OX14 4RN UK

Library and Archives Canada Cataloguing in Publication

CIP data on file with Canada Library and Archives

Library of Congress Cataloging-in-Publication Data

CIP data on file with US Library of Congress

ISBN: 978-1-77491-632-2 (hbk)
ISBN: 978-1-77491-633-9 (pbk)
ISBN: 978-1-03271-959-7 (ebk)

About the Author

K. R. Krishna, PhD
Agricultural Scientist, formerly at the International Crops Research Institute for the Semi-Arid Tropics (ICRISAT), India

K. R. Krishna, PhD (Agriculture), is an Agricultural Scientist. His contributions deal with soil fertility, microbiology, agroecosystems, precision farming, farm robotics, and unmanned aerial vehicle systems. He is a member of the International Society for Precision Agriculture, American Society of Agronomy, Soil Science Society of America, Ecological Society of America, Indian Society of Agronomy, and Soil Science Society of India. He has authored several books on International Agriculture. He is retired from the International Crops Research Institute for the Semi-Arid Tropics (ICRISAT), India. The author has written several books on the topic of agriculture. Some of the notable works include *the Agricultural Sky: A Concept to Revolutionize Farming*, *Aerial Robotics in Agriculture: Parafoils, Blimps, Aerostats, and Kites*, *Unmanned Aerial Vehicles in Crop Production: A Compendium*, *Agricultural Drones: A Peaceful Pursuit*, *Push Button Agriculture: Robotics, Drones and Satellite-guided Soil and Crop Management*, *Agricultural Prairies: Natural Resources and Crop Productivity*, *Agroecosystems: Soils, Crops, Nutrient Dynamics, and Productivity*, *Precision Farming: Soil Fertility and Productivity Aspects*, and *Maize Agroecosystem: Nutrient Dynamics and Productivity*. These books cover a wide range of topics related to agriculture, including innovative farming concepts, the use of aerial robotics in agriculture, unmanned aerial vehicles in crop production, agricultural drones, satellite-guided soil and crop management, natural resources and crop productivity in agricultural prairies, agroecosystems, precision farming, and nutrient dynamics and productivity in maize agroecosystems.

Contents

Acknowledgments

During the preparation of this treatise, several of my friends, colleagues, and organizations have offered scientific reports, published research papers, and photographs. Agencies such as the Food and Agricultural Organization, Rome, Italy; World Meteorological Organization, Zurich, Switzerland; National Aeronautics and Space Agency, Washington, D.C., USA, have offered permission to utilize research publications. Over the years, my efforts in Microbiology have been supported by ICRISAT, Hyderabad, India, and the University of Florida. Following are specific persons and their institutions who have offered permission to use the images and/or research material. A few of them belong to the general public domain. Such institutions have also been acknowledged.

For photographs:

- **Chapter 3:**
  - **Plate 3.1:** National Aeronautics and Space Agency, NASA/ NOAA, Washington, USA: Website National Oceanic and Atmospheric Administration: Washington D.C.
  - **Plate 3.2:** Mr. Sharath Kowligi, Eindhoven, Noord Brabant, Netherlands
- **Chapter 4:**
  - **Plate 4.1:** Dr Schmaltz. Goddard Space Station, National Aeronautics and Space Agency (NASA), Washington, D.C. United States of America
- **Chapter 5:**
  - **Plate 5.5:** World Meteorological Organization (WMO), Sand and Dust Project. Zurich, Switzerland
  - **Plate 5.6:** Mr. Sharath Kowligi and Mrs. Roopashree Basavapatna Narayan, Eindhoven, Noord Brabant, Netherlands
  - **Plates 3.3, 4.2, 5.1 to 5.5:** I wish to appreciate the International Crops Research Institute for the Semi-Arid Tropics (ICRISAT)

Hyderabad India and the ICRISAT Sahelian Centre for the research facilities during the period I served as Scientist.

- **Plates 6.1, 6.2, and 6.3:** Mr. Sharath Kowligi and Mrs. Roopashree Basavapatna Narayan, Eindhoven, Noord Brabant, Netherlands
- **Plate 7.1:** Dr. Kianian Shaharyar, Cereal Rust Laboratory, United States Department of Agriculture, Minneapolis, Minnesota, USA
- **Plate 8.1:** United States Department of Agriculture, Natural Resource Conservation Program, Reno, Nevada, USA; https://www.nrcs.usda.gov/Internet/FSE_DOCUMENTS/nrcs144p2_037150.pdf; Photo in public domain

I wish to thank Dr. Uma Krishna, Bengaluru, India, and offer my best wishes to Mr. Sharath Kowligi, Mrs. Roopashree Basavapatna Narayan, and Miss Tara Bhagirathi Kowligi of Eindhoven, Noord Brabant, Netherlands.

Preface

During the past three centuries, we have made enormous strides in our quest to understand microbes in agrarian regions. A major emphasis has been on their distribution, migration, proliferation, population density, physiological abilities, involvement in geochemical transformations of various essential nutrients, and disease causation on crops, farm animals, and humans (farmers).

During recent decades, we have understood that microbes are well distributed in the troposphere. Microbes thrive at relatively higher population density in the lower troposphere just above crops. They progressively become feeble as we climb the altitudes of the sky. Microbes in the troposphere are derived from the surface horizons of soils, natural vegetation, agricultural crops, water bodies, and animal/human activity. We may also encounter microbes in the extremes of high-altitude troposphere regions and stratosphere. Microbial ability to negotiate extreme conditions of feeble atmosphere may be of great value to us if we can transfer the traits, genetically, to symbiotic and other useful microbes found in agrarian farms. No doubt, the regular deposition of microbe-laden dust/precipitation droplets on farming belts is a natural event of significance to global agriculture. Microbes recycle naturally between agricultural soils/crops and the troposphere.

Inadvertently, we are recycling the microbes between agrarian regions and the stratosphere on a daily basis due to our ability to travel, using aircraft that fly skimming the borders of the lower stratosphere. Recently, we have learned that microbes above agricultural expanses could even be derived from the stratosphere. Or at least microbes that have successfully managed a rocket ride and stayed on the body surface of satellites or space stations.

Specific microbial species, for example, *Pseudomonas syringae, P. fluorescence*, and *P. aeruginosa,* found on aerosols, are attributed to the effect of weather systems above the Earth. They may play a role in nucleation, inducing precipitation (rain), frost, and snow.

There is a lacuna in our knowledge that must be filled to manipulate the migratory trends and paths of aerial microbes for our benefit. Microbes transit all over the globe utilizing the various aeolian processes like wind, tornadoes, hurricanes, and jet streams. Biotic agents like insects and birds aid microbial migration.

Intercontinental drifts of microbes via dust plumes/storms are a major mechanism that decides, to a certain extent, weather and precipitation patterns over agrarian expanses of other regions or even other continents. Microbes can dictate nucleation, formation of droplets, ice crystals, frost, snow, and rain. Any disturbance to a microbial load of the atmosphere and its migratory aspects is a bit risky. It has the potential to alter the agrarian environment drastically.

There are now innumerable examples that reinforce the idea that global epidemics/pandemics on crops are caused by intercontinental weather phenomena like jet streams, high-altitude winds, dust plumes, dust storms, etc. They occur in regions like the Sahara, Sahel (e.g., Harmattan), Gobi (Taklamakan), Atacama (South America), Great Plains of America, etc. Studies on crop diseases like rusts, blights, and blotches that reach epidemic proportions due to such intercontinental weather phenomena have been explained in detail. For example, dust storms aid the migration of rust fungi (e.g., *Puccinia graminis tritici*-Pgt) from Asia to Africa/West Asia, then on to the West Indies and Great Plains wheat belts. 'A dusty sky may also be a sickly sky.'

Microbial migration in the atmosphere is an important natural phenomenon of great significance to Earth systems. Microbial migration is an essential natural aspect of certain plant pathogens that cause epidemics/pandemics on crops (e.g., Fungal Rusts). Aeolian processes are intricately associated with microbial transport in air. It needs attention from farmers. Biotic factors like insects (e.g., aphids, jassids, psyllids, whiteflies, honeybees, locusts), birds, and migrating mammals aid the large-scale dissemination of microbes. They are known to disseminate plant pathogens, resulting in detrimental consequences to important food grains and horticultural crops, as well as forest plantations. There are too many examples of repeated epidemics and pandemics on crops abetted by biotic factors (i.e., vectors). Microbial migration to short distances from one plant to another within a field occurs swiftly. It is essential for the establishment of plant disease.

Microbial flora of the troposphere affects nutrient dynamics in the agroecosystem. Aerial microbes intervene at several steps during physico-chemical transformations of nutrients within the agrarian region. In nature, and while we conduct practical farming, microbes negotiate soils, water bodies, and atmosphere in a continuum.

Nutrient dynamics that ensue due to aerial microbes need consideration. Atmospheric microbes mediate C-fixation, CO_2 evolution, methanogenesis, methanotrophy, and methylotrophy. Also, nitrification, i.e., N_2O and NO_2

emissions, NH_3 emission, etc. Aerial microbes are also capable of biological-N fixation that adds to organic-N in soil and microbial biomass N in the troposphere. Crop burning is practiced in food grain-generating regions. Here, the microbial metabolic processes mediate the transformation of nutrients that occur in smog, particulate matter, and mists. Microbes may detoxify smog.

Agronomic procedures have a profound impact on nutrient elements/microflora found in aerospace above a farm. Soil tillage that aerates the soil profile leads to oxidative conditions. It induces higher activity of aerobic microbes. Enhanced microbial respiration liberates more CO_2 into the agrarian atmosphere. Methanogens in rice culture zones, cattle farms, and other farm enclosures are important. Since they emit methane (CH4), which is a potent greenhouse gas. Methanogens operate under microaerophilic or anaerobic conditions. There are, of course, methanotrophs and methylotrophs, i.e., methane-oxidizers. They could be utilized to reduce CH_4 in the aerospace above crops.

The phyllosphere (i.e., crop's canopy) protrudes into the troposphere of the agrarian sky. The phyllosphere is indeed a vast area twice the size of the land surface on Earth. The phyllosphere encompasses all aerial organs of plants, like leaf lamina, branches, twigs, flowers, and pollen. They harbor and nurture microflora of diverse genetic nature and physiological abilities. Natural phenomena like gaseous exchange, foliar nutrient acquisition, entry of plant pathogens, entry of plant protection chemicals, herbicides, and growth hormones occur at the phyllosphere, i.e., leaf lamina.

Phyllosphere microbes mediate biogeochemical transformations. Phyllosphere microbes are intricately connected with nutrient cycling in agrarian regions. Phyllosphere microbes may detoxify chemicals traced on leaf lamina. Phyllosphere microbes found *in situ* or those applied could be effective biocontrol agents. Extraneous microbial biocontrol agents sprayed onto the phyllosphere do curb disease/epidemics. The phyllosphere is a vast niche with nutrient exudations where biological nitrogen fixation could be enhanced. Overall, the phyllosphere, with its microbial assemblages, is immensely important, although not so well recognized or utilized, even now. Phyllosphere microbes may also utilize exudates to synthesize novel metabolites of significance to farming.

Farmers are prone to utilize foliar mode for the supply of water and fertilizer-based nutrients. Obviously, crop foliage and microbes on it gain importance compared to soil/rhizosphere microbes. Foliage has to manage and mediate more diverse functions. Simultaneously, it has to endure harsh troposphere conditions, if any. No doubt we have to modify the agrarian sky suitably.

We have no idea about the limits to foliar feeding of crops. Actually, in the future, geneticists and plant breeders may have to look for physiological manifestations of crop foliage. Particularly leaf area, cuticular characteristics, stomatal density, and function. Stomata are going to be burdened with greater stress. Resistance to airborne disease too, may reside in the foliage. No doubt, the phyllosphere holds the key to successful crop production.

The agrarian sky has indeed played havoc with crops' survival and productivity. The agrarian sky has bestowed disasters year after year in some regions. Agrarian sky has even forced farmers to abandon their fields and dwelling locations due to epidemics/pandemics on staple crops. The agrarian sky has mediated innumerable epidemics and pandemics on food crops. Fungal pathogens like rust fungi, blights, powdery mildews, and smuts, which devastated cereal food crops, are highly airborne. Spores drift from one region to the other due to aeolian processes like wind, jet streams, precipitation, etc.

The region just above the crop's canopy has harbored, nurtured, and helped several vectors /plant pathogen combinations. Viruses dispersed by aphids, jassids, whiteflies, and mite species are examples. Rapid aerial dispersal of the virus has devasted crops. Perhaps vector management holds the key to the successful reduction of disease severity. Reduction in insect vector population through plant protection chemicals and/or biological control is a good suggestion. Enhancing the crop's inherent resistance to viral multiplication, too, has helped to curb viral pathogens. Avoiding susceptible crop hosts/genotypes in the vector-infested area is suggested. Inventing methods to denature virus particles is also possible. Wheat streak mosaic virus (WSMV) (Virus/*Aceria tosichella*), Cassava mosaic disease (CMD) (Virus/*Bemisia tebaci*), and Maize streak virus (MSV) (virus/ *Cicadulina* sp.) are epidemics of notable severity.

Global surveillance of major food crops for disease onset, particularly those prone to becoming virulent and causing epidemics, has been suggested. There are already a few examples, like Wheat Streak Mosaic disease, which is under aerial watch in different continents, like Australia, Asia, the Middle East, Europe, and North America. It helps to forewarn farmers about the impending disease, its spread, and potential devastation to crops in a region. Perhaps it is time we watched as many crop diseases as possible on a routine basis. Aerostats, drones, and satellite networks with facilities for spectral analysis are being utilized to survey the crops for disease.

Bacterial diseases have affected several crops of importance to farmer's exchequer. *Xanthomonas* sp. caused epidemics that have devastated several crop species. Agrarian sky mediates their spread via wind, turbulence, precipitation, splashing water droplets, and foliar sprays. Farm hygiene is a factor that induces epidemics. Citrus greening (Huanglongbing disease) caused by *Candidatus liberibacter* has reached pandemic proportions in citrus. It is spread by an aerial insect vector – a psyllid (*Diaphorina citri*).

Global farming regions have endured innumerable epidemics caused by microbial plant pathogens. Reduction in exchequer due to epidemics has been severe. For example, Wheat Stripe Rust (*Puccinia striformis*-Pst) resulted in a loss of 2.7 billion USD in 2016. In the Australian wheat belt, the Wheat streak mosaic epidemic (WSMV) resulted in a 1.27 billion A$ loss in 2014. The agrarian sky is to be blamed, to a certain extent.

Farm procedures affect microflora in aerospace. Farm procedures emit dust laden with microbes that get disseminated into the air. Agronomic procedures may disperse microbes that are detrimental to crops. Plus, the agrarian sky may spread human respiratory pathogens and others. Farm animal enclosures may harbor and disseminate pathogens that cause respiratory diseases.

There are too many aspects of global agricultural crop production that are under the influence of microbes adapted to the agrarian sky. There are ample opportunities to improve the advantages derived from aerial microbes. At the same time, deleterious effects can be bestowed on them, too. They need to be somehow curbed.

The take-home message is to look for 'Microbes in the Agrarian sky.' They are important to accomplish several ecosystem functions and achieve stable crop productivity.

Now, about the book *Microbiology of the Agricultural Sky*. There are eight highly informative chapters. They all deal with microbes of the atmosphere above the crop production regions. The first chapter provides glimpses of historical events related to microbiology as applied to global farming. Microbes in the atmosphere, particularly their distribution, diversity, and functions in the troposphere, form the central topic of Chapter 2. Aspects like microbe-mediated nucleation and its impact on crops and earth systems are highlighted.

Microbial migration in the agrarian sky is the focus of Chapter 3. Microbial migration is an essential natural phenomenon that ensures survival, dispersal, and perpetuation. Intercontinental migration or those for short distances provide opportunities for genetic intermixing of microbes. Recombinant microbes may be useful. If they are detrimental, then we have to select them out. Chapter 4 is about the global dust storms, plumes, and mass transfer of microbes from one continent to another. Incidentally, Chapters 3 and 4 both deal in detail with the phenomenon of intercontinental dust storms, plumes, and tornadoes. Their focus is on microbes that take a ride on the storms and disperse all over the globe. It is commensurate with the great inquisitiveness shown in recent times on these topics.

Chapter 5 lists and explains various nutrient transformations that aerial microbes induce in the atmosphere. Aerial microbes affect nutrient dynamics in the agrarian regions. Aerial microbes are intricately linked with greenhouse gas emissions from agrarian fields.

Phyllosphere, which is richly endowed with microbes of diverse metabolic abilities, is the subject matter of Chapter 6. Phyllosphere microbes are vital for crop productivity. Chapter 7 deals with crop disease. Emphasis is on epidemics and pandemics caused by aerial microbes. Discussions on the intercontinental migration of microbes in relation to epidemics/pandemics are interesting. The migratory habit of aerial vectors and virus combinations that induce epidemics has been dealt with. Several examples of epidemics/pandemics that occurred during the recent past have been listed and discussed.

Overall, this treatise provides excellent information about the microbes that thrive in agricultural aerospace and their relevance to agriculture. This book will be useful to the general public. Libraries worldwide will find it to be a useful acquisition. Professors, researchers, and students in universities will like it most as a text and reference source.

—***Krishna Kowligi***
Bangalore, Karnataka, India

CHAPTER 1

Microbes in the Agricultural Sky: Historical and Introductory Aspects

ABSTRACT

This chapter aims to provide glimpses of historical events related to the theme of the book, namely the 'microbiology of the agricultural sky.' At the same time, it introduces various aspects relevant to microbes and their role in the aerospace above agrarian ecosystems.

Historically, our journey to understand microbes began with the discovery of microscopic 'animalcules' by Anton van Leeuwenhoek in 1676 in Holland. Since then, we have gained extensive knowledge about their distribution, genetic nature, physiological aspects, and utility. Studies on the 'Microbiology of the Atmosphere' in general are relatively limited compared to those focusing on soil and water.

Several biologists of the 19th and 20th centuries enhanced our understanding of microbes in the air, aerosols, and crop surfaces. Examination of air samples in the 1850s and 1880s by inquisitive naturalists made it clear that microbes survive, proliferate, metabolize nutrients, absorb moisture, and perpetuate in the air. This is true despite the troposphere and other layers of the atmosphere being identified as a 'fragile or extreme environment.' During the 20th century, there was a brisk search for microbes, their binomial nomenclature, and their physiological function, as well as an understanding of their consequences to crops, animals, and humans. Several reports on microbes traced in the troposphere were published. Air samples drawn from lower altitudes above the soil surface, up to 3 km and even up to 7 km, contained a relatively higher population density. Microbes were traced up to

Microbiology of the Agricultural Sky. K. R. Krishna (Ed)

an altitude of 47 km in the atmosphere. Microbes are recycled between the Earth's surface and the troposphere.

Microbes levitate into the air from the soil surface or water and migrate aerially, eventually getting deposited on a substratum at a distant location. The survival of these microbes depends on the nature of the substratum. Various aeolian processes such as convection currents, low-speed and high-speed winds, jet streams, dust plumes, intercontinental dust storms, clouds, and weather patterns all have an impact on the microbes and their population and diversity in the agrarian sky.

Atmospheric microbes are primarily derived from the soil surface/profile, inland water bodies, oceans, natural vegetation, vast agroecosystems, drylands, and sandy deserts. Over the past three to four decades, focused efforts to study the distribution, migration, and re-establishment of microbes have provided important insights into aerial microbes and their physiological potential in the atmosphere.

Atmospheric microbes mediate the physicochemical transformation of biomolecules found in the air. They play a vital role in regulating the nutrient dynamics of agroecosystems. They partly regulate the dynamics of C, CO_2, CH_4, N_2, NO_2, NH_3, and other gaseous compounds. Major discoveries pertaining to the role of microbes in organic matter degradation, CO_2 and CH_4 evolution, ammonification, NH_3 liberation, nitrification, and biological-N fixation occurred during the latter half of the 19th century. Microbes on Earth are among the largest carbon fixers and emitters. Several heterotrophic and autotrophic microbes, methanogens, methanotrophs, nitrifiers, and NH_3 emitters were identified in the atmosphere, isolated, cultured, and studied during the period from the 1870s to the 1950s. Microbes affect nutrient dynamics in the troposphere by metabolizing organic substrate on particulate matter (PM).

Discoveries in the 1970s and 1980s related to airborne microbes in precipitation, frost, and snow are interesting. We have understood that bacteria like *Pseudomonas syringae* have an important role in nucleation, the formation of frost, snow, and precipitation. Intercontinental dust plumes that carry nucleating microbes, it seems, have an influence on global weather patterns and other Earth systems.

The phyllosphere is a large surface area on Earth, twice the size of the land surface area. It consists of the crop's canopy, including the leaf lamina and other parts. The term 'phyllosphere' was coined by F.T. Last in 1955. The phyllosphere refers to the area above the leaf lamina or branches, with a thickness of about 0.2 mm. It harbors nutrients exuded from the plant tissue and supports a large population of diverse microbes capable of nutrient transformation, such as nitrifiers, methanogens, methanotrophs, biological nitrogen fixers, and detoxifiers of toxins, if present. The phyllosphere also serves as a source of inoculum for airborne plant pathogens.

Our knowledge of airborne crop diseases dates back to ancient times. Detailed studies on airborne microbes causing severe crop diseases were conducted during the 1900s. Throughout history, the agrarian sky has played a role in the dispersal of plant pathogenic microbes, leading to devastating epidemics. In the mid-1800s, Late Blight of Potato caused widespread damage to crops in North America and Europe. Since then, numerous epidemics and pandemics caused by aerial plant pathogenic fungi, bacteria, and viruses have been investigated, with some recurring epidemics.

The agrarian sky provides a niche, albeit feeble, for microbes and facilitates their migration. It also promotes the intermixing of microbial species, potentially leading to recombinants with improved traits. Aerial microbes carrying genes for tolerance to low moisture, desiccation, UV radiation, and unfavorable temperatures can transfer these traits to other microflora, both in the atmosphere and in the soil profile upon deposition. Noteworthy means of intercontinental dispersal include dust plumes, storms, hurricanes, precipitation, high-speed winds, and jet streams. The migratory path of airborne microbial plant pathogens presents an interesting area of study, aiding in the avoidance of susceptible crop hosts along their routes.

Since the 1970s, there have been concerted efforts to unravel the intercontinental migratory routes of plant pathogens. Aeolian processes transport and deposit significant quantities of dust laden with nutrients and microbes over distances ranging from 2 to 15,000 km. Microbes are even capable of circumnavigating the globe. These natural phenomena are attracting increasing attention from microbiologists and meteorologists.

Overall, aerial microbes are intricately linked with atmosphere phenomenon that relate to global farming. The following seven chapters try to provide detailed information and discussions relevant microbes and their role above vast stretches of crops.

Microbes are versatile in nature. They survive on diverse types of substrates and environmental conditions. The discovery of microbes in different media such as air, water, soil, and complex geographical conditions like ocean depths, hot water springs, permanent snow, ice, frost, cold, and frigid atmospheric conditions, marshes, dry desert air, dust storms, etc., has occurred at different times and in unique situations during the past three centuries. Our accumulated knowledge about these microbes and their activity on Earth is immense. Yet, it is perhaps simply too feeble compared to what needs to be deciphered in future years. We only know about a few microbial species that survive and perpetuate in the atmosphere. This mostly pertains to commensals, plant pathogens, or a few others that affect animals and humans on Earth. Microbes in the Agrarian sky and their activity happen to be the central theme of this book.

1.1 DISCOVERY OF MICROBES

There is a general opinion that land and water have supported microbial life since the Precambrian period. In contrast, the atmosphere has primarily provided physical factors for life forms. However, there are reviews that discuss the evolutionary history of airborne microbes. These discussions also include the influence of the atmosphere and aeolian processes on airborne microbes. Establishing a timeline for the evolution of microbes in the troposphere is challenging, as it cannot be easily supported by authenticated fossilized remains, unlike land microbes that have left fossils for examination. Atmospheric processes such as wind, temperature changes, radiation, moisture, and others must have affected the distribution of microbes in the air (Henderson & Salem, 2016).

There are reports on aerosolized microbes from different ages, including ancient periods. It is believed that the existence of microbes in the air and their transmission through aeolian processes was suspected during the medieval period, specifically in the 1540s, by Girolamo Fracastoro. It was discovered that disease spread did not require physical contact; aerial dispersal of disease propagules was sufficient (Morgan, 2021). The earliest unequivocal evidence for the existence of microscopic life, such as organisms (animalcules), was provided by a lens maker named Antonie van Leeuwenhoek in 1676 in Delft, Holland. This was approximately 350 years ago (Gregory, 1971, 1973; Antonie Von Leeuwenhoek, 1941). Furthermore, Antonie van Leeuwenhoek demonstrated that animalcules (microbes) could be transmitted through the air, as wind and air currents dispersed these floating animalcules.

Examination of air/dust samples collected by Charles Darwin in the 1830s from his voyages showed that they contained 'infusoria,' i.e., microbes. It seems that black dust samples collected from the Sahara contained fungal propagules (Darwin, 1845a, b).

Pasteur (1860a) demonstrated that atmospheric conditions for the survival of microbes changed with altitude. As we reached higher altitudes, the population (number) of microbes reduced. For example, Louis Pasteur showed that at an altitude of 2 km on Montan Vert, there were fewer microbes in the air. Cunningham (1873) also detected microbes at higher altitudes of approximately 6 km from the ground surface in the Himalayas.

Later, in the 1890s, the presence of microbes in the lower troposphere was demonstrated in upstate New York by Dyar (1894). Microbes belonging to genera such as *Micrococcus, Bacillus,* and *Sarsina* were detected at lower altitudes in the air above vegetation.

During the early 1900s, studies focused on fungal crop diseases such as rusts, blotches, and blights, primarily addressing the distribution of fungal pathogens in the atmosphere. The objective was to understand the epidemiology of crop diseases. Aerial surveys using aircraft showed that bacteria, yeast, molds, and known fungal pathogenic genera were present at altitudes of 5-6 km (Procter, 1934). Several studies were conducted in North America and European plains to comprehend the aerial behavior of fungal pathogens affecting cereals. Bioaerosols were investigated at different altitudes in the troposphere. One study spanning the North Atlantic from New York to Denmark revealed the presence of microbes such as *Leptospheria, Mycospherella, Trichothecium, Helicosporium,* and *Venturia* even at an altitude of 3 km from the ground surface/agrarian fields (Meier & Lindbergh, 1935). The use of high-altitude balloons to investigate the aerial distribution of microbes, especially plant pathogens, began in the 1930s–1940s in the USA. A few other reports indicated the presence of *Bacillus, Macrosporium, Aspergillus, Penicillium,* and *Rhizopus* in air samples, particularly those drawn above crops and up to an altitude of 3 km (Rogers & Meier, 1936). Soffen (1965) found microbes surviving even at an altitude of 40 km. Perhaps, the particulate matter (PM) provided a niche for microbes to survive in the crevices of organic matter.

Aero-microbiological studies in the 1970s, conducted by Imshenetsky et al. (1976–1979) using balloons, have shown that microbe samples can be collected from higher altitudes in the stratosphere. Samples from the stratosphere can be obtained between 45 and 85 km from the Earth's surface. The tolerance of microbes to extreme conditions, their survival length, and transit are aspects that require further clarification (Griffin, 2004, 2008; Smith et al., 2010; DeLeon-Rodriguez et al., 2013; Cuthbertson & Pearce, 2011). In the 1980s, several studies were conducted to understand the various adaptations and mechanisms found in aerial microbes. Specifically, those that assist microbes in overcoming UV radiation, heat, and desiccating conditions. Sporulation, the production of a thick exosporium, pigmentation, and rapid DNA repair were found to aid microbial survival in harsh atmospheric conditions.

Microbes, especially bacteria, are found in clouds and precipitation (Prenni et al., 2009; Fitt et al., 1989). Snow, frost, fog, and mist also harbor bacteria of diverse genera. For example, ice nucleating *Pseudomonas aeruginosa* and *P. syringae* (Maki, 1974; Sands, 1982). Over the past two decades, smog in agrarian and urban skies has affected crops and humans. The primary cause is fossil fuel burning. Another cause is the periodic burning of crop stubbles and residues in agrarian belts in South and Southeast Asia. Examination of smog has indicated that it contains a large population of inhalable microbes with diverse genetic nature. Inhalation of these microbes is detrimental to humans. Crops may also suffer due to limited photosynthesis, hindrance to gaseous exchange, and toxic effects of pollutants. Additionally, microbial diseases may spread into crop production zones (Cao et al., 2014). Smog undoubtedly aids the migration of microbes in the troposphere. Not only bacteria, but smog may also harbor fungi (including spores), actinomycetes, viruses, and archaea. Bacterial population and diversity traced in dust storms, thunderstorms, and tornadoes that occur above agrarian regions have also been studied and reported.

Plant viruses were discovered 120 years ago. They were reported simultaneously by Dutch microbiologist Martinus Beijerinck and Russian researcher Dmitri Iwanowski while investigating mosaic diseases of tobacco. Undoubtedly, the earliest virus detected was a plant virus (see Beijerinck, 1898; Vander Want & Dijkstra, 2006; Gergerich & Dolja, 2006; Scholthof, 2000, 2001; Zaitlin, 1998). The first known virus is tobacco mosaic virus (TMV). Since then, we have discovered viruses that infect several crop species, fungi, bacteria, farm animals, and even humans (such as farm workers). Most plant viruses are dispersed through aerial vectors or natural factors such as wind and rain. Entry into plant tissue is facilitated by wounds on plant or animal tissue. Currently, there are about 4,000 known, classified, and described viruses, with approximately 1,000 of them being plant viruses. Viruses have been responsible for several epidemics and pandemics in crops. The combination of aerial vectors and viruses in the agrarian sky can cause severe damage to small and large areas of crops.

Initial investigations were aimed at understanding the causal agent (i.e., viruses), their morphological characteristics, epidemiology, and a few ecological aspects. In recent years, most virus research has focused on studying their molecular aspects, disease causation, epidemiology, and crop loss. Aerial vectors and their role in the transmission of viral diseases have been a major focus for plant virologists. These studies have shown us that viral migration is rapid due to vectors such as aphids, thrips, jassids, psyllids, whiteflies, etc. Since the early days of plant viral studies, we have made rapid advancements. Today, we know how to control aerial vectors (insects) to manage the rate of viral dissemination.

During the past century, research efforts on airborne algae have been relatively less compared to other airborne microbial species. The first airborne algal species was discovered by Ehrenberg in 1884 and by Salisbury in 1886. However, the earliest detailed effort to unravel the physiological and ecological aspects of airborne algae was made in the Netherlands in 1935 (Visch van Overeen, 1972). Several other studies have focused on airborne algae in different parts of North America and other continents (Brown et al., 1963; Schlichting, 1969, 1971, 1974). Airborne algae are derived from soil, marshes, lakes, ponds, rivers, other water bodies, and seawater. Annually, about 300 billion tons of seawater with algae escape into the air from the sea (Lee & Eggleston, 1988). A sizeable portion may reach the land surface after drifting via wind and air currents. Airborne algae may contribute immensely to global carbon fixation through their ability for photosynthesis.

There are several reports about archaea found in the troposphere. They are known to play an important role in the biogeochemical cycles on Earth, including the agrarian sky (Wehking et al., 2018; Lehtovirta-Morley, 2018). The population density and diversity of archaea found in the agrarian regions are not well understood. Crop species or soil type preferences are also not clearly known. However, the occurrence of certain species of archaea does point out that soil, water bodies, and crops are major sources of archaea. Particularly, archaea that are found in the particulate matter ($PM_{2.5}$ and PM_{10}). Undoubtedly, archaea are found in the agrarian atmosphere. They have also been traced in high-altitude wind and jet streams (Smith et al., 2013).

In addition to single-celled microbes, microscopic multicellular organisms, insects, and other biotic flora/fauna may also be traced in the troposphere above cropping belts.

1.2 MICROBES IN ATMOSPHERE: HISTORICAL ASPECTS

Morris et al. (2011) state that, for the past two centuries, aerobiology (including Aeromicrobiology) has involved the exploration of the abundance, diversity, survival, and migration of microbes in the atmosphere. This gives the impression that microbes are passive residents of the atmosphere.

However, it has also been observed that the microbial load in the atmosphere can affect atmospheric processes, such as nucleation and snow formation (Morris et al., 2013).

Let us first define the terms 'Aerobiology' and 'Aeromicrobiology.' Aerobiology is a broad term that encompasses the study of the biotic aspects of the atmosphere, including all types of biotic components found in the air. The term 'aerobiology' was first used by Meier in 1936. In greater detail, aerobiology is defined as the study of the aerosolization process, aerial transmission or migration, and deposition of microbes and other life forms (Boehm & Leuschner, 1986; Pepper & Gentry, 2015; Dimmic & Akers, 1969). Minor variations in the definition and meaning of aerobiology have been observed, mainly to suit the purposes of environmental studies, microbiology, crop disease, animal and human respiratory diseases, and the long-distance migration of pathogens (Pepper & Gentry, 2015). Furthermore, Pepper & Gentry (2015) state that aerobiology, as a field of study, is finding increasing usage and applications in agriculture, industry, and environmental aspects.

Aeromicrobiology includes the study of microbes found in the atmosphere, specifically in the air. Particulate matter (PM) present in the atmosphere can also carry microbes, either attached to them or in crevices. These microbes are referred to as 'bioaerosols.' Aerial microbes can also be found on clouds, mists, and fogs that form above crops. The total population, density, and genetic and physiological diversity of aerial microbes can vary significantly. However, compared to soil profiles or other natural environments like forest floors, marshes, lagoons, or rich organic substrates such as peat bogs, the microbial population and diversity encountered in the atmosphere is relatively low. The atmosphere, including the lower and upper troposphere, is considered an 'extreme environment' due to its ability to support microbial growth. Microbes in the troposphere must be well adapted to harsh conditions, including gaseous mixtures, nutritional availability, moisture, and other factors. Nevertheless, microbes are known to bring about a wide range of physicochemical transformations in the atmosphere.

According to a few aerobiologists, the black fungi traced in the dust samples obtained from the Saharan region by Charles Darwin in 1938 are among the earliest recognized efforts to acknowledge the existence of aerial microbes (Darwin, 1845a, b; Weil et al., 2017). Microbes, including bacteria and fungi traced in the Sahara/Sahel, must have endured the extremely harsh conditions presented by the atmosphere. The collected microbes must have overwintered via sporulation or other means of hibernation to overcome the feeble atmosphere (Smith et al., 2011; Chuvochina et al., 2011). Microbes reported to occur in other regions of the Earth, such as the snowy Alps, Mediterranean regions, dry Arabian climate, or even the tropics, must endure the harsh conditions prevalent in the atmosphere. Indirectly, this discovery about fine dust and traced microbes also indicates an early discovery of microbial migration.

The troposphere, with its gaseous and particulate fraction, is a habitat for diverse microbes. It is said that the microbial population of the lower troposphere is around $1 \times 10^4 m^{-3}$. This is much lower than the microbial population encountered in oceans or on agricultural land, which is 100 to 1,000 times higher (Bauer et al., 2002; Burrows et al., 2009; Griffin et al., 2001; Prospero et al., 2005). The concentrations of airborne culturable bacteria were measured earliest by Miquelin in 1883 (Fitt et al., 1989). There are now several methods of measuring the microbial population in the atmosphere.

Sampling and diagnostic procedures are also available to suit the purpose (see Aguilera et al., 2018). Recent methods involve measuring the population and diversity of microbes using molecular methods based on 16s rRNA (Leveau & Tech, 2011; Janda & Abbott, 2007). Otherwise, the most common methods adopted since the mid-1900s have involved dilution plate counting, turbidity measurements, and petri-plate traps with selective medium.

Research during the past two decades has provided us with data on the distribution of microbes in different layers of the atmosphere. Troposphere microbial species are derived from the soil surface or those found on crops/vegetation. This is because atmospheric processes such as wind, turbulence, and precipitation induce the levitation of microbes from these regions. However, we may also encounter microbes that levitate and migrate from distant regions. Several studies have shown that microbes migrate for short distances and even transit intercontinental distances. Therefore, at any given time, the atmosphere serves as a repository of microbes derived *in situ* from the ground or from other locations. Both types of microbes may affect soil processes and crops upon deposition.

Reports on dust plumes, storms, and wind, particularly their ability to support bacteria, are interesting. Finer particles of dust plumes often contain Bacteroidetes, a type of microbe that can tolerate harsh temperatures and desiccating conditions in the troposphere. These finer particles, known as $PM_{2.5}$ or particles smaller than $PM_{2.5}$, can travel longer distances in the troposphere, including intercontinental migration. However, microbes must survive the duration of the aerial journey and the conditions they encounter. Coarse sand grains, dust particles, and organic dust often carry bacterial phyla such as Proteobacteria, Alpha, and Gamma Proteobacteria. These bacterial species are known to play a role in various nutrient transformations. Studies by Tignat-Perrier et al. (2019) and Acosta-Martinez et al. (2015) have suggested that bacterial species found in the troposphere often correspond to the microflora found in landscapes, such as agricultural soils. Additionally, turbulence in oceans can cause moisture drops containing microbes to become airborne, leading to the presence of ocean surface microbes in the troposphere near coasts.

We are concerned about the presence of microbes in the air above crops. Specifically, we are interested in the discoveries related to microbes in the 'agrarian sky.' It has been found that bioaerosols are distributed up to 7 km above the Earth's poles and 20 km above the equator. Polymenakou (2012) emphasizes the importance of tropospheric microbes, as they can have diverse effects on the ecosystem. Interestingly, troposphere microbes are commonly found in clouds (Dillon & Fennel, 2021), fog (Osowski, 2018), snow (Brown & Jamponnen, 2019), precipitation, and water droplets (Prenni et al., 2009; Everts, 2008; Delort et al., 2010). Scharf (2019) suggests that troposphere microbes may influence cloud chemistry, the enucleation process, and precipitation patterns.

Troposphere microbes have the potential to be deposited back into soils, natural vegetation, and cropping belts. Atmospheric processes, such as dust, cloud, and rain cycling, greatly impact microbial diversity in agroecosystems (Bauer et al., 2002, 2003). Fogs, smog, and blizzards also transport microbes, which can re-infect the soil surface and crop canopies (Christner et al., 2012). In other words, the soil is constantly exposed to and enriched with new microbial species from the air.

Annual dust plumes are frequently observed in the vast stretches of the Sahara. These dust plumes carry diverse microbial species, including a large number of bacterial species. Rosselli et al. (2015) have shown that dust plumes migrating above the Mediterranean and into Italian agrarian regions often carry *Alpha* and *Gamma Proteobacteria, Sphingomonas, Paracoccus, Klebsiella, Pantoea, Stenotrophomonas,* etc. Additionally, the dust particles also harbor fungi, protozoa, archaea, and microscopic insects.

Studies on stratosphere microbes are relatively recent (DasSarma & DasSarma, 2018). There is a growing curiosity in understanding the microbial survival mechanisms in extreme climatic conditions. For example, Khodadad (2017) demonstrated that out of 40 million spores released into the stratosphere, only 260 spores survived the harsh conditions.

DasSarma et al. (2020) reported a comprehensive list of microbes, including both prokaryotes and eukaryotes, that were tested and isolated from the stratosphere. Among them, several are spore formers and are susceptible to UV radiation. These microbes have the potential to mutate in the stratosphere and return to the ground. If they intermingle with the soil microflora, they could give rise to novel versions of microbes, which may pose a threat to crops and other biotic components of the soil. Interestingly, it is estimated that over 100,000 flights, including high-altitude long-distance airplane flights, could introduce stratosphere microbes into the soil and troposphere on a daily basis (DasSarma et al., 2020). Many of these microbes could potentially become crop pathogens. In fact, in the future, airplanes and especially low-flying agricultural drones could serve as major vectors for global crop epidemics and pandemics if proper field hygiene practices are not implemented. These findings are based on recent research.

Space flights may have begun in the 1950s/1960s. However, our efforts to study the survival, migration, and physiological activity of microbes in space took much longer. Whether we studied microbes or not, they have been adhering to the surface of vehicles and reaching space. Satellites released by vehicles or space stations may have carried a diverse load of microbes from Earth. Undoubtedly, several species of microbes may have perished, while a few might have survived the journey and returned. Some may have even undergone mutations due to UV and other radiations before returning to the soil as mutants. From 2000 A.D. until now, space microbiologists have discovered several genera of microbes on the surface of space vehicles. For example, *Acinetobacter, Alcaligenes, Bacillus, Corynebacterium, Brevibacterium, Flavobacterium, Micrococcus,* and *Staphylococcus* were reported on the body surface of space rockets in early 2004, then in 2007, and recently in 2018 (Nicholson et al., 2018; La Duc, 2004; Schuerger, 2004). The microbial load on the surface of spacecraft ranged from 1×10^4 to 1×10^8 per vehicle (Nicholson et al., 2018). Fungi discovered on space vehicles belonged to the following genera: *Alternaria, Aspergillus, Bipolaris, Candida, Cladosporium, Fusarium, Mucor,* and a few others (Schuerger, 2004; Nicholson et al., 2018). It is possible that the microbes returned from space may have some effect on agrarian regions if they happen to reach agrarian soils or cropped fields. In 2009, the Indian Space Research Organization (ISRO) reported the discovery of three new bacterial species in the troposphere. They are *Janibacter hoylei, Bacillus isronensis,* and *Bacillus aryabhata* (ISRO, 2009). We have no idea about their ability for nutrient transformations or other physiological manifestations once they have reached agrarian terrain.

1.3 MIGRATION OF MICROBES IN THE AGRARIAN SKY

Motility of microbes in closed systems containing water or other fluid nutrient medium, i.e., nutrient broths, was perhaps known some three centuries ago. Antonie Von Leeuwenhoek first described the movement of 'animalcules' in sewage and other water sources. Mobility is an essential trait for many microbial species. Mobility in moisture droplets or fluid medium occurs via cilia, flagella, or through the production of pseudopodia. Floating, somersaulting, and creeping also aid the migration of microbes, although to small distances. We may encounter all these phenomena on the crop's canopy, particularly on the phyllosphere, on the soil surface, and in the troposphere (snow, fog, frost, mist, dew, precipitation droplets). Such migration may aid the survival and perpetuation of troposphere microbes. However, during recent decades, our focus has been more on the aerial migration of microbes through aeolian processes (convection, wind, tornadoes, hurricanes, jet-streams). The transcontinental journey of microbes and their consequences to crops seems important. Hurricanes and tornadoes also mediate the long-distance movement of microbes in the atmosphere (Yong, 2013; Temkiv et al., 2012).

Microbial migration in the troposphere is an important aspect of aeromicrobiology in agrarian regions. Migration helps microbes find new niches and locate locations with the nutritional factors they require. Migration also aids in avoiding unfavorable and harsh troposphere conditions. Microbes migrate short distances through atmospheric drift, low-speed winds, small-sized dust plumes, and turbulence.

Contamination of the air above crop canopies with microbial spores or propagules is common. This contamination can ultimately facilitate microbial migration and dispersal across farmers' fields. Microbial contamination of irrigation water is equally significant, as it can lead to the dissemination of microbes through the troposphere when farmers adopt aerial sprays. The transport of microbes on pollen, insects such as locusts, honeybees, birds, and other organisms is a clear possibility (Krishna, 2022).

Crop burning is known to induce the levitation of organic dust particles and microbes. Microbial transport in the troposphere also occurs via clouds, frost, wind, and convection currents, among others (Krishna, 2022).

Aerial vehicles such as airplanes, UAVs (drones), parafoils, microlights (triches), sprayer drones, etc., also disperse microbes to varying distances. The goods carried by these vehicles may harbor microbes, which can be dispersed during their movement. Sprayer drones, in particular, are capable of effectively distributing microbes on crops.

Over the past three decades (1990–2020), research on microbial migration in the lower troposphere and its consequences on disease epidemics/pandemics has garnered attention. Significant discoveries have been reported regarding the role of microbes in ice nucleation. Bioaerosols are increasingly being attributed to weather phenomena. Dust plumes that migrate carry important microbial genera like *Pseudomonas* sp., which can induce precipitation or at least alter weather processes (Sands, 1982). Therefore, the long-distance migration of microbes from Asia and the Sahara, along with their nucleation/precipitation properties, makes bioaerosols significant (DiLiberto, 2018; Griffin et al., 2001; Smith et al., 2012; Creamean et al., 2013). African dust plumes embedded with microbes are known to alter weather patterns in the North American Plains (Acosta-Martinez et al., 2015; Brown & Hovmoller, 2002; Murray et al., 2012).

Intercontinental transmission of cereal rust from Central Asia (Mongolia) to the Pacific Northwest via the jet streams was detected in the 1980s and 1990s. There are reports of global circumnavigation of rust propagules via the jet streams (Smith et al., 2012, 2013; Duan et al., 2021; Tang et al., 2018).

Transcontinental transport of microbes, including beneficial microbes, commensals, and pathogens that cause pandemics, is an increasingly important aspect, particularly in the past two decades. Aero-microbiological studies of bioaerosols have clearly demonstrated that long-distance transcontinental aerial transit of microbial pathogens contributes to outbreaks of epidemics and pandemics. A few examples include Wheat Rust fungi (McIntosh et al., 2009; Singh et al., 2005), Sugar cane rust (Purdy, 1985), and cassava mosaic disease (CMD) (Otim-Nape & Thresh, 1998; Legg et al., 2011; Legg & Fauquet, 2004). Vector-mediated viruses also travel intercontinental distances. This occurs when the insect carrying the viral load drifts into the jet streams at higher altitudes (Jones, 2014, 2021).

The migration of microbial pathogens occurs over various distances, ranging from a few meters to longer distances of 5–100 km. Dust storms, plumes, and intercontinental jet-streams aid in the long-distance journey of microbial pathogens. Additionally, high-altitude rapid winds facilitate the transportation of plant pathogenic microbes to different continents. Numerous reports have been published on the migration of microbes in agrarian fields. In recent years, there has been detailed research on the intercontinental migration and spread of microbes (Smith et al., 2012; Brown & Hovmoller, 2002; Acosta-Martinez et al., 2015). Several interesting reports also discuss disease epidemics and pandemics (2 Blades Foundation, 2020). For instance, Acosta-Martinez et al. (2015) states that a significant amount of dust material (organic/inorganic), amounting to several Tg, reaches the American east coast and plains from the Sahara. This dust carries a large biomass of microbes embedded in and on the dust particles. Similar reports exist for the journey of microbes from the Asian desert to the North Pacific of America (Chen, 2005; Tang et al., 2018). Saharan dust plumes are known to transport microbes from North Africa to Southern European agrarian regions and the Alps (Rosselli et al., 2015). One example is the wheat disease-causing fungus *Puccinia graminis tritici* (*Pgt*), which travels in the troposphere from East Africa to West Asia (*Ug 99*) (Brown & Hovmoller, 2002). The sugar cane rust fungus (*P. melancephala*) is also known to migrate from West African regions to Brazil, Cuba, West Indies, and other areas (Anikster et al., 2005; Purdy, 1985).

Intercontinental migration of dust laden with microbes is also known to affect weather patterns and other Earth system functions, particularly those occurring in the troposphere. Long-distance migration from one continent, in fact, affects precipitation and other abiotic factors in a different continent. This aspect of Aeromicrobiology may attract greater attention from meteorologists and agriculturalists in the future.

1.4 AERIAL MICROBES AND NUTRIENT DYNAMICS IN AGRARIAN REGIONS

In the agricultural sky, microbes are versatile with the ability to utilize, transform, release, or remove various gaseous compounds. Collectively, in terms of population and/or physiological diversity, they influence global crop production zones. Their physiological functions in the troposphere and at higher altitudes could be crucial to crop productivity. For example, it seems that about 50% of the O_2 produced on Earth is by microbes. A sizeable portion is contributed by algae/diatoms, etc. About 75% of nitrogen inputs into the atmosphere and 90% of its removal from the troposphere are mediated by soil/aerial microbes. A major share of greenhouse gases (GHGs) like N_2O, which are potent in raising atmospheric temperature, is derived from bacterial activity. CH_4, which is also an important gaseous C compound emitted into the air, is derived from bacteria. They say that 70% of CH_4 emitted from cattle sheds, rice fields, and other farm activities is attributable to microbial activity. A point to note here is that emissions are often caused by microbes traced in soil (surface or in the profile), on animal house floors, and those in the atmosphere (agrarian sky). Of course, not all emissions traced in the atmosphere are derived entirely from agricultural regions. Atmospheric drifts from above the oceans, mountains, and other natural regions are also traced. Historically, these findings about microbial abilities in soil and in the troposphere have improved our understanding of nutrient cycles. Microbes also affect the fate of fertilizer inputs, plant protection chemicals, and other amendments in the air.

The major emphasis in this section is on aerial microbes involved in natural cycles of carbon and nitrogen. Aerial microbes are capable of a series of biogeochemical transformations that could be useful to crops. The distribution of aerial microbes in the troposphere is important, although only partially. Microbial population and genetic/physiological diversity have a direct influence on nutrient transformations. Particulate matter (PM) and gaseous compounds may progressively decrease as we reach higher altitudes. Therefore, microbial activity, particularly nutrient transformations, is expected to decrease based on altitude in the troposphere.

There is another set of transformations that can be detrimental to crops, such as greenhouse gases (GHGs) like CO_2, CH_4, NH_4, N_2O, and NO_2. Loss of soil carbon (C) due to repeated plowing, oxidative conditions, and microbial respiration in the soil may create unfavorable conditions, especially in agrarian regions that are repeatedly exposed to organic matter, nitrogenous fertilizers, and other amendments. Therefore, it is difficult to study and decipher the consequences separately for the agrarian sky. Most often, it is a continuum of soils, soil pores, moisture, air, and the agrarian sky that are intricately linked during the course of crop production. Hence, there are discussions and explanations that pertain to soils, their profile, and nutrient transformations that eventually lead to emissions into the agrarian sky. Vice versa, i.e., the deposition of atmospheric chemical compounds, organic matter, and microbes, has also been emphasized. Indeed, nutrient dynamics in agrarian regions cannot be properly understood if we deal with the 'agrarian sky' in isolation. Natural occurrences in the agrarian sky are intricately linked with the soil phase and crops.

1.4.1 CARBON CYCLE AND GHG EMISSIONS

Microbes in the aerospace above crops are capable of both carbon emission via respiration and fixation via the photosynthetic process. Microbes found at various altitudes in the vertical profile of the troposphere fix carbon through photosynthesis. Photosynthetic bacteria, green algae, blue-green algae (BGA), and diatoms found floating freely in the agrarian sky are capable of carbon fixation. The microbial photosynthetic process enhances aerial biomass. The source of photosynthetic microbes found in the troposphere could be soil, water bodies, high seas, oceans, even snow, and frost. The ocean surface with its waves and turbulence can be an important source of green and red algae for the atmosphere. Atmospheric processes like tornadoes, wind, turbulence, hurricanes, clouds, precipitation, snowfall, and frost breeze can support the migration of photosynthetic microbes from one location to another in the troposphere. Historically, photosynthetic bacteria, algae, and diatoms were traced in aeolian dust about a century ago.

Organic matter degradation, mineralization, and microbial respiration are processes that release carbon dioxide into the troposphere. Such processes were identified long ago, in the early 1900s. Carbon emissions in the form of CO_2 can be significant if the microbial population in the lower troposphere is relatively high. Troposphere microbes act on the organic matter in particulate matter (PM), decomposing microbial mass and soluble carbon compounds present in the moist air. Let's consider a few examples. Cellulose-degrading microbes are found in relatively higher densities in the troposphere. They break down cellulose and release CO_2 into the troposphere. Bacteria such as *Achromobacter, Bacillus, Cellulomonas, Cellovibrio, Pseudomonas,* etc.; Fungi such as *Aspergillus, Fusarium, Penicillium, Rhizoctonia, Trichoderma, Verticillium,* etc.; and Actinomycetes such as *Actinomyces* sp., *Streptomyces* sp. are cellulose degraders. There are several microbial species that have the metabolic ability to degrade hemicellulose. Similarly, the troposphere may support microbes that degrade lignin and phenolic compounds found in organic debris. All of these processes are related to carbon dynamics in the air. Microbes indeed release significant quantities of CO_2, CO, and even CH_4 into the atmosphere. However, it is still much less compared to rice paddy fields or cattle enclosures.

The vast expanses of crops in various regions are major contributors to CO_2 emissions. This is especially true when considering the presence of microbes in agrarian soils, on crop canopies (known as the phyllosphere), and even in the atmosphere above the crops. Heterotrophic microbes, which rely on organic particulate matter (PM) in the troposphere for energy, release significant amounts of CO_2. Agrarian regions also emit carbon in the form of CH_4. Methanogenic microbes found in submerged paddy fields and cattle sheds are the primary sources of CH_4 emissions. Rice production areas with submerged soils that have low redox potential are known to emit CH_4. Additionally, microbes that can tolerate microaerophilic conditions also release CH_4 into the atmosphere. The rice production zones in Southeast Asia are particularly significant contributors to CH_4 emissions due to the anaerobic/micro-aerophilic metabolic processes carried out by these microbes. Lagomarsino et al. (2016) suggest that managing methanogens in both soil and the troposphere could help reduce CH_4 emissions. Proper maintenance of cattle sheds can also help reduce the population of methanogens on surfaces and in the air within the enclosures.

In recent years, the burning of crop residues has led to carbon emissions in agrarian areas. Crop burning releases carbonaceous PM and the attached microbes into the atmosphere. This agricultural practice results in a significant increase in microbial biomass in the air.

Methane (CH_4) emissions from agricultural cropping expanses, especially rice fields, cattle sheds and other sources is an important aspect. It directly influences gaseous composition of agrarian sky. CH_4 is an important GHG liberated into agrarian sky. It is part of carbon cycle in the agroecosystem.

Methanogens are usually confined to anaerobic conditions like submerged paddy fields, cattle shed effluent, marshes, etc. Methanogens are among major causes for CH_4 liberation into atmosphere. Methanogens that belong to bacterial genera and archaea have been identified. Methanogenic bacteria belonging to classes Alpha and Gamma Proteobacteria and Verrucomicrobia (Dedysh & Dunfield, 2011) were identified during past decade. The population and diversity of methanogenic bacteria depends on soil/atmospheric conditions. Rice fields may harbor 10^5 methanogens depending on redox potential, nutritional status, and bacterial species. Methanogenic bacteria and archaea seem to stratify at different depths in the submerged micro-aerophilic conditions, in paddy fields (Vaksmaa et al., 2017). A few examples of methogenic bacteria reported are *Acidobacteria*, *Chloroflexi*, *Proteobacteria*, *Planctomycetes*, and *Actinobacteria*. Methanogenic archaea were reported in paddy soils, recently. A few examples of archaea are *Methanosarcina*, *Methanosaeta*, and *Methanobacerium* (Weber, 2001).

Methanotrophs (CH_4 utilizers) were first identified and isolated in 1906. *Bacillus methanicum* is the earliest species of methanotroph identified. However, currently, over 100 bacterial species with the ability to oxidize CH_4 have been isolated (Jiang et al., 2011; Bowman, 2006). Methanotrophs that metabolize CH_4 emitted into the troposphere play an important role in reducing CH_4 concentration. Methanotrophs actually oxidize C1 compounds such as CH_4 to methanol. CH_4 oxygenase is the enzyme involved in the aerosol microbes (Crombie & Murrell, 2011). Several bacterial species belonging to the phylum Proteobacteria have been identified as methanotrophs (Dedysh & Dunfield, 2011).

During the past decade, it has been shown that fungi (e.g., *Fusarium solani*) are also capable of degrading CH_4. They transform CH_4 emitted into the sky from farms. Laboratory trials have shown that *F. solani* converts about 300 mg m^{-3} h^{-1}. This is similar to the levels achieved by common methylotrophic bacteria such as *Methylomicrobium album* and *Methylocystis* sp. (Vergara-Fernandez et al., 2019).

Methylotrophs are also found in the agricultural atmosphere. Examples of methylotrophs include *Methylomonas, Methylosphaera, Methylomicrobium, Methanocladium,* and *Methylobacter.* These organisms utilize biochemical pathways to capture and detoxify formaldehyde. Obligate methylotrophs, such as *Methylomonas aminofaciens,* have been discovered in the agricultural atmosphere. Additionally, there are methylotrophs that metabolize CH_4 in dry and hot atmospheric conditions, known as 'thermotolerant methylotrophs,' such as *Bacillus brevis* (Yurimoto et al., 2002). Filamentous methanotrophs like *Crenothrix polyspora* (Stoeker et al., 2006) and *Clonothrix fusca* (Vigliotta et al., 2007) have also been identified. Furthermore, there are methanotrophs that can tolerate extremely acidic conditions.

1.4.2 AERIAL MICROBES INVOLVED IN NITROGEN DYNAMICS

In the atmosphere, ammonification is a key process that involves aeolian organic dust, dead microbes, dead insect debris, and certain insoluble/soluble organic matter. All aerial microbes possess enzymes to disintegrate proteins, nucleic acids, and other nitrogenous compounds found in PM. They are also capable of degrading those compounds found in dissolved state in the moisture droplets. The dusty atmosphere and plumes that are generated serve as a good substratum for microbes. The dusty agrarian sky is generally rich in microbes. Ammonification is an aerobic process mediated by aeolian microbes of diverse genera, such as fungi, bacteria, actinomycetes, and archaea. It liberates amino acids followed by gaseous NH_4 compounds and NH_3.

Anaerobic conditions are not easily found in the troposphere above crops. However, anaerobic degradation of nitrogenous compounds is mediated by microbes like *Clostridium* sp. Anaerobic degradation of organic matter is called 'putrefaction.' It results in the loss of NH_4 into the atmosphere.

Ammonia (NH_3) oxidizing microorganisms play a central role in the fate of nitrogen in the environment. They also provide substrate, i.e., N_2O, for denitrification (Lehtovirta-Morley, 2018).

Nitrification is an important biogeochemical transformation encountered in agricultural soil and the tropospheric air above it. It is among the key steps in the nitrogen cycle that occurs in agrarian regions. Pasteur (1862) was the first to recognize the conversion of NH_3 to nitrate as a microbiological process. The requirement of nitrifying microbes was shown by Alexander Muller in Berlin, using sterilized solutions of NH_3 and natural waters containing nitrifying microbes (Muller, 1875). Schloesing & Muntz (1877) provided the first evidence that NH_3 oxidation is a biological process. In 1878, Warington of Rothamsted, Harpenden, England, for the first time, showed that nitrification is actually a two-step process, occurring in both soil and air (Warington, 1878, 1879). The pure culture of nitrifying bacteria was first achieved by Percy Frankland and Grace Frankland in Scotland (Frankland & Frankland, 1890). However, Sergie Winogradsky, in 1890, obtained pure cultures of both NH_3 and nitrite-oxidizing bacteria (*see* Winogradsky, 1890a, b, 1891; Waksman, 1953). In addition to discoveries about nitrification, Winogradsky is known to have discovered chemolithotrophs and chemoautotrophs like sulfur-oxidizing bacteria, i.e., *Beggiatoa alba* (Sedlacek, 2020).

Nitrification has gained importance as anthropogenic nitrogen inputs into agricultural regions have exceeded 120 Tg per year. This leads to the emission of nitrous and nitric oxides with high global warming potential into the agricultural atmosphere (Lehtovirta-Morley et al., 2018). Additionally, nitrification directly affects fertilizer nitrogen use efficiency in farmland and can contribute to water pollution through nitrate leaching and eutrophication.

NH_3-oxidizing microbes are aerobic autotrophic microorganisms. They are classified as ammonia-oxidizing bacteria (AOB), ammonia-oxidizing archaea (AOA), and comammox bacteria (which completely oxidize NH_3 to nitrate) (Lehtovirta-Morley, 2018). Putative NH_3 monooxygenase genes are the main functional genes used to identify NH_3 oxidizers. These genes have been found in the genomic sequences of archaea, such as *Crenarchaeota* (Treusch et al., 2004; Sedlacek, 2020).

At this juncture, it is important to note that the supply of fertilizer-N to crops is a crucial aspect. Equally important, if not more so, is the conversion of fertilizer-N and NH_4 compounds in the soil and in the atmosphere above the crop's canopy. The emission of NH_4 from paddy fields, as well as N_2O and NO_2 from arable soils, holds significant importance. These microbiological processes have an impact on fertilizer-N, affecting N assimilation into crops, use-efficiency, and greenhouse gas emissions. Overall, nitrifying microbes in the soil and air play a vital role in the agroecosystem.

Many of the aforementioned discoveries are highly relevant to the N dynamics that occur in agroecosystems, particularly within the soil profile, water bodies found in farms, and the atmosphere. In the present context, we are specifically interested in the microbiological aspects of N dynamics in the agrarian sky (atmosphere). Since the 1950s, crops have been supplied with fertilizer-N formulations, either to the soil as granules or solution, or directly to the crops as foliar-N in a dissolved state. The aforementioned discoveries hold even greater relevance when considering fertilizer-N utilization, its efficiency, and, more importantly, its consequences to soil quality. Gaseous emissions of N such as NH_4, N_2O, and NO_2 have implications for atmospheric pollution and greenhouse effects.

The microbial contribution to soil nitrogen in agrarian regions that support cereals, legumes, and other crops is carried out by a group of microbes known as Biological Nitrogen Fixers (BNF). There are two types of BNF microbes that can be found in the airspace above crops, on the canopy, and in the roots/rhizosphere. Asymbiotic nitrogen fixers were discovered by Winogradsky (Winogradsky, 1949; Waksman, 1953) and later confirmed by others. On the other hand, rhizobium species are predominantly found in the soil and in the nodules of legume roots, forming a symbiotic relationship with legume crops. They can also be found floating in the aerosol due to turbulence and wind. *Rhizobium* was first recognized as a beneficial organism that contributes to soil nitrogen and plant growth through a symbiotic association, as identified by Beijerinck in 1888. Frank (1889) named the nodule bacterium as *Rhizobium leguminosarum*. The BNF microbes play a vital role in converting atmospheric nitrogen into soil nitrogen and plant nitrogen. Their ability to add nitrogen to the soil is estimated to range from 20 to 60 kg N per season per hectare, depending on the crop, soil types, soil nitrogen status, weather parameters, and species (isolates) of BNF organism.

Asymbiotic biological N fixation that occurs in the troposphere is mediated by microbes such as *Azotobacter*, *Azospirillum*, *Beijerinkia*, *Klebsiella*, and free-floating BGA in the air. Archaea such as *Euryarcheota*, *Crenarcheota*, *Thaumarcheota*, and *Nanoarcheota* also fix atmospheric-N, in the free state, in atmosphere. It is essential that BNF microbes survive the extreme conditions of the troposphere. They must also be able to extract nutrients from the dry or moist air. Most of them are obligate aerobes, requiring optimal levels of O_2. Additionally, for nitrogen fixation to occur, they must possess a mechanism to create microaerophilic or anaerobic conditions, allowing the nitrogenase enzyme to remain active.

The airspace above crops harbors a diverse range of microbial species. For instance, *Actinobacteria, Proteobacteria, Firmicutes, Chlorobi,* and *Bacteroidetes* are commonly encountered. In their review on the evolution of nitrogenase (*nif*) genes, Boyd & Peters (2013) suggest that several genera of atmospheric bacteria may have undergone gene deletions during the evolutionary process.

As mentioned previously, the first known species of rhizobia, *Rhizobium leguminosarum*, was identified in 1889. Most of the research conducted on these microbes has focused on crop and forage legumes such as clover, alfalfa, beans, peas, and soybeans. Typically, these microbes reside in the soil, but they can become airborne due to disturbances caused by farming activities, wind, turbulence, tornadoes, or dust storms. They can float in the troposphere or rise to higher altitudes and migrate over long distances (Acosta-Martinez et al., 2015; Smith et al., 2012).

Mycorrhizal spores can also be dispersed by wind at low altitudes, across soil surfaces, natural vegetation areas, and cultivated fields. Egan et al. (2014) have detected AM fungal spores in the air above crops. Chagnon et al. (2013) and Kivlin et al. (2011) have also reported the presence of propagules of symbiotic fungi, such as arbuscular mycorrhizas (AMs), in the air.

Ectomycorrhizal propagules (spores) have been found in the troposphere as well. Ectomycorrhiza (ECM) species like *Laccaria laccata* and *Laccaria amethystine* produce basidiospores that are resistant to atmospheric conditions. These light spores, capable of withstanding harsh atmospheric conditions, can travel long distances in the air before landing and germinating in new soil (Vincenot et al., 2012).

1.5 MICROBES ON CROP'S CANOPY AND PHYLLOSPHERE: HISTORICAL ASPECTS

Within the context of this volume, the crop's canopy, including the vast phyllosphere surface encountered in agrarian regions, is considered as part of the agricultural sky. The phyllosphere, which protrudes into the agrarian sky, provides a congenial location for microbes with diverse physiological abilities. Microbiologically, the phyllosphere is a very important zone, particularly for microbes that have levitated from the soil due to turbulence. It is also important for microbes and resting structures that have migrated short or long distances independently or on the surface/crevices of plant material. The phyllosphere often provides the most congenial surface for virulent strains of plant pathogens, as they proliferate upon settling after migrating from a different location.

It would be interesting and relevant to consider the historical aspects, particularly the discoveries about phyllosphere microbes and their role in crop production. Several crop production procedures directly affect phyllosphere microbes. Examples include foliar nutrient supply, foliar application of pesticides, fungicides, herbicides, application of biocontrol agents (BCAs), and plant growth-promoting microbes, among others. Now, let us consider a few historical facts about phyllosphere microbes.

Last (1955) was the first to use the term 'phyllosphere.' This occurred while studying the microflora of the leaf surface. He was investigating the etiology and life cycle of the fungal pathogen *Sporobolomyces* sp., which affects wheat and barley (Alklett et al., 2014; Last, 1955). Regarding microbes on other parts of the plant, Boutroux (1884) had discovered microbes on flowers and their parts. More details on phyllosphere microflora and their diverse physiological and ecological functions can be found in Chapter 6. Phyllosphere microbes indeed perform several ecosystem functions.

The phyllosphere refers to the thin region, approximately 0.2 mm thick, on the surface of the leaf lamina and other parts of the plant. Globally, it represents a significantly large surface area. In fact, the phyllosphere is twice as large as the land surface on Earth. Its potential for converting light energy and fixing atmospheric CO_2 through photosynthesis into organic matter is enormous. At the same time, the phyllosphere is a region through which the entire foliar supply of water, nutrients, and chemical amendments must pass before reaching the internal organs of plants. The phyllosphere is one of the major habitats for microbes. Phyllosphere microbes possess diverse physiological abilities in nature, including biological nitrogen fixation (Ruinen, 1956) and other biochemical transformations, the release of hormones to stimulate plant growth, and biocontrol functions based on chemical compounds (e.g., antibiotics, bacteriostatic agents) secreted to inhibit plant pathogens. The microbial population in the phyllosphere varies depending on the crop species and fluctuates based on the season, weather conditions, and physiological status of the crops.

1.5.1 ICE NUCLEATING MICROBES IN AGRARIAN SKY

Ice nucleating agents traced in the atmosphere could be predominantly of biological origin, such as microbes (Murray et al., 2012). Bacterial cells, fungal propagules like hyphae, spores, or other resting structures, protozoa, and archaea may all exhibit ice nucleating properties, particularly when present in clouds in the troposphere. The earliest evidence for ice nucleating properties of bacteria, specifically *Pseudomonas syringae,* was reported by Maki et al. (1974) and Sands in 1982. They were studying the infection and spread of this plant pathogenic bacteria and found high levels of *P. syringae* in ice crystals during the winter months in wheat farming regions of Montana. Researchers at the Max Planck Institute and other centers in Europe have also discovered that spraying *P. syringae* can induce ice nucleation (Max Planck Institute, 2016; Stopelli et al., 2017; Weichner & Frohlich-Nowoisky, 2013; Robbins, 2010; Rowan, 2008; Georgakopoulos & Sands, 1992). Bacteria like *P. syringae* are commonly found on the leaf surface (phyllosphere). While they can be plant pathogens with the potential to cause disease or epidemics, spraying non-pathogenic microbes with nucleation abilities has its advantages. Pathogenic strains can also be used in denatured form to create precipitation. Overall, since 2000 A.D., there has been increasing interest in linking epiphytotic *P. syringae* and a few other species not only with agriculture (diseases) but also with earth system processes such as precipitation and snow formation.

Fungi also possess ice nucleating properties (Pouleur et al., 1992). It has been discovered that fungal species such as *Fusarium avenaceum* and *F. acumiatum* have good ice nucleation capabilities (Sesartic et al., 2013). These capabilities are comparable to well-known ice nucleating bacterial species like *Pseudomonas syringae*. Iannone et al. (2011) found that certain fungi, like *Cladosporium* sp., have hydrophobic proteins on their cell walls that contribute to ice nucleation. Lichens have the ability to nucleate ice at temperatures around – 8°C or higher (Kieft, 1988; Kieft & Ruscetti, 1990; Pouleur et al., 1992; Iannone et al., 2011).

Recently, in 2021, it was shown that viruses, which are known as pathogens on crops, are also associated with weather phenomena. Viruses also possess properties for cloud glaciation and ice crystal formation (Adams et al., 2021). Viruses found in the wind, dust particles, and clouds are known to have ice nucleating properties.

These are important discoveries about microbes and their ability to affect the global precipitation pattern. The combination of PM and nucleating microbes could have significant implications for determining precipitation patterns in specific locations.

1.6 HISTORICAL ASPECTS OF CROP DISEASES, EPIDEMICS, AND PANDEMICS CAUSED BY AERIAL MICROBES

Historical reports related to microbiology, particularly those concerning the aerial spread of crop, animal, and human diseases, indicate that Louis Pasteur was among the earliest to deliberately make well-directed efforts to understand the aerial dispersal of bacteria (Pasteur, 1860b). The transmission of disease-causing microbes over distances of more than 6 km was noted as early as the mid-19th century. In the present context, however, our focus is on crop diseases, especially pathogens harbored or transmitted through the air (i.e., the agrarian sky).

Van der Plank published a comprehensive treatise that deals with epidemics and their control. According to Van der Plank (1963), historically, crop diseases have caused severe losses. For an epidemic to occur, we need at least three natural factors that favor it: a susceptible host, a virulent pathogen, and a congenial environment for the pathogen to multiply and spread (Ameson, 2001; Madden et al., 2007). The timing of infection initiation, multiplication, and disease/devastation caused by the pathogen are aspects to watch. Additionally, crop growth stage and season are factors that control epidemics (Drenth, 2004; O'Grada, 2006). There are monocyclic epidemics caused by soil-borne pathogens, which have only one infection cycle. However, polycyclic epidemics are aided by airborne fungi with repeated infection events. Rust or powdery mildew fungi are good examples. Once again, the agrarian sky plays an important role in regulating these epidemics. *Phytophthora infestans,* the causal agent of the Irish Potato Famine, was actually transmitted via several modes from Mexico (O'Grada, 2006), including aspects like precipitation.

During the past decade, i.e., 2010–2020, there have been suggestions to establish global surveillance for several crop diseases. Especially those with the potential to devastate crops and significantly reduce grain yield. Such surveillance also helps in preventing epidemics/pandemics in their early stages (Carvajal-Yepes et al., 2019). A good example of this is the global monitoring initiative for wheat streak mosaic virus (WSMV) across regions in Australia, Southeast Asia, India, West Asia, Europe, and the North American plains. This initiative began around 2010. Aerial surveys involve spectral analysis of crops using networks of drones, low and high-altitude aerostats, balloons, and satellites (Krishna, 2019).

Over the past two centuries, agricultural experts have accumulated vast knowledge about plant pathogens that thrive in the atmosphere. Detailed understanding of the life cycle, growth of disease-causing fungi/bacteria, epidemiology including aerial migration, virulence, and devastation has been achieved.

Historically, the area known as 'agrarian sky' has had a detrimental impact on crops and their productivity. Agrarian sky has been responsible for numerous epidemics and pandemics, causing devastating diseases such as rusts, powdery mildews, blotches, blights, and blasts. In response to these challenges, significant progress has been made in developing control measures. Research on the intercontinental migration of plant pathogens has been instrumental in understanding the global spread of crop diseases. Let us consider a few notable examples of crop diseases and epidemics. Rusts affecting cereals have been extensively studied, along with other diseases like blights and blotches. Aerial migration of microbes also occurs through vectors such as insects, which is another natural phenomenon influenced by the agrarian sky. This can be particularly devastating when a virulent pathogen, such as a virus, is transmitted by whiteflies, aphids, jassids, psyllids, and other vectors. There are indeed several virus diseases transmitted by vectors that have reached alarming levels of severity. For instance, in recent years, Cassava Mosaic, transmitted by the whitefly (*Bemisia* sp.), has reached epidemic proportions. Numerous crop epidemics have occurred periodically since the 17th century, affecting crops such as wheat, maize, rice, grapes, coffee, cassava, and more (2 Blades Foundation, 2021).

1.6.1 FUNGAL DISEASES: A FEW EXAMPLES

Wheat has been cultivated across various agrarian regions around the world. It is known that wheat planting and harvesting occur almost every day of the year in some regions. Since its cultivation on a larger scale, this crop, like many others, has been susceptible to attacks from biotic factors such as microbes, insects, and birds. The agrarian sky has wreaked havoc on wheat crops for millennia. Mahaffee & Stoll (2016) have conducted a literature survey and compiled information on historical references to crop diseases, particularly epidemics and pandemics affecting wheat and other staple cereals. It appears that the Sumerians and Egyptians recognized wind in the agrarian sky as the source of crop diseases. European cultivators identified that diseases were spread through the wind. The Romans referred to such disease-carrying wind as 'warm wind' or 'hot wind.' Although they did not identify the exact causal agent, they did recognize that the blowing wind carried the 'sickness factor.' The resulting loss of crops and grains could be predicted. Now, we know that such diseases on wheat could be one of the many rusts, blotches, or blights that are dispersed through the agrarian sky (McIntosh, 2009; McIntosh et al., 1995). It seems that ancient wheat cultivators in China and India were aware of blotches and blights as diseases that reduce the crop's potential yield.

Since the 1960s, we have made rapid strides in standardizing methods for collecting air samples, trapping microbes, detecting, identifying, and attributing them to crop diseases, if any. For example, using Hirst traps (Hirst & Stedman, 1963), selective medium, and microscopic analysis, we have confirmed disease-causing microbes. Such investigations allow us to track plant pathogens in the air before they land or settle on foliage. This enables us to plan prophylaxis accordingly. Currently, the trend is to use molecular methods that depend on 16S rRNA sequences to identify microbial species.

Stem rust has been a major problem for wheat-growing farmers for several millennia. In ancient times, it seems that Romans suffered from low wheat grain yield caused by disease-carrying wind and precipitation (Schumann & Leonard, 2000). Therefore, the airborne nature of rust fungi has been known for several millennia. In 1767, Italians Fontana and Tozzetti first explained the parasitic nature of stem rust fungus in wheat (Schumann & Leonard, 2000). Wheat black stem rust is also referred to as black rust, red rust, or red dust disease. It is caused by a basidiomycetous fungus called *Puccinia graminis tritici* (Pgt). In the Indian sub-continent, its repeated occurrence was studied and explained by Professor K C. Mehta (Mehta, 1941). The Pgt overwinters on a collateral host in the southern Indian hills and reinfects the wheat cultivated in the Gangetic plains. Recently, a few virulent strains of *Pgt* have appeared (e.g., Ug-99). They are known to overwinter on collateral hosts and transit intercontinental distances on dust plumes, causing epidemics/pandemics in the Middle East, North European Plains, and other regions (DeBary, 1866; Brown & Hovmoller, 2002; Hovmoller & Justesen, 2007; Hovmoller et al., 2008).

Yellow rust fungus (*P. striformis*) has caused severe disease on wheat grown in various regions of the world. Its epidemics have been observed for the past 5–7 decades. Such epidemics have been attributed to genetic drift of the pathogen, loss of resistance of the host crop, and congenial conditions. Agrarian sky, including dust plumes, wind, jet-stream, and local turbulence due to farm vehicles, has caused the epidemics. From the 1900s until now, this rust fungus has caused epidemics in different agrarian regions such as the North American Great Plains (BGRI, 2022; Brown et al., 2001),

Argentine Pampas (German et al., 2007), Asian wheat belts (Singh et al., 2005), Northeast China (Chen, 2005), the Australian wheat belt (Wellings & McIntosh, 1990), South Africa (Wellings et al., 2007a, b), and Western Europe (Bayeles et al., 1989).

Wheat leaf rust fungus *Puccinia triticina* also exhibits genetic drift and the potential to cause epidemics repeatedly in wheat-growing regions. During the past decade, it has induced epidemics leading to the loss of grains in the North American Plains and Australia (Huerto-Espino et al., 2011; Murray & Brennan, 2009). Agrarian sky supports the long-distance migration of the pathogen via wind, turbulence, and farm vehicles. There is a regular aerial spread of rust fungi from Australia to New Zealand (McKenzie, 1998; Viljanen-Rollinson & Cromey, 2002).

Studies focused on rust fungi affecting wheat, barley, sugarcane, and groundnuts. They aimed to identify aerial survival, viability period, and mode of transmission. In the case of wheat black stem rust, it was found that rust spores levitated into the troposphere and moved via wind into the Gangetic Plains. Mehta (1941) was the earliest to recognize the collateral host and its involvement in aerial spread and re-infection of rust on wheat in the Indian subcontinent. The spread of rust propagules from one continent to another, such as from West Africa to the North American Plains, was identified in the 1970s (Brown & Hovmoller, 2002).

The earliest epidemics of Coffee rust fungus were noticed in Sri Lanka in the 1850s. The spread of the Coffee rust organism *Hemileia vastatrix* was traced to Brazil in 1970 (Bowden, 1971). It has caused epidemics at intervals in Asia and South America (Schieber & Zentmeyer, 1984).

Transmission of sugarcane rust (*P. melanocephala*) from West Africa to Caribbean islands was observed in 1978. The rust propagules are transmitted through low and high-altitude troposphere wind systems, as well as jet streams that carry levitated rust spores (Anikster et al., 2005; Purdy, 1985).

Groundnut rust (*P. arachidis*) spread from tropical America to the African peanut belt in Senegal (known as 'The Groundnut Basin') and other Sahelian countries. It has also been reported in Asia for the past few decades (Savary, 1988).

Plumeria rust, caused by *Coleosporium plumeriae*, has been recognized since 2000 A.D. This fungus is capable of long-distance transmission and has caused epidemics on *Plumeria* trees. Investigations from the 1990s to 2015 have revealed that it first appeared in Central America. Over a span of two decades, it spread from the initial focus to Australia, Southeast Asia, the Indian Subcontinent, and then to East African regions (Kakashima et al., 2017). It is now identified as a fungal disease that is present worldwide. It initially disseminated from locations in Central America (Kakashima et al., 2017).

Phytophthora infestans, the cause of potato blight, has repeatedly affected the crop in North America and Europe, particularly in the mid-19th century (1850s). This devastating disease has also impacted other regions (Ristaino & Gumpretz, 2018; Ristaino et al., 2021; Campos & Ortiz, 2020). The earliest reports on the long-distance transmission of fungal plant pathogens actually revolve around the late blight of potatoes. It was discovered that potato blight spread from Mexico to Ireland, likely facilitated by the movement of biotic factors. The intercontinental transit of people and products further accelerated its spread. The lack of hygiene in cropped fields played a major role in the epidemics, with raindrops, crop residue, and the movement of farm workers contributing to its rapid dissemination. Since then, there have been extensive studies on the long-distance spread of fungal pathogens.

During the 1930s and 1940s, Helminthosporium blight of rice emerged as a major epidemic, spreading throughout the Eastern-Gangetic Plains. This epidemic was caused by *Helminthosporium oryzae (Cochliobolus miyabeanus)* (IRRI, 2021; 2 Blades Foundation, 2021).

Southern corn blight, caused by *Bipolaris maydis*, has induced epidemics in maize cultivation regions in the North American plains, as well as in several other regions of the world. In 1968 and 1970, it devastated the maize crop in North America (Ortega et al., 2020; Gregory, 1971; Agrios, 2005; Chakraborty et al., 2020).

Fusarium head blight (FHB) has devastated wheat crops in North America and other regions across the globe. Epidemics and pandemics of FHB have been recorded in 1915, 1919, 1950, and the 1960s, with the latest outbreak occurring in the 1980s (Schmale & Bergstorm, 2010). FHB effectively utilizes the agricultural sky, including weather patterns, wind, jet streams, and precipitation (Kriss et al., 2012). This pathogen is highly airborne in its ascospore stage and can contaminate the sky above crops extensively. *Fusarium oxysporum var cubens* caused epidemics known as the 'Panama disease of Bananas' in 1868 in Sri Lanka, followed by outbreaks in the 1920s, 1930s, and 1950s in South America (Cardenas-McGill, 2021).

Powdery mildew epidemics on grapes have been known since the mid-18th century. Historically, powdery mildews have caused devastation to grapes and a few other crops. The conidial stage of the powdery mildew fungus, *Erysiphe* spp., is highly airborne. Incidences and epidemics have been reported from agrarian regions all over the world during the past century (Fondevilla & Rubiales, 2012).

In addition to the aforementioned examples of fungal agents that have historically caused severe epidemics, there are downy mildews caused by Oomycetes, smuts caused by *Ustilago* spp., and ergots caused by *Claviceps* sp. These fungal pathogens have induced epidemics in diverse crop species. Epidemics by these aerial fungi have been periodically noted during the past few decades.

1.6.2 BACTERIAL DISEASES OF CROPS

Bacterial epidemics and pandemics have wreaked havoc on crops. Plant pathogenic bacteria have spread through the air and contaminated soil and water resources. Let's consider a few examples. *Xanthomonas* species serves as a prime example. Xanthomonas is a gram-negative, aerobic, short rod-shaped bacterium that belongs to the family Pseudomonadaceae. The intercontinental movement of plant pathogenic microbes, including *Xanthomonas* spp., has been extensively documented between the 1970s and 2010. *Xanthomonas campestris pv. glycine* has attacked vast soybean fields in North America and other regions, causing multiple epidemics of soybean blotch throughout the past century. Similarly, *X. campestris pv. phaseoli* has induced epidemics of bacterial blotch in lima beans, with its spread facilitated by rainwater and wind. Periodic reports of *X. campestris pv. vesicatoria* epidemics on capsicum, tomato, brinjal, and other crops have been documented over the past 50 years. *X. campestris* is also known to affect Brassica species (Kuan et al., 1986).

Bacterial pathogens on wheat with the ability for higher severity are *X. translucens pv. cerealis, X.t. translucens,* and *X.t secalis* (Hagborg, 1974). Epidemics of bacterial blight on rice have been known for years (Saha, 2015). In the 1960s, the dwarf rice variety IR8 experienced a serious attack by virulent strains of *Xanthomonas oryzae.* The bacterial pathogen is generally endemic. It causes low levels of infection, mutates, and turns virulent. Environmental conditions such as congenial relative humidity (RH), warm temperature, wind, precipitation, and nutritional status aid its rapid spread as an epidemic. The epidemic spread to almost all rice production areas in Southeast Asia and South Asia (Bangladesh, India, Pakistan, and Sri Lanka). Processes in the agrarian sky, movement of rice grains and its products, humid and windy conditions were responsible for the spread of the bacterial pathogen.

The role of clouds, precipitation, and splashing droplets in the dissemination of fungi and bacteria was first described by Pierre Miquel in 1888 (Fitt et al., 1989). Faulwetter (1917) experimentally demonstrated the dispersal of Xanthomonas, which causes 'angular leaf spot of cotton,' mediated by wind-blown rain and water droplets. *Xanthomonas* sp. is, in fact, one of the most notorious bacterial pathogens. There are species of *Xanthomonas* that attack and cause epidemics/pandemics on crops such as cotton (angular leaf blotch) and vegetables. Historically, the genus *Xanthomonas* has repeatedly affected several crops in different regions of the earth. For example, a report of Xanthomonas leaf blight of onion caused by *Xanthomonas axonopodis* was first made in 1978 in Hawaii. Today, it has rapidly spread to the Caribbean, United States Plains, South America, South Africa, Asia, and a few European nations (Shwartz & Ghent, 2011). Obviously, the 'agricultural sky' has been congenial to the bacterial pathogen throughout the seasons and years, particularly when the disease intensity has been severe.

Epidemics of Citrus canker occurred in the 1990s in Florida, affecting the quality of fruit yield. The cause was identified as *Xanthomonas campestris var citri.*

Outbreaks of Tomato leaf spot, caused by *Xanthomonas perforans,* have been reported in several vegetable-producing regions across all continents. Over the past two decades, molecular epidemiology has been effectively utilized for diagnosis and identification of epidemic/pandemic-causing strains of the pathogenic organism. Novel sequences of 16S rRNA on the bacterial pathogens have been employed to identify the specific strains responsible for inducing these epidemics. Additionally, strain differences based on geographic location have been identified through molecular analysis. For instance, strains of Xanthomonas species causing tomato or capsicum leaf spot in Iran (Osdaghi et al., 2017), Ethiopia (Kebede et al., 2014), and Australia (Roach et al., 2018) have been identified using molecular methods (Abrahmian et al., 2019).

Pseudomonas sp. has caused epidemics on few crops during past 5 decades (McInnes et al., 1988). They have reduced crop yield. These are aerially transmitted. Plus, they get disseminated through splashing rainwater, precipitation, snow, frost, and other aeolian processes.

Citrus greening is among the recently identified diseases with potential to achieve epidemic proportions. It is aerially transmitted by a Psyllid insect vector – *Diaphorina citri.* The disease has moved from Southeast Asia to Africa, then to Europe. Now, it is spreading rapidly in the Florida's citrus belt (da Graca, 1991, 2004). Studies on spread of epidemics and pandemics continues.

The list of plant pathogens capable of long-distance transmission is increasing. For example, during the past decade (2010-2020), diseases like citrus greening have been identified as capable of long-distance transmission via an insect vector and wind. *Candidatus liberibacter* is the bacterial species that induces an epidemic on citrus.

Despite advancements in disease/epidemic detection, prophylaxis methods, development of genetically resistant crop stocks, insect vector control, and agronomic procedures that aim to avoid infection and its spread, we still encounter numerous diseases that reach epidemic proportions. The presence of the "Agrarian sky" can contribute to the occurrence, spread, and severity of epidemics and pandemics (*see* 2 Blades Foundation, 2020). Undoubtedly, the aerospace above crops plays a role in the causation of epidemics, their spread, and their severity.

1.6.3 AERIAL INSECT VECTOR DISPERSED VIRAL DISEASES OF CROPS

Historically, the agrarian sky has facilitated the spread of aerial vectors and viruses, leading to devastating outbreaks of epidemics. The agrarian sky plays a crucial role in the long-distance migration of vectors, particularly insects. In this case, we are dealing with viruses that are transmitted by airborne insect vectors. The vector and the virus effectively utilize the agrarian sky to initiate and propagate diseases, which may eventually develop into epidemics under favorable conditions. Commonly encountered airborne vectors of plant viruses include whiteflies (*Bemisia* sp.) and aphids such as *Aphis cracivora, Myzus persicae, Toxoptera citricia, A. gosssypi, Pentalonia nigranervosa,* and others. Jassids and Psyllids are additional insect vectors of viruses, while mites also serve as vectors of plant viruses. Let us consider a few examples to illustrate the devastating impact of the 'vector plus virus' combination when the agrarian sky is conducive.

Jones (2014, 2020, 2021) discusses a series of viruses and their vectors that have historically caused epidemics/pandemics on food crops. These epidemic outbreaks have occurred multiple times throughout the past century in various geographical regions. Furthermore, Jones lists certain principles and reasons for viruses that lead to epidemic outbreaks. For instance, the introduction of a crop into a new area or the introduction of an insect vector/

virus combination in a new area creates conditions for outbreaks. Genetic drift (mutation) towards virulence also contributes to epidemics.

Let us consider a few crop diseases caused by viruses that have reached epidemic proportions. WSMV was first observed in Nebraska in the 1920s. Its spread is facilitated by a mite known as *Aceria tosichella* (Skoracka et al., 2018). This vector can also transmit WSMV infection to other crops, primarily cereals such as pearl millet, foxtail millet, maize, echinocloa, and a few grasses. The presence of collateral hosts provides viral inoculum that could develop into an epidemic on wheat (Chalupnikova et al., 2017; French & Stenger, 2003; Singh et al., 2018).

Maize mosaic virus (MMV) is a disease with the potential to reach epidemic proportions at any time, given congenial conditions for the aerial vector-leaf hopper (*Peregrinus maidis*) and virus duplication. Historical data from the 3rd to 9th century indicates that maize grown in Mexico and Guatemala were susceptible to MMV. It appears that the collapse of the Mayan kingdom in the 'Peten' region during the 9th century was attributable to MMV. The devastation of the maize crop was caused by the combination of the aerial virus and vector. Resistance genes were absent to combat the aerial viral attacks. It seems that viruliferous leaf hoppers spread from the Caribbean to Mexico during the 8th century through aeolian processes. The severity of the disease was so great that the 'Peten tribes' did not resettle in that region for a long period of time. Historically, this is a clear example of the combination of aerial microbes and insect vectors with the potential to cause crop devastation (Brewbaker, 1979). Of course, it was aided and abetted by the 'agrarian sky.'

Cassava mosaic virus, vectored by the whitefly *Bemisia* sp., has caused epidemics in various agrarian regions over the past three decades (Jones, 2020, 2021). The earliest detection of the virus disease was in Sub-Saharan tropical Africa (Nigeria) in the 1980s (Thresh et al., 1994).

1.7 FARM PROCEDURES AND AERIAL MICROBES

Soil tillage is a farm operation that has been practiced for several millennia. It has evolved from being solely a human activity to one supported by farm animals. Since the mid-1850s, it has been conducted using steam-powered or petroleum-fueled tractors with heavy discs or tines. The emission of PM and microbes has steadily increased due to soil disturbance (Cassel et al., 2003; Ottl et al., 2005; Peterfalvi et al., 2018). Soil tillage has consistently altered the microbial population and diversity in both the soil profile and the airspace above the open field. It has also caused microbial migration across fields and over longer distances. As it affects microbes, it also impacts various physicochemical transformations attributed to them. In addition to tillage, there are a series of other field and post-harvest processes that result in PM emissions (Arslan & Aybek, 2013). These processes also release dust and microbes into the air.

Tillage disturbs the surface layers of soil, inducing higher levels of microbial activity, respiration, and other activities. This leads to a higher oxidative state of the soil, resulting in an increased rate of microbial respiration and the release of more CO_2 into the atmosphere. Ploughing also helps to mix and turn the soil, promoting uniformity in pH, redox potential, moisture, and nutrient availability for crop roots and microbes. Conventional tillage releases a greater quantity of CO_2 into the atmosphere, while reduced or zero tillage helps conserve organic matter, dust, and microflora. As a result, carbon in the soil is conserved by reducing CO_2 emissions. In-season procedures such as weeding, earthing up, and pruning may also impact the microflora in the upper layers of the open field (Arslan & Aybek, 2013).

Microbes resting on PM levitate from soil and crop's canopy into the troposphere. It is normal to find soil bacteria or those resting on large PM/dust on foliage/crop's canopy and in the lower troposphere. While finer PM helps microbes migrate for longer distances, larger PM may travel shorter distances and settle down on soil along with microbes. For example, large PM (PM >10) often supports *Pseudomonas aeruginosa,* while finer particles have different genera of microbes on them (Lighthart & Schaffer, 1994; Schaffer & Lighthart, 1997).

Farm inputs such as fertilizer-N, P, K, gypsum, organic matter, pre-emergent herbicides, and insecticides also have an impact on soil microflora. Certain farm operations, like the application of biological-N fixers, directly influence microflora. Similarly, agronomic procedures that involve the use of bio-control agents for spraying alter the microbial load and diversity above the crop's canopy (Usta, 2013; Frampton et al., 2012). Bacteriophage-based formulations also affect bacterial diversity and population in the air above the crop (Jagdale & Kapudnis, 2012). Combine harvesting and grain processing procedures also affect the density of organic dust in the air and have an impact on microflora above cropped fields (Arslan & Aybek, 2011; Arslan et al., 2010; Arslan & Aybek, 2013).

Farmhouses, animal enclosures, feedlots, and grain processing regions are known to harbor a variety of microbes. The aeolian dust inside and surrounding the enclosures also harbors microbes. The basic assumption here is that the aerial space within farm processing houses, sheds, and animal enclosures is part of the 'agrarian sky.' The microbiology of aeolian space inside the farm enclosures needs attention. Microflora needs to be characterized. Often, the microflora inside the farmhouses, animal enclosures like cattle sheds, poultry sheds, and piggery homes may be in a continuum with those outside, in open fields (Wilson et al., 2002). Microbes do escape or get spilled and mixed with those above-cropped fields. *Vice versa,* i.e., the intrusion of microbes from aerospace above crop fields into farm animal enclosures and processing units is possible. Historical facts about farmhouse hygiene, dust/PM emission, and their consequences on microbial distribution need attention. It has caused several epidemics among livestock. Respiratory diseases in farm animals are often mediated by airborne microbes. Farm workers too are affected by airborne disease-causing agents found inside farmhouses (Cassey et al., 1980; Gibbs et al., 2010).

We may observe that farm animals are exposed to microflora above their heads within their enclosures. This airspace accumulates dust, chemicals, and microbes from the ground. Activities such as animal movement, feed lots, farmer's work, and open fields contribute to the suspension of microbes. For instance, samples taken from litter and the airspace in poultry enclosures revealed the presence of *Paecilomyces varioti, Trichoderma* spp., *Aureobasidium pullulans,* and *Hyalodendron lignicola* (Dennis & Gee, 1973). Pathogenic microbes that have the potential to cause respiratory disorders in farm animals and workers are commonly found inside enclosures (Cassey et al., 1980; Gibbs et al., 2010). Historical evidence suggests that simple devices like scrubbers, air filters, dust filters, and smoke screens can help prevent the unnecessary movement of dust, gases, and microbes, particularly between the inside and outside of the farmhouse.

The following seven chapters of this volume discuss various important aspects related to microbial species found in the agrarian sky. The microbiology of the agrarian sky has been explained in great detail. Furthermore, their role in nature, specifically in the aerospace domain, has been thoroughly delineated. It is worth acknowledging the significance of microorganisms and their diverse functions in the atmosphere above global agroecosystems.

KEYWORDS

- **aerial microbes**
- **aerial vectors**
- **aeromicrobiology**
- **Anton van Leuwenhoek**
- **atmosphere**
- **bio-geochemical cycles**
- **crop diseases**
- **discovery**
- **infusoria**
- **Louis Pasteur**
- **nutrient dynamics**
- **phyllosphere**

REFERENCES

2 Blades Foundation, (2021). *Major Crop Disease Outbreaks: Disease Epidemics are an Ongoing Concern* (pp. 1–12). https://2blades.org/major-crop-disease-outbreaks/ (accessed on 19 January 2024).

Abrahmian, P., Timilsina, S., Minsavage, G. V., Potnis, N., Jones, J. B., Goss, E. M., & Vallad, G. E., (2019). Molecular epidemiology of *Xanthomonas* outbreaks in tomato plants from transplant as determined by single-nucleotide polymorphism analysis. *Applied and Environmental Microbiology, 85*(18), e01220-19. doi10.1128/AEM.01220-19/.

Acosta-Martinez, V., Van, P. S., Moore-Kucera, J., Baddock, M. C., & Zobeck, T. M., (2015). Microbiology of wind eroded sediments: Current knowledge and future. *Aeolian Research, 18*, 99–113.

Adams, M., Atanasova, N. S., Sofieva, S., & Murray, B. J., (2021). Ice nucleation by viruses and their potential cloud glaciation. *Biogeosciences, 18*, 4431–4444. https://doi.org/10.5194/bg-18-4431-2021/.

Agrios, G. N., (2005). Plant diseases caused by fungi. *Plant Pathology* (5th edn., pp. 467, 468). Elsevier Academic, Amsterdam, Netherlands.

Aguilera, A., De Diego-Castilla, G., Osuna, S., Bardera, R., Sor Mendi, S., Blanco, Y., & González-Toril, E., (2018). *Microbial Ecology in the Atmosphere: The Last Extreme Environment*. Intech Open. doi: http://dx.doi.org/10.5772/intechopen.81650/.

Aleklett, K., Hart, M., & Shade, A., (2014). The microbial ecology of flowers: An emerging frontier in phyllosphere research. *Botany, 92*, 253–266. doi: 10.1139/cjb-2013-0166?.

Ameson, P. A., (2001). Plant disease epidemiology: Temporal aspects. *Plant Health Instructor*. American Phytopathological Society. doi: 10.1094/PHI-A-2001-0524-01.

Anikster, Y., Eilam, T., Bushnell, W. R., & Kosman, E., (2005). Spore dimensions of *Puccinia* species of cereal hosts as determined by image analysis. *Mycologia, 92*, 474–484.

Antonie, V. L., (1941). *The Collected Letters of Antoni Van Leeuwenhoek* (Vol. 2). Sets and Zeitlinger Co. Amsterdam, the Netherlands. https://www.amazon.in/Collected-Letters-Antoni-van-Leeuwenhoek/dp/9026500416/ (accessed on 19 January 2024).

Arslan, S., & Aybek, A., (2011). PM10, PM2.5 and PM1 concentrations in cotton ginners. 13th international congress and agricultural mechanization and energy, Istanbul, Turkey, Tarim Makinalan Bilimi Dergisis. *Journal of Agricultural Machinery Science, 7*, 113–119.

Arslan, S., & Aybek, A., (2013). *Particulate Matter Exposure in Agriculture* (pp. 1–33). Intech Open. http://dx.doi.org/10.5772/50084/.

Arslan, S., Aybek, A., & Ekerbiçer, H., (2010). Measurement of personal PM10, PM2.5 and PM1.0 exposures in tractor and combine operations and evaluation of health disturbances of operators. *Journal Agricultural Sciences, 16*, 104–115.

Bauer, H., Giebl, H., Hitzenberger, R., Kasper-Giebl, A., Reischl, G., Zibuschka, F., & Puxbaum, H., (2003). Airborne bacteria as cloud condensation nuclei. *Journal of Geophysical Research, 108*, 4658. doi: 10.1029/2003JD003545.

Bauer, H., Kasper-Giebl, A., Löflund, M., Giebl, H., Hitzenberger, R., Zibuschka, F., & Puxbaum, H., (2002). The contribution of bacterial and fungal spores to the organic carbon content of cloud water, precipitation, and aerosols. *Atmospheric Research, 64*, 109–119.

Bayeles, R. A., Channell, M. H., & Stigwood, P., (1989). *Yellow Rust of Wheat* (pp. 11–17). United Kingdom Pathogen Virulence Survey- 1989 Annual Report.

Beijerinck, M. W., (1898). **Concerning a Contagium vivum fluidum as cause of the spot disease of tobacco leaves**. In: Johnson, J., (ed.), *Classics* (Vol. 7, pp. 33–52). American Phytopathological Society, St. Paul.

BGRI Archive Site, (2022). *About Borlaug Global Rust Initiative* (p. 118). https://globalrust.org/about-bgri/ (accessed on 19 January 2024).

Boehm, F., & Leuschner, R. M., (1986). *Advances in Aerobiology Proceedings of International Conference on Aerobiology*. Birkhuuser Verlag, Berlin, Germany.

Boutroux, L., (1884). Sur la conservation des ferments alcooliques dans la nature. *Annales Des Sciences Naturales: IV Botanique, Paris, France, 17*, 145–209.

Bowden, J., Gregory, P. H., & Johnson, C. G., (1971). Possible wind transport of coffee leaf rust across the Atlantic Ocean. *Nature, 229*, 500, 501.

Bowman, J. P., (2006). The methanotrophs- the families *Methylococcaceae* and *Methylocystaceae*. *Pro, 5*, 266–289.

Boyd, E. S., & Peters, J. W., (2013). New insights into the evolutionary history of biological nitrogen fixation. *Frontiers in Microbiology*. https://doi.org/10.3389/fmicb.2013.00201/.

Brewbaker, J. L., (1979). Diseases of maize in the wet low land tropics and the collapse of the classical Maya civilization. *Economic Botany, 33*, 101–110.

Brown, J. K. M., &, Hovmoller, M. S., (2002). Aerial dispersal of pathogens on the Global and Continental scales and its impact on plant disease. *Science, 297*, 537–541.

Brown, R. M. Jr., Larson, D. A., & Bold, H. E., (1963). Airborne algae: Their abundance and heterogeneity. *Science, 143*, 583–585.

Brown, S. P., & Jumpponen, A., (2019). Microbial ecology of snow reveals taxa-specific biogeographical structure. *Microbial. Ecology, 77*, 946–958.

Brown, W. M., Hill, J. P., & Velasco, J. R., (2001). Barley yellow rust in North America. *Annual Review of Phytopathology, 39*, 367–384.

Burrows, S. M., Elbert, W., Lawrence, M. G., & Posch, U., (2009). Bacteria in the global atmosphere- part 1. Review and synthesis of literature data for different ecosystems. *Atmospheric Chemistry and Physics, 9*, 9263–9280.

Campos, H., & Ortiz, O., (2020). *The Potato Crop: Its Agricultural, Nutritional and Social Contribution to Humanoids* (pp. 351–430). Springer Open Book. https://doi.org/10.1007/978-3-030-28683-5/.

Cao, C., Jiang, W., Fang, J., Lang, J., Tian, G., Jiang, J., & Zhu, T. F., (2014). Inhalable microorganisms in Beijing's PM2.5 and PM10 pollutants during a severe smog event. *Environmental Science and Technology, 48*, 1499–1507.

Cardenas-McGill, S., (2021). *Plant Pandemics Threaten Future Food Supply* (pp. 1–7). Futurity. https://www.futurity.org/plant-pandemics-crops-food-supply-2571952-2/ (accessed on 19 January 2024).

Carvajal-Yepes, M., Cardwell, K., Nelson, A., Garrett, K. A., Giovani, B., Saunders, D. G. O., Kamoun, S., et al., (2019). A global surveillance system for crop diseases (PDF). *Science, 364*(6447), 1237–1239. American Association for the Advancement of Science (AAAS). doi: 10.1126/science.aaw1572. ISSN 0036-8075. PMID 31249049. S2CID 195750384/.

Cassel, T., Trzepla-Nabaglo, K., & Flocchini, R., (2003). PM10 emission factors for harvest and tillage of row crops. *International Emission Inventory Conference 'Emission Inventories – Applying New Technologies*. San Diego. https://www3.epa.gov/ttn/chief/conference/ei12/poster/cassel.pdf/ (accessed on 19 January 2024).

Cassey, J. A., Curriero, F. C., Cosgrove, S. E., Nachman, K. E., & Schwartz, B. S., (1980). High-density livestock operations, crop field application of manure, and risk of community-associated methicillin-resistant staphylococcus aureus infection in Pennsylvania. *JAMA International Medicine, 173*(21), 1980–1990. doi, 10.1001/jamainternmed.2013.10408/.

Chagnon, P., Bradley, R. L., Maherali, H., & Klironomos, J. N., (2013). A trait-based framework to understand life history of mycorrhizal fungi. *Trends in Plant Science, 18*, 484–491. doi:10.1016/j.tplants.2013.05.001/.

Chakraborty, P., Chakraborty, A., & Gupta-Bhattacharya, S., (2020). Dispersal of airborne pathogenic conidia of *Bipolaris oryzae* inciting brown spot disease of paddy in West Bengal, India. *Journal of Tropical Agriculture, 58*, 197–208.

Chalupnıkova, J., Kundu, J. K., Singh, K., Bartakova, P., & Beoni, E., (2017). Wheat streak mosaic virus: Incidence in field crops, potential reservoir within grass species and uptake in winter wheat cultivars. *Journal of Integrated Agriculture, 16*, 60345–60357.

Chen, X. M., (2005). Epidemiology and control of stripe rust (*Puccinia striiformis f sp. tritici*) on wheat. *Canadian Journal of Plant Pathology, 27*, 314–337.

Christner, B. C., Cai, R., Morris, C. E., McCarter, K. S., Foreman, C. M., Skidmore, M. L., Montross, S. N., & Sands, D. C., (2012). Geographic, seasonal, and precipitation chemistry influence on the abundance and activity of biological ice nucleators in rain and snow. *Proceedings of National Academy of Science, USA, 105*, 18854–18859.

Chuvochina, M. S., Marie, D., Chevallier, S., Petit, J. R., Normand, P., & Alekhina, I. A., (2011). Community variability of bacteria in Alpine snow (Mont Blanc) containing Saharan dust deposition and their snow colonization potential. *Microbes and Environment, 26*, 237–247.

Creamean, J. M., Suski, K. J., Rosenfeld, D., Cazorla, A., DeMott, P. J., Sullivan, R. C., White, A. B., et al., (2013). Dust and biological aerosols from the Sahara and Asia influence precipitation in the Western U.S. *Science, 6127*(339), 1572–1578.

Crombie, A., & Murrell, J. C., (2011). Methods in methane metabolism: Methanotrophy. *Methods in Enzymology, 495*, 119–133.

Cunningham, D. D., (1873). Microscopic examinations of air (Calcutta: Superintendent of government printing). Quoted in: Earth's stratosphere and microbial life. DasSarma et al., 2020 *Current Issues in Molecular Biology, 38*, 197–244. https://doi.org/10.21775/cimb.038.197.

Cuthbertson, L., & Pearce, D. A., (2017). Aeromicrobiology. In: Margesin, R., (ed.), *Psychrophiles: From Biodiversity to Biotechnology* (pp. 41–55). Cham, Switzerland: Springer International Publishing.

Da Graça, J. V., & Korsten, L., (2004). Citrus huanglongbing: A review; present status and future strategies. In: Naqvi, S. A. M. H., (ed.), *Diseases of Fruits and Vegetables* (Vol. I, pp. 229–245). Kluwer Academic Press, Dordrecht, The Netherlands.

Da Graça, J. V., (1991). Citrus greening disease. *Annual Review of Phytopathology, 29*, 109–136.

Darwin, C. R., (1845a). *Journal of Research into Natural History and Geology of the Countries Visited During the Voyage of HMS Beagle Round the World*. Under the Command of Captain Fitzroy, R.N. John Murray, London.

Darwin, C., (1845b). An account of the fine dust which often falls on vessels in the Atlantic Ocean. *Quarterly Journal of Geological Society of London,2*, 26–30.

DasSarma, P., Antunes, A., Simoes, M. F., & DasSarma, S., (2020). Earth's stratosphere and microbial life. *Current Issues in Molecular Biology, 38*, 197–244. https://doi.org/10.21775/cimb.038.197.

DasSarma, S., & DasSarma, P., (2018). Survival of microbes in Earth's stratosphere. *Current Opinion in Microbiology, 43,* 24–30.

DeBary, A., (1866). *Neue Untersuchungen uber die Uredineen insbesondere die Entwicklung der Puccinia Graminis und den Zusammenhang Duselben Mit Aecidium Berberis* (pp. 15–20). In: Monalsber. K. Preuss. Akademi. Wiss. https://rusttracker.cimmyt.org/wp-content/uploads/2011/11/rustdiseases.pdf/ (accessed on 19 January 2024).

Dedysh, S. N., & Dunfield, P. F., (2011). Methods in methane metabolism part B: Methanotrophs. *Methods in Enzymology, 495*, 31–44.

DeLeon-Rodriguez, N., Lathem, T. L., Rodríguez, R. L., Barazesh, J. M., Anderson, B. E., Beyersdorf, A. J., Ziemba, L. D., et al., (2013). Microbiome of the upper troposphere: Species composition and prevalence, effects of tropical storms, and atmospheric implications. *Proceedings of the National Academy of Sciences of the United States of America, 110*(7), 2575–2580.

Delort, M., Vaitilingom, M., Amato, P., Sancelme, M., Parazols, M., Maihot, G., Laj, P., & Deguillaume, L., (2010). A short review of the microbial population in clouds: Potential roles in atmospheric chemistry and nucleation processes. *Atmospheric Research, 98*, 249–260. https://doi.org/10.1016/j.atmosres.2010.07.004/ (February, 17th, 2022).

Dennis, C., & Gee, J. M., (1973). Microflora of broiler house litter and dust. *Journal of General Microbiology, 78*, 101–107.

Di Liberto, T., (2018). *Dust from the Sahara Desert Stretches Across the Tropical Atlantic Ocean in Late June/Early July 2018*. www.climate.gov/news-features/event-tracker/dust-sahara-desert-stretches-acrosstropical-atlantic-ocean-late (accessed on 19 January 2024).

Dillon, K., & Fennel, D. E., (2021). *Bacteria and Blue Green Algae Get Rides on Clouds* (pp. 1–4). School of Environment and Biological Sciences, Rutgers University, New Brunswick, NJ, USA. http://astrobiology.com/2021/02/bacteria-and-algae-get-rides-in-clouds.html (accessed on 19 January 2024).
Dimmic, R. L., & Akers, A. B., (1969). *An Introduction to Experimental Aerobiology* (p. 494). Wiley Interscience, New York.
Drenth, A., (2004). Fungal epidemics-does spatial structure matter. *New Phytologist, 163*, 4–7. doi: 10.1111/j.1469-8137.2004.01116.x.
Duan, X., Guo, C., Zhang, C., Li, H., Zhou, Y., Gao, H., Xia, X., et al., (2021). Effect of East Asian atmospheric particulate matter deposition on bacterial activity and community structure in the oligotrophic Northwest Pacific. *Environmental Pollution, 283*, 117088, https://doi.org/10.1016/j.envpol.2021.117088/.
Dyar, H. G., (1894). XII.— On certain bacteria from the air of New York. *Annals of the New York Academy of Sciences, 8*(1), 322–380.
Egan, E., Li, D. W., & Klironomos, J., (2014). Detection of arbuscular mycorrhizal fungal spores in the air across different biomes and ecoregions. *Fungal Ecology, 12*,1–7. doi: 10.1016/j.funeco.2014.06.004 /.
Ehrenberg, C. G., (1844). *Bericht Iiber die zur Bekanntmachun Geeigneten Verhandlungen der Konigl* (Vol. 9, pp. 194–197). Preuss. *Akad. Wiss. (Berlin)*.
Everts, S., (2008). *Bacteria in Clouds* (Vol. 86, No. 15, pp. 23–28). Chemical Engineering News. https://cen.acs.org/articles/86/i15/Bacteria-Clouds.html/ (accessed on 19 January 2024).
Faulwetter, R. C., (1917). Dissemination of the angular leaf spot of cotton. *Journal of Agricultural Research, 8*, 457–475.
Fitt, B. D. L., McCartney, H. A., & Walklate, P. J., (1989). The role of rain in dispersal pathogen inoculum. *Annual Review of Phytopathology, 27*, 24–70.
Fondevilla, S., & Rubiales, D., (2012). Powdery mildew control in pea: A review. *Agronomy for Sustainable Development, 32*, 401–409.
Frampton, R. A., Pitman, A. R., & Fineran, P. C., (2012). Advances in bacteriophage-mediated control of plant pathogens. *International Journal of Microbiology, 2012*, 326452. doi: 10.1155/2012/326452/.
Frank, A. B., (1889). Über die pilzsymbiose der leguminosen. *Berichte der Deutschen Botanischen Gesellschaft, 7*, 332–346.
Frankland, P. F., & Frankland, G. C., (1890). The nitrifying process and its specific ferment. *Part I Philosophical Transactions of Royal Society B., 181,* 107–128.
French, R. C., & Stenger, D. C., (2003). Evolution of wheat streak mosaic virus dynamics of population growth within plants may explain limited variation. *Annual Review of Phytopathology, 41*(1), 199–214. doi: 10.1146/annurev.phyto.41.052002.095559/.
Georgakopoulous, S., & Sands, D. C., (1992). Epiphytic populations of *Pseudomonas syringae* on barley. *Canadian Journal of Microbiology, 38*, 111–114.
Gergerich, R. C., & Dolja, V. V., (2006). Introduction to plant viruses, the invisible foe. *The Plant Health Instructor*. doi: 10.1094/PHI-I-2006-0414-01.
German, S., Barcellos, A., Chaves, M., Kohli, M., Campos, P., & De Viedma, L., (2007). The situation of common wheat rusts in the Southern Cone of America and perspectives for control. *Australian Journal of Agricultural Science, 58*, 620–630.

Gibbs, S. G., Green, C. F., Tarwater, P. M., Mota, L. C., Mena, K. D., & Scarpino, P. V., (2006). Isolation of antibiotic-resistant bacteria from the air plume downwind of a swine confined or concentrated animal feeding operation. *Environ. Health Perspect., 114*(7), 1032–1037.

Gregory, P., (1971). The Leeuwenhoek lecture-1970. Airborne microbes: Their significance and distribution. *Proceedings of Royal Society B- Biology, 177*, 469–483.

Gregory, P., (1973). *The Microbiology of the Atmosphere* (p. 377). Leonard Hill, Aylesbury, UK.

Griffin, D. W., (2004). Terrestrial microorganisms at an altitude of 20,000 m in Earth's atmosphere. *Aerobiologia, 20*(2), 135–140.

Griffin, D. W., (2008). Non-spore forming eubacteria isolated at an altitude of 20,000 m in Earth's atmosphere: Extended incubation periods needed for culture-based assays. *Aerobiologia, 24*(1), 19–25.

Griffin, D. W., Garrison, V. H., Herman, J. R., & Shinn, E. A., (2001). African desert dust in the Caribbean atmosphere: Microbiology and public health. *Aerobiologia, 17*(3), 203–213.

Hagborg, W. A. F., (1974). Notes on bacterial diseases of cereals and some other crop plants. *Canadian Plant Disease Survey, 54*, 129–151.

Henderson, T. J., & Salem, H., (2016). The atmosphere: Its developmental history and contributions to microbial evolution and habitat. In: *Aerobiology: The Toxicology of Airborne Pathogens and Toxins* (pp. 1–41). doi: 10.1039/9781849737913-00001.

Hirst, J. M., & Stedman, O. J., (1963). Dry liberation of fungal spores by raindrops. *Journal of General Microbiology, 33*, 335–344.

Hovmøller, M. S., & Justesen, A. F., (2007). Rates of evolution of avirulence phenotypes and DNA markers in a northwest European population of *Puccinia striiformis* f. sp. *tritici*. *Molecular Ecology, 16*, 4637–4647.

Hovmoller, M. S., Yahaouii, A. H., Milus, E. A., & Justessen, A. S., (2008). Rapid global spread of two aggressive strains of a wheat rust. *Molecular Ecology*. doi: 10.1111/j.1365-294X.2008.03886.x/.

Huerta-Espino, J., Singh, R. P., Germán, S., McCallum, B. D., Park, R. F., Chen, W. Q., et al., (2011). Global status of wheat leaf rust caused by *Puccinia triticina*. *Euphytica, 179*, 143–160. doi: 10.1007/s10681-011-0361-x.

Iannone, R., Chernoff, D. I., Pringle, A., Martin, S. T., & Bertram, A. K., (2011). The ice nucleation ability of one of the most abundant types of fungal spores found in the atmosphere. *Atmospheric Chemistry and Physics, 11*(3), 1191–1201. doi: 10.5194/acp-11-1191-2011.

Imshenetsky, A. A., Lysenko, S. V., & Kazakov, G. A., (1978). Upper boundary of the biosphere. *Applied and Environmental Microbiology, 35*(1), 1–5.

Imshenetsky, A. A., Lysenko, S. V., & Lach, S. P., (1979). Microorganisms of the upper layer of the atmosphere and the protective role of their cell pigments. In: *Life Sciences and Space Research, XVII: Proceedings of the Open Meeting of the Working Group on Space Biology of The Twentieth Plenary Meeting of COSPAR* (pp. 105–110).

Imshenetsky, A. A., Lysenko, S. V., Kasakov, G. A., & Ramkova, N. V., (1977). Resistance of stratospheric and mesospheric micro-organisms to extreme factors. *Life Sciences and Space Research* (Vol. 15, pp. 37–39). Paper presented at: COSPAR Life Sciences and Space Research. Pergmon Press, Philadelphia, Pennsylvania, USA.

Imshenetsky, A. A., Lysenko, S. V., Kazakov, G. A., & Ramkova, N. V., (1976). On micro-organisms of the stratosphere. *Life Sciences and Space Research, 14*, 359–362.

IRRI, (2012). *Rice Knowledge Bank.* International Rice Research Institute, Manila, Philippines http://www.knowledgebank.irri.org/training/fact-sheets/pest-management/diseases/item/brown-spot/ (accessed on 19 January 2024).

ISRO, (2009). *New Microbes Discovered in Earth's Stratosphere* (pp. 1–3). Indian Space Research Organization, Ahmadabad, India https://www.sciencedaily.com/releases/2009/03/090318094642.htm (accessed on 19 January 2024).

Jagdale, S., & Kapudnis, B., (2012). Bacteriophage liquid formulation: A potential green tool for the management of pomegranate bacterial blight. *Biological Control, 158,* 104597 https://doi.org/10.1016/j.biocontrol.2021.104597 /.

Janda, J. M., & Abbott, S. L., (2011). 16S rRNA gene sequencing for bacterial identification in the diagnostic laboratory: Pluses, Perils and Pitfalls. *Journal of Clinical Microbiology, 45,* 2761–2764. doi10.1128/JCM.01228-07.

Jiang, H., Chen, J. C., Murrell, P., Jiang, C., Zhang, X., Xing, T., & Smith, T. J., (2011). A brief review of methanotrophs. *Comprehensive Biotechnology* (2nd edn., Vol. 6, pp. 249–262). https://www.sciencedirect.com/topics/biochemistry-genetics-and-molecular-biology/methanotroph/ (accessed on 19 January 2024).

Jones, R. A. C., (2014). Plant virus ecology and epidemiology: Historical perspectives, recent progress, and future prospects. *Annals Applied Biology, 164,* 320–347.

Jones, R. A. C., (2020). Disease pandemics and major epidemics arising from new encounters between indigenous viruses and introduced crops. *Viruses, 12,* 1388. doi: 10.3390/v12121388 www.mdpi.com/journal/viruses/ (accessed on 19 January 2024).

Jones, R. A. C., (2021). Global plant virus disease pandemics and epidemics. *Plants, 10,* 233 https://doi.org/10.3390/plants10020233/.

Kakashima, M., Ji, J. X., Zhao, P., Wang, Q., Li, Y., & McKenzie, E. H. C., (2017). Geographic expansion of a rust fungus on *Plumeria* in Pacific and Asian countries. *New Zealand Journal of Botany, 55*(2), 178–186. doi: 10.1080/0028825X.2017.1281316/.

Kebede, M., Timilsina, S., Ayalew, A., Admassu, B., Potnis, N., Minsavage, G. V., Goss, E. M., et al., (2014). Molecular characterization of *Xanthomonas* strains responsible for bacterial spot of tomato in Ethiopia. *European Journal of Plant Pathology, 140,* 677–688. doi: 10.1007/s10658-014-0497-3/.

Khodadad, C., Gregory, W. M., Leandro, J. M., Thakrar, P. J., Micheal, L. A., Catechis, J. A., & Smith, D. J., (2017). Stratosphere conditions inactivate bacterial endospores from a mars spacecraft assembly facility. *Astrobiology, 17*(4), 337–350. doi:10.1089/ast.2016.1549/.

Kieft, T. L., & Ruscetti, T., (1990). Characterization of biological ice nuclei from a lichen. *Journal of Bacteriology, 172,* 3519–3523.

Kieft, T. L., (1988). Ice nucleation activity in lichens. *Applied Environmental Microbiology, 54,* 1678–1681.

Kivlin, S., Hawkes, C. V., & Treseder, K. K., (2011). Global diversity and distribution of arbuscular mycorrhizal fungi. *Soil Biology and Biochemistry, 43,* 2294–2303. doi:10.1016/j.soilbio.2011.07.012.

Krishna, K. R., (2019). *Aerial Robotics in Agriculture: Parafoils, Blimps, Aerostats and Kites* (p. 498). Apple Academic Press Inc. Palm Bay, Florida, US.

Krishna, K. R., (2022). *The Agricultural Sky: A Concept to Revolutionize Farming* (p. 588). Apple Academic Press Inc. Palm Bay, Florida, USA.

Kriss, A. B., Paul, P. A., & Madden, L. V., (2012). Variability in *Fusarium* head blight epidemics in relation to global climate fluctuations as represented by the El-Nino-Southern

Oscillation and other atmospheric patterns. *Phytopathology, 102*, 55–64. doi: 10.1094/PHYTO-04-11-0125.

Kuan, T. L., Minsavage, G. V., & Schad, N. W., (1986). Aerial dispersal of *Xanthomonas campestris* pv. *campestris* from naturally infected *Brassica campestris. Plant Disease, 70*, 409–413.

La Duc, M. T., (2004). Microbial monitoring of spacecraft and associated environments. *Microbia lEcology, 47*, 150–158.

Lagomarsino, A., Agnelli, A. E., Pastorelli, R., Pallara, G., Rasse, D. P., & Silvennoinen, H., (2016). Past water management affected GHG production and microbial community pattern in Italian rice paddy soils. *Soil Biology and Biochemistry, 93*, 17–27. doi: 10.1016/j.soilbio.2015.10.016/.

Last, F. T., (1955). Seasonal incidence of *Sporobolomyces* on cereal leaves. *Transactions of British Mycological Society, 38*(3), 221–239. doi: 10.1016/S0007-1536(55)80069-1.

Lee, T. F., & Eggleston, P. T., (1988). Airborne algae and cyanobacteria. *Grana, 28*, 63–66.

Legg, J. P., Jeremiah, S. C., Obiero, Maruthi, M. N., Ndvetabula, I., Okao-Okubja, Bouwenmeester, H., et al., (2011). Comparing the regional epidemiology of the cassava mosaic and cassava brown streak virus pandemics in Africa. *Virus Research, 159*, 161–170. doi: 10.1016/j.virusres.2011.04.018/.

Legg, J., & Fauquet, C., (2004). Cassava mosaic Gemini viruses in Africa. *Plant Molecular Biology, 56*, 585–599.

Lehtovirta-Morley, L. E., (2018). Ammonia oxidation: Ecology, physiology, biochemistry and why they must all come together. *FEMS Microbiology Letters, 365*, fny058. doi, 10.1093/femsle/fny058/.

Leveau, J. H. J., & Tech, J. J., (2011). Grapevine microbiomics: Bacterial diversity on grape leaves and berries revealed by high-throughput sequence analysis of 16S rRNA amplicons. *Acta Horticulture, 905*, 31–42. doi: 10.17660/ActaHortic.2011.905.2/.

Lighthart, B., & Schaffer, B. T., (1994). Bacterial flux from chaparral into the atmosphere in mid-summer at a high desert location. *Atmospheric Environment, 28*, 1267–1274.

Madden, L., Huges, G., & Van, D. B. F., (2007). Study of plant disease epidemics. *American Phytopathological Society Get Web Address the Study of Plant Disease Epidemics* (p. 238). American Phytopatological Society, Pilot Knob, Minnesota, USA. https://doi.org/10.1094/9780890545058/.

Mahaffee, W. F., & Stoll, R., (2016). The ebb and flow of airborne pathogens: Monitoring and use of disease management decisions. *Phytopathology Review, 106*, 420–431.

Maki, L. R., Galyan, E. L., & Caldwell, D. R., (1974). Ice nucleation induced by *Pseudomonas syringae. Applied and Environmental Microbiology, 28*, 456–459.

Max Plank Institute, (2016). *The Effect of Bacterial Ice Nuclei*. Mainz, Germany. https://www.mpg.de/10470442/ice-formation-bacteria-syringae/ (accessed on 19 January 2024).

McInnes, T. B., Gitatis, R. D., McCarter, S. M., Jaworski, C. A., & Phatak, S. C., (1988). Airborne dispersal of bacteria in tomato and pepper transplant fields. *Plant Disease, 72*, 575–579.

McIntosh, R. A., (2009). History and status of the wheat rusts. In: McIntosh, R. A., (ed.), *Proceedings of the Borlaug Global Rust Initiative* (pp. 11–23). Technical workshop BRI Cd Obregon, Mexico.

McIntosh, R. A., Wellings, C. R., & Park, R. F., (1995). *Wheat Rusts an Atlas of Resistance Genes* (pp. 1–200). CSIRO Publications Victoria, Australia.

Mckenzie, E. C. H., (1998). Rust fungi of New Zealand—An introduction, and list of recorded species. *NewZealand Journal of Botany, 36*, 233–271. doi: 10.1080/0028825X.1998.9512564/.

Mehta, K. C., (1941). The wheat rust problem of India. *Current Science, 10*, 357–361.

Meier, F. C., & Lindbergh, C. A., (1935). Collecting micro-organisms from the arctic atmosphere: With field notes and material. *Scientific Monthly, 40*, 5–20.

Meier, F. C., (1936). *Paper, US Army Air Corps Stratosphere Flight of 1935 in Balloon Explorer II* (pp.152, 153). National Geographic Society; Washington DC.

Morgan, W., (2021). *The Physician Who Presaged the Germ Theory of Disease Nearly 500 Years Ago* (p. 105). https://www.scientificamerican.com/article/the-physician-who-presaged-the-germ-theory-of-disease-nearly-500-years-ago/ (accessed on 19 January 2024).

Morris, C. E., Monteil, C. L., & Berge, O., (2013). The life history of *Pseudomonas syringae*: Linking agriculture to earth system processes. *Annual Review of Phytopathology, 51*. doi: 10.1146/annurev-phyto-082712-102402/.

Morrris, C. E., Sands, D. C., Bardin, M., Jaenicke, R., Vogels, B., Leyronas, C., Ariya, P. A., & Psenners, R., (2011). Microbiology and atmospheric processes: Research concerning the impact of airborne micro-organisms on the atmosphere and climate. *Biogeosciences, 8*, 17–25. doi: 10.5194/bg-8-17-2011/.

Müller, A., (1875). *Ammoniakgehalt des Spree- und Wasserleitungs Wassers in Berlin* (pp. 121–123). Fortsetzung der Vorarbeiten zu einer zukünftigen Wasser-Versorgung der Stadt Berlin ausgeführt in den Jahren 1868 und 1869.

Murray, B. J., O'Sullivan, D., Atkinson, J. D., & Webb, M. E., (2012). Ice nucleation by particles immersed in super cooled cloud droplets. *Chemical Society Reviews, 41*, 6519–6554. doi: 10.1039/c2cs35200a.

Murray, G. M., & Brennan, J. P., (2009). Estimating disease losses to the Australian wheat industry. Australasia. *Plant Pathology, 38*, 558–570.

Nicholson, W. L., Race, M. S., & Schuerger, A. L., (2018). Migrating microbes and planetary protection. *Trends in Microbiology: Science and Society* (pp. 389–392). doi: 10.1016/j.tim.2009.07.001: https://www.researchgate.net/publication/26785528/ (accessed on 19 January 2024).

O'Grada, C., (2006). *Irelands Great Famine* (p. 212). University College, Dublin, Ireland. https://www.ucdpress.ie/display.asp?K=9781904558583&aub=Cormac%20O%20Grada/ (accessed on 19 January 2024).

Ortega, S. F., Ferrocino, I., Adams, I., Silvestri, S., Spadaro, D., Gullino, M. L., & Boonham, N., (2020). Monitoring and surveillance of aerial mycobiota rice paddy through DNA metabarcoding and qPCR. *Journal Fungi, 6*, 372–378.

Osdaghi, E., Taghavi, S. M., Hamzehzarghani, H., Fazliarab, A., & Lamichhane, J. R., (2017). Monitoring the occurrence of tomato bacterial spot and range of the causal agent *Xanthomonas perforans* in Iran. *Plant Pathology, 66*, 990–1002. doi: 10.1111/ppa.12642.

Osowski, V., (2018). *Microbes Catch a Ride on Fog but Change During the Trip* (pp. 1–3). Cary Institute of Ecosystem Studies, Michigan State University, East Lansing, Michigan, USA. https://www.futurity.org/fog-microbes-1876442/ (accessed on 19 January 2024).

Otim-Nape, G. W., & Thresh, J. M., (1998). The current pandemic of cassava mosaic virus disease in Uganda. In: Jones, G., (ed.), *The Epidemiology of Plant Diseases* (pp. 423–443). Kluwer: Dordrecht, The Netherlands.

Ottl, D., Funk, R., & Sturm, P., (2005). PM emission factors for farming activities. In: *Proceedings of the 14th Symposium Transport and Air Pollution* (Vol. 85, pp. 411–419). Technical University Graz, Austria.

Pasteur, L., (1860a). Expériences relatives aux générations dites spontanées. *Comptes rendus Hebdomadaires des Séances de l'Académie des Sciences, 50,* 303–307. (English translation: Pasteur, L., (1860a). Experiences relating to so-called spontaneous generations. *Weekly Reports of the Sessions of the Academy of Sciences, 50*, 303–307.

Pasteur, L., (1860b). Suite a une pr ` ec´ edente communication relative aux ´ gen´ erations dites spontan ´ ees, Comptes rendus hebdomadaires ´ *Des Seances de l'Acad'emie des Sciences, 51*, 675–678. (English translation: Pasteur, L., (1862). Etudes sur les mycoderme. *C. R. Academy Science, 54,* 265–270. https://en.wikipedia.org/wiki/Nitrification/(accessed on 19 January 2024).

Pepper, I. L., & Gentry, T. J., (2015). Earth environments. Aeromicrobiology. In: Pepper, I. L., Gerba, C. P., & Gentry, T. J., (eds.), *Environmental Microbiology* (3rd edn., pp. 89–111). doi: http://dx.doi.org/10.1016/B978-0-12-394626-3.00005-3.

Peterfalvi, N., Keller, B., & Magyar, M., (2018). PM10 emission from crop production and agricultural soils. *Agro chemistry and Sol Science, 67*, 143–159. doi: 10.1556/0088.2018.67.1.10/.

Polymenakou, P. N., (2012). Atmosphere: A source of pathogen or beneficial microbes? *Atmosphere, 3*, 87–102. doi: 10.3390/atmos3010087.

Pouleur, S., (1992). *Applied Environmental Microbiology, 58*, 2960–2964.

Prenni, A., Petters, M., Kreidenweis, S., Heald, C., Martin, S., Artaxo, P., Garland, R., et al., (2009). Relative roles of biogenic emissions and Saharan dust as ice nuclei in the Amazon basin. *Nature, 2*, 402–405. doi: 10.1038/NGEO517/.

Proctor, B. E., (1934). The microbiology of the upper air. I. In: *Proceedings of the American Academy of Arts and Sciences* (Vol. 69, No. 8, pp.315–340). American Academy of Arts and Sciences.

Prospero, J. M., Blades, E., Mathison, G., & Naidu, R., (2005). Inter hemispheric transport of viable fungi and bacteria from Africa to the Caribbean with soil dust. *Aerobiologia, 21*, 1–19. doi: 10.1007/s10453-004-5872-7/.

Purdy, L. H., (1985). Sugarcane rusts. In: Roelfs, A. P., & Bushncll, W. R., (eds.), *The Cereal Rusts* (pp. 237–256). Academic Press, Orlando.

Ristaino, J. B., & Gumpertz, M. L., (2018). New frontiers in the study of dispersal and spatial analysis of epidemics caused by species in the genus *Phytophthora. Annual Review of Phytopathology, 38,* 541–576.

Ristaino, J. B., Anderson, P. K., Bebbard, D. P., Braumane, K. A., Cunniffef, N. K., Fedorof, N. V., Finegold, C., et al., (2021). The persistent threat of emerging plant disease pandemics to global food security. *Proceedings of National Academy of Sciences USA, 118*(23), 1–9, e2022239118. https://doi.org/10.1073/pnas.2022239118/.

Roach, R., Mann, R., Gambley, C. G., Shivas, R. G., & Rodoni, B., (2018). Identification of *Xanthomonas* species associated with bacterial leaf spot of tomato, capsicum and chilli crops in eastern Australia. *European Journal of Plant Pathology, 150*, 595–608. doi: 10.1007/s10658-017-1303-9.

Robbins, J., (2010). *From Trees and Grass, Bacteria that Cause Snow and Rain* (pp. 1–3). New York Times. N. Y., USA. https://www.nytimes.com/2010/05/25/science/25snow.html (accessed on 19 January 2024).

Rogers, L. A., & Meier, F. C., (1936). *National Geographic Society, Technical Papers* (pp. 146–151). Quoted in: Earth's Stratosphere and microbial life. DasSarma et al. (2020). *Current Issues in Molecular Biology, 38*, 197–244. https://doi.org/10.21775/cimb.038.197.
Rosselli, R., Fiamma, M., Deligios, M., Pintus, G., Pellizrro, G., Canu, A., Duce, P., et al., (2015). Microbial immigration across the Mediterranean via airborne dust. *Scientific Reports, 5*, 16306. doi: 10.1038/srep16306.
Rowan, K., (2008). *Snowflakes Aren't Innocent and Fluffy- They Are Bacteria Bombs from the Sky*. https://www.discovermagazine.com/planet-earth/snowflakes-arent-innocent-and-fluffytheyre-bacteria-bombs-from-the-sky/ (accessed on 19 January 2024).
Ruinen, J., (1956). Occurrence of *Beijerinckia* species in the phyllosphere. London. *Nature, 177*, 220–221.
Saha, S., Garg, R., Biswas, A., & Rai, A. B., (2015). Bacterial diseases of rice: An overview. *Journalof Pure and Applied Microbiology, 9*, 725–726.
Salisbury, J. K., (1886). On the cause of intermittent and remittent fevers, with investigations that tend to prove that these affections are caused by certain species of *Palmellae*. *American Journal of Medicine Science, 51*, 51–75.
Sands, D. C., Langhans, V. E., Scharen, A. L., & De Smet, G., (1982). The association between bacteria and rain and possible resultant meteorological implications. *Journal of Hungarian Meteorology Service, 86*, 148–152.
Savary, S., Bosc, I. P., Noirot, M., & Zadoks, I. C., (1988). Peanut rust in West Africa: A new component in a multiple pathosystem. *Plant Disease, 72*, 1001–1009.
Schaffer, B. T., & Lighthart, B., (1997). Survey of culturable airborne bacteria at four diverse locations in Oregon: Urban, rural, forest, and coastal. *Microbial Ecology, 34*, 167–177.
Scharf, C. A., (2019). *The Atmospheric Microbiome* (pp. 1–3). Scientific American Unlimited-Life Unbounded. https://blogs.scientificamerican.com/life-unbounded/the-atmospheric-microbiome/ (accessed on 19 January 2024).
Schieber, E., & Zentmeyer, G. A., (1984). *Coffee Rust in the Western Hemisphere Plant Disease, 68*, 89–93.
Schlichting, H. E. Jr., (1969). The importance of airborne algae and *Protozoa*. *Journal of Air Pollution Contr. Assoc., 19*, 946–995.
Schlichting, H. E. Jr., (1971). A preliminary study of algae and *Protozoa* in seafoam. *Bot. Mar., 14*, 24–29.
Schlichting, H. E., (1974). Periodicity and seasonality of airborne algae and *Protozoa*. In: Helmut, L., (ed.), *Phenology and Seasonality Modeling* (pp. 620434). Springer-Verlag, Heidelberg.
Schloesing, J. J. T., & Muntz, A., (1877). Sur la nitrification par les ferments organizes. *C R Academy of Science, LXXXIV*(30), 1–3.
Schmale, III. D. G., & Bergstrom, G. C., (2010). *Fusarium* head blight in wheat. *The Plant Health Instructor* (pp. 1–8). doi: 10.1094/PHI-I-2003-/.
Scholthof, K. B. G., (2000). Tobacco mosaic. *The Plant Health Instructor.* doi: 10.1094/PHI-I-2000-1010-01. http://www.apsnet.org/edcenter/intropp/lessons/viruses/Pages/TobaccoMosaic.aspx/ (accessed on 19 January 2024).
Scholthof, K. B. G., (2001). 1898 – the beginning of virology…time marches on. *The Plant Health Instructor.* doi: 10.1094/PHI-I-2001-0129-01. http://www.apsnet.org/edcenter/intropp/lessons/viruses/Pages/TobaccoMosaic.aspx/ (accessed on 19 January 2024).
Schuerger, A. C., (2004). Microbial ecology of the surface exploration of mars with human-operated vehicles. In: Cockell, C. S., (ed.), *Martian Expedition Planning* (pp. 363–386).

American Astronautical Society publication AAS03-322, Univelt Publishers, Santa Barbara, California, USA.

Schumann, G. L., & Leonard, K. J., (2000). *Stem Rust of Wheat (Black Rust).* American Phytopathology Society *the Plant Health Instructor.* doi: 10.1094/PHI-I-2000-0721-01/.

Sedlacek, C. J., (2020). It takes a village: Discovering and isolating the nitrifiers. *Frontiers in Microbiology, 11*, 1900. PMCID: PMC7431685. doi10.3389/fmicb.2020.01900/.

Sesartic, A., Lohmann, U., & Storelvmo, T., (2013). Modeling the impact of fungal spore ice nuclei on clouds and precipitation. *Environmental Research Letters, 8*, 014029. doi: 10.1088/1748-9326/8/1/014029/.

Shwartz, H., & Gent, D. H., (2011). *Xanthomonas Leaf Blight* (pp. 1–4). Fact Sheet 2951 Colorado State University Extension Service. https://extension.colostate.edu/docs/pubs/garden/02951.pdf (accessed on 19 January 2024).

Singh, K., Wegulo, S. N., Skoracka, A., & Kumar, J., (2018). Wheat streak mosaic virus: A century old virus with rising importance worldwide. *Molecular Plant Pathology, 19*, 2193–2206.

Singh, R. P., William, H. M., Huerta-Espino, J., & Rosewarne, G., (2005). Wheat rust in Asia: Meeting the challenges with old and recent technologies. In: *New Directions for a Diverse Planet, Proceedings of the 4th International Crop Science Congress* (pp. 1–13). Brisbane, Australia.

Skoracka, A., Rector, B. G., & Hein, G. L., (2018). The interface between wheat and the wheat curl mite, *Aceria tosichella*, the primary vector of globally important viral diseases. *Frontiers in Plant Science, 9*, 1098. doi: 10.3389/fpls.2018.01098/.

Smith, D. J., Griffin, D. W., & Schuerger, A. C., (2010). Stratospheric microbiology at 20 km over the Pacific Ocean. *Aerobiologia, 26*, 35–46.

Smith, D. J., Griffin, D. W., McPeters, R. D., Ward, P. D., & Schuerger, A. C., (2011). Microbial survival in the stratosphere and implications for global dispersal. *Aerobiologia, 27*(4), 319–332.

Smith, D. J., Jaffe, D. A., Birmele, M. N., Griffin, D. W., Schuerger, A. C., Hee, J., & Roberts, M. S., (2012). Free tropospheric transport of microorganisms from Asia to North America. *Environmental Microbiology, 64*, 973–985.

Smith, D. J., Timonen, H. J., Jaffe, D. A., Griffin, D. W., Birmele, M. N., & Perry, K. D., (2013). Intercontinental dispersal of bacteria and archaea by transpacific winds. *Applied Environmental Microbiology, 79*, 1134–1139.

Soffen, G. A., (1965). Atmospheric collection at 130,000 feet. *Proceedings of the Atmospheric Biology Conference*. Minneapolis, Minnesota, USA Jet Propulsion Document No. 65N24000. https://www.jstage.jst.go.jp/article/bss/23/3/23_3_151/_article/-char/ja/ (accessed on 19 January 2024).

Stoecker, L., Bernd, B., Bjorn, S., Per, H. N., Jeppe, L. N., Christian, B., et al., (2006). Cohn's Crenothrix is a filamentous methane oxidizer with an unusual methane monooxygenase. *Proceedings of National Academy of Sciences USA, 103,* 2363–2367. www.pnas.orgcgidoi10.1073pnas./ (accessed on 19 January 2024).

Stopelli, E., Conen, F., Guilbaud, C., Zopfi, J., Alwell, C., & Morris, C. E., (2017). Ice nucleators, bacterial cells and *Pseudomonas syringae. Biogeosciences, 14*, 1189–1196. doi: 10.5194/bg-14-1189-2017/. www.biogeosciences.net/14/1189/2017/ (accessed on 19 January 2024).

Tang, K., Huang, Z., Huang, J., Maki, T., Zhang, S., Shimizu, A., Ma, X., et al., (2018). Characterization of atmospheric bioaerosols along the transport pathway of Asian dust during the dust-bioaerosol-2016 campaign. *Atmospheric Chemistry and Physics, 18*, 7131–7148. https://doi.org/10.5194/acp-18-7131-2018.

Temkiv, T. S., Finster, K., Hansen, B. M., Nielsen, N. W., & Karlson, U. G., (2012). The microbial diversity of a storm cloud as assessed by hailstones. *FEMS Microbiology Ecology, 81*, 684–695.

Thresh, J. M., Fargette, D., & Otim-Nape, G. W., (1994). Effects of African cassava mosaic Gemini virus on the yield of cassava. *Tropical. Science. London, 34*, 26.

Tignat-Perrier, R., Dommergue, A., Thollot, A., Keuschnig, C., Magnad, O., Vogel, T. M., & Larose, C., (2019). Global airborne microbial communities controlled by surrounding landscapes and wind conditions. *Scientific Reports, 9*, 14441. https://doi.org/10.1038/s41598-019-51073-4/.

Treusch, A. H., Kletzin, A., Raddatz, G., Ochsenreiter, T., Quaiser, A., Meurer, G., et al., (2004). Characterization of large-insert DNA libraries from soil for environmental genomic studies of archaea. *Environmental Microbiology, 6*, 970–980. 10.1111/j.1462-2920.2004.00663.x/.

Usta, C., (2013). *Microorganisms in Biological Pest Control: A Review* (pp. 287–317). Intech Open. http://dx.doi.org/10.5772/55786/.

Vaksmaa, A., Van, A. T. A., Ettwig, K. F., Lupotto, E., Valè, G., Jetten, M. S. M., & Lüke, C., (2017). Stratification of diversity and activity of methanogenic and methanotrophic microorganisms in a nitrogen-fertilized Italian paddy soil. *Frontiers in Microbiology, 8*, 2127. doi: 10.3389/fmicb.2017.02127/.

Vanderplank, J. E., (1963). *Plant Disease: Epidemics and Control* (p. 349). Academic Press, New York.

VanderWant, J. P. H., & Dijkstra, J., (2006). **A history of plant virology** *Archives of Virology, 151*, 1467–1498.

Vergara-Fernandez, A., Morales, P., Scot, F., & Guerrero, S., (2019). Methane biodegradation and enhanced methane solubilization by the filamentous *Fusarium solani. Chemosphere, 226*, 24–35. doi: 10.1016/j.chemosphere.2019.03.116/.

Vigliotta, G., Nutricati, E., Carata, E., & Tredici, S. M., (2007). *Clonothrix fusca* Roza 1896, a filamentous, sheathed, methanotrophic proteobacterium. *Applied and Environmental Microbiology, 73*, 3556–3565. doi: 10.1128/AEM.02678-06.

Viljanen-Rollinson, S. L. H., & Cromey, M., (2002). Pathways of entry and spread of rust pathogens: Implications for New Zealand's biosecurity. *New Zealand Plant Protection.* doi: 10.30843/nzpp.2002.55.3911/.

Vincenot, L., Nara, K., Sthultz, C., Labbé, J., Dubois, M. P., Tedersoo, L., Martin, F., & Selosse, M. A., (2012). Extensive gene flow over Europe and possible speciation over Eurasia in the ectomycorrhizal basidiomycetes *Laccaria amethystina* complex. *Molecular Ecology, 21*, 281–299. https://doi.org/10.1111/j.1365-294X.2011.05392.x.

Visch, V. O. M. A., (1972). Onderzoekingen naar goene organismen en protozoen in de atmosfeer. In: *Aerobiologie* (pp. 54–57).

Waksman, S., (1953). *Sergei N. Winogradsky his Life and Work* (p. 150). New Brunswick, New Jersey: Rutgers University Press.

Warington, R., (1878). IV.—On nitrification. *Journal of Chemistry. Society. Transactions, 33, 44–51.* doi:10.1039/CT8783300044.

Warington, R., (1879). XLIX.—On nitrification. (part II.). *J. Chem. Soc., Trans.,* ***35,*** *429–456.* doi:10.1039/CT8793500429. ISSN0368-1645.

Weber, S., Lueders, T., Friedrich, M. W., &. Conrad, R., (2001). Methanogenic populations involved in the degradation of rice straw in anoxic paddy soil. *FEMS Microbiology and Ecology, 38*, 11–20.

Wehking, J., Pickersgill, D. A., Bowers, R. M., Teschner, D., Poschl, U., Frohlich-Nowoisky, J. F., & Desperes, V., (2018). Community composition and seasonal changes of archaea in coarse and fine air particulate matter. *Biogeosciences, 15*, 4205–4214. https://doi.org/10.5194/bg-15-4205-2018/.

Weichner, T., & Frohlich-Nowoisky, J., (2013). *The Effect of Bacterial Ice Nuclei* (pp. 1–3). Max-Planck Gessellshaft, Germany. https://www.mpg.de/10470442/ice-formation-bacteria-syringae (December, 12th, 2021).

Weil, T., De Filippo, C., Albanese, D., Donati, C., Pindo, M., Pavarini, L., et al., (2017). Legal immigrants: Invasion of alien microbial communities during winter occurring desert dust storms. *Microbiome, 5*, 32. doi: 10.1186/s40168-017-0249-7/.

Wellings, C. R., & McIntosh, R. A., (1990). *Puccinia striiformis* f. sp. *tritici* in Australia: Pathogenic changes during the first 10 years. *Plant Pathology, 39*, 316–325.

Wellings, C. R., (2007a). *Puccinia striiformis* in Australia: A review of the incursion, evolution, and adaptation of stripe rust in the period 1979–2006. *Australian Journal of Agricultural Research, 58*, 567–575.

Wellings, C. R., (2007b). *Puccinia striiformis f. sp. tritici* in Australasia: Pathogenic changes during the first 10 year. *Plant Pathology, 39*, 316–325.

Wilson, C. S., Morrow-Tesch, J., Straus, D. C., & Cooley, J. D., (2002). Airborne microbial flora in a cattle feedlot. *Applied and Environmental Microbiology, 68*, 3238–3242. doi:10.1128/AEM.68.7.3238-3242.2002/ (September, 10th, 2022).

Winogradsky, S., (1890). A recherches sur les organismes de al nitrification. *Annals of Institute of Pasteur, 4,* 213–331.

Winogradsky, S., (1890). B Sur les organisms de la nitrification. *Ann. Inst. Pasteur, 4*, 257–275.

Winogradsky, S., (1891). Sur les organisms de la nitrification. *Ann. Inst. Pasteur, 5*, 577–616.

Winogradsky, S., (1949). *Microbiologie du Sol* (p. 136). Problèmes et méthodes. Cinquante Années de recherches. Ceuvres Complètes. Masson & Cie. Paris.

Yong, B., (2013). *Frequent-Flier Microbes Ride Hurricanes and Live in Clouds* (pp. 1–3). National Geographic. https://www.nationalgeographic.com/science/article/frequent-flier-microbes-ride-hurricanes-and-live-in-clouds (accessed on 19 January 2024).

Yurimoto, H., Hisashi, R. H., Yasueda, Y. R., Sakai, Y., & Kato, Y., (2002). The ribulose monophosphate pathway operon encoding formaldehyde fixation in a thermotolerant methylotroph, *Bacillus brevis* S1. *FEMS Microbiology Letters*. https://doi.org/10.1111/j.1574-6968.2002.tb11345.x/.

Zaitlin, M., (1998). The discovery of the causal agent of tobacco mosaic disease. In: Kung, S. D., & Yang, S. F., (eds.), *Discoveries in Plant Biology* (pp. 105–110). World Publishing Co. Ltd. Hong Kong.

CHAPTER 2

Microbes in Atmosphere: An Agrarian Perspective

ABSTRACT

The atmosphere, specifically the troposphere, is one of the largest repositories of microbes with diverse genetic constitution and physiological abilities. The troposphere is also home to microbes that are immediately relevant to agricultural expanses. This chapter emphasizes the survival, distribution, and population density of microbes in different layers of the atmosphere. The primary sources of microbes found in the troposphere are agricultural soils, water bodies, crop canopies, and phyllosphere. Farm animals, vehicles, operations, and general farm activities release microbes into the lower troposphere. Compared to soil or water, the atmosphere is generally considered an 'extreme environment' for microbes. However, microbes can survive and proliferate in even harsher environments such as the fringes of volcanoes, hot water springs, snow, and arid desert air. The stratosphere extends from 10 km to 50 km altitude from agrarian terrain. It occasionally supports microbes that drift or levitate from the troposphere and cross over. The stratosphere is a very cold region for microbes to tolerate, and microbial population is generally weak, if present at all. Mesosphere, Thermosphere, and Ionosphere are inhospitable to microbial life forms. Occasionally, microbes have been found attached to spacecraft. Some of them have survived space journeys as clumps, where the outer layer of microbial cells perishes but the inner core survives as spores.

Atmospheric factors such as solar radiation, ambient temperature, relative humidity (RH), and moisture affect microbial activity. Aeolian processes such as wind, jet streams, tornadoes, precipitation (rainfall), snow, frost,

Microbiology of the Agricultural Sky. K. R. Krishna (Ed)

mist, fog, and smog affect microbes found in the troposphere above cropping expanses. The atmosphere aids in the transport of teragrams (Tg) of dust laden with microbes each year from one continent to another, via jet streams, dust storms, and plumes. A large biomass of microbes shifts each year in the troposphere. The microbial load of dust particles is an important factor.

During the past two decades, research on intercontinental drifts of aeolian dust, the occurrence of large dust plumes, dust storms, and their involvement in microbial migration has received priority. The phenomenon of nucleation, which aids in the formation of ice crystals, snow, and rainfall, is being intensely explored. Microbes found in the troposphere, such as *Pseudomonas syringae* and a few others, are attributed to affecting Earth's systems functions. In fact, microbiological ice nucleation, known as 'bioprecipitation,' is supposedly a major rain-creating mechanism in farming regions.

Crop residue burning is gaining acceptance among farmers. It releases particulate matter, organic dust, microbes, and gaseous matter into the troposphere. This phenomenon affects photosynthesis, respiration, and other normal physiological processes of crops. Farm animals and workers also suffer from smog and toxic compounds that settle on the soil surface periodically.

Climate change affects the survival and rates of several physiological processes of microbes in the troposphere. Conversely, airborne microbes play a role in GHG emissions such as CO_2, CH_4, and climate change processes.

The atmosphere harbors a large population of diverse plant pathogenic microbes. The troposphere plays a crucial role in the aerial migration and spread of crop diseases. Tropospheric factors such as wind, turbulence, precipitation, snow, frost, etc., facilitate the spread of plant pathogens. Additionally, the troposphere supports various aerial insect vectors that transmit diseases to crops, not only within the same field but also over longer distances. Several epidemics/pandemics have been attributed to vector-transmitted plant pathogenic viruses.

Troposphere also supports aerial microbes involved in the biological control of crop diseases and insect pests. The interactions of biocontrol microbes, in situ phyllosphere microbes, and plant pathogens are worth studying in detail. In addition, there are beneficial microbes like biological N fixers, mycorrhizal fungi, and plant growth-promoting species that thrive for a period of time in the troposphere. They are dispersed via troposphere mechanisms like wind, turbulence, clouds, and precipitation.

Overall, atmospheric layers, particularly the lower and upper troposphere, play a vital role in supporting microbial activity in agroecosystems.

2.1 ATMOSPHERE ABOVE AGRARIAN REGIONS: BACKGROUND

Atmospheric features specific to agrarian regions are the focus here. Special features relevant to microbes, their survival, growth, perpetuation, and activity require attention. There are several ways to define the atmosphere based on purpose and context. There are a few distinct layers of the atmosphere that are easily identifiable. These layers have specific traits that aid in their demarcation. Understanding the vertical profile of the atmosphere, as well as the spatial-temporal variation of characteristics/factors in different layers of the atmosphere, is necessary. Variations in atmospheric traits can be observed above each agrarian location. These features can be very specific and dependent, to some extent, on the terrain.

By definition, 'Agricultural sky' refers to the region just above the ground surface of agricultural regions on Earth. It excludes the sky (atmosphere) above oceans or permanent snow regions in the Arctic or Antarctic. Additionally, it does not include the sky above mountainous non-agricultural zones. The agricultural sky encompasses the atmosphere immediately above crops, as identified and described by Krishna (2022). Now, let us consider the physical features of the atmosphere exclusively. The atmosphere consists of several layers that surround the Earth. These layers can be distinguished based on characteristics such as temperature, pressure, radiation effects, moisture, and the distribution of microbes and other life forms. These layers are the troposphere, stratosphere, mesosphere, thermosphere, and exosphere. The exosphere leads us into outer space.

The troposphere is the layer closest to the Earth's surface, encompassing terrain, crop belts, water bodies, and infrastructure. It extends from the ground surface to an altitude of 8–15 km (5–9 miles) (NASA, 2018; Zell, 2017). The troposphere holds great significance for global agriculture, as well as for microbial life and their physiological activity in agrarian regions. Microbes are found ubiquitously in the troposphere (Polymenakou, 2012). Specifically, the troposphere above agricultural zones extends from the soil surface to 33,000 ft into the atmosphere, which is equivalent to 10 km or 6.2 miles from the agrarian soil surface. Weather phenomena occurring in the troposphere, such as clouds, water vapor, snow, frost, haze, and precipitation, greatly influence all global agricultural zones. These conditions support diverse species of microbes (Amato et al., 2007a, b; Sattler et al., 2001; Bauer et al., 2003; Deleon-Rodriguez, 2013). As we ascend to higher altitudes in the troposphere, the ambient temperature and pressure decrease. Wind and air become cooler in the upper troposphere. Microbes are able to survive and remain metabolically active even in the extremely cold conditions of moving clouds (Sattler et al., 2001). Organic substrates such as dust particles, smog, and soot become less abundant as we ascend to higher altitudes. As a result, microbial diversity and population are also sparse in the higher altitudes of the troposphere. The conditions in the troposphere near the ground surface are often most favorable for crop growth as well as microbial life forms. Microbes are known to facilitate numerous biogeochemical transformations, some of which are important ecosystem functions carried out by troposphere microbes. They play a role in maintaining natural elemental cycles in the region. The atmosphere consists of various gases, with nitrogen and oxygen being the most important, constituting about 99% of the gaseous compounds in the atmosphere. Other trace gases include carbon dioxide (CO_2), hydrogen, argon, helium, neon, and ozone. Atmospheric CO_2 is directly responsible for biomass accumulation by crops, green and blue-green algae, and other photoautotroph's.

Next, the stratosphere extends from the edge of the troposphere at an altitude of 10 km to 50 km. It is a region that surrounds the Earth. Microbial life is weak, if present at all, in the stratosphere. However, there are a few reports about microbes in the stratosphere. The ozone layer is located in the stratosphere region. Ozone absorbs high-energy UV radiation and converts it into heat. As the altitude from the ground surface increases, the stratosphere becomes warmer. Microbial life is either absent or negligible in this region. The impact of stratospheric phenomena on agrarian regions, including crops and microbes, is weak, if any. At higher altitudes, ranging from 1,500 m to 10,000 m above the ground surface, we encounter jet streams of wind. This natural phenomenon is believed to facilitate the long-range dispersal of microbes above agrarian regions (Horneck et al., 2010).

Mesosphere is the region immediately above the stratosphere. It begins at 50 km and extends to 85 km (53 miles) from the ground surface. Practically, it has little to no influence on crops, natural vegetation, or fauna. Most importantly, in the present context, it has a weak effect on microbial life in different cropping ecosystems. The temperature is extremely cold for life forms, ranging from –90°C to –120°C. Temperature decreases as we ascend to higher altitudes. The mesosphere has thin air, with atmospheric pressure below 1% of that encountered at the sea surface. As mentioned earlier, the mesosphere becomes even colder at higher altitudes, making it difficult for microbes to survive. Tolerance to super cool temperatures and microaerophilic conditions is essential. However, such cold-tolerant microbes are scarce, and it is challenging to trap and culture them on artificial medium (Horneck et al., 2010).

The thermosphere extends from 636 km (373 miles) to 1,000 km (621 miles) above the Earth's surface. The width of the thermosphere fluctuates. The thermosphere is almost devoid of air and has little to no nitrogen, carbon, or oxygen. It does not normally support life forms, including microbes, due to the high energy X-rays and UV radiation present. The temperature in the upper thermosphere ranges from 500°C to 1,700°C or even higher. The thermosphere supports the occurrence of Northern and Southern lights. The influence of the thermosphere on microbes in agrarian regions (soil, crops) and the troposphere should be low, if any.

The ionosphere is not a well-demarcated region with distinct boundary layers. It may extend into parts of the mesosphere and thermosphere, where high-energy radiation is observed. This radiation creates ions, which are electrically charged particles.

The exosphere is the final layer of the atmosphere and the frontier of Earth's gaseous envelope. Air is negligible in the exosphere, and it may leak or disperse into outer space. The exosphere extends from 100,000 m (62,000 miles) to 190,000 km (120,000 miles). It begins at an altitude of 500 km from the ground surface. The outer fringe of the exosphere marks the beginning of space.

In addition to the aforementioned layers, due to Earth's magnetic behavior, we have a magnetosphere. This magnetosphere concentrates electrons and protons into distinct bands at altitudes of 3,000 km and 16,000 km from the ground surface. This phenomenon is known as the 'Van Allen Radiation Belt.'

In the above paragraphs, we have described various layers of the atmosphere extending from the ground (i.e., soil surface of agrarian regions) to the exosphere. Among these layers, it is the troposphere that plays a vital role in sustaining agrarian regions on Earth. Now, let us discuss the details about the agrarian sky and its variations as they relate to microbial flora and its activity. When making inferences about microbes on the soil surface and in the air above crops, it is important to consider intricate details. Despite the large variations encountered in troposphere conditions, especially regarding substrates, microbes may be almost omnipresent in the troposphere. They can be found in every corner of the troposphere, particularly in layers closer to farming regions. The agricultural sky still imposes certain vital properties for microbes to navigate in order to survive, interact, and perpetuate. Microbes frequently transition from the soil to the troposphere and *vice versa*.

The atmosphere is a challenging environment for microbes to grow and multiply. Due to the limited nutrient availability and desiccation, microbes struggle to thrive in the atmosphere. This applies to various settings such as above crops, agroecosystems, natural vegetation, general terrain, hills, oceans, rivers, and lakes. Unlike soil or other substrates like marshes, oceans, rivers, lakes, crop surfaces, or animal body surfaces, the atmosphere provides significantly fewer nutrients. As a result, microbial populations in the atmosphere remain small and do not develop into dense populations easily. Additionally, the uncongenial conditions in the atmosphere restrict biological activity in general, including microbes. However, there are certain microbial species that can form resting structures. These microbes can survive in the atmosphere for a limited time until the resources are depleted or suppressive factors become dominant. These resting bodies of microbes overwinter. Microbial spores overcome unfavorable conditions to rejuvenate later when and if conditions are optimal. For example, bacteria sporulate, fungi also form spores and resting structures. Certain protozoa form cysts. Atmospheric factors such as relative humidity, aridity, desiccation, temperature, and oxygen tension limit the survival period of microbes. Other factors such as UV irradiance, pollutants like SO_2, H_2O_2, O_3, ionic content, high-speed wind, etc., also affect microbial activity. Factors like severe cold, snow, frost, moisture deficiency, and heat induce atmospheric microbes to hibernate for varying lengths of time. Microbial adaptations are crucial for their survival and perpetuation.

At this juncture, we can easily speculate that bioaerosols above agricultural zones may harbor microbes. These microbes are closely associated with agrarian soils, crops, and other aspects of the ecosystem. The agrarian sky may specifically contain microbes originating from agricultural fields. In contrast, in other regions, the microflora, dust particles, and even gaseous composition may be more influenced by the general nature of the terrain. For instance, the atmosphere above large water bodies such as dams, lakes, rivers, and oceans may differ from that encountered above agrarian regions. It is this difference that is intended to be emphasized in this chapter. For example, the dust particles, gaseous emissions, and microbes present in the atmosphere above a semi-arid farmland in Sahel may be completely different from those in the tropical cropping zones of southern Nigeria. Similarly, the microbes emanating from the dry arable Great Plains atmosphere may differ from those in the temperate moist forests of the Pacific Northwest. The microflora found above European plains may also differ from that observed in the West Asian drylands. It is important to identify and investigate these microbes and their impact on crops. It should be noted that the inter-continental migration (dispersal) of microbes contributes to a certain degree of mixing and uniformity in the microbial flora throughout the global atmosphere.

2.2 OCCURRENCE AND DISTRIBUTION OF MICROBES IN THE ATMOSPHERE

Historically, the occurrence of microbes in the air was first identified by Spallanzani in the mid-18th century. Later, in the mid-19th century, Louis Pasteur's works provided greater insights into microorganisms in the ambient air, including their abundance, diversity, and role (Capanna, 1999; Ehrenberg, 1830, 1849). The discovery of the existence of microbes in the air immediately above the ground and at slightly higher altitudes was made in France (Ehrenberg, 1830; Pasteur, 1860). According to a review by Morris et al. (2011), aerobiological investigations have been popular for the past two centuries. Aerobiological studies began around 1847-1850 with reports about the organic nature and life on atmospheric dust. These studies have focused on the abundance, diversity, survival, and movement of microbes in different layers of the atmosphere. Microbes have generally been recognized as inhabitants of harsh atmospheres. Aerosol microbes continue to attract attention. Recent findings about ice nucleating microbes, the biological control of plant pathogens residing on the phyllosphere, the inter-continental transit and pathway of plant pathogens in the atmosphere, and the induction of snow, frost, and rain have emphasized the importance of aerial microbes in agricultural pursuits. Morris et al. (2011) and Morris & Sands (2017) further state that 'Biometeorology' requires attention. We should strive to decipher more about the interactive effects of the atmosphere and microbes. Perhaps, there is a lack of knowledge regarding the importance of the 'Agricultural sky' and its various manifestations related to physicochemical processes, biotic components like microbes, insects, birds, etc. (Krishna, 2022). Aerial microbes, for instance, may be vital to atmospheric processes relevant to farming.

Aguilera et al. (2018) identify the atmosphere as a region with extreme conditions for microbes to survive. It can be compared to harsh arid deserts, ocean depths with high pressure, high altitude permanent snow regions with extremely low temperatures, and more. The microbes that manage to survive in the atmosphere and continue to exist are known as extremophiles (Rothschild & Mancinelli, 2009; Madigan & Marrs, 1997).

First, let us explore the sources of microorganisms that contribute to the microbial load of the atmosphere. The atmosphere receives microbial inoculum in both small and large population sizes. The majority of its microbial flora comes from soil, natural vegetation, forest stands, water bodies, crops, domestic animals in farms, or wild animals in sanctuaries. There are other sources of microbes in the atmosphere, such as clouds, rain, dust storms, and tornadoes. These phenomena actually transport microbes from one location to another. Sometimes, the source of microbes could be from a far-off location on Earth (even different continents).

The canopy of plants, including branches, leaves, inflorescence, flowers, and fruits, serves as a significant source of microbes. Microbes present on the canopy of crops can easily enter the atmosphere through mild breezes, winds, or turbulence. Domestic and wild animals may carry a diverse range of microbes on their body surfaces, further contributing to the microbial load in the atmosphere. It is important to note that a crop infected with contagious microbial inoculum can have disastrous consequences. Plant pathogens that can be easily separated from the plant canopy can rapidly spread themselves and diseases throughout entire fields via atmospheric processes.

Agrarian soils often contain loose sand and dust particles that carry a variety of microbes. High-speed winds, accompanied by turbulence, can propel these dust particles and microbes into the atmosphere. They can then levitate, float, and reach different altitudes within an agroecosystem. The repeated plowing, loosening, and aeration of agrarian soils can release a large population of microbes into the atmosphere. Additionally, various agronomic procedures such as discing, clod crushing, interculture, earthing-up, weeding, sprinkler irrigation, and combine harvesting conducted in the outfield can also liberate microbes into the atmosphere. Turbulent water sources, droplets, and moisture from clouds and snowflakes also contribute to the presence of microbes in the atmosphere.

A diverse assemblage of microbes can be found in the atmosphere. However, their exact importance in conducting ecosystem functions is still not fully understood (Smets et al., 2016; Rothschild & Mancinelli, 2001). Most microbes are found in the troposphere, particularly the lower troposphere. Natural vegetation, crops, and forests contribute a large number of microbes to the lower troposphere, while a few may reach the stratosphere. Microbes from the soil may even reach altitudes higher than 41 km, although their survival becomes difficult. The atmosphere is primarily an extremophilic environment due to the scarcity of organic matter. Factors such as desiccation, low relative humidity, extreme temperatures, low oxygen levels, and high UV radiation at higher altitudes make the troposphere inhospitable for microbes (Aguilera et al., 2018). For single-celled organisms, the Earth's atmosphere serves as a refuge and potentially a habitat (Scharf, 2019).

Methods for studying such 'extremophiles' are available. Initially, most studies on microbes in the high-altitude atmosphere were conducted by reaching mountain tops or tall buildings. However, there are undoubtedly constraints when studying microbes that reside at higher altitudes of the atmosphere. Various means such as aircraft, balloons, parafoils, rockets, satellites, and more recently drones (Unmanned Aerial Vehicles) have been used to collect authentic air samples. Different microbial sample collection systems have been adopted, including active filtration of air, impingement of air, use of selective medium, enrichment medium, and complex nutrient media. Traps with nutrient media have been employed to capture microbes in the upper troposphere, tropopause, and low stratosphere (Bowers et al., 2009; Griffin, 2008, 2010; Lacey & West, 2006).

Wommack et al. (2010, 2015) argue that the atmosphere and its microbial aspects have immediate consequences for crops, domestic animals, and humans. Therefore, we should have already accumulated detailed knowledge about these aspects quickly, just as we have done with soils, water, or crops. The atmosphere should not be considered merely as a medium for the transit of microbes. It not only aids in the short or long-distance dispersal of microbes but also serves as a habitat where microorganisms dwell, grow, multiply, and mediate numerous bio-geochemical reactions relevant to crops and agro-climate in general.

The atmosphere is not merely a repository of microbes. In fact, the troposphere, in particular, serves as a habitat for microbes. According to Wommack et al. (2010), the atmosphere is not the most inhospitable or extreme environment for microbes to thrive. There are numerous locations on Earth that are even *harsher* for microbial survival. However, microbes naturally adapt to these challenging conditions and flourish. For instance, microbes encounter environments with fewer nutrients and moisture compared to the atmosphere. They also thrive in regions with unfavorable conditions such as relative humidity, temperature, moisture, pH, and oxygen tension.

Interestingly, during certain seasons, crops including stubbles and weeds contribute to the lower troposphere by releasing more particulate matter (PM), nutrients, organic debris, microbes, and moisture. This occurs during activities such as plowing, pollen shedding, harvesting, and post-harvest stubble clearing.

During their existence in the troposphere, microbes should adapt to the ecosystem. The survival time and general longevity are important. For example, certain bacteria (e.g., *Bacillus* sp. and *Clostridium* sp.) may survive for up to 200 years in a dried state or as spores. On the other hand, a few *Salmonella* spp. found in the atmosphere may perish in a short while, say, 6–10 minutes after duplication. Pigmentation, if present, may help in the survival of microbes by reducing the ill effects of UV radiation (Shomrat, 2013).

The atmosphere in agricultural regions may often support microbes derived from the soil surface, crop residue, organic matter, and the crop's canopy. Farming practices, soil characteristics, wind, and turbulence mediate the transport of microbes from the crop's canopy to the atmosphere. For example, the microbes present in the troposphere immediately above a rice paddy would be related to the puddled submerged soils. Rice cultivation involves the alternation of submergence and dry arable conditions, resulting in varying microbial populations and diversity in the atmosphere. The microbial composition above temperate crops will differ significantly from that above tropical crops. Wetland crops like rice may emit gaseous compounds that are different from those found in dry arable soils, which in turn affects bioaerosols. Submerged paddy fields accumulate microbes that thrive in microaerophilic or anaerobic conditions, as well as microbes that can survive in chemically reduced soils. Therefore, the lower troposphere above a paddy field may harbor these same species of microbes.

Crops are also sources of aerosol microbes. It is expected that aerosol microbes will vary depending on the crop species (Linderman & Upper, 1985; Linderman et al., 1982). There are examples of fungi, such as *Alternaria* sp., that are specifically associated with crop species. The canopy and the air above it are infested with fungal spores (Mitakakis et al., 2001). Therefore, the diversity of aerosol microbes is determined by crops and cropping systems.

Ruiz Gill et al. (2020) have provided clear evidence that atmospheric microflora differs based on terrain, land features, water bodies, urban or rural locations. Furthermore, it is important to acknowledge that within an agrarian region, atmospheric microbes (bioaerosols) vary depending on soil type. This is in addition to the influence of crop species and cropping systems adopted. More detailed information is needed regarding how soil, crops, and agronomic procedures influence the presence of microbes on dust particles and free-floating ones (Lighthart, 1984). Let us consider the factors that determine microbes in the aerosol. Ruiz-Gill et al. (2020) state that the physicochemical characteristics of PM are crucial. Aerosols containing pollutants may support an entirely different set of microbes (Liu et al., 2019). Similarly, the size of PM (dust) plays a significant role. Microbes found on fine particles (< 2.5 μm) and large-sized particles (> 2.5 μm) appear to differ (Ruiz-Gill et al., 2020; Liu et al., 2018). The localization of dust particles at different altitudes also affects the microbial species traced in the atmosphere (Ruiz-Gill et al., 2020). The sun's radiation influences the microbes found in the air. Differences in aerosol microbes based on the diurnal period can be observed (Lighthart, 1997, 1984). Overall, microbes in aerosols may depend on agricultural activity on the terrain. Specific agrarian practices may emit dust and microbes related to the agrarian region into the atmosphere.

The chemical nature of dust particles is influenced by the source, such as terrain, vegetation, soils, crops, organic composts used, water resources, and weathering processes. In response, the microflora present in the atmosphere also fluctuates based on the chemical composition of the substrate. Fine particles of dust or water droplets containing dissolved compounds like d- and l-amino acids, carboxylic acids, and sugars may support specific microbes. By analyzing the chemical characteristics of dust particles and the microbial species present, we can even identify the source and location from where the microbes originated. In fact, the chemical markers and microbial diversity of dust particles allow us to identify the source of microbes in the troposphere (Feltracco et al., 2021). If a similar concept is applied to farms, we may be able to accurately trace the source of plant pathogens. Specific farms or crops that release dust laden with plant pathogens could also be identified.

Detailed investigations by Flemming & Wuertz (2019) suggest that bacteria and archaea found on Earth (1.2×10^{30} cells) perpetuate in the 'big five habitats' (deep ocean, subsurface of oceans, upper oceanic sediments, soil surface, and oceans). The remaining may be traced in the atmosphere, phyllosphere, and groundwater. Most microbes are found in biofilms, in groups, rather than as single cells in the atmosphere. Biofilms offer protection to microbes from radiation, desiccation, and other unfavorable conditions. Furthermore, it is inferred that microbes in these biofilms mediate most biogeochemical cycles in the atmosphere. Microbes in biofilms might play a vital role in nutrient transformations within agroecosystems. Nutrient cycling stages in agrarian regions may depend on microbial diversity and population in the biofilms.

2.2.1 TROPOSPHERE: A HABITAT FOR AGRICULTURALLY RELEVANT MICROBES

The focus of this section is on the microbes that localize and perform various ecosystem functions at the confluence of soil, water, and the lower troposphere. The troposphere, including the soil surface in agrarian regions, has played a crucial role in the evolution of cropping systems and has partially determined crop productivity. The troposphere is also home to numerous crop-disease causing microbes, while beneficial organisms rely on troposphere conditions for their survival and dispersal in agrarian regions. Additionally, the troposphere harbors a wide range of microbes that mediate geochemical reactions and cycles, such as carbon mineralization, biological nitrogen fixation, and nitrate reduction. A significant amount of biochemical substrates undergo transformation in the troposphere region above crops. Notably, carbon fixation by photo-autotrophic green algae, blue-green algae, and plants is a process closely associated with the troposphere.

We should also highlight the distribution, genetic diversity, and population of microbes in the atmospheric layers. The troposphere contains numerous phyla of microbes that are relevant to agricultural crop production. There are several commensal microbes observed in the agrarian atmosphere, many of which are severe plant pathogens that require immediate and careful attention, especially if the farmer intends to achieve a reasonable harvest. However, there are also a few beneficial microbes that can move from the soil phase to the troposphere, particularly the lower troposphere. These beneficial microbes, such as Bradyrhizobia, arbuscular mycorrhizas (AMs), and PGPR, can be advantageous if they encounter a suitable host. Bioaerosols are distributed up to 7 km at the Earth's poles and 20 km above the equator. The troposphere region extends from the soil surface to the bottom of the stratosphere. Biological aerosols contribute to cloud formation, which in turn affects the Earth's hydrological cycle. These natural hydrological phenomena have immediate significance for crop production trends. Microbes in the troposphere may impact cloud chemistry, enucleation process, and precipitation patterns (Scharf, 2019).

A review by Polymenakou (2012) highlights the immediate effects that tropospheric microbes can have on the ecosystem. In the troposphere, microbes are generally found incorporated into clouds, fog, snow, and water droplets. Dust particles also support aeolian microbes. They are exposed to the vagaries of atmospheric conditions. Free-floating individual cells or groups or colonies of microbes are also traced in the troposphere. Such microbes could be deposited back to soils, natural vegetation, and cropping belts. This process of natural cycling of dust, clouds, and rain can alter the microbial diversity immensely in an agroecosystem (Bauer et al., 2002). Fogs, smog, and blizzards also carry microbes and inoculate them into new soil surfaces and onto crop canopies (Christner et al., 2008). The soil microflora is constantly impinged and added with new microbe species. Hence, microbial diversity and the total population are transient. This way, the atmosphere may even control the distribution and recycling of plant pathogenic microbes or introduce fresh epidemic or pandemic-causing organisms into agrarian fields. In addition to affecting microflora, tropospheric processes simultaneously influence the physicochemical nature of the soil. The atmospheric chemistry generally gets influenced simultaneously with the microbial population (Ariya & Amyot, 2004; Christner et al., 2008; Mohler et al., 2007). Together, these may affect cropping zones (Polymenakou, 2012).

There is an opinion that no microbial species are native to the troposphere. Most of them are derived from Earth's surface features. Microbes that inhabit the troposphere are mostly derived from soils, vegetation, and water bodies. The tropospheric microflora is also determined by geographic location. Microbes like *Brevundimonas* and *Sphingomonas* are found in the tropospheric region of the Antarctic sky (Gonzalez-Toril, 2009). The troposphere in Australian locations harbors microbes derived from saline lakes, sediments, soil crusts, sand, and crops (Busse et al., 2003; Abed et al., 2012; de Deckker et al., 2008). The movement of microbes from water to air is also common (Baylor et al., 1977).

In the Sahelian zone, the troposphere contains both dust and attached microbial load. The primary sources are the sandy soil surface, dust storms, crops, pollen, and other factors. The microbial diversity can also be influenced by farming operations and composting. Therefore, it is common to find microbes that are present in higher numbers in the soil, crops, and lofted into the troposphere (Chen et al., 2020). Additionally, marine and riverine regions can contribute to tropospheric microbes (Gao et al., 2017; Gilbert et al., 2012). It appears that microbes from aquatic systems are frequently detected in the atmosphere. The presence of dust, fog, or smog in the air can indicate the presence of thriving microbes in the troposphere. The microbial composition of the air above compost pits, waste treatment plants, polluted waters, soil, and waste heaps is largely derived from the ground surface (Bru-Adan et al., 2009; Le Goff et al., 2010; Han et al., 2012). Extensive crop areas can also contribute to microbial diversity, mainly consisting of phyllosphere microbes, pathogens, symbionts, and commensals found in the soil. Pollen released into the troposphere can also introduce microbes found on its surface. It is important to note that pollen shredding is a short-lived phenomenon that occurs during a specific stage of the crop.

The presence of toxic pollutants such as SO_2, CH_4, and organic and mineral PM can have an impact on tropospheric microbes and their survival. These contaminants and pollutants can curtail their physiological activity. Contaminants like SO_2, H_2O_2, polycyclic aromatic hydrocarbons, as well as solar radiation and UV-irradiance are detrimental factors that reduce microbial survival in the troposphere. Dust storms, thunderstorms, wind, and strong drifts can transport microbes in the troposphere. While water (moisture), water-soluble ions, PM, and optimum temperature are favorable conditions for the survival and multiplication of microbes, even in the troposphere (Chen et al., 2020).

Microbes are continuously introduced into the 'agricultural sky' (troposphere) from various sources such as open soil, mountains, cropland, forests, natural vegetation, and marshes, among others. The dispersal of microbes at low altitudes is significant in farming regions. Some microbes only travel short distances, while others can travel long distances and settle on soils, plants, or even inhospitable rocky terrain. We are interested in studying the diversity of microbial species that reach agrarian regions, including crops, plantations, pastures, etc. (Burrows et al., 2009a, b; Griffin, 2007, 2008; Smith et al., 2013). Microbes from deserts also contribute to the existing diversity above crops. It is possible that numerous potentially pathogenic microbial strains are constantly reaching the atmosphere above cropping zones.

Long-term ecological studies of the atmosphere have shown that air masses and aerosols play a crucial role in determining microbial load and diversity. Let's consider an example. Researchers in Northeast Spain have analyzed air, rain, and snow samples for seven years, covering all seasons, using 16s and 18s rRNA sequences. Understanding the mechanisms of microbial dispersal is important for comprehending the factors influencing microbial diversity in the atmosphere. It appears that the microbial component received from other regions increases annually, driven by the global movement of microbes. In fact, a significant population of microbes is exchanged between different ecosystems, potentially altering microbial diversity and the rates of biochemical transformations in the troposphere (Zhu, 2007). The survival rate of microbes may be determined by the substrate, as not all microbes can persist for extended periods in the troposphere. Many of them are transient dwellers in the air. For instance, Caliz et al. (2018) identified approximately

1,200 bacterial species, but found that only 12 genera were common in 90% of the analyzed air samples. Some of the ubiquitous genera included *Sphingomonas, Methylobacterium, Massila, Pseudomonas, Acidiphillium, Ramlibacter, Mucilaginbacteer, Hymenobacter,* and *Noviherbaspirrilum.* Among fungi, 32% of the samples contained members of Ascomycota and 26% had Basidiomycota. Archaea were present in 4% of the samples, predominantly in summer, spring, and winter. Microbial eukaryotes were detected in only 10% of the air samples above the Pyrenees. Overall, the microbial load was found to be influenced by the season.

Several strains of bacteria, fungi, and archaea found in the lower troposphere are not just residents; they are metabolically active. They mediate several biogeochemical transformations. In fact, a few of the microbiological transformations of soil nutrients occur in the lower troposphere. For example, nitrogen compounds are transformed by nitrifiers and denitrifiers. Atmospheric diatomic nitrogen (N_2) is fixed into microbial cells by nitrogen fixers (e.g., diazotrophs in the air). Organic compounds are utilized as an energy source. Microbes rapidly respire and release CO_2 into the troposphere. This is similar to their activity in arable oxidative soils found in farms. Sulfur compounds are also transformed in the atmosphere. Sulfur compounds are often found with smog. SO_2 is the chief contaminant. *Thiobacillus thiooxidance* and *Acidobacillus thiooxidance* are examples of microbes involved in the chemical transformations of S compounds in the air. We may also encounter viruses that affect crops. Bacteriophages are also detected in the troposphere. Bacteriophages that infect plant pathogenic bacteria may be relevant to disease control during crop production.

Tropospheric microorganisms are known to multiply and go through several generations. Obviously, natural selection of versatile microbes occurs in the troposphere. Several microbial species and their biotypes that specifically tolerate the harsh climate prevalent in the troposphere are known. Such natural selection pressure by lower tropospheric conditions will be applied to microbes relevant to agricultural crop production.

Now, let us discuss fungi found in the troposphere. Generally, we encounter a high degree of fungal diversity in the aerosols. Particularly, in the lowermost layers of the troposphere closer to cropping belts and urban human inhabitations. This is especially true in the tropical farm sky (Prasla et al., 2013; Ravikala & Nagalakshamma, 2016; Ezike et al., 2016; Ibanez et al., 2001). There is a need to monitor the fungal genera commonly occurring in the air. The fungal flora in the aerosols may vary with season, atmospheric conditions, native vegetation, crops, wind velocity, etc. Such information should be updated periodically. It will allow us to take prophylactic measures against crop diseases. There are several microbe collection and identification methods (Almaguer et al., 2013, 2018). Soil, organic matter, crop surfaces, residues, and wind can serve as sources for new fungi in a region. Monitoring aerosols in the agrarian sky has direct advantages, particularly in terms of disease forecast and control measures (Frenguelli, 1998; Levetin et al., 2016).

Consideration of fungal diversity and population in a tropical agrarian region can be exemplified. In the Caribbean region, a survey of fungal genera in the lower atmosphere was conducted a few years ago. However, the situation may have changed over time due to farming practices, crops, and domestic animals. Almaguer et al. (2018) conducted a survey of aerosols and discovered fourteen new fungal genera. They employed volumetric air sampling procedures and microscopic examination. Among the fungal species identified, six produced ascospores (ascomycetes) and eight produced conidia. These fungal types accounted for 56% of the total fungal flora in the air sample. In a previous sampling in the same area, 30 fungal genera were found, with *Cladosporium, Aspergillus,* and *Penicillium* being the predominant fungi. It was observed that relative humidity (RH) in the atmosphere played a significant role in influencing both fungal diversity and population.

Several other fungi were noticed in the aerosols. Some of the ascospore-producing fungal types included *Chaetomium, Paraphaeosphaeria, Pleospora,* and *Sporormiella.* Conidial fungi detected by Almaguer et al. (2018) included *Alternaria, Beltrania, Bipolaris, Epicoccum, Fusarium, Gliomastix, Helicoma, Helicomyces, Pestalotia, Pithomyces, Pseudocercospora, Pyricularia, Spegazzinia, Sporidesmium, Stemphylium, Tetraploa,* and *Torula.* Several fungi were recorded repeatedly during various surveys. Some newly traced genera were *Diaporthe, Didymosphaeria, Massarina*, and *Helminthosporium.* There are diurnal fluctuations in fungal diversity and population in aerosols. This holds true for any location within agrarian regions, especially if the farm zone supports large-scale crop production, which would host a diverse range of fungal pathogens and commensals. The counts of fungal spores also vary depending on the month of the year. For instance, in the tropical Caribbean, fungal spore density (37 spores m^{-3}) in the troposphere near crops is high during dry months, particularly from November to April. Additionally, Almaguer et al. (2018) state that a higher degree of sporulation occurs in only a few fungal species, not all. Fungal genera such as *Cladosporium, Coprinus, Aspergillus/Penicillium, Curvularia, Nigrospora,*

Periconia, Ganoderma, Leptosphaeria, Xylariaceae, Venturia, Cercospora, and *Monodictys* produce spores. Overall, studies on aerosols and their biotic components are relevant to crop management, particularly in terms of disease control and prophylaxis. They help trace the long-distance dispersal of pathogenic fungi and the contamination of the troposphere above crops. This often occurs due to high-speed winds, high-altitude jet streams, dust storms, thunderstorms, snow, frost, and rain. These studies provide a warning about fungal load and its potential destructive effects on crops.

Aerosolization is an important natural process that is relevant to many agrarian regions. Soils containing dust, loose sand, or organic particulates are prone to forming aerosols in the troposphere. This phenomenon is particularly prevalent at the atmospheric boundary layer. Aerosols play a role in transporting bacteria and virus particles over long distances. Microbes can deposit anywhere on crops in different regions or continents. Reche et al. (2018) observed that the deposition rates of viruses are much higher than those of bacterial cells. The downward flux measured for viruses ranged from 0.26×10^9 to $> 7 \times 10^9\ m^{-2}\ day^{-1}$. Viral deposition rates were 9–461 times greater than those noticed for bacteria. Viral deposition rates occurred via organic particles $< 7\ \mu m$ in size, while bacterial deposits were slightly related to larger PM $> 0.7\ \mu m$. Bacterial deposit rates were enhanced during rain events (Reche et al., 2009). Studies on viruses and/or bacteria at the atmospheric boundary layer have immediate relevance when forecasting diseases in crops.

Upper troposphere microbes may also affect agrarian regions. Microbes found at high altitudes of 8–11 miles above the ground surface were similar to those encountered on the soil surface. The proportion of microbes in the dust sampled at 8 km altitude, i.e., the upper troposphere, was high. Most of the microbes sampled at higher altitudes were alive and in an active metabolizing condition (Main, 2013). It is believed that moisture may make the dust heavier, causing it to fall to the ground. It appears that microbes transiently reside in the upper troposphere. The microbial flora is undoubtedly dynamic, with certain types of microbes being more abundant at certain times. The reasons for these successional changes in diversity need to be accurately understood. It is possible that physically harsh climates, sparse nutritional status of dust particles, and competition from other microbes better adapted to the atmosphere may contribute to differences in microflora over space and time.

Wade (2013) suggests that there are advantages to understanding the microbial flora that exists in a nearly permanent state. It is also important to monitor transient microbes and those that are weak visitors. Analysis of air samples has shown that approximately 17 microbial species were present in all samples from the upper troposphere. This suggests that there are microbes capable of absorbing nutrients, facilitating nutrient transformations, harnessing energy, and surviving at high altitudes in the troposphere, specifically at altitudes ranging from 8-14 km above the soil surface. The role of these microbes, if any, within agroecosystems and natural vegetation remains to be understood.

2.2.2 STRATOSPHERE MICROBES AND THEIR INFLUENCE ON VEGETATION AND CROPS

Microbes are commonly encountered at higher densities in the lower troposphere, closer to farming regions. The same is not true for the stratosphere, as it is an extremely harsh environment for microbes to thrive. Currently, we have no data to suggest that the stratosphere above vast crop production zones differs in any way from those above other types of terrain, such as dry deserts, snow, or industrial regions. It is important for us to gather this information. It is plausible to speculate that cropped fields could release microbes that eventually reach the stratosphere. Conversely, the influence of stratosphere microbes (if any) descending onto vegetation also requires further investigation.

The colonization of microbes in the stratosphere is dependent on natural factors and energy sources. The microbial population in the stratosphere is weak or negligible. Factors such as irradiance, lack of moisture, and supercool conditions limit the number, density, distribution, survival, and perpetuation of microbes. Heist et al. (2020) have demonstrated that stratosphere microbes can be detected up to an altitude of 32 km from the ground surface of agrarian regions. They are able to withstand the extreme conditions created by factors such as cold, desiccation, and UV irradiance.

Microbes are extremely abundant in agricultural soil. Some of them have adaptations for aerial dispersal (Papke & Ward, 2004; Kellogg & Griffin, 2006). Microbial dispersal into the stratosphere may occur due to volcanic eruptions (Jacob, 1999), Brewer–Dobson atmospheric circulation, dust storms (Kellogg & Griffin, 2006), monsoons, electrostatic forces created by thunderstorms (Dehel et al., 2008), and rocket launches (Griffin, 2004).

There is relatively little literature about the natural occurrence of microbes in the stratosphere. Studies on their significance to aeolian processes are also negligible. Exploring the strategies that stratosphere microbes adopt for survival is worth considering. Knowledge about their utilization of energy sources, biochemical transformations they mediate, and tolerance to harsh ambient conditions is indeed limited. After all, stratosphere microbes may play a role when they descend onto terrestrial or cropping ecosystems.

Efforts to understand the effects of the stratosphere on microbes began during the early Apollo Space missions in the 1960s and 1970s. Bacteria, fungi, and algae were all found to be suitable for space flight and survival. However, during a later period, space effects were frequently studied using bacterial samples. Initially, the National Aeronautics and Space Agency (NASA) employed high-altitude balloons to study microbes in the stratosphere. The balloons placed microbes at an altitude of 30 km in the stratosphere, which was possible after a 9-hour journey. Fungi and their spores were found to survive the harsh stratosphere environment, including extreme temperatures, radiation, and low pressure (Bartscheidt, 2020).

Bryan et al. (2019) used helium balloons to sample dust particles and microbes at altitudes ranging from 3 to 29 km. Some aerosol samples were collected at an altitude of 38 km in the stratosphere. These samples were then analyzed for adenosine triphosphate (ATP) concentrations in the aerosols. The detection of bacterial cells indicated that certain microbial species thrive at lower altitudes in the stratosphere. The vertical distribution of microbes was remarkably consistent, at approximately $2–5 \times 10^5$ cells m^3. Similar findings regarding stratosphere microbes were reported by Yang et al. (2005, 2008), Vaitilingom et al. (2012), Sattler et al. (2001), and Smith et al. (2013, 2013b).

Smith (2011) mentioned a decade ago that there has been an increasing interest in studying microbes that reside in atmospheric layers, particularly those in the stratosphere (Narlikar et al., 2003; Griffin, 2004, 2008; Shivaji et al., 2006, 2009; Yang et al., 2008; Smith et al., 2009). This interest is driven by both curiosity and the practical value of microbes that are resistant to extreme environmental conditions. The intermixing of bacteria normally found in agricultural soils with those in the stratosphere may offer certain advantages. Microbes found in agrarian soils and other regions could harbor various species of microbes, some of which may have suitable adaptations to survive in the stratosphere, albeit for a limited period. There is also a clear possibility of troposphere microbes floating near the tropopause and entering the stratosphere, where they can survive for a certain length of time. The microbes acclimatized to the troposphere could be genetically related, potentially giving rise to new strains with a mixed genetic nature. The stratosphere may then select species/strains that possess the ability to withstand harsh UV radiation, as well as extremes of relative humidity, temperature, and pressure.

There are reports about microbes with adaptations that enable them to survive in the stratosphere. However, reports about stratosphere microbes with the ability to adapt to various substrates and their nutrient composition, metabolic activity, and biochemical transformations are scarce (Smith et al., 2011). These reports are important if we wish to gauge the ability of soil microbes, plant pathogens, and commensals to be transported from agrarian soil to the extremities of the tropopause or stratosphere. Microbes that transition from the tropopause to the stratosphere also need suitable genetic traits to adapt.

Microbes in the stratosphere tolerate extreme environments. Smith et al. (2011) point out that most inferences about microbes found in the upper layers of the atmosphere attribute their tolerance to genetic resistance. Genetic resistance involves genes for active DNA repair mechanisms. Other adaptations include the production of photo-protective pigments, sporulation, and more. Intermating, followed by genetic hybridization and natural selection, may enhance bacterial strains with the ability to survive in the stratosphere for longer periods. Therefore, we should exercise extra caution before introducing genetically hybrid microbes that are tolerant to stratospheric conditions. We must ensure that nothing goes astray! Additionally, we are uncertain about their potential effects on agrarian regions, especially when they descend and settle on agrarian soils and crop canopies.

There are numerous reports about the movement of vegetative cells and spores of microbes in the stratosphere. Therefore, the stratosphere may aid in the dispersal of microbes that cause diseases in crops. Some of the dispersed microbes could be beneficial microbes (DasSarma et al., 2020).

Sometimes, it is easier to examine microbes by simulating the conditions prevalent in the stratosphere. Smith et al. (2011); and (Smith 2013a, b), for example, evaluated *Bacillus subtilis* for survival, metabolic activity, multiplication, and dispersal. They simulated the stratosphere conditions by creating reduced pressure, low temperature, high desiccation, and intense UV irradiation. They compared the performance of *B. subtilis* strains isolated from the stratosphere with those obtained from conditions of normal agricultural soil. They evaluated both sets of *B. subtilis* strains at –70°C, 10–12% RH, UV irradiation at 315–400 nm wavelength, and low atmospheric pressure. They found that after 6 hours in stratosphere conditions, *B. subtilis* spores were killed because of UV irradiation. It is generally believed that *Bacillus* strains sporulate to overcome extremes of the environment.

Viability was not affected by other parameters such as pressure, temperature, and RH in the absence of UV irradiation. Therefore, it was concluded that the survival of sporulating bacteria or other organisms during long-distance dispersal depends on UV irradiation. Vegetative cells of microbes like *B. subtilis* are often more susceptible to extreme conditions and UV radiation. In agricultural zones, the survival of stratospheric microbes depends on weather parameters and the effects of radiation. Dry desert dust may harbor bacteria, but for long-distance travel, they must avoid UV radiation, possibly by seeking shelter in the crevices of dust particles. Once stratospheric microbes have descended onto agricultural soils, they may rejuvenate, grow, and multiply rapidly. Spores can easily germinate on the soil surface, and stratospheric microbes may also penetrate the soil profile to different depths, perpetuating their presence.

There have been several efforts made to study stratosphere microbes *in situ*. Large balloons (filled with liquid neon) were launched to collect air samples from 20–41 km above the Earth's surface. Approximately a decade ago, at least three bacterial species were identified as normal residents in the stratosphere using 16s rRNA sequencing. These species are *Janibacte-rhoylei, Bacillus isronensis,* and *Bacillus aryabhata* (ISRO, 2009). These bacteria have been found in the stratosphere but not in the terrain or soil profiles of agrarian regions. Stratosphere air samples were collected at various times throughout the year. According to their reports, approximately 12 bacterial species and 6 fungal species were discovered in the stratosphere air samples. The mechanism of microbial survival in the stratosphere is still not fully understood. There are stratosphere microbes that are tolerant to UV radiation. Furthermore, the biochemical transformations mediated by these stratosphere microbes are not yet known. It is worth noting that the Earth's surface, including vast agrarian regions, is constantly exposed to various microbial species such as bacteria, fungal propagules, and a large number of viral particles. Some of these may originate from the stratosphere.

2.2.3 MICROBES IN THE UPPER LAYERS OF THE ATMOSPHERE: MESOSPHERE, THERMOSPHERE, AND IONOSPHERE

Mesosphere begins at a height of 95–120 km altitude from the ground surface. The temperature, at –85°C, is totally inhospitable for any kind of microbes to survive or transit along with inert material. The temperature cools down further as we move to higher altitudes within the mesosphere. Therefore, the possibility for life forms like microbes to survive and conduct physiological activity is remote. Bryan (2017) mentions the ability of microbes to survive in the upper stratosphere and lower mesosphere. These inferences may need support from rigorous testing of microbes by re-exposing them to the mesosphere. This will help in proving Koch's postulates for the survival of bacteria in the mesosphere.

It seems that dust particles laden with microbes, particularly in the size range from 1 to 100 µm, levitate due to 'photophoretic forces.' This might be the reason for the occurrence of microbial cells in the mesopause and lower mesosphere regions (Rohatschek, 1984, 1996; Imshenetsky et al., 1977, 1978). Pigmentation of spores/cells adds to tolerance to radiation ill-effects.

The thermosphere above the mesopause is an even more challenging region for any life forms. Temperatures may reach 1,700°C. It is cloudless, with no moisture for life forms. The International Space Station (ISS) is supposed to be located in this region, approximately 350–450 km from Earth's surface. Yet, there are reports of bacterial clumps being able to survive the rigors of this region.

The exosphere is a space that does not support microbes. Here, we find man-made satellites orbiting Earth and reflecting signals. Solar winds, harmful radiations, and lack of water make it impossible for life forms.

2.2.4 MICROBES DETECTED IN SPACE AND ON SPACE CRAFTS

Lambert (2020) mentions microbes that were exposed to the conditions of space. The bacteria were placed on the outside of the ISS. It appears that bacteria survived the harsh conditions when they were in clumps. The bacteria in the center of the colony thrived during this ordeal. The DNA of cells in the outer layers of colonies was destroyed by radiation. Clumps of bacteria survived for over three years in the harsh climate of space (e.g., *Deinococcusspp.*). An outer layer of dead microbes protects the bacteria nestled inside. Therefore, it is argued that bacteria may survive for longer periods when traveling on the outer surface of spaceships. It is also possible that bacteria can drift between two planets. Now, how does this relate to cropping ecosystems on Earth? How do they react to atmospheric conditions in the troposphere, closer to soils and crops, if allowed to flourish? These are questions that will be answered in the future. We have no idea how they affect individual crop species. Microbes in the stratosphere may also have a significant impact on the agrarian ecosystem. It is crucial to understand their physiological activity under favorable troposphere conditions, especially those relevant to crops and crop production. We are aware of the long-distance horizontal migration of soil and epiphytotic bacteria. It is important to determine how quickly or frequently stratosphere microbes descend upon crops, if they ever reach cropping regions or soils. Some aerobiologists believe that space and interplanetary travel may introduce new microbes that can tolerate harsh space conditions.

Nicholson et al. (2009) have studied the trans-continental migration of microbes and suggest that microbes may contaminate interplanetary space. Similarly, the migration of space microbes via spaceships is a clear possibility. It is necessary to avoid a few detrimental microbes and regulate their entry into the Earth's atmosphere. This avoidance of stratospheric or space microbes should be based on their potential activities in the Earth's atmosphere and agrarian soils. Furthermore, interplanetary contamination of atmospheres should also be taken into consideration.

Dust particles from other planets that are incorporated with microbes may contaminate Earth's atmosphere to varying degrees. The study of microbial incursions into the troposphere and/or soil depends on our ability to first identify such microbes.

There are countless plant pathogenic microbes that are normal inhabitants of agricultural croplands and the sky above them. These microbes may temporarily reach high altitudes and enter the stratosphere or even layers beyond that are inhospitable. Plant pathogens can indeed rise to high altitudes in the atmosphere and may even reach planetary boundary layers. Eventually, they descend and get deposited on the soil or crop canopies, causing diseases in important crops (Maldonado-Ramirez et al., 2005; Horneck et al., 2010). Similar observations have been made with *Gibberellazeae* and *Fusarium* sp. spores, which reach boundary layers and disperse over long distances. This high-altitude long-distance transmission may initiate epidemics or pandemics in major cropping zones. However, further verification is needed for this aspect.

Spacecrafts and stations may release microbes and other biological material into the atmosphere. One of the projects conducted under the International Space Station (ISS) involved growing lettuce under space conditions. The goal was to assess various nutritional contents and microflora on the leaves and other parts of the lettuce. The ultimate intent was to cultivate food crops *in situ* in space. Lettuce (*Lactuca sativa*) grown in space was compared with lettuce grown on Earth. Leaf tissue samples from space flight, the ISS, and the ground exhibited differences in Fe, K, Na, P, S, and Zn content, total phenolic levels, and anthocyanin. The microbial load and quality also differed between lettuce leaf samples from the space station and those from the ground. Leaves from the space station harbored a higher number of microbes (Khodadad et al., 2020). Potential human pathogens such as *Salmonella, Escherichia,* and *Pseudomonas* were absent in the space station samples. Crops grown in the ISS should also be screened for plant pathogens, as they may return to the ground along with astronauts or spacecraft and potentially affect agroecosystems in an adverse manner. Space microbes may mutate to higher virulence due to exposure to harmful radiation. Generally, lettuce samples grown in the ISS showed a diverse range of microbes. Lettuce had a higher microbial population on the root surface and inside the tissue compared to the shoot system. In the future, it may be necessary to monitor the microbiological quality of plant material transported to the space station and upon its return. Microbes on the ISS may be exposed to harmful radiation, which can lead to mutations in surface microbes.

Microbes from the stratosphere or space, including the ISS, space crafts, and balloons, may or may not have immediate drastic deleterious effects on agricultural soils. However, we must remain vigilant regarding mutations or genetic drifts that may occur after space microbes have reached agrarian soils, if such a phenomenon has indeed taken place. As space travel becomes more frequent and microbes frequently hitchhike on space journeys, we cannot discount the possibility of microbial contamination (see ISRO, 2009; Khodadad et al., 2020; Lambert, 2020; Horneck et al., 2010). Space vehicles and crew should be monitored for space microbes that may be spilled or spread into agricultural zones. Routine quarantine checks for biological material shipped out or returned from space flights should be adopted. We cannot easily dismiss the potential detrimental effects of space microbes (particularly mutated ones) on the vast food generating systems on Earth. Agrarian soils are highly conducive to space microbes. Hence, they could potentially become virulent and prolific.

2.3 ATMOSPHERIC FACTORS INFLUENCE MICROBES AND THEIR ACTIVITY IN THE AGRICULTURAL SKY

The origin of microbes thriving in the lowermost altitudes of the troposphere is mostly agricultural soils, crops, trees, and other geographical features of the terrain. Wind, breeze, moving dust, clouds, precipitation, birds, insects, and moving ground animals also inoculate microbes into the air. Knowledge about the source or mode of entry of microbes found in the troposphere is important. They all have to negotiate and withstand the rigors of extreme physical conditions prevalent in the lower layers of the atmosphere. There are a few factors operating in the troposphere that microbes have to mandatorily endure and survive. Microbes have to be metabolically active and perpetuate. They have to withstand prevalent relative humidity, temperature, UV irradiance, ozone, pollution, etc.

It seems that extensive investigations have been conducted on the microbial content of soil and aquatic ecosystems (Gusareva, 2019). However, efforts to study the dynamics of microbes in the atmosphere have been limited. Scientific investigations on airborne microbes in the tropical lower troposphere are relatively scarce. Our knowledge about microbes in the troposphere above tropical forestry and agrarian regions is indeed limited (Wommack et al., 2010, 2015). Gusareva et al. (2019) found that the tropical atmosphere supported a larger share of fungi, fungal spores, and propagules (hyphae). Bacterial genera accounted for only 14.5% of the diversity, while Archaea accounted for 2.6%. The presence of crops' pollen and minute seeds was negligible.

Investigations by Guserevaa et al. (2019) revealed that microbes exhibit distinct diurnal cycles in the lower troposphere. These cycles involve changes in population, diversity, and activity within the troposphere, similar to microflora found in soil or water bodies. The fluctuations in aerosol microbes within a single day are significantly greater compared to variations observed over weeks or months. Molecular techniques used to examine microbial taxa have shown that microbes do exist in the atmosphere over long periods of time. For instance, bacteria and fungi have been detected in the troposphere for up to 13 months. Aerosol microbes can both disperse and decline in population within a specific region, while others migrate into the atmosphere from remote areas. It is generally believed that tropospheric microbes are either local or originate from distant regions through wind dispersal (Smith, 2012; Mayoll et al., 2017; Burrows et al., 2009a, b). However, the long-term presence of microbial species depends on tropospheric factors such as solar irradiance, temperature, relative humidity, oxygen levels, wind patterns, and more. The temperature within the atmosphere predominantly influences microbial diversity, population, and activity.

Diurnal changes and their influences on microbial diversity and population density have immediate applications in agriculture. The lower troposphere harbors and mediates the rapid dispersal of plant pathogens. Prophylactic sprays of fungicides, bactericides, antibiotics, and other chemicals can be timed accurately based on changes in microbial flora. We can time the sprays to avoid the build-up of pathogens. Inferences are based on migratory and cyclical trends of airborne pathogens. We usually consider the weather pattern above crops prior to spraying. Migratory trends of microbes will also depend on diurnal changes. The simplest suggestion would be to change the cropping pattern. It avoids the build-up of pathogens in a region. Prophylactic sprays could also be decided based on such information.

2.3.1 SOLAR RADIATION AFFECTS MICROBIAL LIFE IN THE ATMOSPHERE

Microbes floating in the air above agrarian regions are exposed to various radiations. Commonly, microbes are exposed to solar UV radiation. UV radiation is among the most potent radiations that affect various living organisms, including microbes. The UV radiation can be particularly devastating on microbes dispersed in the air. Such radiation affects DNA synthesis and duplication. There are reports about the quantitative aspects of UV radiation and its influence on atmospheric microbes. The effect of solar radiation has been studied both under controlled conditions (simulations) and in open cropped fields (Madronich, 2018). The inactivation of aeolian microbes does not occur only due to UV radiation. Several other detrimental factors might also operate on the microbial life above agrarian locations. Solar UV radiation might affect the normal life cycle processes in the agrarian regions. Microbial functions relevant to agro ecosystems may get repressed because irradiation affects DNA and its proper functioning. Under severe conditions, microbes in the agrarian regions may perish. The UV radiation may also induce mutants of atmospheric microbes. In the open fields, UV radiation inactivates microbes found on soil or crop surfaces (Ben-David & Sagripanti, 2010).

Generally, microbes are not resistant to harmful radiation. They perish under constant exposure to harmful radiation. Often, harmful radiations restrict the proliferation of microbes in the air. However, certain microbes found in the aerosol are resistant to UV radiation. They possess enzymes to rapidly repair the damage to DNA. Pepper & Garba (2015) list several mechanisms that help microbes overcome harmful radiation. These include microbes living inside larger airborne particles, the presence of pigments, high relative humidity (RH), and cloud cover. These tend to absorb or shield bioaerosols from radiation. Microorganisms also have mechanisms for repairing the DNA damage caused by UV radiation. For example, *Dienococcus radiodurans* possesses a radiation resistance mechanism. *Dienococcus radiodurans* is a soil bacterium and is considered a radiation-resistant organism. An important component of its radiation resistance is the ability to enzymatically repair the damage caused to chromosomal DNA.

Oxygen is essential for aerobic microbes found in the troposphere. Generally, the troposphere immediately above the crop's canopy is a highly oxygenated region. However, crops, epiphytotic microbes, and free-living atmospheric microbes may all encounter oxygen radicals that are uncongenial. Pepper & Garba (2015) mention that there are open air factors like oxygen that have a direct influence on the survival of diverse types of microbes. Reactive forms of oxygen can repress microbes in the atmosphere. Atmospheric microbes may encounter other reactive forms like superoxide, peroxides, and hydroxide. In the presence of chlorine, nitrogen, or sulfur, they can form harmful compounds. Such a process is often mediated in the agricultural sky by lightning, water, gases, and radiation. Reactive ions distributed in the air can inactivate or suppress microbial activity. They may lead to the physical decay of atmospheric microbes. We need to understand the effect of various radiations, lasers, and chemical sprays above the crop canopy. We have to monitor the phylloplane/phyllosphere for microbes that interact with radiation and chemicals. For example, high levels of hydrocarbons and ozone can inactivate many microbial species.

Ozone (O_3) is an air pollutant. It is generated in the troposphere through a photochemical reaction. Ozone is created by the action of sunlight on volatile organic compounds (VOCs) or oxides of nitrogen emitted. It is detrimental to life forms such as natural vegetation, crops, and forest stands, including microbes floating in the air. Additionally, it affects those residing in the phyllosphere, crop canopy, and other surfaces on a farm. There are numerous reports on the deleterious effects of ozone on crops and thriving microbes in agrarian regions. For instance, the use of ozone can lead to the perishing of microbes like bacteria, molds, and archaea found on crop surfaces. Ozone application has been shown to reduce airborne microbes to undetectable levels (Dohrmann & Tebbe, 2005; Grulke & Heath, 2019). Common bacteria such as *E. coli* and plant pathogens like *Pseudomonas syringae*, *P. aeruginosa,* and *Bacillus* sp. have all been found to be susceptible to ozone.

2.3.2 AMBIENT TEMPERATURE AFFECTS THE SURVIVAL, ECOSYSTEM FUNCTIONS, AND PROLIFERATION OF MICROBES IN THE AGRARIAN ATMOSPHERE

The temperature in the ambient troposphere controls, to a certain extent, the rate of physiological processes in microbes. Microbes found on the soil surface, crops, farm vehicles, and in the air are all affected by ambient temperature. Atmospheric microbes exhibit specific adaptations. Examples of these adaptations include modifications in enzyme activity, sporulation, and hibernation. When the temperature exceeds the tolerance level of microbes, they become inactive and may even die. The inactivation of microbes due to high temperature is attributed to desiccation, which is the loss of water from cells. Both functional and structural proteins may become denatured. Similarly, temperatures lower than a certain threshold are also detrimental to microbes in the atmosphere. Microbes have a slower life cycle and longer generation time in low temperature conditions. For instance, the generation time for *Bacillus* sp. is 25 to 30 minutes at a normal temperature of 25°C–28°C. However, it increases to 45–60 minutes when the incubation temperature is lowered to 15°C or below. Additionally, when the agricultural sky experiences freezing temperatures, microbial cells perish due to ice crystal formation in the cytoplasm and cell wall surface. The interaction between relative humidity (RH) and diurnal temperature is a crucial aspect that affects microbial population, diversity, and specific ecosystem functions of microbes.

We can detect cold-loving microbes (psychrophiles) in the temperate sky. Mesophiles and thermophiles also exist depending on the location of agrarian regions. For instance, the atmosphere in arid desert-like conditions with sparse vegetation/crops in the Sahel can support microbes that tolerate higher temperatures, reaching up to 45°C–50°C. These microbes may form spores to survive unfavorable temperatures. The atmosphere in temperate agrarian regions harbors numerous microbes that can withstand lower temperature regimes. Cold tolerance is essential for all beneficial microbes (e.g., Bradyrhizobia). Plant pathogens exposed to air or those that migrate through wind or dust particles will inevitably require temperature tolerance. Microbes residing in agrarian atmospheres are known to contribute to nucleation, snow formation, and precipitation induction (rain).

Temperature can significantly impact phyllosphere microbes, including their population, diversity, and activity. On the phylloplane, microbes respond to leaf surface characteristics, exudates, nutrient availability, moisture, ambient temperature, and other interacting microbial species.

2.3.3 RELATIVE HUMIDITY (RH) IN THE ATMOSPHERE INFLUENCES BIOAEROSOL MICROBES

To stay within the context of the book, we will only consider soil microbes from agrarian regions, airborne commensals, plant pathogens, and their ability to thrive at different relative humidity (RH) levels in the atmosphere.

Soil microbes thrive at varying moisture levels. Even when dispersed into the air, soil microbes may still encounter RH levels similar to those found in soil pores. RH changes in the agrarian troposphere occur based on factors such as season, precipitation levels, evaporation, ambient temperature, and cropping systems adopted. Microbial species that overwinter as resting spores may have a better tolerance for adverse RH conditions. In fact, several bacterial genera are known to enter sporulation based on environmental signals, such as low RH, desiccation, high temperature, freezing, and adverse moisture. Agricultural microbes experience simultaneous changes in RH and temperature. In response, fungal pathogens often sporulate and release spores into the atmosphere. Spores with thick walls and low metabolic activity have a better survival rate under adverse RH conditions. For example, spores of rust fungi are more tolerant to desiccating conditions in the troposphere.

A study by Peccia et al. (2001) suggests that commonly aerosolized bacteria, such as *Serratia marcescen, Bacillus subtilis,* and *Escherichia coli,* absorb a certain amount of water from the atmosphere. This occurs between RH values of 20% to 95%. The absorption of moisture increases the weight of microbes, but it may not result in an expansion of cell size. In nature, various atmospheric factors affect agriculturally beneficial microbes, plant pathogenic bacteria, and commensals. These factors operate at different intensities and durations. Aerial microbes, including bioaerosols, invariably respond to changes in RH in the atmosphere. They may also be influenced by exposure to the atmosphere during transit, migration on clouds, precipitation droplets, snow, and dust particles. Bacteria may respond differently to these combined effects. For instance, RH can impact beneficial microbes like nitrogen fixers (Peccia et al., 2001). The natural fluctuations of RH over agricultural areas require regular attention. By creating unfavorable RH conditions, it may be possible to hinder the growth of plant pathogens in crop canopies.

As mentioned previously, the microbes present in the troposphere primarily originate from soils, crop surfaces, harvested products left in the fields, stubbles, water bodies, and farm structures. Their ability to survive in the air greatly depends on the relative humidity (RH) they encounter. Optimum metabolic activity of microbial cells requires minimum levels of RH. Microbes thrive in the agricultural sky when RH is within the optimal range of 30 to 70%. However, they perish and decompose when exposed to saturation levels of RH at 100% (Wells, 1955; Riley & Kaufman, 1972; Fuzzi et al., 1997). In extremely arid regions, deserts, and other areas with scarce moisture, microbes cannot survive due to the extremely low RH levels of 5-10%. Bacteria, fungi, and archaea simply become desiccated and perish. It appears that gram-negative bacteria found in aerosols are more resistant to desiccation and have a higher survival rate. Gram-positive bacteria, on the other hand, are more adept at retaining moisture (Mohler et al., 2007; Pepper & Garba, 2015). They also have a better chance of survival at higher RH levels. RH also plays a role in the survival of virus particles in the atmosphere. Viruses with nucleocapsids and protein envelopes have a higher chance of survival in the atmosphere. They can endure at low (<50%) or variable RH levels.

The troposphere is periodically exposed to foliar sprays of plant protection chemicals. These sprays are water-based. High-tech farms often use center-pivot irrigation, which temporarily alters the relative humidity (RH) above the canopy. This change in RH may promote the multiplication of bacteria, fungi, and other microorganisms. Additionally, dust particles may settle on the soil due to the sprays, providing protection to the microbes against desiccation.

We could potentially explore applications based on the 'RH preference' of atmospheric microbes. For instance, we could conduct aerial sprays of water to temporarily increase RH and stimulate microbial growth. Alternatively, we could create an unfavorable RH environment for the microbes. Increasing RH could also enhance the effectiveness of certain fungicides, as bacteria and fungi absorb moisture from the atmosphere.

The phylloplane and phyllosphere are exposed to the vagaries of the atmosphere. However, maintaining a congenial relative humidity (RH) in the phyllosphere region is essential. Microbes require optimum RH in the crop's canopy. We can restore lost water on the leaf surface through aerial sprays (sprinklers). Phylloplane conditions, such as ambient temperature and RH, are crucial for the survival and perpetuation of microbes.

2.3.4 MICROBES AND PRECIPITATION

The clouds interfere with the interception of light by natural vegetation and crops. Clouds also mediate certain chemical transformations because they are made of water. Clouds play a role in the hydrological cycles in the troposphere. However, in nature, they serve as a major habitat for microbes in the atmosphere. They facilitate the short and long-distance transport of aerosol microbes. Clouds act as a cycle for microbes between the soil and troposphere. Therefore, understanding the microflora of clouds is important. Amato et al. (2017) reported the occurrence of complex communities of bacteria, fungi, and archaea in the clouds. Some examples include alpha- (Sphingomonadales, Rhodospirillales, and Rheobases), beta-, and gamma-proteobacteria (Pseudomonadales). In addition to understanding microbial diversity, it is essential to comprehend the functional aspects of various microbial species encountered in the clouds and precipitation.

Mineral dust, particulate organic matter, and sand from the soil surface were thought to be highly effective in clumping raindrops and allowing them to fall on crop fields. Currently, the major enucleators and rain creators are believed to be biological agents such as bacteria, fungi, spores, pollen, etc. It has been observed that wind, which kicks up dust combined with microbes, induces precipitation (Welsh, 2011). In the Nordic region, for instance, bacteria and other microbes that initiate precipitation in the atmosphere have been identified. Danish meteorologists believe that the reduced accuracy of precipitation forecasts, to some extent, could be attributed to microbial flora. It appears that microbes play a key role in the formation and distribution of precipitation (Brix, 2015).

Atmospheric bacteria induce enucleation and cause precipitation to occur (Schiermeier, 2008). Undoubtedly, precipitation events replenish agrarian soil with bacteria traced in the atmosphere. Therefore, there are suggestions to spray enucleating microbes above the clouds. Microbes are to be sprayed whenever there is a shortage of dust and PM. This is because dust particles are invariably required to induce precipitation. Microbes that are not pathogenic to crops, animals, and humans are utilized as enucleators. Freshwater organisms could also be used as enucleators. They could be sprayed periodically as needed. Spraying microbial cultures is definitely cost-efficient compared to silver grains, solidified CO_2 particles, powders, sand, or other sources of mineral PM.

In fact, Shiermeier (2008); Christner et al. (2008, 2012), and a few others express that soil microbes that get lofted often become part of aeolian microflora. They are important in completing the hydrological cycle on Earth and play a vital role in regulating moisture in agrarian regions. Microbes may be key natural factors affecting precipitation patterns experienced in many agrarian regions. Indirectly, crop productivity may, after all, be influenced by the enucleation via microbes leading to rain. There are suggestions to use aerial bacterial sprays to determine the location where it should rain (Ravilious, 2016). This could be done by preferentially applying enucleating microbes over the clouds. For example, rain-bearing clouds may miss a particular arid agrarian region because of a lack of enucleators. Then, microbial cultures could be sprayed at certain altitudes above the clouds.

Furthermore, reports suggest that bio-precipitation may actually become common in the future, particularly to overcome water deficits in agrarian regions. There are indications that microbes located at far-off locations in other continents may actually be controlling weather and precipitation patterns in a different continent. Although evidence for it is still feeble, the migratory aspects of microbes, both at local and inter-continental ranges, need careful consideration. Drastic interference in this natural phenomenon may not be wise.

There is a suggestion to enrich organic soils with microbes, particularly commensals and non-pathogenic strains of *P. syringae* or similar genera. The blowing wind and turbulence can carry them into the atmosphere. Therefore, a rich soil with stabilized microflora is considered a good option to supply enucleators. Afterward, we wait for nature to induce rain. Some farmers have expressed that enucleators need to be inoculated in the right spot, perhaps around 50 km ahead, taking into account wind velocity (Blomfield, 2015). Simulations of raindrops and natural precipitation events have shown that a single drop of water can support multiple enucleators (Garcia et al., 2019). For instance, a single drop may contain *Pseudomonas viridiflava, P. flourascence, P. agglomerans, Exserohilumturcicum,* and *Xanthomonas campestris*. This means that, for our assurance, we can try inoculating clouds with multiple genera and strains of enucleating microbes. Several microbial genera that reside on the foliage of natural vegetation and crops may also find their way into raindrops through clouds, potentially contributing to precipitation.

2.4 MICROBES IN ATMOSPHERIC DUST AND ICE NUCLEATION

It is not easy to freeze water. All forms of water, such as clouds, water vapor, and droplets, require nucleating agents. Nucleating particles are tiny solid particles that aid in the freezing process (Cox et al., 2013; Vali, 1995). In the troposphere, these nucleating agents are typically dust particles, bacteria, fungal spores, soot, organic matter, dust, and sand particles. Aerosols that carry biological particles (bacteria, fungi, archaea), inorganic dusts (e.g., kaolinite crystals, sand particles), and organic matter play a significant role in the formation of rain-bearing clouds over agroecosystems (*see* Boucher, 2013; Tobo et al., 2012, 2014). Bioaerosols can have a major impact on global hydrological cycles, particularly those relevant to farming enterprises. The interaction between aerosols laden with biological material and moisture is crucial for ice nucleation processes.

Farming regions developed in dry areas with sandy or loose surface soils are prone to creating dust plumes. Fields with loose soil left open without cover crops are vulnerable to wind, which generates dust plumes. Dust plumes often occur during the fallow season. Arid deserts also generate large-scale dust plumes/storms. Such dust, along with biological material, creates ice and rain-bearing clouds (Ginoux et al., 2012). There is clear evidence regarding the ice nucleating properties of agrarian soils from different continents (O'Sullivan et al., 2014; Tobo et al., 2014).

The ice nucleation properties of dust created in agricultural regions differ in terms of efficiency. For instance, Steinke et al. (2016) conducted a study on the variations in aerosol formation above farming regions in Argentina, Germany, and Mongolia. They assessed the ice nucleation activity of surface soil and dust particles from these agrarian areas. The researchers discovered that the ice nucleation properties of bioaerosols from the three farming locations were ten times higher than those of dust from deserts. It appears that the differences in ice nucleation properties among soil dust samples decreased below a temperature of 247 K (–26.5°C). The ice nucleation properties were also influenced by the atmospheric relative humidity. The biochemical nature of ice nucleating dust should also be taken into consideration.

There are a few different modes of ice nucleation processes traced in the atmosphere. They are:

i. **Immersion/Condensation:** It is initiated in the cloud droplets.
ii. **Deposition Nucleation:** It involves water vapor being directly converted into ice.
iii. **Contact Freezing:** It occurs when dry interstitial aerosols mix with supercooled aerosol particles (Marcolli, 2014).

Murray et al. (2012) state that ice nucleating agents found in the atmosphere, including those above agrarian regions, are predominantly of biological origin. These agents can be bacteria, fungal hyphae, spores, protozoa, archaea, and viruses. The earliest detection of bacteria, specifically *Pseudomonas syringae,* in rainmaking clouds and on nucleating dust particles was reported by Maki et al. (1974) and Sands et al. (1982). Bacteria, such as *P. syringae,* are known to be highly efficient nucleators, especially at temperatures above 14°C (Rowan, 2008). It is clear that we should continue our search for more efficient nucleators.

Let us now consider the mechanisms involved during microbe-mediated ice nucleation. As stated earlier, freezing pure water is not easy; it requires –37°C. However, ice nucleating bacteria can help freeze water at just 0°C. These bacteria have proteins with specific amino acid sequences that aid in ice nucleation. Studies conducted at the Max Planck Institute in Mainz, Germany, have shown that bacteria like *Pseudomonas syringae,* which possess ice nucleating proteins, are required for ice nucleation. The surface proteins remove heat from water droplets and induce crystallization. A better understanding of bacteria-mediated ice nucleation may help us take measures against frost/snow-induced damage to crops. Spraying denatured cells of *Pseudomonas syringae* at higher altitudes in agricultural areas could induce rain in the agrarian zones (Max Planck Institute, 2016; Stopelli et al., 2017; Weidner & Frohlich-Nowoisky, 2013). Previous studies on rain, snow, and frost formation have given us the idea that inert particles made of organic matter, carbonaceous soot, and inorganic sand crystals aid in nucleation.

Microbiologists have examined clouds, ice crystals, and snowflakes in the Montana farm belts. They traced a bacterial ice nucleating agent in 70% of atmospheric samples. *Pseudomonas syringae* was the most common nucleating agent (Sands et al., 1982; Robbins, 2010; Rowan, 2008). Incidentally, these ice nucleating microbes, like *P. syringae* and several other species, are standard inhabitants of leaf surfaces. They are frequently found as plant pathogens in cropped fields. The ice nucleating bacteria actually damage plant cells through ice crystals that fall on the leaf surface. Pandey et al. (2016) have stated that microbes such as *P. aeruginosa* are well-known plant pathogens. They have a role in crop disease and related ecological aspects in agricultural zones. They attack crops by nucleating ice crystals. Additionally, these microbes with ice nucleation ability (INA) affect cloud formation, precipitation, and snowfall. We need to exploit this trait appropriately.

There are several microbial species that induce enucleation and ice crystal formation. Trials are being conducted to control ice-nucleating and plant pathogenic bacteria, such as *P. syringae*. The formation of ice crystals can be harmful to leaf tissue, as it facilitates the entry and development of diseases like *P. syringae*. Agricultural experts in Californian vine growing regions are considering the application of plant protection chemicals containing copper as foliar sprays, using overhead sprinklers, in the future. These plant protection sprays help reduce the population of enucleating bacteria *P. syringae* on the surface of grape foliage (Northcut, 2017; UC Cooperative Extension Service, 1997; UC Delivers, 2019). Additionally, they are exploring microbial sprays that can deter the buildup of *P. syringae* on leaves. This approach aims to prevent frost and ice nucleating bacteria, such as *P. syringae*. The underlying rationale is that if the bacteria can be controlled with antibiotics or copper compounds, the risk of frost can be reduced by up to 3°F.

The biological ice nucleation is the major mechanism for creating rain in most farming regions. It is referred to as 'bio-precipitation.' Dust, inorganic particles, or any other particulate surface can also induce rainfall. A global study was conducted to explore the relationship between clouds, rain, snow, and nucleating microbes. In agrarian regions, various types of microbes have been found to act as rain inducers. In fact, one suggestion is to spray bacteria/fungi on the phyllosphere (leaf), soil surface, and terrain to induce the crystallization of water droplets and trigger rainfall. The application of bacterial solutions may actually help in generating rain in dry areas (Louisiana State University, 2008). Furthermore, it is now evident that ice nucleating bacteria play a significant role in natural precipitation cycles. Atmospheric bacteria do contribute to regulating hydrological cycles to some extent in all regions, including agricultural zones.

Microbial species with the capability for ice nucleation are frequently found on crop canopies and natural vegetation. A few examples identified using 16S rRNA sequencing are *Pseudomonas aeruginosa, P. viridiflava, P. fluorescence,* and species of the genus *Xanthomonas.* Among them, a few are plant pathogens. They could be denatured before dispersing in the sky. Most are commensals in agrarian soils and on crop surfaces.

Bacterial ice nucleation is a natural process that occurs in the atmosphere. It involves catalyzing the phase transition of water molecules by a living organism. Perhaps, the Ice Nucleating Activity (INA) of bacteria could be utilized by farmers to seed clouds and induce rain (Morris et al., 2004; Morris & Sands, 2017). The INA of bacteria seems to be related to certain outer membrane and cell wall proteins that catalyze the nucleation of moisture. Studies involving DNA sequence homology, particularly that of 16s rRNA, have shown that INA genes have been transferred to different bacterial species. There are horizontal gene transfer systems from a common ancestral bacterial species (Gurian-Sherman & Lindow, 1993; Edwards et al., 1994; Kajava & Lindow, 1994). The INA protein has highly conserved repeats of 4, 8, and 48 amino acids. It seems that the central core of the INA protein induces hydrogen bonding between the protein and neighboring water molecules.

Hill et al. (2014) attempted to identify and quantify the INA genes using qRT-PCR methods. They reported that, in addition to *P. aeruginosa,* other species such as *P. putida, P. auricularis,* and *P. po* possess INA genes. They discovered 10^8 *ina* genes per gram of cereal leaf blades and 10^5-10^6 *ina* genes per gram on legume/broad-leaved crops. Mutants with high INA activity or those lacking *ina* genes may aid in proper experimentation, comparison, and understanding the role of *ina* genes in precipitation formation in agrarian regions.

In nature, such INA is believed to assist bacteria in accessing water and nutrients even in unfavorable conditions. It aids in the disruption of plant cells, releasing cell contents and nutrients from damaged plant cells. INA can be detrimental to plants such as wine grapes, olives, tea, apricots, and beans. Several major cereals, legumes, oilseed crops, and others sown in regions with severe winters may also suffer frost damage due to ice nucleation by atmospheric bacteria. The population of bacterial cells on the surface of plant material varies from thousands to tens of millions of cells per gram of tissue.

Evidence regarding bioaerosols and their ice nucleating agents has revealed the presence of several other microbial species that also perform similar ice nucleating functions (Sesartic et al., 2013). Lichens have demonstrated the ability to nucleate ice at temperatures as low as *Cladosporium* 8°C or higher (Kieft, 1988; Kieft & Ruscetti, 1990). Similarly, the first reports about the ice nucleating ability of spores from fungi such as *Fusarium acuminatum* and *Fusarium avenaceum* were made in the 1990s (Pouleur et al., 1992). Haga et al. (2013) discovered that rust and bunt fungal spores differed in their potential and efficiency in ice nucleation in the atmosphere. Spores of *Cladosporium* sp. were found to be less efficient in nucleating moisture, which was attributed to certain hydrophobic proteins on the spore surface. Undoubtedly, the quantification of the ice nucleating property of tropospheric bacteria and/or inert particles is necessary (Joly et al., 2014). The efficiency of ice nucleating microbes could be related to the quantitative expression of specific cell wall proteins on ice nucleating bacteria/fungi. This aspect must be thoroughly investigated.

The ice nucleating activity of microbes is a phenomenon that occurs in the agricultural sky. It could be advantageous to apply this during crop production. Standardizing procedures may be necessary. The most efficient ice nucleating strains of bacteria or fungal spores must be identified. The ability to easily multiply the microbes on semi-synthetic or natural substrates and in large quantities is crucial for large-scale application. The microbial formulation should be sprayed over cropped fields and entire agrarian regions. The timing of application, considering factors such as crop stage, precipitation needs, wind velocity, and water quantity required for crops, needs to be determined. The economy of spraying microbes should also be considered, ensuring efficiency as an agronomic procedure that can be repeated frequently. Currently, there are agricultural drones (UAVs) equipped with sprayers. These drones can easily reach higher altitudes in the troposphere (above clouds). With the use of sprayer drones, it should be possible to uniformly distribute bacteria or fungal spores possessing ice nucleating activity across a region in the sky.

Microorganisms are intricately linked with glaciers, snow, frost, haze, and fog. Ice-nucleating bacteria and dust mediate the formation of snowflakes in the temperate atmosphere. Additionally, glaciers support several species of pigmented algae. Green algae deposited on ice sheets/snow fix carbon. There are viruses that infect algae and release carbon and other nutrients that are cycled (Hodson et al., 2008; Edwards, 2021). Reports suggest that aeolian algae and wind-blown mineral particles deposit on snow. There are seasonal variations in this natural process. The snow algae are primary producers because many of them photosynthesize and fix carbon (Umino & Takeuchi, 2012).

Aeolian dust that deposits on snow induces certain changes in the snow. The microbial diversity of the snow may depend on the origin of dust particles, snow structure, and the survival of microbes during transit. Recent deposits of microbe-laden dust are known to exhibit a greater diversity of microbes. Photoautotrophs, such as green algae, appear to colonize snow more than bacterial heterotrophs (Courville et al., 2020; Brown & Jumpponen, 2019; Dastrup et al., 2018). Depending on the source of dust/microbes, we may even encounter plant pathogens. The ability of plant pathogens to overwinter, hibernate, become activated, and infect temperate crops should be monitored.

Smog, haze, and pollution affect the microbes that drift into the atmosphere. In China, smog has reduced air quality over cropping zones. A similar situation is repeatedly encountered in the Indo-Gangetic plains. The total airborne microorganisms (TAM) found in the atmosphere is a good measure of the air quality. In some areas, TAM in snowfall is low (Xie et al., 2018). This may reduce the chances of crop health deterioration if plant pathogens are also proportionately less.

2.5 MICROBES AND CLIMATE CHANGE

The microbial aspects of climate change have been highlighted in many reviews. Most of them emphasize the role of microbes in terrestrial and aquatic ecosystems (Dutta & Dutta, 2016). However, in the present context, we are interested in the influence of climate change on microbes residing in the troposphere. Undoubtedly, climate change affects the physiological functions, nutrient transformation, and survival of aerial microbes. On the other hand, it is important to understand in detail how microbes can induce climate change in an agro-ecoregion.

The airborne microbiome consists of microbes injected from the terrain. In the lower troposphere, microbes are usually derived from soil, crops, and natural vegetation. Caliz et al. (2018) point out that microbial assemblages in such microbiomes are dynamic, with prominent changes in microflora being common. Diurnal and seasonal changes may occur in the microflora of the agrarian troposphere. Aeroplankton changes can be traced based on the altitude in the troposphere. Furthermore, climate changes also affect the microbial assemblages in the atmosphere (Zhu et al., 2017; Smith et al., 2013). The effects of climate change on microbes can be observed prominently in the boundary layer, i.e., nearer to the crop's canopy. In other words, microbes should be monitored at low altitudes of the troposphere (Bowers et al., 2011, 2013; Barberan et al., 2014, 2015; Fierer et al., 2008). Microbes mediate greenhouse gas (GHG) emissions. At the same time, there are soil and aeolian microbes that utilize GHGs. Aerial microbes convert GHGs into compounds that are not detrimental. In fact, climate change effects may result in the disruption of many of the usual geochemical transformations (Abetenh et al., 2018). Overall, climate change may affect microflora and their activity in the troposphere.

Climate change influences several agricultural operations and productivity. Climate change undoubtedly affects microbes thriving in agrarian regions, as they play an active role in various soil-related transformations of nutrients. The microbe-mediated release of greenhouse gases (GHG) into the troposphere needs to be monitored. Not only do soils contribute to this, but tropospheric conditions above crops may also aid in the further conversion of substrates into GHG. Rice paddies alone contribute about 20% of CH_4 emissions from cropland, despite occupying only 10% of global cropland (Cavicchioli et al., 2019). Under anaerobic or microaerophilic conditions, as found in paddy fields, microbes are known to release methane (CH_4). Additionally, ruminant microbes also contribute to CH_4 emissions in the atmosphere.

Climate change affects several types of microbes residing in soil and air. It has an impact on N_2 dynamics in the soil and air. Microbes play a role in transforming N compounds, including processes such as organic matter decomposition, N-mineralization, nitrification, and denitrification, all of which are affected by climate change. N_2O is released as a result of nitrate reduction. Nitrifiers, denitrifiers, N-fixers, and N decomposers are involved in these N transformations. Agricultural soils are major contributors of N_2O and NO_2 gas emissions into the atmosphere. Additionally, microbes transiently residing in the troposphere also mediate N transformations. Furthermore, several transformations related to sulfur (S) and SO_2 in the atmosphere are also mediated by microbes.

Climate change affects the survival and activity of microbes in the troposphere. Aridity induces the desiccation of microbes, reducing their chances of survival. Aspects such as acid rain, smog, and ozone pollution also have an impact on microbes. The atmosphere is affected by crop burning, pollution from SO_2, ozone, and dust/smog. Poor air quality undoubtedly reduces the survival of microbes in aerosols. In recent years, crop burning has become a common and cost-effective agricultural practice. Farmers who grow cereals find it quick and easy to burn the stubbles of the previous crop, resulting in large quantities of dust, smog, and pollutants covering the agrarian sky. Microbes are also carried from the crop fields into the troposphere. Microbes undergo severe stress in the atmosphere, and nutrient transformations that occur in a clean sky (atmosphere) are hindered due to pollutants. Crop production procedures also contribute greenhouse gases and pollutants to the troposphere. Clearly, field procedures have an impact on aerial microbes.

2.5.1 CROP RESIDUE BURNING, SMOG, AND MICROBES

Burning crop residues, stubbles, and weeds in the open field is a common practice in many agrarian regions (Bhattacharya et al., 2021). Crop residue burning is prevalent in the Indo-Gangetic belt where rice-wheat or sugarcane-fallow cropping systems are practiced (Sidhu et al., 1998; Singh & Sidhu, 2014; Sarkar et al., 2018; Goswami et al., 2019). Crop residue burning is also widespread in other agrarian regions such as the Great Plains of North America, Northeast China, and the Pampa of Argentina. In all cases, this agronomic procedure affects the atmosphere through pollution with dust particles, smog, and GHG emissions. Crop residue burning increases the concentration of CO_2 in the atmosphere. Such GHG emissions may impact microbial flora and climate parameters such as temperature, CH_4, CO_2, and NO_2 concentrations. Crop residue burning clears stubbles from fields but pollutes atmosphere. Polluted atmosphere can spread/diffuse to neighboring regions. This can happen via wind, clouds, and drifting smog. Microbial load, diversity, and activity in the atmosphere is affected deleteriously. Microbes' resident in the dust particles emanating from stubbles may however survive for a certain period. Crop residue burning ultimately induces global warming. There could be alternative methods to overcome deterioration of microbial population in the atmosphere.

Bhattacharya et al. (2021) have examined the effects of crop residue removal, crop residue burning, and *in situ* application of stubble decomposing microbes. Crop burning causes a significant loss of nutrients (N, P, K, and S), in addition to carbon. By using cellulolytic and ligninolytic microbes on stubbles of rice, wheat, or other crops, mineral nutrients and carbon are preserved in the soil profile. This method also helps protect atmospheric components, including microbial load (Borah et al., 2009). In fact, crop residue burning severely deteriorates air quality (Grover & Chaudhary, 2019). According to a report, burning one ton of crop stubble leads to a loss of 5.5 kg N, 2.3 kg P, 25 kg K, and more than 1 kg of S, in addition to organic carbon.

Crop residue burning releases large quantities of PM. Carbonaceous dust carries microbes with diverse functional abilities. Such stubble burning events in the Indo-Gangetic Plains and other regions also release SO_2, nitrous oxides (N_2O), ozone, CH_4, and water vapor (Kumari et al., 2018; Jain et al., 2014). This is a combination of PM, microbes, pollutants, and chemical compounds that contribute to the formation of smog (Gadde et al., 2009). Microbes attached to dust particles can reach higher altitudes in the troposphere. These microbes on dust particles can be transported by high altitude air jet-streams. At this altitude in the troposphere, microbes can migrate longer distances and then descend on far-off agrarian regions. Low altitude dust, along with microbes, may contaminate nearby crop fields. Smog and dust plumes can carry various microbial species that are plant pathogens. For example, when wheat crop residue (stubbles) is burnt, it may release propagules of rust fungi into the troposphere. Such smog and dust plumes will definitely spread the disease to other regions situated at short and/or long distances (Gadde et al., 2009; Gupta et al., 2004).

Wildfires in standing forest plantations and natural vegetation have become increasingly common in recent years. Consequently, it is important to consider the consequences of burning natural vegetation on microbial population and diversity in the atmosphere. This specific example is from the natural vegetation zones in Florida, USA. The burning of crops, weeds, and vegetation releases particulate matter (PM) and microbes into the atmosphere. It appears that the troposphere above these wildfire zones contained 5 times more microbial cells ($6.7 \pm 1.3 \times 10^4$ cells m^{-3}) and biological ice nucleating particles (INPs) ($2.4 \pm 0.91 \times 10^3$ INPs m^{-3} active at temperatures $\geq -15°C$). Investigations into microbial diversity indicated that they originated from the burnt vegetation. Microscopic and cultural studies revealed that 80% of aerosolized microbes were viable. Therefore, Moore et al. (2021) concluded that burning vegetation contributes to the microbial population in the smog that forms in the troposphere.

Stefanski & Shivakumar (2016) state that an effective warning system is needed to detect trans-continental dust storms, sand drifts, and tornadoes. This system would help farmers protect their crops through suitable prophylaxis. We can apply this suggestion to aerobiology above crops. Similar periodic monitoring is necessary for microbial pathogens that travel across continents on dust particles. This will provide us with an idea of the pathogenic microflora present in the atmosphere. It is crucial to prevent the occurrence of a sick or disastrous agricultural sky caused by aeolian microbial pathogens.

2.6 AEROSOL MICROBES AND CROP-MICROBE ASSOCIATIONS

2.6.1 PHYLLOSPHERE MICROBES CONSTANTLY INTERACT WITH ATMOSPHERIC COMPONENTS AND CROPS

The phyllosphere is a region on the plant leaf, extending to a thickness of 0.2 mm across the entire leaf surface. It is considered a favorable habitat for microbes within the overall area of the crop's canopy, which is part of the troposphere above the soil surface. This aerial part of the crops is densely inhabited by microbes and plays a vital role in the interaction between the crop and the atmosphere (Bringel & Couvee, 2015). However, the phyllosphere can also be a challenging environment for microbes to survive, grow, and thrive. The ability of microbes to thrive in the phyllosphere depends on factors such as crop species, leaf exudates, season, and cultivation region. It is important to note that phyllosphere microbes are in constant interaction with atmospheric gases, particulate matter (PM), and other microbes. Microbes on the leaf surface also interact with volatile organic compounds (VOCs). The microbial flora of the phyllosphere plays a crucial role in regulating plant pathogens that attempt to enter the host through the leaf surface. Interestingly, phyllosphere microbes are commonly found in the lower troposphere. The use of standardized molecular methods, such as 16s rRNA analysis, has provided opportunities for large-scale analysis of phylloplane and phyllosphere microflora. These methods enable rapid identification of the diverse microbial community present on the leaf surface and in the phyllosphere (Krishna, 2022; Bringel & Couvee, 2015).

Airborne microbes that settle on the phyllosphere may be suppressed by native flora present on the leaf. Conversely, airborne microbes may also act antagonistically towards the phyllosphere microflora. Another possibility is the biological control of pathogens. This can be achieved by applying antagonistic microbes to outcompete, suppress, or even eradicate the pathogens.

2.6.2 AERIAL MICROBES AS PLANT PATHOGENS

Aeolian microbes and their ability to cause diseases on food crops such as wheat, barley, or rice have been known since ancient times. The air, known as "disease-borne wind" by the Romans, carries disease propagules and was identified as the cause of fungal epidemics that destroyed their wheat and barley crops. Airborne diseases, particularly fungal diseases like *Puccinia* rusts, have been extensively studied (Sache, 2000; Madden, 1992; Aylor, 1990, 2003; Aylor et al., 2008; Hirst & Stedman, 1963; Rowell & Romig, 1966; Sache, 2000). We have gathered detailed evidence and developed several preventive measures to control airborne rust fungi (Nagarajan & Joshi, 1985; Nagarajan & Singh, 1990). The troposphere is inhabited by numerous microbial species that can potentially infect botanical species. Many of these pathogens produce light spores with specific adaptations such as thick spore walls, thick exosporium, striations on the spore body, spines, and tolerance to heat, desiccation, and even UV radiation. These adaptations allow them to withstand adverse conditions and travel long distances. They spread rapidly through wind and water, including precipitation, foliar sprays, irrigation channels, and snow. Believe it or not, certain aeolian microbes are dreaded by farmers. This is because they spread epidemics and pandemics. For example, rust fungi can travel long distances and initiate new outbreaks of disease. Light breezes, turbulent winds, dust storms, and jet streams serve as vectors for plant pathogens. There are numerous examples of diseases caused by aeolian microbes, including fungi, bacteria, and viruses. More detailed information and discussions on aeolian microbes as plant pathogens can be found in Chapters 4 and 6. Here are a few examples of airborne diseases that destroy crops. Rust fungi (*Puccinia* spp.), blotches (*Xanthomonas campestris*), and viruses like Cassava Streak Virus can rapidly spread at different altitudes in the troposphere through wind. Dust storms that travel at higher altitudes and longer distances of over 15,000 km can also carry these airborne plant pathogens and initiate new disease outbreaks. In some cases, plant pathogens can travel intercontinental distances when they reach established jet streams (see Krishna, 2022; Kellogg & Griffin, 2006; Griffin et al., 2003, 2006).

It is important to note that African dust storms also contain several plant pathogenic species. Fungal spores traced at high altitudes in jet streams are known to induce major diseases upon descending on crop canopies (Hau & Vallavieille-Pope, 2006; Schmale & Ross, 2015). The aeolian dust harbors enough fungal propagules to cause disease at far-off locations. Troposphere microbes are omnipresent. Researchers have noted that the microbial load (fungal spores) in each dust particle is crucial. Each dust particle may carry between 10,000 and 15,000 spores of rust fungi. The troposphere above crops may contain millions of tons of dust. Therefore, we encounter a large mass of microbes with diverse capabilities. They are all usually viable and have the potential to initiate disease or symbiotic associations, depending on the crop species.

Actually, microbes inside floating dust could be commensals or pathogens. Reports about symbionts that assist crops with biological nitrogen fixation (BNF) are also available. For instance, aeolian dust or wind may carry Bradyrhizobia from the Sahel region all the way to the eastern coastal states of North America and Brazil. This phenomenon is beneficial to farmers who grow legumes. Sugarcane rust is another example of Trans-Atlantic migration of aeolian dust, as it carries rust propagules. The wind facilitates the transportation of rust fungus across the oceans, from the Sahel to North America. Propagules of bacterial pathogens, actinomycetes, archaea, and viruses have also been detected in the wind. It is not sufficient to only understand the dust cycles that occur in various locations of the Sahel. Microbes found on dust with the potential to cause diseases (such as rust, downy mildews, blotches, streaks/mosaics caused by viruses, etc.) are equally important (Imbler, 2020).

Airborne microbes may also be traced on farm vehicles, farm animals, implements, and irrigation channels, among others. Microbes thrive on dust that resides on farm vehicles and get distributed with the turbulence caused by the wind. Airborne microflora that settles into irrigation channels can also get dispersed with water.

Fusarium head blight (FHB) is a disease caused by *Fusarium fujiense*. This fungal pathogen is spread via airborne spores. Inflorescences (panicles) previously infected with spores serve as good sources for the rapid and long-range dispersal of FHB. Viable conidia can be easily detected in the airspace above crops (Del Ponte et al., 2003). Spores of powdery mildew can also become airborne, with short-distance spread possible through shattering spores. Other examples of airborne plant pathogens include Downy Mildew fungus (*Mycosphaerella* spp.), Neck Rot of Cereals fungus (*Gibberellazeae*), and Apple Scab (*Ventureainequalis*). Atmosphere-mediated epidemics/pandemics caused by such fungal pathogens may occur, depending on the aerospace conditions above the crop's canopy (Aylor, 1990, 1992, 2003; Aylor et al., 2008).

2.6.3 AERIAL MICROBES AS BIOLOGICAL CONTROL AGENTS

Aerial microbes of diverse genetic constitution are found in the phyllosphere. Phyllosphere organisms constantly interact with aeolian intruders that have landed on the phylloplane. These organisms are known to antagonize and destroy intruding pathogenic strains (Ritpitakphong et al., 2016). The native phyllosphere microflora is also known to physically block the entry points of disease-causing organisms, thus suppressing the pathogens. The atmosphere also contains biocontrol agents (BCAs) that suppress various fungal pathogens. Airborne bacteriophages also play a role as aeolian microbes, as they can suppress host bacteria, which may be plant pathogens. It is important to explore the atmosphere and identify microbial interactions that can be useful in crop disease control. There may be virus/phage particles in the air that reduce microbial diseases.

Microbes involved in the biological control of important insect pests, such as leaf eaters, head miners, pod borers, and army worms, may be present in the aerosol. The airspace above a cropped field that has been sprayed with biological control formulations (powders, granules, liquid cultures) typically contains corresponding microbes in the immediate atmosphere above. During spraying, microbes may escape to higher altitudes and be carried by wind, convective currents, and turbulence. It is important to note that upon spraying, the atmospheric population of biological control organisms like *B. thuringiensis* or *Metarrhizium* or virus particles (e.g., NPV) may be more concentrated. Over time, their population decreases as the formulation and microbes settle onto the crop canopy or soil. It should be acknowledged that biological control microbes have to compete for space in the phyllosphere after spraying. Microbes on insect cadavers may also become airborne. We need to understand the viability, shelf life, and tolerance of biological control microbes. In the air, these microbes must withstand harsh temperatures, desiccation, radiation, and other factors. Biocontrol microbes may even reach jet streams and undergo long-distance migration to other agrarian regions. For example, there is a possibility that *B. thuringiensis* spores may levitate and travel long distances in the higher altitudes of the atmosphere.

Biological control microbes are known to disperse to neighboring farms and impact pests. Residual activity is also crucial. Biological control organisms can reside on the phyllosphere, crop residue, and even surface layers of soil. They can be carried into the atmosphere whenever wind and turbulence occur. Aerial biocontrol microbes play a vital role in managing insect pests, helping to keep them below the threshold. Prominent microbes known to control lepidopterous pests include *Bacillus thuringiensis, Serratia marsesans, Pseudomonas aeruginosa, Metarrhizium* (fungus), and nuclear polyhedrosis virus (occluded virus).

Airborne *B. thuringiensis* cells and spores can drift for varying distances after a spray event in crop fields. The spread of *B. thuringiensis* or *Metarrhizium* sp. through the air, along with infected insect pests, should be noted. Airborne insects themselves can also spread the biological control agents in the troposphere. The spread of virus particles of NPV via airborne insects, wind, irrigation sprays, and droplets is also possible.

Recent reports suggest that agricultural drones may have a better chance of being used against locusts (Short-horned grasshoppers). Drones can regularly monitor the sky and provide relevant information about locusts and the build-up of *Metarrhizium* within the swarms (*see* Krishna, 2018, 2020). There are sprayer drones that can directly enter the 'grasshopper clouds' and apply biological control agents, such as formulations of *Metarrhizium acridum,* an entomophagous fungus. Aerial sprays of fungal BCAs are being regularly employed to control locusts in the Sahara and Sahel. The population of BCA (fungus) increases with the locust swarms and spreads as they migrate from one location to another. The fungal spores also drift with air currents, carried by infected short-horned grasshoppers. An agricultural sky with a significant build-up of *Metarrhizium* spores is detrimental to locusts, particularly during their airborne transit. Biocontrol microbes will also need to resist any antagonists they encounter in nature.

The phyllosphere microbes are known to impart a certain degree of resistance against pathogens traced on the leaf surface. For example, phyllosphere microbes may reduce pathogenesis caused by powdery mildew fungus. The microbial flora on the leaf surface may play an important role in the resistance to *Botrytis cinerea* (Ritpitakphong et al., 2016; Leveau & Tech, 2011). However, the impacts of pathogens on the phyllosphere microbiome are largely underexplored.

2.6.4 MICROBES BENEFICIAL TO CROPS TRACED IN ATMOSPHERE

Microbes beneficial to crops, such as symbionts, are often found in the airspace above cropping regions. Fields left fallow without crops may lose certain species of microbes to the atmosphere, along with dust particles. Many of these microbes could be symbiotic bacteria (Bradyrhizobia), AM, ectomycorrhiza (ECM), free-living biological nitrogen-fixers (BNF), and plant growth promoting bacteria (PGPR).

Aerospace above crops may aid in the dispersal of beneficial microbes. Schmale & Ross (2015) believe that there are global highways at higher altitudes in the atmosphere. These jet streams may transport propagules of beneficial organisms as well. For example, Bradyrhizobia, Azospirillum, Azotobacter, AM fungi, ECM fungi, and PGPR may travel long distances through the air. We cannot rule out the hybridization of native soil microflora containing BNF, AM fungi, ECM fungi with those introduced via dust storms, low altitude winds, and jet streams. Perhaps, the selection pressure for efficient symbiotic efficiency is crucial, especially for those who want to take advantage of the atmospheric dispersal of beneficial microbes. The aerial transit actually replenishes the loss of microbial propagules and incessantly adds new genotypes of beneficial microbial species. We need to study this phenomenon in greater detail and harness its advantages.

Ectomycorrhizal fungi produce spores with traits that allow them to withstand tropospheric conditions. The ECM fungal spores could be windborne. They are exposed to harsh sunlight, UV radiation, and desiccation, just like any other microbe. The small size of spores, pigmentation, spiny outgrowth, dryness, modifications in spore wall, and light weight help ECM to transit in the air with fairly good survival. Small bits of hyphae also act as ECM propagule. They are often airborne during dispersal. ECM fungi are known to transit long distances in the atmosphere. Basidiospores found in the surface soil are prone to reach the atmosphere. For example, basidiospores of size 4.5 to 8.5 µm belonging to *Laccaria laccata* are easily lofted into the sky due to turbulence in the wind. They may survive in the atmosphere for a long duration. The long-distance dispersal, in fact, is known to aid genetic crossing among similar species found in different locations. This phenomenon maintains gene flow within that ECM genus (*Laccaria laccata* and *L. amythesia*) (Vincenot et al., 2012).

Small-sized spores of VA mycorrhizas are easily transported through the air. High-speed winds, turbulence, and dust storms can carry AM fungal spores into the atmosphere. These spores can float for extended periods without losing viability. However, factors such as light, UV irradiance, and desiccation can be detrimental, reducing the survival period of AM spores. Pigmentation may help protect against the suppressive effects of light and UV irradiance. Large-sized spores of AMs are not uncommon, such as *Gigaspora gigantia*. Fruiting bodies of *Scleroscystis* sp., similar in size to mustard grains and lightweight, can also become airborne. For example, sandy soils in the Sahel region support large-sized *Gigaspora* species, which can potentially be carried by sand or dust storms from one place to another.

Spores may be rich in energy sources due to oil droplets. Sandy soils prone to dust storms can transport AM spores of different sizes and viability. AM mycorrhiza can reach higher altitudes and travel longer distances before settling onto crops in distant locations. Precipitation, clouds, and moist atmosphere can facilitate the descent of light spores of AM. Overall, the viability, transit distance, and endurance in the air may depend on various characteristics of AM fungal hyphae and spores (Lebanon Turf, 2022).

There is a possibility for long-distance transit of mycorrhizal spores, as they are similar to other fungi. The transit of mycorrhizal spores and/or hyphal bits may occur when they become trapped in dust particles. Inside these particles, mycorrhizal spores are protected from the harsh factors of the atmosphere. There is no good reason to believe that mycorrhizal spores do not travel short or long distances in the atmosphere. AM fungi are promiscuous and can colonize the roots of most food crop species. Therefore, upon landing after a transit through the troposphere, they have a good chance of landing on the soil surface and infecting an appropriate host crop.

KEYWORDS

- **agrarian regions**
- **agrarian sky**
- **atmosphere**
- **bioaerosol**
- **climate change**
- **ice nucleation**
- **microbial diversity**
- **microbial load**
- **particulate matter**
- **precipitation**
- **relative humidity**
- **solar radiation**
- **stratosphere**
- **troposphere**

REFERENCES

Abed, R. M., Ramette, A., Hubner, V., De Deckker, P., & De Beer, D., (2012). Microbial diversity of aeolian dust sources from saline lake sediments and biological soil crusts in arid Southern Australia. *FEMS Microbiology Ecology, 80*, 294–304.

Abetenh, E., Gizaw, E., Tsegaye, Z., & Tefera, G., (2018). Microbial function and climate change. *Pollution and Climatology, 2*, 1. doi: 10.4172/2573-458X.1000147.

Aguilera, A., De Diego-Castilla, G., Osuna, S., Bardera, R., Sor Mendi, S., Blanco, Y., & González-Toril, E., (2018). *Microbial Ecology in the Atmosphere: The Last Extreme Environment.* Intech Open. doi: http://dx.doi.org/10.5772/intechopen.81650/.

Almaguer, M., Aira, M. J., & Rodríguez-Rajo, F. J., & Rojas, T. I., (2013). Study of airborne fungus spores by viable and non-viable methods in Havana, Cuba. *Grana., 52*, 289–298.

Almaguer, M., Aira, M. J., Rojas, F. J. R., & Fernandez-Javier, R. R., (2018). New findings of airborne fungal spores in the atmosphere of Havana, Cuba, using aerobiological non-viable methodology. *Annals of Agricultural and Environmental Medicine, 25*, 349–359. doi: 10.26444/aaem/89738.

Amato, P., Demeer, F., Melaouhi, A., Fontanella, S., Martin-Biesse, A., Sancelme, M., Laj, P., & Delort, A., (2007a). A fate for organic acids, formaldehyde, and methanol in cloud water: Their biotransformation by micro-organisms. *Atmospheric Chemistry and Physics, 7*, 4159–4169. doi: 10.5194/acp-7-4159-2007.

Amato, P., Joly, M., Besaury, L., Oudart, A., Taib, N., & Monea, A. I., (2017). Active microorganisms thrive among extremely diverse communities in cloud water. *PLoS One, 12*(8), e0182869. https://doi.org/10.1371/journalpone.0182869/.

Amato, P., Parazols, M., Sancelme, M., Laj, P., Mailhot, G., & Delort, A., (2007b). Microorganisms isolated from the water phase of tropospheric clouds at the Puy de Dôme: Major groups and growth abilities at low temperatures: Microorganisms from the water phase of tropospheric clouds. *FEMS Microbiology Ecology, 59*, 242–254. doi: 10.1111/j.1574-6941.2006.00199.x.

Ariya, P. A., & Amyot, M., (2004). New directions: The role of bioaerosols in atmospheric chemistry and physics. *Atmosphere and Environment, 38*, 1231–1232.

Aylor, D. E., (1990). The role of intermittent wind in the dispersal of fungal pathogens. *Annual Review of Phytopathology, 28*, 73–92.

Aylor, D. E., (1992). Release of *Venturia inaequalis* ascospores during unsteady rain: Relationship to spore transport and deposition. *Phytopathology, 82*, 532–540. http://dx.doi.org/10.1094/Phyto-82-532./.

Aylor, D. E., (2003). Spread of plant disease on a continental scale. Role of aerial dispersal of pathogens. *Ecology, 84*, 1989–1997.

Aylor, D. E., Schmale, E. J., Shields, E. J., Newcombe, M., & Nappo, C. J., (2008). Tracking the potato late blight pathogen in the atmosphere, using unmanned aerial vehicles and Lagrangian modeling. *Agricultural Forestry and Meteorology, 151*, 251–260.

Barberan, A., Henley, J., Fierier, N., & Casamavor, (2014). Structure, interannual recurrence, and global scale connectivity of airborne microbial communities. *Science of the Total Environment, 487*, 187–195.

Barberan, A., Ladau, J., Leff, J. W., Pollard, K. S., Menninger, H. L., Dunn, R. R., & Fierer, N., (2015). Continental-scale distributions of dust associated bacteria and fungi. *Proceedings of National Academy of Science U. S. A., 112*, 5756–5761.

Bartscheidt, P., (2020). *Which Will Survive? A Microorganism Zoo in the Stratosphere* (pp. 1–8). MARSBOX Project, German Aerospace Centre and NASA of USA. https://www.dlr.de/content/en/articles/news/2020/01/20200124_a-microorganism-zoo-in-the-stratosphere.html/ (accessed on 19 January 2024).

Bauer, H., Giebl, H., Hitzenberger, R., Kasper-Giebl, A., Reischl, G., Zibuschka, F., & Puxbaum, H., (2003). Airborne bacteria as cloud condensation nuclei. *Journal of Geophysical Research, 108*, 4658. doi: 10.1029/2003JD003545.

Bauer, H., Kasper-Giebl, A., Löflund, M., Giebl, H., Hitzenberger, R., Zibuschka, F., & Puxbaum, H., (2002). The contribution of bacterial and fungal spores to the organic carbon content of cloud water, precipitation, and aerosols. *Atmospheric Research, 64*, 109–119.

Baylor, E. R., Peters, V., & Baylor, M. B. J. S., (1977). Water-to-air transfer of virus. *Science, 197*, 763–764.

Ben-David, A., & Sagripanti, J., (2010). A model for inactivation of microbes of microbes suspended in the atmosphere by solar ultraviolet radiation. *Photochemistry and Photobiology, 86*, 895–905. https://doi.org/10.1111/j.1751-1097.2010.00738.x/.

Bhattacharya, S., Sahu, A., & Phalke, D. H., (2021). *In situ* decomposition of crop residues using lignocellulolytic microbial consortia: A viable alternative to residue burning. *Environmental Science and Pollution Research, 28*, 32416–32433.

Blomfield, K., (2015). *The Microbial Rain Makers* (pp. 1–6). The Conscious Farmer. http://www.theconsciousfarmer.com/microbial-rainmakers/ (accessed on 19 January 2024).

Borah, N., Barua, R., Nath, D., Hazarika, K., Phukon, A., Goswami, K., & Barua, D. C., (2016). Low energy rice stubble management through *in-situ* decomposition. *Proceedings of Environmental Science, 35*, 771–780.

Boucher, O., (2013). Clouds and aerosols. In: Stocker, T. F., et al., (eds.), *IPCC Fifth Assessment Report 2013: The Physical Science Basis. Contribution of Working Group I to the Fifth Assessment Report of the Intergovernmental Panel on Climate Change* (pp. 571–657). Cambridge Univ. Press, Cambridge, U.K.

Bowers, R. M., Clements, N., Emerson, J. B., Wiedinmyer, C., Hannigan, M. P., & Fierer, N., (2013). Seasonal variability in bacterial and fungal diversity of the near-surface atmosphere. *Environmental Science and Technology, 47*, 12097–12106.

Bowers, R. M., Lauber, C. L., Wiedinmyer, C., Hamady, M., Hallar, A. G., & Fall, R. R., (2009). Characterization of airborne microbial communities at a high-elevation site and their potential to act as atmospheric ice nuclei. *Applied and Environmental Microbiology, 75*(15), 5121–5130.

Bowers, R. M., McLetchie, S., Knight, R., & Fierer, N., (2011). Spatial variability in airborne bacterial communities across land-use types and their relationship to the bacterial communities of potential source environments. *ISME J., 5*, 601–612.

Bringel, F., & Couvee, I., (2015). Pivotal roles of phyllosphere microorganisms at the interface between plant functioning and atmospheric race gas. *Frontiers in Microbiology Review.* Article https://doi.org/10.3389/fmicb.2015.00486.

Brix, L., (2015). *Bacteria in the Atmosphere Cause Rain* (pp. 1–4). Nordic Science. https://sciencenordic.com/climate-denmark-videnskabdk/bacteria-in-the-atmosphere-cause-rain/1423395/ (accessed on 1 February 2024).

Brown, S. P., & Jumpponen, A., (2019). Microbial ecology of snow reveals taxa-specific biogeographical structure. *Microbial. Ecology, 77*, 946–958. doi: 10.1007/s00248-019-01357-z.

Bru-Adan, V., Wery, N., Moletta-Denat, M., Boiron, P., Delgenes, J. P., & Godon, J. J., (2009). Diversity of bacteria and fungi in aerosols during screening in a green waste composting plant. *Current Microbiology, 59*, 326–335.

Bryan, N. C., (2017). *Microbial Distribution and Survival in the Troposphere and Stratosphere* (p. 172). PhD Dissertation, Louisiana State University at Monroe, Louisiana, USA.

Bryan, N. C., Christner, B. C., Guzik, T. G., Granger, D. J., & Stewats, M. F., (2019). Abundance and survival of microbial aerosols in the troposphere and stratosphere. *The ISME Journal, 13*, 2789–2799. https://doi.org/10.1038/s41396-019-0474-0.

Burrows, S. M., Elbert, W., Lawrence, M. G., & Pöschl, U., (2009a). Bacteria in the global atmosphere—Part 1: Review and synthesis of literature data for different ecosystems. *Atmospheric Chemistry and Physics, 9*, 9263–9280.

Burrows, S. M., Elbert, W., Lawrence, M. G., & Pöschl, U., (2009b). Bacteria in the global atmosphere—Part 2: Modeling of emissions and transport between different ecosystems. *Atmospheric Chemistry Physics, 9*, 9281–9297.

Busse, H. J., Denner, E. B. M., Buczolits, S., Salkinoja-Salonen, M., Bennasar, A., & Kampfer, P., (2003). *Sphingomonas aurantiaca* sp. *Nov., Sphingomonas aerolata* spnov and *Sphingomonas faeni sp Nov.*, air- and dust borne and Antarctic, orange pigmented, psychrotolerant bacteria, and emended description of the genus *Sphingomonas*. *International Journal Systematic Evolution Microbiology, 53*, 1253–1260.

Caliz, J., Triardo-Margaritz, X., Camerero, L., & Casameyer, E. O., (2018). A long-term survey unveils a strong seasonal pattern in the airborne microbiomes coupled to general and regional atmospheric circulations. *Proceedings of National Academy of Science, 115*, 12229–12234.

Capanna, E., (1999). Lazzaro Spallanzani: At the roots of modern biology. *Journal of Experimental Zoology, 285*, 178–196.

Cavicchioli, R., Ripple, W. J., Kenneth, N., Timmis, K. N., Azam, F., Bakken, L. R., Baylis, M., et al., (2019). Scientists' warning to humanity: Microorganisms and climate change. *Nature Reviews: Microbiology, 17*, 569–586. www.nature.com/nrmicro; https://doi.org/10.1038/s41579-019-0222-5/ (accessed on 19 January 2024).

Chen, X., Kumari, D., & Varenyam, A., (2020). A review on airborne microbes: The characteristics of sources, pathogenicity, and geography. *Atmosphere, 11*, 919. doi: 10.3390/atmos11090919.

Christner, B. C., Cai, R., Morris, C. E., McCarter, K. S., Foreman, C. M., Skidmore, M. L., Montross, S. N., & Sands, D. C., (2012). Geographic, seasonal, and precipitation chemistry influence on the abundance and activity of biological ice nucleators in rain and snow. *Proceedings of National Academy of Science, USA, 105*, 18854–18859.

Christner, B. C., Morris, C. E., Foreman, C. M., Cai, R., & Sands, D. C., (2008). Ubiquity of biological ice nucleators in snowfall. *Science, 319*. doi: 10.1126/science.1149757/.

Courville, Z. R., Lieblappen, R. M., Thurston, A. K., Barbato, R. A., Fegyveresi, J. M., Farnsworth, L. B., Derry, J., et al., (2020). Microorganisms associated with dust on alpine snow. *Frontiers in Earth Science, 8*, 122. doi: 10.3389/feart.2020.00122.

Cox, S. J., Raza, Z., Kathmann, S. M., Slater, B., & Michaelides, A., (2013). The microscopic features of heterogenous ice nucleation may affect the macroscopic morphology of atmospheric ice crystals. *Faraday Discussions, 167*. doi: https://doi.org/10.1039/C3FD00059A.

DasSarma, P., Antunes, A., Simoes, M. F., & DasSarma, S., (2020). Earth's stratosphere and microbial life. *Current Issues in Molecular Biology, 38*, 197–244. https://doi.org/10.21775/cimb.038.197.

Dastrup, D. B., Carling, G. T., Collins, S. A., Nelson, S. T., Fernandez, D. P., & Tingey, D. G., (2018). Aeolian dust chemistry and bacterial communities in snow are unique to airshed locations across northern Utah, USA. *Atmos. Environ., 193*, 251–261. doi: 10.1016/j.atmosenv.2018.09.016.

De Deckker, P., Abed, R. M. M., De Beer, D., Hinrichs, K. U., O'Loingsigh, T., Schefuß, E., Stuut, J. B. W., et al., (2008). Geochemical and microbiological fingerprinting of airborne dust that fell in Canberra. *Geochemistry and Geophysics Geosystems,* 9.

Dehel, T., Lorge, F., & Dickinson, M., (2008). Uplift of microorganisms by electric fields above thunderstorms. *Journal of Electrostatics, 66*, 463–466.

Del Ponte, E. M., Shah, D. A., & Bergstrom, G. C., (2003). *Spatial Patterns of Fusarium Head Blight in New York Wheat Fields Suggest Role of Airborne Inoculum.* Online. Plant Health Progress. doi: 10.1094/PHP-2003-0418-01-RS.

DeLeon-Rodriguez, N., Konstantinidis, K. T., Lathem, T. L., Rodriguez, R. L. M., Barazesh, J. M., Anderson, B. E., Beyersdorf, A. J., et al., (2013). Microbiome of the upper troposphere: Species composition and prevalence, effects of tropical storms, and atmospheric implications. *Proceedings of National Academy of Science U.S.A., 110*, 2575–2580. doi: 10.1073/pnas.1212089110.

Dohrmann, A. B., & Tebbe, C. C., (2005). Effect of elevated troposphere ozone on the structure of bacterial communities inhabiting the rhizosphere of herbaceous plants native to Germany. *Applied and Environmental Microbiology*, 7750–7758.

Dutta, H., & Dutta, A., (2016). The microbial aspect of climate change. *Energy Ecology Environment 1*(4), 209–232. doi 10.1007/s40974-016-0034-7/.

Edwards, A. R., Van, D. B. R. A., Wichman, H. A., & Orser, C. S., (1994). Unusual pattern of bacterial ice nucleation gene evolution. *Molecular Biology and Evolution, 11*, 911–920.

Edwards, A., (2021). *Microbes in the Icy Ecosystem* (pp. 1–12). https://microbiologysociety.org/publication/current-issue/life-on-a-changing-planet/article/microbes-in-icy-ecosysems.tml/ (accessed on 19 January 2024).

Ehrenberg, C. G., (1830). New observations on blood-like phenomena in Egypt, Arabia, and Siberia, together with an overview and criticism of those previously known. *Annals of Physics, 94*, 477–514. doi: 10.1002 and p.18300940402.

Ehrenberg, C. G., (1849). *Passatstaub und Blutregen, Abhandlungen der K¨oniglichen Akademie der Wissenschaftenzu Berlin* (pp. 269–460).

Ezike, D. N., Nnamani, C. V., Ogundipe, O. T., & Adekanmbi, O. H., (2016). Airborne pollen and fungal spores in Garki, Abuja (North-Central Nigeria). *Aerobiologia, 32*, 697–707.

Feltracco, M., Barbaro, A., Hoppe, C. J. M., Wolf, K. E., Spolaor, A., Layton, R., Keushnig, C., et al., (2021). *Airborne Bacteria and Particulate Chemistry Capture Phytoplankton Bloom Dynamics in an Arctic Fjord* (pp. 1–7). https://doi.org/10.1016/j.atmosenv.2021.118458.

Fierer, N., Liu, Z., Rodriguez-Hernandez, M., Knight, R., Henn, M., & Hernandez, M. T., (2008). Short term temporal variability in airborne bacterial and fungal populations. *Applied Environmental Microbiology, 74*, 200–207. doi: 10.1128/AEM.01467-07.

Flemming, H., & Wuertz, S., (2019). Bacteria and archaea on earth and their abundance in biofilms. *Nature Reviews Microbiology, 17*, 247–252.

Frenguelli, G., (1998). The contribution of aerobiology to agriculture. *Aerobiologia, 14*(2, 3), 95–100.

Fuzzi, S., Mandrioli, P., & Perfetto, A., (1997). Fog droplets—An atmospheric source of secondary biological aerosol particles. *Atmospheric Environment, 31*, 287–290.

Gadde, B., Bonnet, S., Menke, C., & Garivait, S., (2009). Air pollutant emissions from rice straw open field burning in India, Thailand, and the Philippines. *Environmental Pollution, 157*, 1554–1558.

Gao, J. F., Fan, X. Y., Li, H. Y., & Pan, K. L., (2017). Airborne bacterial communities of PM2.5 in Beijing-Tianjin-Hebei megalopolis, China as revealed by Illumina MiSeq sequencing: A case study. aerosol. *Air Quality Research, 17*, 788–798.

Garcia, E. B., Hanlon, R., Makris, M. R., Powers, W. G., Sanchez, C. J., Karatum, O., Marr, L. C., et al., (2019). Microbial diversity of individual rain drops from simulated and natural precipitation events. *Atmospheric Environment, 209*, 112–115. https://doi.org/10.1016/j.atmosenv.2019.04.023Getrightsandcontent/ (accessed on 19 January 2024).

Gilbert, J. A., Steele, J. A., Caporaso, J. G., Steinbruck, L., Reeder, J., Temperton, B., Huse, S., et al., (2012). Defining seasonal marine microbial community dynamics. *ISME Journal, 6*, 298–308.

Ginoux, P., Prospero, J. M., Gill, T. E., Hsu, N. C., & Zhao, M., (2012). Global-scale attribution of anthropogenic and natural dust sources and their emission rates based on MODIS deep blue aerosol products. *Reviews in Geophysics, 50*, RG3005. doi: 10.1029/2012rg000388.

Gonzalez-Toril, E., Robert, A. R., Delmas, J., Petit, J., Komarek, J., & Elster, J., (2009). Bacterial diversity of autotrophic enriched cultures from remote, glacial Antarctic, Alpine and Andean aerosol, snow, and soil samples. *Biogeosciences (European Geoscience Union), 6*, 33–44.

Goswami, S. B., Mondal, R., & Mandi, S. K., (2019). Crop residue management options in rice-rice system: A review. *Archives in Agronomy and Soil Science*. https://doi.org/10.1080/03650340.2019.1661994.

Griffin, D. W., (2004). Terrestrial microorganisms at an altitude of 20,000 m in earth's atmosphere. *Aerobiologia, 20*, 135–140.

Griffin, D. W., (2007). Atmospheric movement of microorganisms in clouds of desert dust and implications for human health. *Clinical Microbiology Reviews, 20*, 459–477.

Griffin, D. W., (2008). Non-spore forming eubacteria isolated at an altitude of 20, 000 m in Earth's atmosphere: Extended incubation periods needed for culture-based assays. *Aerobiologia, 24*, 19–25.

Griffin, D. W., (2010). Observations on the use of membrane filtration and liquid impingement to collect airborne microorganisms in various atmospheric environments. *Aerobiologia.* doi: 10.1007/s10453-010-9173-z.

Griffin, D. W., Kellogg, C. A., Garrison, V. H., Lisle, J. T., Borden, T. C., & Shinn, E. A., (2003). Atmospheric microbiology in the northern Caribbean during African dust events. *Aerobiologia, 19*, 143–157. https://doi.org/10. 1023/B:AERO.0000006530.32845.8d.

Griffin, D. W., Westphal, D. L., & Gray, M. A., (2006). Airborne microorganisms in the African desert dust corridor over the mid-Atlantic ridge, ocean drilling program, leg 209. *Aerobiologia, 22*, 211–226. https://doi.org/10.1007/s 10453-006-9033-z.

Grover, D., & Chaudhry, S., (2019). Ambient air quality changes after stubble burning in rice-wheat system in an agricultural state of India. *Environmental Science Pollution Research, 26*, 20550–20559.

Grulke, N. E., & Heath, R. L., (2019). Ozone effects on plants in natural ecosystems (pp. 1–26). doi: 10.1111/plb.12971.

Gupta, P. K., Sahai, S., Singh, N., Dixit, C. K., Singh, D. P., & Sharma, C., (2004). Residue burning in rice-wheat cropping system: Causes and implications. *Current Science, 87*, 1713–1715.

Gurian-Sherman, D., & Lindow, S. E., (1993). Bacterial ice nucleation: Significance and molecular basis. *FASEB Journal, 7*, 1338–1343.

Gusarevaa, E. S., Acerbia, E., Laua, K. J. X., Luhunga, I., Premkrishnana, N. V., Kolundzijaa, S., Purbojatia, R. W., et al., (2019). Microbial communities in the tropical air ecosystem follow a precise diel cycle. *Proceedings of National Academy of Science, 116*(46), 23299–23308 www.pnas.org/cgi/doi/10.1073/pnas.1908493116.

Haga, D. I., Lannone, M. J., Wheeler, R., Mason, E., Polishchuk, T., Fetch, B., Van, D. K. I., et al., (2013). Ice nucleating properties of rust and bunt fungal spores and their transport at high altitudes, where they can cause heterogenous freezing. *AGR Atmospheres, 118*, 7260–7272.

Han, Y., Li, L., Liu, J., & Zhang, M., (2012). Microbial structure and chemical components of aerosols caused by rotating brushes in a wastewater treatment plant. *Environmental Science and Pollution Research International, 19*, 4097–4108.

Hau, B., &Vallavieille-Pope, S., (2006). Wind dispersal of disease. In: *The Epidemiology of Plant Disease* (pp. 387–416). Springer.

Heist, A., Litchman, R., & Litchman, F., (2020). *Every Little Thing* (pp. 1–4). https://www.biographic.com/invisible-nature-life-in-the-clouds/ (accessed on 19 January 2024).

Hill, T. C. J., Moffett, B. F., DeMott, P. J., Georgeakopoulos, D. G., Stump, W. L., & Franca, G. D., (2014). Measurement of ice nucleation-active bacteria on plants and in precipitation by quantitative PCR. *Applied and Environmental Microbiology, 80*, 1256–1267.

Hirst, J. M., & Stedman, O. J., (1963). Dry liberation of fungal spores by raindrops. Journal of *General Microbiology, 33*, 335–344.

Hodson, A., Anesio, A. M., Tranter, M., Fountain, A., & Osborn, M., (2008). Glacial ecosystems. *Ecological Monographs, 78*, 41–67.

Horneck, G., Klaus, D. M., & Mancinelli, R. L., (2010). Space Microbiology. American society of microbiology. *Microbiology and Molecular Biology Reviews*, 121–156. doi: 10.1128/MMBR.00016-09.

Ibanez, V., Villegas, G. R., & Nolla, J. M. R., (2001). Airborne fungi monitoring in Santiago, Chile. *Aerobiologia, 17*, 137–142.

Imbler, S., (2020). *A Giant Dust Storm is Heading Across the Atlantic* (pp. 1–4). https://www.theatlantic.com/science/archive/2020/06/saharan-dust-storms-giving-earth-life/613441/ (accessed on 19 January 2024).

Imshenetsky, A. A., Lysenko, S. V., & Kazakov, G. A., (1978). Upper boundary of the biosphere. *Applied and Environmental Microbiology, 35*, 1–5.

Imshenetsky, A. A., Lysenko, S. V., Kasakov, G. A., & Ramkova, N. V., (1977). Resistance of stratospheric and mesospheric micro-organisms to extreme factors. *Life Sciences and Space Research* (Vol. 15, pp. 37–39). Paper presented at: COSPAR Life Sciences and Space Research (Philadelphia, Pennsylvania, USA: Pergamon Press, A. Wheaton & Co. Exeter).

ISRO, (2009). *New Microbes Discovered in Earth's Stratosphere* (pp. 1–3). Indian Space Research Organization, Ahmedabad, India https://www.sciencedaily.com/releases/2009/03/090318094642.htm (accessed on 19 January 2024).

Jacob, D. J., (1999). *Introduction to Atmospheric Chemistry* (pp. 1–77). Princeton: Princeton University Press.

Jain, N., Bhatia, A., & Pathak, H., (2014). Emission of air pollutants from crop residue burning in India. *Aerosol Air Quality Research, 14*, 422–430.

Joly, M., Amato, P., Deguillame, L., Monier, M., Hoose, C., & Delort, A. M., (2014). Quantification of ice nuclei active at near 0°C temperatures in low-altitude clouds at the Puy de Dôme atmospheric station. *Atmospheric Chemistry Physics, 14*, 8185–8195. www.atmos-chem-phys.net/14/8185/2014/. doi: 10.5194/acp-14-8185-2014 (accessed on 19 January 2024).

Kajava, A. V., & Lindow, S. E., (1994). A model of the three-dimensional structure of ice nucleation proteins. *Journal Molecular Biology, 232*, 709–717.

Kellogg, C. A., & Griffin, D. W., (2006). Aerobiology and the global transport of desert dust. *Trends in Ecology and Evolution, 21*, 638–644.

Khodadad, C. L. M., Hummerick, M. E., Spencer, L. L. E., Diit, A. R., Richards, J. T., Romeyn, M. W., Smith, T. M., et al., (2020). Microbiological and nutritional analysis of lettuce crops grown on the international station. *Frontiers in Plant Science, 6*. https://doi.org/10.3389/fpls.2020.00199.

Kieft, T. L., & Ahmadjian, V., (1989). Biological ice nucleation activity in lichen mycobionts and photobionts. *Lichenologist, 21*, 355–362.

Kieft, T. L., & Ruscetti, T., (1990). Characterization of biological ice nuclei from a lichen. *Journal of Bacteriology, 172*, 3519–3523.

Kieft, T. L., & Ruscetti, T., (1990). Characterization of biological ice nuclei from a lichen *Journal of Bacteriology, 172*, 3519–3523.

Kieft, T. L., (1988). Ice nucleation activity in lichens. *Applied and Environmental Microbiology, 54*, 1678–1681.

Krishna, K. R., (2018). *Agricultural Drones: A Peaceful Pursuit* (p. 376). Apple Academic Press Inc. Waretown, new Jersey, USA.

Krishna, K. R., (2020). *Unmanned Aerial Vehicles in Crop Production* (p. 683). Apple Academic Press Inc. Palm Bay, Florida, USA.

Krishna, K. R., (2022). *The Agricultural Sky: A Concept to Revolutionize Farming* (p. 589). Apple Academic Press Inc. Palm Bay, Florida, USA.

Kumari, S., Verma, N., & Lakhani, A., (2018). Tropospheric ozone enhancement during post-harvest crop-residue fires at two downwind sites of the Indo-Gangetic plain. *Environmental Science and Pollution Research, 25*, 18879–18893.

Lacey, M. E., & West, J. S., (2006). *The Air Spora: A Manual for Catching and Identifying Airborne Biological Particles* (p. 377). Springer: Dordrecht, The Netherlands.

Lambert, J., (2020). If bacteria stick together, they can survive for years in space. *Frontiers in Microbiology*. https://www.snexplores.org/article/space-bacteria-survival-years-astrobiology/ (accessed on 19 January 2024).

Le Goff, O., Bru-Adan, V., Bacheley, H., Godon, J. J., & Wery, N., (2010). The microbial signature of aerosols produced during the thermophilic phase of composting. *Journal of Applied Microbiology, 108*, 325–340.

Lebanon Turf, (2022). *Effects of Light on Mycorrhizal Fungi* (pp. 1–12). Lebanon Corporation Inc. Lebanon, PAA, USA. https://www.lebanonturf.com/education-center/biological-plant-treatments/effects-of-light-exposure-on-mycorrhizal-fungi/ (accessed on 19 January 2024).

Leveau, J. H. J., & Tech, J. J., (2011). Grapevine microbiomics: Bacterial diversity on grape leaves and berries revealed by high-throughput sequence analysis of 16S rRNA amplicons. *Acta Horticulture, 905*, 31–42. doi: 10.17660/ActaHortic.2011.905.2/.

Levetin, E., (2016). Aerobiology of agricultural pathogens. In: Yates, M., Nakatsu, C., Miller, R., & Pillai, S., (eds.), *Manual of Environmental Microbiology* (4th edn.). ASM Press, Washington, DC.

Lighthart, B., (1984). Microbial aerosols: Estimated contribution of combine harvesting to an airshed. *Applied and Environmental Microbiology, 47*, 430–432.

Lighthart, B., (1997). The ecology of bacteria in alfresco atmosphere. *FEMS Microbiology Ecology, 23*, 263–274.

Linderman, J., & Upper, C. D., (1985). Aerial dispersal of epiphytotic bacterial over bean plants. *Applied and Environmental Microbiology, 50*, 1229–1232.

Linderman, J., Constantinidou, H. A., Barchet, W. R., & Upper, C. D., (1982). Plants as sources of airborne bacteria, including ice nucleating activity bacteria. *Applied and Environmental Microbiology, 44*, 1059–1063.

Liu, H., Hu, Z., Zhou, M., Hu, J., Yao, X., Zhang, H., Li, Z., et al., (2019). The distribution variance of airborne microorganisms in urban and rural environments. *Environmental Pollution, 247*, 898–906.

Liu, H., Zhang, X., Zhang, H., Yao, X., Zhou, M., Wang, J., He, Z., et al., (2018). Effect of air pollution on the total bacteria and pathogenic bacteria in different sizes of particulate matter. *Environmental Pollution, 233*, 483–493.

Louisiana State University, (2008). *Evidence of 'Rain-Making' Bacteria Discovered in Atmosphere and Snow* (pp. 1–3). Science News. https://www.sciencedaily.com/releases/2008/02/080228174801.htm (accessed on 19 January 2024).

Madden, L. V., (1992). Rainfall and the dispersal of fungal spores, Advances of. *Plant Pathology, 8*, 39–79.

Madigan, M. T., & Marrs, B. L., (1997). Extremophiles. *Scientific American, 276*(4), 82–87.

Madronich, S., Bjorn, L. O., & MKenzie, R. L., (2018). Solar UV radiation and microbial life in the atmosphere. *Photochemical and Photobiological Sciences*.

Main, D., (2013). *Loads of Microbes Found High in Atmosphere* (pp. 1–4). Live Science. https://www.livescience.com/26645-microbes-in-the-sky.html/ (accessed on 19 January 2024).

Maki, L. R., Galyan, E. L., & Caldwell, D. R., (1974). Ice nucleation induced by *Pseudomonas syringae*. *Applied and Environmental Microbiology, 28*, 456–459.

Maldonado-Ramirez, S. L., Schmale, D. G. I., Shields, E. J., & Bergstrom, G. C., (2005). The relative abundance of viable spores of *Gibberellazeae* in the planetary boundary layer suggests the role of long-distance transport in regional epidemics of *Fusarium* head blight. *Agricultural and Forest Meteorology, 132*, 20–27.

Marcolli, C., (2014). Deposition nucleation viewed as homogeneous or immersion freezing in pores and cavities. *Atmospheric Chemistry and Physics, 14*, 2071–2104. https://doi.org/10.5194/acp-14-2071-2014.

Max Plank Institute, (2016). *The Effect of Bacterial Ice Nuclei*. Mainz, Germany. https://www.mpg.de/10470442/ice-formation-bacteria-syringae/ (accessed on 19 January 2024).

Mayol, E., Arrieta, J. M., Jimenez, M. A., Martinez-Asensio, A., Garcias-Bone, N., Dachs, J., Gonzalez-Gaya, B., et al., (2017). Long-range transport of airborne microbes over the global tropical and subtropical ocean. *Nature Communications,8*, 201, 202.

Mitakakis, T. Z., Clift, A., & McGee, P. A., (2001). The effect of local cropping activities and weather on the airborne concentration of allergenic *Alternaria* spores in rural Australia. *Grana, 40*, 230–239.

Mohler, O., DeMott, P. J., Vali, G., & Levin, Z., (2007). Microbiology and atmospheric processes: The role of biological particles in cloud physics. *Biogeoscience Discussions, 4*, 2559–2259.

Moore, R. A., Bomar, C., Kobziar, L. N., & Christner, B. C., (2021). Wildland fire as an atmospheric source of viable microbial aerosols and biological ice nucleating particles. *The ISME Journal, 15*, 461–472.

Morris, C. E., & Sands, D. C., (2017). Impacts of microbial aerosols on natural and agro ecosystems: Immigration, invasions, and their consequences. In: Delort, A. M., & Amato, P., (eds.), *Microbiology of Aerosols* (p. 428). John Wiley and Sons Inc. Hoboken, New Jersey, USA.

Morris, C. E., Georgakopoulos, D. G., & Sands, D. C., (2004). Ice nucleation active bacteria and their potential role in precipitation. *Journal of Physics IV*. France. doi: 10.1051/jp4:2004121004.

Morris, C. E., Sands, D. C., Bardin, M., Jaenicke, R., Vogel, B., Leyronas, C., Ariya, P., & Psennet, R., (2011). Microbiology and atmospheric processes: Research challenges concerning the impact of airborne micro-organisms on the atmosphere and climate. *Biogeosciences, 8*, 17–25. www.biogeosciences.net/8/17/2011/; doi: 10.5194/bg-8-17-2011 (accessed on 19 January 2024).

Murray, B. J., O'Sullivan, D., Atkinson, J. D., & Webb, M. E., (2012). Ice nucleation by particles immersed in supercooled cloud droplets. *Chemical Society Reviews, 41*, 6519–6554. doi: 10.1039/c2cs35200a.

Nagarajan, S., & Joshi, L. M., (1985). Epidemiology in the Indian subcontinent. In: Roelfs, A. P., & Bushnell, W. R., (eds.), *The Cereal Rusts; Diseases, Distribution, Epidemiology, and Control* (Vol. 2, pp. 371–402). Orlando, FL, USA, Academic Press.

Nagarajan, S., & Singh, D. V., (1990). Long distance dispersion of rust pathogens. *Annual Review of Phytopathology, 28*, 139–153.

Narlikar, J. V., Lloyd, D., Wickramasinghe, N. C., Harris, M. J., Turner, M. P., & Al-Mufti, S., (2003). A balloon experiment to detect microorganisms in the outer space. *Astrophysics andSpace Science, 285*(2), 555–562.

NASA, (2018). *AIR: Our Atmosphere* (p. 7). www.nasa.gov (accessed on 19 January 2024).

Nicholson, L. W., Schuerger, & Race, M. E., (2009). Migrating microbes and planetary protection. *Trends in Microbiology, 17*(9), 389–392. doi: 10.1016/j.tim.2009.07.001.

Northcut, G., (2017). *Researchers Move Closer to Preventing Frost Damage in Vineyards by Managing Bacteria on the Vines* (pp. 1–8). Farm Progress. https://www.farmprogress.com/marketing/market-brings-plenty-opportunity (accessed on 19 January 2024).

O'Sullivan, S., Murray, B. J., Malkin, T. L., Whale, T., Umo, N. S., Atkinson, J. D., Price, H. C., et al., (2014). Ice nucleation by soil dusts: Relative importance of mineral dust and biogenic components. *Atmospheric Chemistry and Physics Discussions, 13*, 20275–20317. doi: 10.5194/acpd-13-20275-2013.

Pandey, R., Usui, K., Livingstone, A. R., Fisher, S. A., Pfaendtner, J., Backus, E. H. G., Nagata, Y., et al., (2016). Ice nucleating bacteria control the order and dynamics of interfacial water. *Scientific Advances*, *2*(4), e1501630. doi: 10.1126/sciadv.1501630.

Papke, R. T., & Ward, D. M., (2004). The importance of physical isolation to microbial diversification. *FEMS Microbiology Ecology, 48*, 293–303.

Pasteur, L., (1860). Expériencesrelatives aux générationsditesspontanées. *Comptes Rendus Hebdomadaires des Séances de L'académie des Sciences, 50*, 303–307. (English translation:

Pasteur, L., (1860). Experiences relating to so-called spontaneous generations. *Weekly Reports of the Sessions of the Academy of Sciences, 50,* 303–307).

Peccia, J., Werth, H. M., Miller, S., & Hernandez, M., (2001). Effects of relative humidity on the ultraviolet induced inactivation of airborne bacteria. *Aerosol Science and Technology, 35*, 728–740.

Pepper, I. L., & Gerba, C. P., (2015). Aeromicrobiology. In: *Environmental Microbiology* (3rdedn., pp. 89–111). University of Arizona, USA. doi: http://dx.doi.org/10.1016/B978-0-12-394626-3.00005-3.

Polymenakou, P. N., (2012). Atmosphere: A source of pathogen or beneficial microbes? *Atmosphere 3*, 87–102. doi: 10.3390/atmos3010087.

Pouleur, S., (1992). *Applied Environmental Microbiology, 58*, 2960–2964.

Prasla, I., Duman, K., Ciochetto, Z., Burman, A., Mahon, A., & Park, S., (2013). Significant heterogeneity in airborne mold quantities on the Caribbean Island of St. Kitts: Health implications and impact on food preservation. *Virology Mycology, 3*, 1–5.

Ravikala, K. L., &Nagalakshamma, K. V., (2016). Survey on outdoor airborne fungal spores of Tumakuru, Karnataka state, India. *International Journal of Pharmacology and BioSciences 7*(1), 575–577.

Ravilious, K., (2016). *Rain Makers: How High-Flying Bacteria Could Control the Clouds* (pp. 1–2). https://www.newscientist.com/article/mg23030690-400-rain-makers-how-highflying-bacteria-could-control-the-clouds/ (accessed on 19 January 2024).

Reche, I., D'Orta, G., Mladenov, N., Winget, D. M., & Suttle, C. A., (2018). Deposition rates of viruses and bacteria above the atmosphere boundary layer. *ISME Journal, 12*, 1154–1162. https://doi.org/10.1038/s41396-017-0042-4.

Reche, I., Ortega-Retuerta, E., Romera, O., Pulido-Villena, E., Morales-Baquero, R., & Casamayor, E. O., (2009). Effect of Saharan dust inputs on bacterial activity and community composition in Mediterranean lakes and reservoirs. *Limnology and Oceanography, 54*, 869–879.

Riley, R. L., & Kaufman, J. E., (1972). Effect of relative humidity on the inactivation of airborne *Serratia marcescens* by ultraviolet radiation. *Applied Microbiology, 23*, 1113–1120.

Ritpitakphong, U., Falquet, L., Vimoltust, A., Berger, A., Métraux, J. P., & L'Haridon, F., (2016). The microbiome of the leaf surface of *Arabidopsis* protects against a fungal pathogen. *NewPhytologist, 210*(3), 1033–1043. doi: 10.1111/nph.13808/.

Robbins, J., (2010). *From Trees and Grass, Bacteria that Cause Snow and Rain* (pp. 1–3). New York Times. N. Y., USA. https://www.nytimes.com/2010/05/25/science/25snow.html (accessed on 19 January 2024).

Rohatschek, H., (1984). The role of gravitophotophoresis for stratospheric and mesospheric particulates. *Journal of Atmospheric Chemistry,1*, 377–389.

Rohatschek, H., (1996). Levitation of stratospheric and mesospheric aerosols by gravito-photophoresis. *Journal of Aerosol Science, 27*, 467–475.

Rothschild, L. J., & Mancinelli, R. L., (2001). Life in extreme environments. *Nature, 409*, 1092. doi: 10.1038/35059215.

Rowan, K., (2008). *Snowflakes Aren't Innocent and Fluffy- They are Bacteria Bombs from the Sky*. https://www.discovermagazine.com/planet-earth/snowflakes-arent-innocent-and-fluffytheyre-bacteria-bombs-from-the-sky/ (accessed on 19 January 2024).

Rowell, J. B., & Romig, R. W., (1966). Detection of urediospores of wheat rusts in spring rains. *Phytopathology, 56*, 807–811.

Ruiz-Gill, T., Jacquelinne, J., & Acunab, J. J., Fujiyoshib, S., Tanakaf, D., Nodae, J., Fumito, M., & Jorquerab, M. A., (2020). Airborne bacterial communities of outdoor environments and their associated influencing factors. *Environment International, 145*, 106156. https://doi.org/10.1016/j.envint.2020.106156.

Sache, I., (2000). Short distance dispersal of wheat rust spores by wind and rain. *Agronomie, 20*, 757–787.

Sands, D. C., Langhans, V. E., Scharen, A. L., & De Smet, G., (1982). The association between bacteria and rain and possible resultant meteorological implications. *Journal of Hungarian Meteorology Service, 86*, 148–152.

Sarkar, S., Singh, R. P., & Chauhan, A., (2018). Crop residue burning in northern India: Increasing threat to Greater India. *Journal of Geophysics Research-Atmosphere*. https://doi.org/10.1029/2018JD028428.

Sattler, B., Puxbaum, H., & Psenner, R., (2001). Bacterial growth in supercooled cloud droplets. *GeophysicsResearch Letters, 28*, 239. doi: 10.1029/2000GL011684.

Scharf, C. A., (2019). *The Atmospheric Microbiome* (pp. 1–3). Scientific American Unlimited-Life Unbounded https://blogs.scientificamerican.com/life-unbounded/the-atmospheric-microbiome/ (accessed on 19 January 2024).

Schiermeier, Q., (2008). 'Rain-making' bacteria fund around the world. *Nature*, 1–3. https://www.nature.com/articles/news.2008.632/ (accessed on 19 January 2024).

Schmale, D. G., & Ross, S. D., (2015). Highways in the sky: Scales of atmospheric transport of plant pathogens. *Annual Review of Phytopathology, 53*, 591–611. doi: 10.1146/annurev-phyto-080614-115942.

Sesartic, A., Lohman, U., & Storelmov, T., (2013). Modeling the impact of fungal spore ice nucleating and precipitation. *Environmental Research Letters, 8*, 014029. doi: 10.1088/1748-9326/8/1/014029.

Shivaji, S., Chaturvedi, P., Begum, Z., Pindi, P. K., Manorama, R., & Padmanabhan, D. A., (2009). Isolation of three novel bacterial strains, *Janibacterhoylei sp. Nov., Bacillus isronensis sp. Nov.* and *Bacillus aryabhattai* sp. Nov. from cryotubes used for collecting air from upper atmosphere. *International Journal of Systematic Evolution in Microbiology*. doi: 10.1099/ijs.0.002527-0.

Shivaji, S., Chaturvedi, P., Suresh, K., Reddy, G. S. N., Dutt, C. B. S., & Wainwright, M., (2006). *Bacillus aerius sp. Nov., Bacillus aerophilus sp. Nov., Bacillus stratosphericus sp. Nov.* and *Bacillus altitudinis sp. Nov.,* isolated from cryogenic tubes used for collecting air samples from high altitudes. *International Journal of Systematic and Evolutionary Microbiology, 56*, 1465–1473.

Shomrat, A., (2013). *Distribution and Source of Microbes in Air* (pp. 1–4). https://plantlet.org/distribution-and-sources-of-air-microbes/ (accessed on 19 January 2024).

Sidhu, B. S., Rupela, O. P., Beri, V., & Joshi, P. K., (1998). Sustainability implications of burning rice and wheat straw in Punjab. *Economic and Political Weekly, 33*, 163–168.

Singh, Y., & Sidhu, H. S., (2014). Management of cereal crop residues for sustainable rice-wheat production system in the Indo-Gangetic plains of India. *Proceedings of Indian National Science Academy, 80*, 95–114.

Smets, W., Moretti, S., Denys, S., & Lebeer, S., (2016). Airborne bacteria in the atmosphere: Presence, purpose, and potential. *Atmospheric Environment, 139*, 214–221. doi: 10.1016/j.atmosenv.2016.05.038.

Smith, D. J., (2012). Free tropospheric transport of microorganisms from Asia to North America. *Microbiology and Ecology, 64*, 973–985.

Smith, D. J., (2013a). Aeroplankton and the need for a global monitoring network. *Bioscience, 63*, 515–516.

Smith, D. J., (2013b). Microbes in the upper atmosphere and unique opportunities for astrobiology research. *Astrobiology, 13*, 981–990. doi: 10.1089/ ast.2013.1074.

Smith, D. J., Griffin, D. W., & Schuerger, A. C., (2009). Stratospheric microbiology at 20 km over the Pacific Ocean. *Aerobiologia, 26*, 35–46.

Smith, D. J., Griffin, D. W., McPeters, D. R., Ward, P. D., & Schuerger, A. C., (2011). Microbial survival in the stratosphere and implications for global dispersal. *Aerobiologia, 27*, 319–332. doi: 10.1007/s10453-011-9203-5.

Smith, D. J., Timonen, H. J., Jaffe, D. A., Griffin, D. W., Birmele, M. N., & Perry, K. D., (2013). Intercontinental dispersal of bacteria and archaea by transpacific winds. *Applied Environmental Microbiology, 79*, 1134–1139.

Stefanski, R., & Sivakumar, M. V. K., (2009). Impacts of sand and dust storms on agriculture and potential agricultural applications of a SDSWS. *IOP Conference Series: Earth and Environmental Science,7*: 012016wheat.

Steinke, I., Funk, R., Busse, J., Iturri, A., Kirchen, S., Leue, M., Mohler, O., et al., (2016). Ice nucleation activity of agricultural soil dust aerosols from Mongolia, Argentina, and Germany. *Geophysical Research-Atmosphere*, 121. doi: 10.1002/2016JD025160.

Stopelli, E., Conen, F., Guilbaud, C., Zopfi, J., Alwell, C., & Morris, C. E., (2017). Ice nucleators, bacterial cells and *Pseudomonas syringae*. *Biogeosciences, 14*, 1189–1196. www.biogeosciences.net/14/1189/2017/; doi: 10.5194/bg-14-1189-2017(accessed on 19 January 2024).

Tobo, Y.,DeMott, P. J.,Hill, T. C. J.,Prenni, A. J.,Swoboda-Colberg, N. G.,Franc, G. D., & Kreidenweis, S. M., (2014). Organic matter matters for ice nuclei of agricultural soil origin. *Atmospheric Chemistry and Physics,14*, 8521–8531. doi:10.5194/acp-14-8521-2014 .

Tobo, Y., DeMott, P. J., Raddatz, M., Niedermeier, D., Hartmann, S., Kreidenweis, S. M., Stratmann, F., & Wex, H., (2012). Impacts of chemical reactivity on ice nucleation of kaolinite particles: A case study of levoglucosan and sulfuric acid. *Geophysics Research Letters, 39*, L19803. doi: 10.1029/2012GL053007.

UC Cooperative Extension Service, (1997). *Foliar Sprays for Frost Protection of Young Citrus and Avacado- 1995* (pp. 1–3). http://ceventura.ucanr.edu/Com_Ag/Subtropical/Publications/Frost/Foliar_Sprays_for_Frost_Protection_of_Young_Citrus_and_Avocado_-_1995_/ (accessed on 19 January 2024).

UC Delivers, (2019). *Biological Frost Control Strategies Reduce Crop Damage* (pp. 1–5). UC Sonoma County, CA, USA. https://ucanr.edu/blogs/blogcore/postdetail.cfm?postnum=29663/ (accessed on 19 January 2024).

Umino, T., & Takeuchi, N., (2012). *The Spatial and Seasonal Variations in Mineral Particle Composition on the Snow Surface and Their Possible Effect on Snow Algae in the Tateyama Mountains, Japan.* American Geophysical Union Fall Meeting Abstract id. C13B-0609/.

Vaïtilingom, M., Attard, E., Gaiani, N., Sancelme, M., Deguillaume, L., & Flossmann, A. I., (2012). Long-term features of cloud microbiology at the puy de Dôme (France). *Atmospheric Environment, 56*, 88–100.

Vali, G., (1995). Principles of ice nucleation. In: Lee, Jr. R., Warren, G. J., & Gusta, L. V., (eds.), *Biological Ice Nucleation and Its Applications* (pp. 1–28). American Phytopathological Society, St. Paul, 25 Mn, USA.

Vincenot, L., Nara, K., Sthultz, C., Labbé, J., Dubois, M. P., Tedersoo, L., Martin, F., & Selosse, M. A., (2012). Extensive gene flow over Europe and possible speciation over

Eurasia in the ectomycorrhizal basidiomycete *Laccariaamethystina* complex. *Molecular Ecology, 21*, 281–299. https://doi.org/10.1111/j.1365-294X.2011.05392.x.

Wade, L., (2013). *Microbes Survive, and Maybe Thrive, High in the Atmosphere* (pp. 1–5). Science Magazine. https://www.sciencemag.org/news/2013/01/microbes-survive-and-maybe-thrive-high-atmosphere/ (accessed on 19 January 2024).

Weidner, T., & Frohlich-Nowoisky, J., (2013). *The Effect of Bacterial Ice Nuclei* (pp. 1–3). Max-Planck Gessellshaft, Germany. https://www.mpg.de/10470442/ice-formation-bacteria-syringae (accessed on 19 January 2024).

Wells, W. F., (1955). *Airborne Contagion and Air Hygiene: An Ecological Study of Droplet Infections* (p. 344). Harvard University Press, Cambridge, MA, USA. 10.1093/ajcp/25.11.1301/.

Welsh, J., (2001). *Live Bacteria Help Create Rain, Snow and Hail* (pp. 1–7). https://www.livescience.com/14299-bacteria-create-rain-snow-hail.html/ (accessed on 19 January 2024).

Womack, A. M., Artaxo, P. E., Ishida, R. C., Muellers, R. C., Saleska, S. R., Wiedmann, K. T., Bohhannan, B. J. M., &, & Green, L. J., (2015). Characterization of active and total fungal communities in the atmosphere over the Amazon rainforest. *Biogeosciences, 12*, 6337–6349.

Womack, A. M., Brendan, J. M., & Green, J. L., (2010). Biodiversity and biogeography of the atmosphere. *Philosophical Transactions of Royal Society B, 365*, 3645–3653.

Xie, Z., Li, Y., Lu, R., Li, W., Fan, C., Liu, P., Wang, J., & Wang, W., (2018). Characteristics of total airborne microbes and air quality. *Journal of Aerosol Science, 116*, 57–65.

Yang, Y., Itahashi, S., Yokobori, S. I., & Yamagishi, A., (2008). UV-resistant bacteria isolated from upper troposphere and lower stratosphere. *Biological Science- Space, 22*, 18–25.

Yang, Y., Yokobori, S., Kawaguchi, J., Yamagami, T., Iijima, I., & Izutsu, N., (2005). *Investigation of Cultivable Microorganisms in the Stratosphere Collected by Using a Balloon in 2005* (pp. 35–42). JAXA Research and Development Report, JAXA-RR-08-001. Ibaraki. https://global.jaxa.jp (accessed on 19 January 2024).

Zell, H., (2017). *Earth: Earth's Atmospheric Layers* (pp. 1–3). NASA/Goddard Space Centre, Ohio, USA.

Zhu, Y. G., (2017). Microbial mass movements. *Science, 357*, 1099–1100.

CHAPTER 3

Aerial Migration of Microbes: Influence on Agricultural Ecosystems

ABSTRACT

The migration of microbes is a natural phenomenon observed in agrarian regions. Several ecosystem functions depend on the aerial migration of microbes. Migration helps microbes find a suitable environment or niche when they encounter inhospitable substrates, limited nutrient availability, or challenging environmental conditions. This holds true for the microbes found in the agrarian sky. The focus of this chapter is on the migration of microbes in the agrarian troposphere and its impact on crops.

The migration of aerial microbial assemblages is among the points to consider. Microbial diversity, density, and physiological abilities such as nutrient acquisition from the medium (air/clouds), adaptations to survive harsh radiations, relative humidity (RH), desiccation, and temperature fluctuations are crucial for the survival of microbes during short/long-range dispersal. The nature of the substratum at the point of deposition and the physiological versatility of microbes to establish themselves at the point of location are important factors. Microbial traits such as thick cell walls, sporulation, dormancy, hibernation, pigmentation, etc., may aid in migration. Attention should be given to the vertical dispersal of microbes to higher altitudes above agrarian regions, which occurs due to convection currents in the troposphere profile.

Microbes levitate into the troposphere and other layers of the sky due to low-level breezes on the farm soil surface, mild and strong winds, turbulence, convection currents, splashing of precipitation water, etc. In dry regions, dust storms, large plumes, drifting snow, fog, frost, and rain induce

Microbiology of the Agricultural Sky. K. R. Krishna (Ed)

microbial migration. There are several abiotic factors that contribute to aerial migration of microbes with diverse genetic nature. Microbial survival during levitation, migration, and upon deposition in a remote substratum determines the successful spread of microbes. Clouds support microbial migration and provide a habitat for microbes. Clouds also facilitate the aerial transit of microbes. Precipitation, such as rain, aids the microbial descent to the earth's (soil) surface. Frost, fog, mist, blizzards, and tornadoes also induce the migration of aerial microbes.

The nature of dust streams, dust particles, and microbial survival on particulate matter (PM) are aspects to consider. The size, weight, and buoyancy of aeolian dust particles in relation to microbial load are worth noting. Larger dust particles may only transport microbes for short distances, due to gravitational descent and deposition on soil or substratum. On the other hand, smaller PM particles can float and transit for longer durations, allowing for migration over longer distances, even intercontinental ones. Screening microbial pathogens residing on smaller particles at the point of levitation can help reduce disease intensity on crops.

Several examples of long-distance migration of microbes are discussed in this chapter. Intercontinental dust plumes, storms, and jet streams facilitate the dispersal of several teragrams (Tg) of particulate matter (PM) laden with microbes. In fact, a significant amount of microbial biomass is transferred from one continent to another. For instance, it is estimated that annually, approximately 8–12 Tg of dust containing diverse microbes migrates from North African dry farming regions to southern European farms and the Alpine region. Larger amounts of dust and microbial mass are known to migrate from the Sahara in Africa and the Gobi in Central Asia to other regions. This migration is known to enhance the genetic diversity of the locations where the dust/microbes settle and deposit. They may also undergo genetic hybridization with the local microbes. Aerial microbes carrying genes for tolerance to harsh atmospheric conditions may ultimately transfer them to the local microbial flora on the soil surface and crop canopies. The migration of plant pathogens is also observed. Historically, several epidemics/pandemics on food and commercial crops have been attributed to the migration of pathogenic/virulent strains utilizing intercontinental dust storms. Aerial insects that rapidly migrate from crop canopies to others also induce epidemics. Long-distance migration of wheat rust fungi (*Puccinia graminis tritici* (Pgt), *P. striformis* (Pst)) from Asia/Africa to the North American Plains occurs due to intercontinental dust storms/plumes. Similarly, the sugarcane rust fungus (*P. melanocephala* (Pm)) migrates from the African continent to Brazil, Cuba, and West India, providing excellent examples of intercontinental dispersion of fungal plant pathogens. Microbes migrate at 3,000 ft utilizing jet streams above agrarian regions and oceans for distances of 9,000 to 15,000 km. They disperse from the Atlantic coast in West African farms to the West Indies and on to the Great Plains. In the other direction, microbes, including plant pathogens, transit from the Gobi and Taklamakan in Central Asia to the Pacific Northwest. There are excellent examples of microbial migration from North Africa to West Asia and European food production zones. Dust plumes and storms that aid the migration of microbes within the Australian continent have been recorded. Interestingly, the global circumnavigation of microbes with the potential to cause crop disease has been identified.

There are highways and definite routes of migration for microbes, especially plant pathogens, in the intercontinental sky. Examples include annual dust storms and jet streams. This information is useful for avoiding the planting of susceptible cultivars in the path of dust plumes that carry pathogens. In fact, it is suggested to use resistant genotypes of crops and suitable cropping systems to prevent plant pathogens from affecting favorable substrates.

Aerial biotic factors that aid the migration of microbes require attention. Insects, birds, flying mammals, pollen, and other elements support the migration of microbes within agroecosystems. The agrarian sky harbors a diverse array of aerial insect species that act as vectors for microbial dispersal above major crop production zones. Aerial vectors such as aphids, jassids, psyllids, thrips, whiteflies, and bugs are known to cause significant damage to crops, initially as sap-sucking pests. However, their role as carriers of deadly plant viruses is equally important. Aphids, for instance, found above crops, are capable of transmitting approximately 150 different types of viruses. Beetle species are also known to transmit over 35 different viruses among crops, some of which can lead to devastating epidemics. The migratory behavior of aphids, including take-off, flight distance, flight pattern, flight behavior, and floating in the troposphere due to wind, determines the rate of virus spread.

It implies that the migration of microbes in the aerospace above agrarian regions is worth observing. Knowledge about microbial migration trends has several useful applications in crop production, particularly in forecasting and alerting farmers about impending diseases. It also has relevance to biogeochemical transformations in the agrarian atmosphere and soil profile. In the recent past, we have learned that aerial microbes and their migration are relevant to several earth system phenomena, such as weather patterns, even in distant locations, i.e., different continents.

3.1 AERIAL MIGRATION OF MICROBES IN AGRARIAN REGIONS: BACKGROUND

The migration of microbes in the agricultural sky is an essential natural phenomenon. Migration helps microbes in several ways. It is a phenomenon that helps microbes disperse and find better niches. Migration also helps them move away from harsh conditions. Additionally, migration allows microbes to intermix, mate, and improve genetic diversity.

Migration may also be an inevitable activity that microbes experience as a natural phenomenon. For example, when microbes are lifted into the atmosphere due to high-speed winds, turbulence, tornadoes, or sand/dust storms, their levitation and transit are inevitable. Migration may become a necessity when microbes are exposed to harsh conditions.

We are concerned here with the microbiology of the agricultural sky and its relevance to crop growth and productivity. The agricultural sky is a repository of microbes that are commensals, pathogenic to crops, or symbionts. In the atmosphere, all three types of aeolian microbes need mobility, to a certain extent, for survival and perpetuation. Aerial mobility differs based on microbial species. The ability of microbes to garner nutritional requirements sometimes depends on mobility. 'Migration' is also frequently referred to as the 'natural dispersal of microbes.' Migration utilizing different modes is essential for a few microbial species, basically to spread. Such microbes later rapidly multiply in the new location. However, it depends on the nutritional and environmental conditions of the new location, as well as the occurrence of natural detriments, antagonistic organisms, and toxins in the medium. Most studies suggest that bioaerosols disperse via a natural vehicle (vector) such as dust particles (aerosols), organic dust, clouds, water droplets, raindrops, wind, tornadoes, storms, etc.

Microbes found in agrarian regions are no exception to the abovementioned aspects. There are dreaded fungi of great significance to the global production of cereals. These fungi migrate as a routine natural phenomenon, moving from one field to the next, from one cropping belt to another, and even from one continent to another. Migration is an essential characteristic for most aerial microbes. Microbes in temperate regions transit via snow, fog, frost, clouds, cold breeze, wind, and even high-altitude jet streams found in the upper troposphere. Even for microbes that are sedentary or stuck to a substrate, migration may still be required, at least until they find a suitable surface. Globally, there are innumerable microbial phyla found in bioaerosols. Bioaerosol microbes assist in nucleation and the formation of snow and rain droplets, depending on atmospheric conditions. Indirectly, the migration of microbes influences nucleation and precipitation patterns in a location. Bioaerosols may be involved in long-distance migration, consequently influencing the nucleation properties of global rain-bearing clouds, snow, fog, and rain.

Microbial migration via the atmosphere, such as through wind, precipitation, and ice crystals, is an important phenomenon that facilitates the rapid spread of crop pathogens from one field or cropping zone to another. Numerous observed epidemics and pandemics have demonstrated that microbial migration, survival, and perpetuation are crucial factors in the causation and devastation of diseases affecting crops.

It is worth noting that microbial migration to nearby locations can occur due to the shattering of sporocarps. Short-distance migration may occur as a result of disturbances to the crop's canopy or splashing water. Longer-distance movement occurs through wind dispersal. All of these mechanisms contribute to the enhanced survival of microbes within an agroecosystem.

Aerosols are particles found in the atmosphere, varying in composition. They can be biological (bioaerosols) or mineral matter. Their size ranges from 0.001 μm to 100 μm. The shape and mass of aerosols also differ (Kulkarni et al., 2011). In recent years, aerosols have gained increasing attention as significant agents of climate change (Solomon et al., 2011). For instance, arid regions such as deserts are major sources of mineral dust particles in the atmosphere. These dust particles are lifted by storm activity and transported, sometimes reaching altitudes of 5,000 meters above sea level (m.a.s.l.) (Prospero et al., 2005). The Sahara Desert region is one of the main sources of aerosols (Goudie & Middleton, 2001, 2006), contributing 630–710 Mt of dust to the atmosphere annually (D'Almeida, 1986). Approximately 60% of Saharan dust emissions are transported southward, 25% westward to the Atlantic, 5% eastward to the Middle East, and 10% northward to Europe (Shao et al., 2011). Undoubtedly, microbes closely attached to or within aerosols can be disseminated proportionately.

Schmale & Ross (2015) have remarked that microbial transport in the atmosphere may have clear highways. Knowledge about such routes helps us detect microbes, their diversity, and potential to cause disease if they happen to be pathogenic. Firstly, detailed knowledge about vertical and horizontal migration needs to be accrued. Deciphering microbial routes at lower altitudes and the upper troposphere is essential to initiate anti-fungal or anti-bacterial sprays on crops (Brown & Hovmoller, 2002; Smith et al., 2018; Morris & Sands, 2017). Accurate methods for collecting air samples in the transit routes are also important. The migratory routes of Brown stem rust, Black stem rust, and Wheat streak mosaic are good examples of dispersal routes. It seems migratory pathways are specific even at the isolate level (e.g., Ug 99 of *Puccinia graminis tritici*). As prophylaxis, we may plant wheat varieties resistant to Ug99 or spray fungicides on crops found in its route.

Smith et al. (2018) point out that sampling, identifying, and analyzing microbes transiting in the upper troposphere region is not easy due to their origin. The vertical transit of microbes to higher altitudes is of great importance. Therefore, aircraft collector campaigns were conducted at altitudes ranging from 0.3 km to ~12 km. For instance, microbes belonging to Rumonococcacea, Lachnospiraceae, Pseudomonadaceae, etc., were detected in samples collected from the troposphere. Perhaps drones could also be utilized to collect air samples at different altitudes. In fact, the lack of easy, efficient, and rapid methods is the reason for the limited progress in Aeromicrobiology research above agrarian regions. The earliest efforts to collect, identify, and quantify the propagules (spores) of plant pathogens in the atmosphere were reported by Hirst (1959) and Hirst et al. (1967). These efforts have contributed to our understanding of the long-distance (inter-continental) migration of microbes.

Behzad et al. (2018) have identified several major dust storm generating areas around the world. These include the Great Basin in North America, the Atacama in South America, the Sahara and Sahel in West Africa, the Kalahari and Namibian deserts in Southern Africa, the Syrian, Egyptian, and Arabian deserts in West Asia, the Thar in South Asia, the Gobi and Taklamakan in the Mongolian region, and the Great Victorian Desert in Western and Central Australia. They have also identified regions that generate dust storms capable of aiding the long-distance migration of microbes. Additionally, they have identified regions that induce shorter journeys of microbes in the atmosphere. In these cases, microbial transport is relatively lower compared to the massive Saharan dust storms. Practically, there are atmospheric phenomena in each continent that aid microbial migration. Microbial migration in the lower and upper troposphere is an essential natural event. It aids several other natural phenomena in the atmosphere, such as weather patterns, precipitation trends, nutrient dynamics, and the spread of crop disease-causing organisms. Tampering with the long-distance migratory routes of microbes could be risky, as it can affect global climate and weather patterns. This, in turn, has a bearing on global crop harvests.

As mentioned earlier, the Sahara/Sahel region generates the largest amount of dust annually. During cultivation, the migration of microbes can be observed on crops, leading to diseases. The Harmattan dust carries microbes from the east to the west of the Sahara. The diverse ecosystem functions are constantly expressed by these microbes. This region is characterized by extremely high temperatures throughout the year, with a long and intense dry season from October to May. A brief and irregular rainy season occurs during the West African monsoon. The survival of microbes is a crucial aspect for perceptible migration to occur.

The Sahelian atmosphere supports a huge population of microbes throughout the year. There is a routine yearly weather pattern of microbial migration from Libya to Chad, then to Niger, Burkina Faso, and finally to the Senegalese Coast (Atlantic Coast). This pattern usually occurs from November, after staple crops have been harvested and bare soil is exposed to wind force. The aeolian wind carries vast amounts of dust and microbes from east to west. This has a significant impact on the crops grown during the next season and requires careful monitoring and study. Rainfall in the region is monomodal and occurs from June to October. Precipitation is also a major means of transporting aeolian microbes. This aspect also requires greater attention (Geagea et al., 1999).

The Sahel experiences a semi-arid climate. The rainy season lasts for two to three months in the north and about four months in the south. During the winter months (November-March), the dry, microbe-laden dust storms known as 'Harmattan' and 'Haboob' blow from the northeast to the southwest. These storms cause the migration of dust and sand (USAID, 2018). Understanding microbial migration within the Sahel is equally important, especially when compared to trans-continental movement through dust storms. It is possible that we can assess the onset of crop diseases, epidemics, and pandemics based on trends in microbial migration.

The migration of microbes in the troposphere above agrarian regions, as well as other types of terrain, involves three essential aspects (Pepper & Garba, 2015). These aspects include the natural conditions in the troposphere, the morphological and physiological characteristics of microbes, and the environmental conditions prevalent in the agricultural field. These aspects determine the migration pattern. The essential steps for migration itself are: (i) launching of bioaerosols (microbes), (ii) transportation, and (iii) deposition of microbes at a distant location.

Firstly, let us consider the launch of microbes from terrestrial locations such as open soils, crop canopies, natural vegetation, irrigation sources, animals, farm vehicles, etc. The conditions for launching are not the same for all microbes or dust particles. Launching microbes or dust into higher altitudes may involve several mechanisms. For example, the brisk activity of farm animals, vehicles, and humans can induce the launching of microbes into the atmosphere. High-speed winds and turbulence (e.g., tornadoes) can also induce the launching of soil microbes. There are microbe-related mechanisms that facilitate easy launching into the troposphere. For instance, there are fungi that release spores with a significant force into the sky, causing levitation and floatation in the air. The dispersed spores from the sporocarp can reach a few meters away from the parent fungal colony. The number of microbes launched at a specific point, in one event, into the air is crucial. It is estimated that annually, about 7.6×10^3 to 3.5×10^{23} microbe-laden dust particles reach the upper layers of the troposphere (Burrows et al., 2009a, b). The intensity of microbes launched from a location is key to the perpetuation of microbes found in bioaerosols.

Now, let us consider the transport of bioaerosols. Dust storms, clouds, and raindrops may mediate the transport of microbes from one location to another. In agrarian regions, it is often the breeze, wind, and dust storms that support short or long-distance migration. Spores that are light and small can easily levitate, remaining in the tropospheric air for a longer time. This allows them the opportunity to travel longer distances. However, if the spore size, weight, and moisture conditions of the atmosphere are unfavorable, transport may be limited to shorter durations and distances. Higher altitudes enable spores to remain in the atmosphere for longer durations, increasing the possibility of longer journeys. Microbes can also be transported at high altitudes in the troposphere due to jet streams. These jet streams can carry spores thousands of kilometers from their place of origin, reaching different continents. The survival and perpetuation of these spores depend on the natural conditions during transit and the environment where they are deposited.

Microbes found on the soil surface, crop canopy, phyllosphere, and fruits/pods often undergo sub-microscale or mesoscale transport. This involves the movement of bioaerosols for distances ranging from 100 m to 1 km from their place of origin or levitation. Mesoscale transport is common when dealing with longer journeys of microbes in the troposphere. For example, over 100 km from the origin of spore inoculum. The horizontal movement of microbes in the atmosphere is determined by factors such as air turbulence, viscosity, mean wind velocity, dimensions of spores, and other adaptations. Linear air flow is perhaps important for the longer journey of spores in the troposphere. However, in nature (i.e., agroecosystems), microbes floating in the air experience both vertical and linear wind movements.

Deposition of microbes occurs due to several different natural conditions in the atmosphere. The mechanisms employed by bacteria, fungi, actinomycetes, archaea, and viruses are noteworthy. The primary mechanism associated with deposition is the gravitational settling, where particles are affected by gravity and settle down. Another mechanism is downward diffusion. Surface impaction occurs when microbial spores floating in the air collide with crops, canopies, trees, leaves, branches, or other structures on a farm. Downward deposition of bioaerosols (microbes) can also occur due to rain droplets and electrostatic condensation. Microbes with a negative charge may associate with particles that have a positive charge, resulting in condensation and settlement on soils.

Airborne bacteria above fields are prone to disperse to other fields located at short or relatively longer distances. Intrafield dispersal of microbes is also of great significance to crop production. The uncontrolled spread of pathogenic bacteria within a field and their buildup can be a precursor to epidemics. McInnes et al. (1988) found that aerosols above crops contained microbes such as *Pseudomonas syringae* and *Xanthomonas campestris,* which are pathogens on crops. They tracked these bacterial species in samples collected early in the morning at 07:00, 12:00, 15:30, and 18:30. Microbes were detected at the highest intensity in samples collected between 15:30 and 17:30. Crop production procedures such as clipping, earthing-up, and fungicide application affected the microbes in the air above crops. Certain agronomic procedures that disturbed crops released more microbes into the air. The breeze above crops caused microbes to travel short distances until they settled in a different field. Bactericidal and fungicidal sprays had a definite effect on the microbes detected in the aerosol above crops. Pathogens like *P. syringae* reduced in population due to bactericides, although it did not influence several other genera. Microbes may survive and transit to neighboring farms despite sprays with Cu-containing plant protection chemicals. At this juncture, we should note that all microbes with suitable physio-ecological adaptations could survive, transit, and disperse above the cropped fields.

About a decade ago, Burrows et al. (2009a) expressed that we have investigated the lower troposphere more extensively in terms of bacterial diversity. We have also gained insights into the short-distance transit of bacteria, which impact nutrient dynamics in the atmosphere and soil upon deposition. When these bacteria strike a crop field susceptible to disease, they can become potent pathogens. However, not much attention has been given to the genetic diversity of microbes in the upper troposphere. These microbes are believed to travel longer distances and deposit at farther locations. Their contribution to soil and above-ground ecosystem functions requires greater attention.

In the case of fungi, the timing of sporulation and its release into the atmosphere is crucial. It can determine various aspects of fungal migration, including dispersal, deposition, and *in situ* proliferation. Oneto et al. (2020) have compiled evidence and supporting reports to demonstrate the significance of fungal sporulation, spore release into the atmosphere, wind transit, and return to the soil or crop canopy. It appears that fungal spores released during the daytime have the ability to float in the atmosphere and survive for longer periods. Furthermore, Oneto et al. (2020) state that fungal sporulation and release follow a rhythmic pattern, occurring at specific times of the day or night (Gilbert & Reynolds, 2005). This pattern may be influenced by temperature regimes and seasonal cycles. It is said that the fungal spore release pattern plays a crucial role in the rapid and efficient migration in the air. For instance, fungal plant pathogens such as *Mycosphearella fujiense, Rhizoctonia solani,* and many others selectively release spores in the morning (Meredith et al., 1973; Burt, 2002; Zhang et al., 2005). The shattering and forceful ejection of spores occur during the morning hours. On the other hand, some fungi release spores during the night. In the case of a few other fungi, spore release is controlled by circadian rhythms (Heintzen & Liu, 2007). Fungi also release spores intermittently to ensure better survival (Aylor, 1990). Such releases may also induce waves of disease on crops. It is believed that the selective timing of spore release helps fungi overcome the harsh atmospheric conditions. Fungal spores and hyphae avoid desiccation and low relative humidity until they reach their destinations (Jones & Harrison, 2004). In the case of certain fungi like *Helminthosporium maydis,* the spores are loosely held, so even a light breeze or mild turbulence is enough to lift them into the agricultural sky (Aylor & Lukens, 1974; Waggonners, 1974).

Oneto et al. (2020) have opined that spore fitness during the dispersal of a fungal species is important. The ability to overcome selection pressure applied by the substrate, at the point where they settle after a short or long journey in the troposphere, is crucial. This is particularly true for fungal species that are prone to long-distance spread in the air. Spore adaptations such as easy levitation (buoyancy), resistance to UV radiation, ability to withstand desiccation, low RH, and ability to imbibe moisture once they settle on a substrate are important. Interestingly, fungi that accomplish long-distance migration seem to adapt to the new environment by altering their sporulation patterns and timing of their release in the new site (Jones &

Harrison, 2004; Kramer & Pady, 1968). They adjust to the diurnal pattern of the new site. The rapid build-up of the population of fungal plant pathogens depends on a few modifications. The genetic diversity of pathogens helps them survive the harsh journey in the atmosphere (Springer, 1993; Mundt et al., 2009). Such a wide genetic pool helps fungi traverse long distances and establish themselves in new agrarian regions.

Burrows (2009b) suggests that the long-distance migration of microbes from arid, sandy deserts or regions with loose sandy soil (e.g., Sahel) could be intermittent. Seasonal additions of microbes into distant locations may be conspicuous. However, we have to realize that dust generation from ecosystems with natural vegetation and crops could be observed in all seasons. The microbial load of PM may differ based on seasons. Inter-regional migration of microbes is almost routine throughout the year, occurring in tropical vegetation zones. The congenial ambient conditions allow microbes from wet tropical ecosystems to survive and multiply, resulting in a higher microbial load on each dust particle compared to the relatively lower number noted for deserts.

Viruses are among the most abundant microbes on Earth. They are well distributed in the troposphere above the terrain (open land), vegetation zones, agricultural zones, urban areas, and oceans. Viruses show remarkable biodiversity. Their size allows them versatility in terms of levitation into the atmosphere and rapid transit across long distances. Their dispersal is mediated by several atmospheric factors such as clouds, precipitation, wind, convection currents, dust storms, etc. Viruses play a significant role in disease causation in agroecosystems. According to Weinbauer (2004) and Reche et al. (2017), when aerosolization of soil, dust, and organic PM occurs, it includes microbes. This initiates long-range dispersal in the atmosphere. However, there are impediments to free long-range transit for microbes, including bacteria, fungi, and viruses. Viruses become attached to organic dust particles smaller than 0.7 μm. It appears that the deposition rates of viruses are positively correlated with those of organic dust particles. This allows viruses to survive for longer periods of time, as they can endure harsh conditions by remaining within the organic particles. Additionally, once they are nestled in the crevices and folds of dust particles, they can avoid harmful radiation. The long-distance migration of viruses seems to enhance genetic diversity. Furthermore, Reche et al. (2017) suggest that we have some understanding of viral migration in the lower troposphere, but we need to better comprehend their deposition rates. The deposition of viruses, the conditions of the substrate in soil/crops, and their rapid multiplication inside a host crop are crucial factors. A viral pathogen that finds the right substrate with a suitable host can induce disease in crops. Virus deposition rates have been estimated to range from 0.26×10^9 to $>7 \times 10^9$ m^{-2} per day. In comparison, bacterial deposition rates were much lower, ranging from 0.3×10^7 to $>8 \times 10^7$ m^{-2} per day.

Vertical dispersal of bacteria from soil and crop canopies improves their chances of survival, but not always. Firstly, it causes bacteria to levitate. Then, it provides an opportunity for horizontal movement along the breeze and strong winds. Both surface layers of soil and crop canopies serve as vast repositories of microbes with diverse biochemical and physiological abilities. Surface soils induce the formation of aerosols containing bacteria (Graham & Harrison, 1975; Graham et al., 1977; Bovallius et al., 1978a, b; Vinette & Kennedy, 1975). Let us consider an example related to common bacteria found in fields with cereal and vegetable crops. Lindemann et al. (1982) have shown that soil structure, canopy height, wind shear, and strength are important factors that cause bacteria to become airborne from the soil and crops. Generally, the upward flux of microbes into the air was higher in fields with crops compared to bare soils. Fields with alfalfa had four times more bacteria than dry fallow soils. The vertical gradient of bacterial population could be observed up to 10 meters above the crop canopy. The vertical distribution directly affects the dispersal of bacteria in the vicinity and nearby cropped fields, and also provides conditions for long-range dispersal.

Mayol et al. (2017) state that approximately 42% of the microbes found in the atmosphere could originate from terrestrial sources. Another 25% could come from the surface of oceans or freshwater bodies. However, the remaining portion appears to be contributed through long-distance migration. Despite this, many estimates of microbes in the atmosphere assume that the contribution from oceans is negligible. Bacterial emissions from land to the atmosphere are estimated to be between 40–1,800 Gg dry weight per year (Burrows et al., 2009a, b). Mayol et al. (2017) further mention that during migration, microbial populations can be lost due to deposition or death in dust storms or plumes. Typically, when sampling the atmosphere about 1,000 km away from the origin, only 20% of the microbes can be traced compared to the original location. Their calculations have shown that it may take 3.5 to 17 days for microbes (prokaryotes/eukaryotes) to travel a distance of 1,000 km in the atmosphere (Finlay, 2002). Undoubtedly, the inter-continental migration of microbes has significance for agrarian regions, oceans, and the atmosphere itself. The agrarian region between the origin and final deposition of microbes at a distance may also be influenced. This is due to microbial deposition during transit at various points along the journey route.

3.2 ATMOSPHERIC FACTORS AND MIGRATION OF MICROBES

Wind (including jet streams), clouds, and precipitation are the most important factors that affect microbial migration in the atmosphere. Migratory microbes are deposited at distant locations on the Earth's surface. It appears that on a daily basis, wind efficiently picks up microbes such as bacteria, fungi, viruses, and archaea. These microbes can be dispersed through wind over long distances, ranging from 10 to 100 km. Microbes are deposited due to moisture, precipitation, rain, snow, or fog. Additionally, gravitational force also plays a role in causing microbes to settle in agrarian regions. According to Wehner (2018), approximately 800 million virus particles can travel down and deposit on vegetation on Earth within a few hours.

Wind and heavy rainfall induce soil erosion, leading to the loss of soil (sand, silt, and clay) and organic matter. Microbes are dispersed along with the eroded particles. Fields that are left fallow may experience a loss of microbial symbionts due to wind-induced erosion. Additionally, eroded soil can distribute plant pathogens. Wind erosion is particularly common in locations with loose soil, such as drylands and arid conditions. Breezes, low-speed winds, and high-speed winds are prevalent in plains regions, easily causing turbulence that can lift microbes and disperse them into the lower atmosphere. The horizontal movement of microbes can range from a short distance of a few meters to 10,000 meters or even a few kilometers. Regulating the wind-mediated dispersal of microbial pathogens is challenging, as it may lead to crop epidemics. However, the use of wind breaks can help prevent the spread of soil/dust loaded with microbes. Agronomic practices like deep plowing, which forms clods and aggregates, can also help reduce erosion (Shepherd, 2016). Sanitation measures during sprinkler irrigation are essential to prevent the drifting of pathogenic fungi, bacteria, and viruses along with mist and droplets.

Wind tunnel experiments help us decipher patterns of aerial movement of soil particles, particularly sand grains and dust that are emitted after a wind or dust storm event. Soil erosion caused by wind leads to the release of sand particles, dust, and various microbes such as bacteria, fungi, protozoa, and archaea into the atmosphere. Reports indicate that finer particles carry Bacteroidetes, which have a higher resistance to desiccation and therefore survive better. These particles can stay in the atmosphere for longer durations, allowing them to migrate over longer distances. On the other hand, coarser particles eroded from the soil contain *Proteobacteria,* which are believed to play a role in soil nutrient transformations related to nitrogen and carbon in the atmosphere (USDA-REE, 2013; Acosta-Martinez et al., 2015). Therefore, conserving soil against erosion may also help maintain a diverse population of microbes within fields. According to Tignat-Perrier et al. (2019), the microbial load and diversity in the atmosphere may depend on the landscape (soil, natural vegetation, crops) from which dust particles originate.

Jet streams are high-altitude winds generated in the atmosphere. Jet streams are formed when air masses from different regions converge. Jet streams found at 60°N latitude are called 'polar jet streams,' while those traced at lower latitudes are called 'subtropical jet streams' (Plate 3.1). They can be observed in the upper troposphere of the Earth in both the northern and southern hemispheres. Jet streams flow in waves, with prominent regions of 'troughs and ridges.' Jet streams blow faster in the winter due to the large temperature difference in the air masses. Jet streams have an impact on weather patterns. They can cause dust storms, as dust microbes are easily carried to higher altitudes of the atmosphere when there are fast winds with turbulence. Microbes that reach the jet streams can be transported over long distances. Jet streams also move large masses of dust particles between continents (*see* Acosta-Martinez et al., 2015; NASA, 2021; Slijepcevic, 2020; Plate 3.1). It is expected that a significant amount of dust, as well as a large population/mass of microbes, is transported.

'Microbial highways' in the sky were discovered about a decade ago (Schmale & Ross, 2015). Millions of microbes (bacteria, fungi, viruses, and archaea) were observed in the upper troposphere (8–15 km). The jet streams found in the upper troposphere are known to carry dust particles laden with microbes over long distances. There is evidence to show that microbes migrate at high altitudes in the atmosphere. For example, jet streams above the Mediterranean Sea occur at an altitude of 8 km and are known to transport microbes via dust plumes. Approximately 80 Tg of dust flows to European destinations annually, and we should expect a proportionate quantity of microbial mass to be transported along with the dust plumes. The amount of microbial mass transported depends on the microbial load of the dust particles (Rosselli et al., 2015). The evidence gathered to date proves that microbes causing diseases/pandemics in crops do migrate on jet streams (e.g., rust fungi) (Brown & Hovmoller, 2002; Mahaffee & Stoll, 2016; Acosta-Martinez et al., 2015).

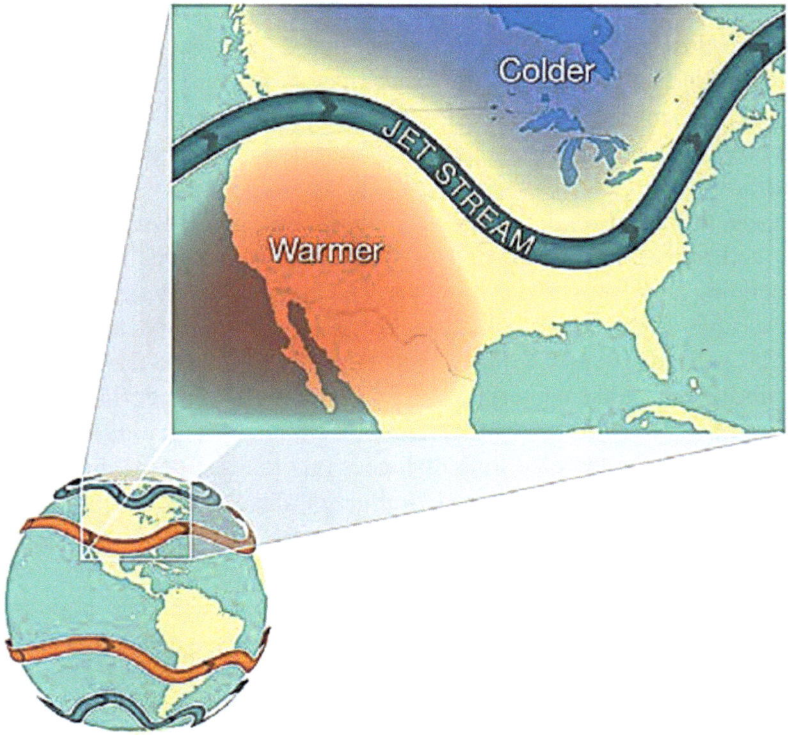

PLATE 3.1 Jet streams of wind at high altitude in the atmosphere.

Note: There are four jet streams (two polar and two subtropical) that encircle the Earth, with average speeds of 110 km/h^{-1}. They can reach high speeds between 200 and 300 km/h^{-1}. Jet streams form wherever cooler and warmer air masses meet. These powerful air currents transport billions of tons of dust particles across continents. Additionally, they facilitate the rapid movement of diverse genetic and physiologically capable microbes on a large scale. The transport of disease-causing fungal propagules via jet streams is responsible for global epidemics and pandemics in major food crops such as wheat, maize, and sugarcane. Furthermore, viral particles carried by jet streams contribute to several viral pandemics. Microbes can migrate over oceans and reach a different continent in as little as 7–12 days when riding the jet streams (e.g., from Sahel to North America)..

Source: NASA/NOAA, Washington, USA: Website of National Oceanic and Atmospheric Administration: Washington D.C. SciJinks; https://scijinks.gov/jet-stream/.

Clouds harbor diverse species of bacteria, fungi, viruses, and archaea (Amato, 2012). The presence of moisture in clouds helps to keep microbes hydrated. Certain microbes, such as *Bacillus* spp. and *Cladosporium,* are known to possess the ability to resist or tolerate extreme conditions. Cloud microbes utilize adenosine triphosphate (ATP) to meet their energy requirements. Additionally, cloud microbes may utilize organic matter found in dust particles for their carbon and energy needs. These microbes can float and travel along with cloud masses for varying distances. *Pseudomonas aeruginosa,* a well-known pathogenic bacteria, possesses unique properties that enable it to induce nucleation and precipitation. Under natural conditions, nucleation assists the bacterial pathogen in lysing leaf cells and entering tissues. *Pseudomonas aeruginosa* has specialized proteins on its cell wall that clump together ice crystals. Nucleation aids the bacterium in descending and depositing onto crops. The migratory patterns and nucleation of microbes like *P. aeruginosa* may have a significant impact on weather and rainfall patterns (Everts, 2008; Delort et al., 2010; Plate 3.2).

Studies on cloud bacteria and algae in Central France indicate several genera of algae as normal inhabitants of clouds. A few algal species were suspected to have traveled long distances on clouds. For example, members of Chlorellaceae, blue-green, red, and green algae, and diatoms were found in the cloud sample collected at altitudes ranging from 2,200 ft to 4,800 ft in the atmosphere (Dillon & Fennel, 2021).

Clouds support microbes of diverse genetic nature. The population of microbes resting in cloud masses and moving along with them in the atmosphere depends on a few factors. These factors include the source of microbes, such as terrain, soil, and vegetation, as well as the ability of microbes to adapt to cold freezing conditions and multiply rapidly. Microbes in the clouds get dispersed to different agrarian regions. Additionally, microbes in the clouds aid nucleation by providing a particulate surface on which water molecules aggregate and deposit on soil, crops, and natural vegetation as rainfall. Furthermore, microbes in the clouds allow the formation of water droplets (rain) at slightly higher temperatures. They may also induce nucleation in low floating clouds.

At the same time, clouds are known to harbor disease-causing organisms. They transport plant pathogens into different cropping zones (Hennings & Lynch, 2018). Perhaps we should refer to them as 'contagious clouds.' They have a detrimental effect on crops and farm animals. Monitoring cloud microbiota may help us detect crop pathogens such as bacteria, fungal spores, hyphal bits of actinomycetes, and archaea. This can provide advance warning, especially if the density of propagules of pathogenic microbes is higher than the threshold. Agrarian regions with a short cloudy season and low microbial density may not benefit from microbial migration via clouds. It is worth noting that microbial migration in the agrarian sky could affect weather and precipitation patterns, not only locally but also in distant locations.

For microbes, fog is a congenial ecosystem to survive, interact, and manifest different biogeochemical transformations. Fog is supposedly a good medium for microbes to float and migrate from one location to another (Osowski, 2018; Plate 3.2). Fog may actually serve to feed terrestrial ecosystems with microbes. This occurs when microbes in the fog deposit at a specific location. Agroecosystems do receive microbes from fogs that pass above their canopy. However, fog may not be a suitable vector for all types of microbes. It is important to identify the preferred species and understand their effects on natural vegetation and crops in the area. Fogs that occur in agrarian regions, deserts, coastal areas, or cold temperate regions do aid in the transport of microbes. While fogs can move over longer distances, they are not as widespread as intercontinental phenomena like high altitude winds, jet streams, or other means.

Fog is an important atmospheric phenomenon in agrarian regions. It disperses moisture, dust, and microbes across cropped fields. Fog provides a habitat for bioaerosols (microbes). Additionally, depending on the geographic location, fogs can aid in the dispersal of microbes over longer distances (Michigan State University, 2020). Estimations suggest that fog is a more significant source of bacterial and fungal propagules compared to raindrops. However, since fog can transport pathogen-laden dust particles, it can also contribute to the spread of diseases. Plant pathogens (spores) can remain suspended in the fog for extended periods. Furthermore, microbes present in the fog can induce nucleation and precipitation.

The microbial diversity of snow in the cropping zones of different continents was examined by Brown & Jumponnen (2019). Snow is an important atmospheric factor in Scandinavia, the European Plains, North America, and several parts of the southern hemisphere. Snow at higher altitudes may support a certain degree of microbe migration. Brown & Jumponnen (2019) compared the diversity of microbes in snowy regions of Europe and North America. They found that bacterial and fungal diversity was comparatively lower than that of snow algae. The dispersal of bacteria and fungi depended more on the size of the microbe-laden snow particles. Snow algae could be endemic to certain locations where periodic snowfall occurs throughout the year. Snow experienced during the crop season or in the fallow period has an impact on microbial diversity and activity. Snow can severely affect microbial survival and diversity in plantation zones. The combination of a snowy cold front and a pathogen can be devastating to plantations (Plate 3.3).

PLATE 3.2 Clouds above the Northern European wheat growing regions.

Note: Clouds are said to act as oases for aerial microbes. While the atmosphere in general is considered an 'extreme environment,' clouds provide favorable conditions for aeolian microbes to survive, travel, and reproduce. Clouds also play a crucial role as dispersal agents in the troposphere above wheat fields. Microbes can migrate over long distances while perched on clouds.

Source: Sharath Kowligi, Eindhoven, Noord Brabant, Netherlands.

PLATE 3.3 Snow aids the migration of microbes in the Pacific Northwest of the USA.

Note: The above image depicts a natural forest stand of *Pinus* sp. near Crater Lake in Oregon, Pacific Northwest America. Snow is an important atmospheric factor that transports significant amounts of moisture, microbes, and nutrients. The formation of snow particles is influenced by nucleating microbes such as *Pseudomonas aeruginosa*. Snow serves as an effective dispersal agent for microbes that have the potential to cause disease or initiate symbiosis. It can also carry plant pathogenic organisms like *Pseudomonas aeruginosa*. Additionally, snow contributes nutrients to the soil. Snow deposits are a major source of irrigation water and dissolved nutrients in the northern Great Plains of America. Simultaneously, snow disperses microbes into farming zones.

Source: Dr. K. R. Krishna, Personal Collection, Bangalore, India.

A study aimed at deciphering the role of rain in the migration and spread of bacterial populations in the atmosphere has been conducted. It was found that rains can disturb the soil surface and disrupt soil aggregates encapsulated with bacteria. Under tropical temperatures, rains create a mist/splash that spreads bacteria to a shorter distance from the origin. Rain itself is a good vector for bacteria. Raindrops are known to spread microbes quickly and over longer distances. The migration of bacteria from one cropped field to another in the vicinity or even a few kilometers away from the point of generation is mediated by precipitation events. Both natural rain and sprinklers may create mist, which easily picks up plant pathogens or commensals found in the atmosphere. This mist may then spread to entire cropped fields. Rains contribute to the rapid spread of plant pathogens. It appears that microbial diseases like blotches, blights, and spots on leaves are easily spread via mist/precipitation. Researchers have attempted to decipher the impact of raindrops using high-resolution continuous imaging methods. Rains actually create aerosols containing microbes. Estimates by researchers at MIT (Cambridge, MA, USA) indicate that globally, precipitation may account for 1-25% of microbes displaced as aerosols and placed in the atmosphere. This makes it suitable for the migration of microbes to different distances from the point of origin.

Microbes found on terrain, crop's canopy, and leaf surface are supposedly dispersed to different distances by the impact of raindrops. Storms, heavy rain, and lashing droplets may all aid the movement of microbes to different distances from the place of origin (Geagea et al., 1999; Chu, 2017). In this regard, experiments by Kim et al. (2019) using dry spores of *Puccinia triticina* have suggested that fungal spores move from the surface (leaf lamina) to a certain distance outside the leaf due to the vortex created by splashing raindrops. When rain drifts, it causes microbes in the air to disperse to a farther distance. Furthermore, Kim et al. (2019) used high-resolution photography, scanning electron microscopy, and spore counts to assess the impact of rain on fungal spores. A combination of air vortex, splashing droplets, and levitation into the atmosphere aids the migration of microbes to longer distances. Hurricanes and tornadoes with moisture can cause large-scale movement of microbes in the atmosphere (Yong, 2013). Temkiv et al. (2012) opine that the atmosphere, being an 'extreme environment,' may actually select microbes (survival of the fittest). So, those species that resist the harsh conditions of transport, in fact, add to biodiversity both in the clouds and on the terrain. They also studied the microbial diversity of clouds and hailstorms by actually examining the hailstone. Hailstone analysis using culturing and non-culturing 16S rRNA methods showed that hurricanes and rapidly moving clouds can carry microbes. Hailstones affect the microbial composition of the terrain. Hurricanes may carry commensals or pathogens and disperse them to distant agrarian zones.

There are studies on atmospheric relative humidity (RH) and its influence on microbes. We are particularly interested in how RH affects the migratory patterns of microbes. Microbial survival is crucial for microbes to transition, deposit, and establish themselves in new environments. Undoubtedly, atmospheric RH plays a significant role during transit. Let's consider an example. A study conducted in the 1950s by Brown (1953) found that most microbes, including the three tested microbial genera *Achromobacter, Pseudomonas,* and

Micrococcus, had better survival rates at RH values of 45 to 90%. Microbes had the lowest mortality rates when the atmospheric RH was 65–70%. These microbes migrate at high altitudes, in cold temperatures, and under desiccating conditions. Currently, several microbial genera and species that can withstand low RH, desiccation, and dryness in the atmosphere have been identified. Discussions on other atmospheric factors such as temperature, UV radiation, oxygen tension, pollutants, etc., are already available in Chapter 2.

Monteil et al. (2014) state that microbes such as *Pseudomonas aeruginosa* and *Botrytis cinerea* serve as good models, particularly in understanding the role of aerial microbes in the water cycle, nucleation, and disease causation in crops. According to their experiments, the occurrence and spread of *P. aeruginosa* are actually linked to the electrical conductivity, pH, physical and chemical constituents of rain, and the movement of air masses. The distance traveled by fungal spores of *B. cinerea* depends on atmospheric conditions such as exposure temperature and solar radiation. These factors, along with the characteristics of the terrain, crops, and precipitation patterns, contribute to the dissemination of these microbes.

3.3 VECTORS FOR AERIAL MIGRATION OF MICROBES IN THE AGRARIAN REGIONS

In this section, we focus on the migration of microbes using biological vectors. Specifically, we examine insect vectors such as whiteflies, aphids, psyllids, bugs, and others, which spread crop pathogens. Several examples highlight microbes that have the potential to cause epidemics in crops, resulting in significant crop loss. Additionally, there are man-made vectors of microbes found in agricultural farms, including farm vehicles and drones (unmanned aerial vehicles), which can also contribute to the migration of microbes.

3.3.1 AERIAL BIOLOGICAL VECTORS OF MICROBES RELEVANT TO AGRICULTURE

Pollen drifts in the air due to wind, turbulence, mist, fog, snow, and rain. Pollen is also carried longer distances by bees, insects, birds, and ground animals. Microbes attached to pollen can rapidly spread into agrarian regions. Clearly, pollen is an important vector for transmitting microbes during pollen shed. The migration of plant pathogenic microbes on pollen grains can be detrimental to crop areas. Card et al. (2007) points out that invertebrates, Spiroplasmas, and Phytoplasmas are not generally spread via pollen dispersion. However, several fungal plant pathogens that deposit into or infect floral parts are dispersed by pollen. The aerial spread of fungal diseases can be mediated to some extent by the intensity of pollen shedding and drifting. According to Card et al. (2007), there are several plant viruses that hitch a ride on pollen and infect neighboring crops. Potyviruses, Alphacryptoviruses, Ilaviruses, and Nepoviruses have been identified as being dispersed by pollen, acting as a vector (Aparcio et al., 1999). In total, approximately 39 viruses have been documented as being spread by pollen in agricultural fields. The distance over which viral transmission or migration occurs in these fields depends on factors such as pollen morphology, release pattern, ability to float in the atmosphere, and survival conditions like relative humidity, temperature, and wind speed. Additionally, a few viroids that are pathogenic to crops are also dispersed through pollen. For instance, tomato apical stunt viroid (TASVd) migrates by attaching to pollen (Antignus et al., 2007). Avocado sunblotch viroid (ASVd) also migrates to neighboring avocado fields via pollen (Desjardins et al., 1984). Similarly, Blueberry leaf mottle virus (Boylan-Pett et al., 1991), Prunus necrotic ringspot ilarvirus, Barley streak virus, Barley stripe virus (Caroll & Mayhew, 1976), Grapevine viruses (Cory & Hewitt, 1968), and many others are transmitted through pollen.

Insects belonging to several families assist in the aerial transmission of viruses, bacteria, and fungi in agricultural zones. These insects migrate short distances between plants and sometimes move from one region to another, spreading plant viruses. Certain families of insects play a role in disseminating various types of disease-causing viruses, some of which can be devastating to food crops. Let's consider a few examples. Aphids (Aphidoidea) transmit over 150 different plant viruses. Leafhoppers (Cicadellidae) are known to spread more than 80 different viruses that can cause diseases. They are also capable of transmitting phytoplasmas, spiroplasmas, and mycoplasma-like organisms (MLOs). Plant hoppers (Fulgoroidea) aid in the migration of over 20 different types of viruses and a few MLOs in agrarian regions. They are responsible for transmitting several cereal virus diseases. Additionally, there are tree hoppers (Membracidae) that act as aerial vectors.

There are a few more examples. Whiteflies (Aleurodidae) transmit viruses with the potential to cause diseases on legumes, vegetables, cotton, potatoes, and several others. At least 20 different virus diseases are spread by whiteflies. Psyllids (Psyllidae), Thrips (Thripidae), Mealybugs (Pseudococcidae), dipteran flies like *Rhagoletis pomonella,* Cabbage maggots (*Delia radicum*) (Anthomyiidae), Leaf miners (Agromyzidae) also act as vectors.

Coleopteron beetles like leaf beetles (Chrysomellidae) are known to transmit at least 35 different plant viruses. Bark beetles (Scolytidae) aid in the dispersal of fungal pathogens like *Ceratocystis ulmi*. It spreads Dutch Elm disease. *Conotrachelus nenuphar* (Curculionidae) is a vector for *Sclerotinia fructicola,* which causes brown rots on fruit crops. Honeybees (*Apis melliferae* – Apidae) are vectors for *Erwinia amylovora*. This bacterial species causes fire blight of fruits. This list is not exhaustive. Plus, several new reports may appear periodically, in short intervals.

Overall, there are numerous examples of microbes being spread by aerial insects. These insects include bugs, flies, aphids, mites, birds, and animals that carry microbial pathogens or commensals on their body surfaces and transport them. Factors such as the life cycle events and migratory patterns of vectors, the season, the crop stage, as well as the pathogen and its virulence, should be considered. The activity of aerial vectors, particularly in the canopy and a few meters above, can be crucial. There are also situations when vectors carrying pathogenic microbes have been lifted and transmitted via high altitude winds and jet streams. When this occurs, we can expect the aerial transmission of microbes over long distances. This can lead to epidemics and pandemics in distant locations, especially in the absence of disease-resistant crop cultivars.

Insect vectors of microbes, particularly those causing epidemics or pandemics on crops, have received considerable attention from researchers and farmers. Let us consider a few examples to demonstrate the importance of insect control or management in preventing the migration of viruses. Cassava mosaic virus disease (CMVD) and cassava brown streak virus disease (CBVSD) serve as apt examples. In these cases, the microbes are dispersed by insects (Legg et al., 2011; Alicai et al., 2019). These two viruses spread rapidly, causing epidemics in the tropical cassava production zones of East and Central Africa. The viruses migrate across the canopy via the insect vector – whitefly (*Bemisia tabaci*). During the mid-1980s, these viruses quickly spread to several nations where cassava cultivation zones exist. The vector, *Bemisia tabaci,* migrated and spread the viral disease (Legg et al., 2014). The aerial spread of viruses could be halted if we effectively control the whitefly. Whitefly control should be implemented before the onset of epidemics. Atmospheric factors have an impact on both the vector and the virus (pathogen). A favorable atmosphere leads to greater devastation of the cassava crop. The vector flies from plant to plant and field to field, spreading the virus. The whitefly is active within a range of a few meters from the focal center, yet it rapidly covers the entire field and beyond. Data collected between 2004 and 2017 shows that the whitefly (*B. tabaci*) has caused pandemics in East and Central Africa. The vector has facilitated the migration of the virus from Kenya, Tanzania, and Mozambique to Uganda and Cameroon (Alicai et al., 2019).

Aphids are a class of insects that migrate in the atmosphere. They hover within and surround the canopy of crops. They are among the world's most potent vectors of crop diseases. The 'cereal aphids' are particularly deadly insects that significantly reduce global food grain production. These cereal aphids directly affect crops through their sucking behavior. However, their impact on crops is immense as they also transmit deadly cereal viruses. According to Parry (2013), although there are over 5,000 species of aphids in agriculture, only a few species affect cereal and legume crops. When cereal aphids take flight or are carried by high-speed winds, viral pathogens migrate with them. In essence, 'aphid movement in the air is microbial movement.' Several cereal aphids are known to assist in the migration of microbes such as bacteria, fungi, and viruses. Some examples of aphid species acting as vectors of cereal crop pathogens include the Rose-Grain aphid (*Metopolophium dirhodum*), Grain aphid (*Sitobion avenae*), Bird Cherry-Oat aphid (*Rhopalosiphum padi*), Corn aphid (*Rhopalosiphum maidis*), Russian wheat aphid (*Diuraphis noxia*), Indian Grain aphid (*Sitobion miscanthi*), Rice root aphid (*Rhopalosiphum rufiabdominalis*), Apple grass aphid (*Rhopalosiphum insertum*), Blackberry-cereal aphid (*Sitobionfragariae*), Fescue (or grass) aphid (*Metopolophium festucae*), and Greenbug (*Schizaphis graminum*) (Irwin & Tresh, 1988; Parry, 2013; Blackman & Eastop, 2000; Brown et al., 1993).

Parry (2013) defines aphid migration as the periodic flight of insects beyond the boundaries of their old habitats. The migratory behavior of insects differs significantly. Furthermore, knowledge of migratory behavior is important because it directly affects feeding patterns and the dissemination of viral pathogens in crop fields. Strategies for viral disease and vector management may depend on the migratory patterns of aphids (Irwin et al., 2007). In the case of aphid migration, aspects such as initiation, path, distance, and duration are determined by wind velocity. Aphids take off when the wind is mild, but they float randomly when it is gusty and fast (Johnson, 1954, 1969). Once the aphid vector reaches an altitude of 1 m in the atmosphere, the wind takes over the movement of the aphid insects (Compton, 2002; Taylor, 1974; Haine, 1955). Usually, flights below 1 m in height are referred to as 'voluntary appetite flights.' During these flights, aphids disseminate viruses while feeding. Aphids engage in appetite flights when the wind speed is less than 8 km/h (Kennedy, 1990; Kennedy & Booth, 1963; Thomas & Vevai, 1940; Bottenberg & Irwin, 1991). The flight path of aphids can be influenced by a wind velocity as low as 2 km/h. At wind velocities up to 4.8 km/h, aphids are known to be in continuous flight with intermittent alighting/feeding for 4-8 hours. This is more than sufficient to rapidly spread an epidemic of viral disease. Aphid migration primarily occurs during the day, resulting in the dispersal of microbes in agricultural fields or beyond. Parry (2013) mentions that there are two peaks of flights and microbial dissemination via aphids in a day: one in the early morning and another in the mid-afternoon when aphids are active. Migration patterns are influenced by the season, with aphids flying out for 1 to 3 hours during the normal migratory period (Loxdale et al., 1993; Kennedy & Booth, 1963; Nottingham & Hardie, 1989; Nottingham et al., 1993). In some cases, these flights may last only 30 minutes, while there are instances when aphids fly nonstop for 6-9 hours in the atmosphere. During these flights, aphids spread viral pathogens. Cereal fields are generally free of aphid vectors during nights with cold temperatures.

Aphid take-off, in-flight speed, flight path, terrain, vegetation, crops, and landing are crucial aspects. These aspects determine the extent to which microbes can be dispersed. Aphids have the ability to fly at higher altitudes and reach lower jet streams, enabling long-distance movement. This can result in intra or even inter-continental migration. Temperatures in the atmosphere affect aphid migration when they are at higher altitudes (Isard et al., 1990). Precipitation also affects the migratory pattern of aphids and, in turn, the extent of viral dissemination (Wallin & Loonan, 1971; Mann & Harrington, 1998).

Long-distance migration of viruses can also occur through insect vectors. This phenomenon is attributed to the updraft of aphids, which carry them into high-speed winds and jet streams (Wallin & Loonan, 1971; Robert, 1987; Lewis & Taylor, 1964). This is a random and less frequent method of microbial migration. When in flight, aphids are known to land based on terrain characteristics such as vegetation, crops, pastures, fallows, etc. Aphids are induced to land in crop fields or farms by yellow, orange, and green wavelengths around 540 nm (Hendrie et al., 1985). Distracting aphids with materials that reflect different wavelengths has practical applications in agriculture. For instance, landing can be avoided by avoiding blue and ultraviolet range radiation. Reflective silver-colored ribbons, commonly used for birds, could also be utilized to prevent aphid landing (Brown et al., 1993; Compton, 2002; Loxdale et al., 1993).

There have also been several simulation studies conducted to determine the rate of viral spread via vectors. Reports indicate that in 3 to 5 hours of flight, aphids may cover over 10 km distance in a cropped field. Therefore, it only takes a few hours for viruses to spread from one field to another. Additionally, their ability to drift along with jet streams increases infection. Within a few days, aphids can disperse the viral pathogens to different agrarian regions within a continent.

Citrus Greening, also known as 'Huanglongbing disease,' has reached epidemic proportions. It has spread to different continents through an aerial insect vector. The disease is caused by a bacterium called *Candidatus liberibacteria* (Teixeira et al., 2005). This bacterium is transmitted aerially by a psyllid insect known as *Diaphorina citri.* It was initially discovered in Asian citrus plantations. Over the past 2–3 decades, this combination of aerial vector and bacterium has affected large areas of the citrus belt in Brazil and Florida (Bove, 2006; da Garca, 1991; da Garca & Korsten, 2004). There are two types of bacterial pathogens: one heat-sensitive and the other heat-tolerant. It is the heat-tolerant bacterial species that survives and spreads rapidly through the vector (Brlansky & Rogers, 2007).

Let us consider a couple more examples of bacterial diseases that are spread by aerial insects (vectors) – Psyllids. However, these examples are about diseases on vegetables, fruits, and other crop species. Since the causal agent (bacterial species) is vectored via a rapidly spreading Psyllid, we can expect the microbes to spread quickly if prophylaxis is not practiced. The Zebra disease of Potato is already a serious disease in the American and Australian continents. The bacterium Candidatus *Liberibacter solanacearum* causes it. But the pathogen is spread by a Psyllid fly, namely *Bactericera cockerelli* (Department of Environment, Food and Rural Affairs, 2014). Symptoms on potatoes include stunting of crops in the field, chlorosis, leaf curls, and zebra spots. There are a few strains of bacteria that cause zebra spots. Perhaps, it is a matter of time before the bacterium is spread via an aerial vector and reaches epidemic proportions. The flight pattern, life cycle, and bacterial virulence are a few important aspects to monitor.

Phytoplasmas migrate in apple orchards by utilizing aerial insects (vectors) such as Psyllid flies. Reports suggest that at least three species of Phytoplasmas, namely *Candidatus Phytoplasma mali, Ca.P. pyri,* and *Ca.P. prunorum,* are known to spread among fruit trees such as apples, pears, and stone fruit (Sousa et al., 2012). The pathogen spreads rapidly through a vector – a Psyllid known as *Cacopsylla picta, C. pyri,* and *C. pruni.* To prevent epidemics, it is important to monitor the life cycle of the Psyllid fly, its flight pattern, distance covered, and the virulence of the pathogen.

3.4 AERIAL MIGRATION OF MICROBES BENEFICIAL TO CROP PRODUCTION: BIOLOGICAL NITROGEN FIXERS, MYCORRHIZAS, AND BIOCONTROL AGENTS (BCAS)

There is a vast literature on Biological N-Fixation (BNF) by symbionts that form nodules on the roots of leguminous crops (e.g., Bradyrhizobia on soybean, cowpea, peanuts, etc.). These microbial species are applied to the soil as liquid formulations or granules. Alternatively, they can be seed treated. In the absence of live-host roots, they can survive in the soil in an independent state. However, turbulence in the wind can lift them into the atmosphere from the surface layer of the soil. In aerosol form, they may migrate short or long distances and become available for the next crop in a different location. Bradyrhizobial species can migrate as bioaerosol from one field to another depending on atmospheric conditions such as wind, turbulence, precipitation events, fogs, and smog. Even farm vehicles that cause drifts of surface soil may disperse rhizobium, particularly those residing in surface soil, on plant debris, or crop residue.

Similarly, asymbiotic microbes capable of biological nitrogen fixation (BNF) can be dispersed as aerosols. These bioaerosols can migrate over long distances. It is common to find asymbiotic N-fixing microbes such as *Azotobacter, Azospirillum,* and others contaminating nearby fields. This contamination occurs when bioaerosols land in neighboring fields. Additionally, there are phyllosphere bacteria that fix atmospheric nitrogen in an asymbiotic state (free-living). These bacteria can utilize organic energy sources present on leaf surfaces to fix nitrogen. For instance, there are commercial formulations of asymbiotic N-fixers like Envita, produced by Azotic North America in Raleigh, North Carolina, USA, which are sprayed onto cereal and legume crops. These N-fixing microbes can disperse to different distances when sprayed using sprinklers. Fertigation with asymbiotic N-fixers and irrigation water, using center-pivot sprinklers, is also possible. The aerosols containing N-fixers can deposit on the crop's canopy, including leaf lamina, branches, and stems. They can also remain suspended and migrate over long distances. When asymbiotic N-fixers in surface soils and those suspended in foliar sprays reach high altitudes and jet streams, they can migrate even further.

Blue-green algae (BGA) are used in rice production. BGA cultures are added to submerged paddy fields to take advantage of atmospheric nitrogen fixation. There are several reports on the amount of nitrogen fixed by BGA, ranging from none to 20–40 kg N ha^{-1} $season^{-1}$. Common genera of BGA found in rice fields across Asia include *Nostoc, Anabaena, Aulosira,*

Cylndrospermum, etc. (FAO, 1981; Roger & Kulasooriya, 1982; Roger & Watanabe, 1982). BGA may reach higher altitudes in the troposphere due to low troposphere wind, turbulence, and atmospheric convection. The migration of BGA through irrigation water has been partially studied, but the migration of BGA through air/wind in the troposphere requires further understanding. The distribution of BGA in the troposphere above crops, individual fields, or cropping regions has an impact on nitrogen fixation. Aerial dispersal of BGA from one region to another can be beneficial in various ways and may contribute to improving their genetic diversity.

Aerial distribution, population density, survival, tolerance to factors that operate in the atmosphere, and their role in agricultural conditions have been well reported for prokaryotes such as fungi, bacteria, viruses, and protozoa (Brown et al., 1964; Lee & Eggleston, 1989; Sahu & Tangatur, 2015; Sharma et al., 2007). However, the same may not be true for algal distribution in the agricultural sky. Studies documenting the vertical distribution of blue-green and green algae in the atmosphere are scarce, if any. Some of the blue-green algae are well-known free-living biological nitrogen fixers. Their ability to migrate through wind, precipitation, or other means in the atmosphere is important. The distribution of these algae in soil and water bodies in agrarian regions may depend, to some extent, on their ability to disperse in the atmosphere. Understanding the aerosolization of blue-green algae is crucial (Sharma & Singh, 2010). Long-distance migration of blue-green algae has significant consequences for biological nitrogen fixation. Aerial dispersal can replenish algal species with the ability to fix atmospheric nitrogen. However, it is important to note their ability to tolerate harsh atmospheric conditions. Efforts to trace and cultivate blue-green algae that can tolerate aerial conditions and dispersal are worthwhile.

The arbuscular mycorrhizal (AM) fungi have a worldwide distribution in soils and crop roots. They can be found in all agrarian regions, facilitated by natural migration. Aerial dispersal of AM spores through convection currents, wind, jet-streams, precipitation, and flowing surface water cannot be ruled out. AM fungal spores and hyphal fragments can also be dispersed through irrigation water. Furrow or sheet irrigation, as well as foliar sprays, may transport AM propagules from one location to another. Dispersal can occur through birds, animals, and even farm vehicles that come into direct contact or carry soil/root debris (e.g., disc harrows, tined soil tillers, hoes, clod crushers, furrow/ridge makers). Aerial migration of AM fungal spores has its advantages. It enhances their population in new farming locations where they are deposited, providing an opportunity for fungal isolates to cross and enhance genetic diversity. Long-distance dust plumes may carry multiple AM fungal species. The addition of new isolates/genotypes into a soil ecosystem can offer better advantages from AM symbiosis to crops. However, it is important to unequivocally demonstrate that high altitude, long-distance dust plumes do carry AM fungal spores/propagules. The number, volume, and shape of spores may be correlated with dispersal ability in fungi (Chagnon et al., 2013; Kivlin et al., 2011).

The arbuscular mycorrhizal (AM) fungi primarily inhabit the soil. They can be found in the soil profile either freely or attached to roots. Various agricultural activities such as plowing, discing, soil loosening, intercultivation, or weed uprooting can expose them to wind and turbulence. Therefore, it is reasonable to find AM propagules in tropospheric dust at low altitudes. AM spores may even reach higher altitudes and be carried by jet streams. This phenomenon is similar to the dispersal of other soil microbes at different altitudes in the troposphere. The morphology, weight, buoyancy, radiation tolerance, and desiccation resistance of spores are crucial factors. Regions that generate dust storms, such as deserts (like the Sahara) or areas with loose sandy soils (like the Sahel), contain AM fungal spores (based on personal observations). High-speed winds, rainstorms, or dust storms can lift and transport AM fungal spores. Further research is needed to understand the dispersal of AM fungal spores through tropospheric winds. Fungi are ubiquitous and spread throughout agrarian regions worldwide via the troposphere. Undoubtedly, these symbiotic fungi exhibit significant migratory trends.

Research on the aerial migration of arbuscular mycorrhizal (AM) fungi is scarce. In 2014, Egan et al. expressed that our understanding of the aerial dispersal of AM fungal hyphae or spores is minimal, if any. Mycorrhizal spores can be carried by wind at low altitudes, across soil surfaces, natural vegetation areas, and cultivated fields. Egan et al. (2014) detected AM fungal spores in the air above agricultural crops. They investigated the air above various cropping regions in the Great Plains and other areas. Even a slight breeze, turbulence, or uplift can trigger the dispersal of AM spores. Almost any farm machinery that disturbs the soil surface can generate a small dust plume. The AM fungal propagules become airborne during activities such as tillage, earthing up, weed uprooting, complete plant harvesting, uprooting, and stubble removal. Soil tillage dislodges AM spores from dry roots and soil aggregates, propelling them into the troposphere. When there is enough lift, they may reach the upper troposphere and even jet streams. Several soil fungi and their spores can levitate to the upper troposphere and migrate. Such phenomena also occur with AM fungal spores in fields. So, AM fungi can also move via jet streams and over long distances. However, we need authentication through clear experiments. Perhaps, collecting aeolian dust after long-distance transit and growing AM plants using the aerial sample of dust will clarify the situation. Egan et al. (2014) and Kivlin et al. (2011), for example, have identified spores of AM genera like *Glomus* sp. in air samples. Furthermore, soils usually contain a more diverse species of AM fungi compared to the air above the crop's canopy. They have also opined that troposphere wind may not be the major mode of dispersal (Plate 3.4).

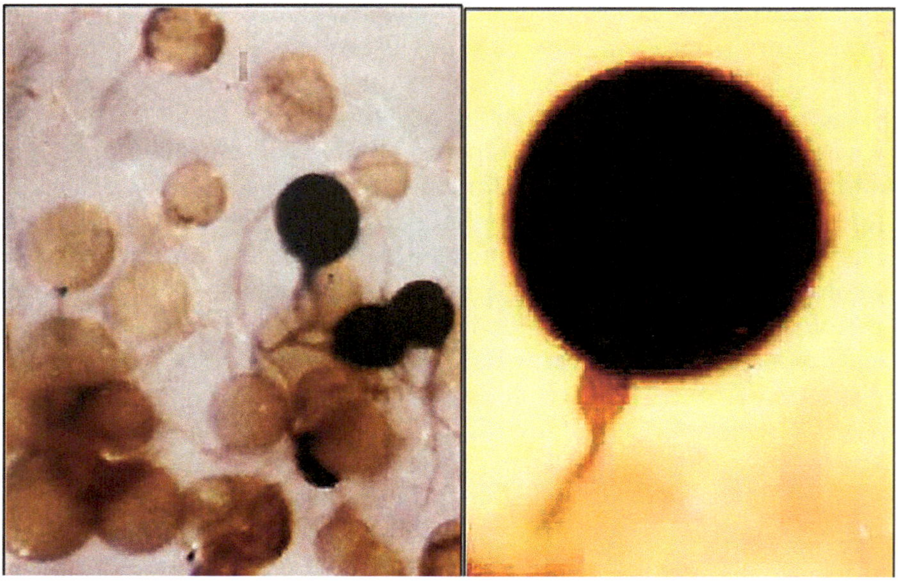

PLATE 3.4 Arbuscular mycorrhizal spores reach different altitudes and migrate in the troposphere due to aeolian processes such as wind and turbulence. [Left: *Glomus aggregatum* from sandy desert soil at the fringes of the Thar desert, near Nagaur, in Rajasthan, Northwest India. Right: A large spore of *Gigaspora gigantea* from sandy Oxisols of Sahelian West Africa].

Note: These spores were isolated from regions cultivating Pearl millet/cluster bean in Northwest India and Pearl millet/cowpea in the Sahel on sandy soils. They are easily prone to reaching the troposphere through wind, dust storms, or turbulence. Fungal spores are light in weight, small in size, and buoyant. Therefore, there is a possibility for them to be transported over short or long distances via aeolian dust. They travel in the air, similar to particulate matter or other recorded microbes. Often, rain-bearing clouds carry a significant amount of particulate matter before they precipitate at a location. Such storms may contain spores of arbuscular mycorrhizal fungi. It is comparable to dust storms carrying spores of plant pathogenic fungi (*Glomus aggregatum* spores are ~ 200 µm and *Gigaspora gigantea* spores are ~ 240–410 µm in diameter).

Source: Dr. K. R. Krishna, International Crops Research Institute for the Semi-arid Tropics, Hyderabad, India.

Basidiomycetous ectomycorrhizal fungi exist. They produce spores on fruiting bodies. For instance, mushrooms that form symbiotic mycorrhizal associations produce lightweight spores compared to those of AM fungi. Some species disperse spores through a shattering mechanism, spreading them over relatively shorter distances. Additionally, spores have ornamentations, thick walls, and various aerodynamic shapes, enabling rapid lift and migration in the atmosphere. Ectomycorrhizal spores that migrate in the upper troposphere often possess characteristics to withstand harsh conditions such as desiccation, low relative humidity, unfavorable temperatures, and radiation. Spore pigmentation helps avoid the detrimental effects of radiation. Spines, wings, or other protrusions aid in floating in the air (Lebenonturf, 2013). Several basidiomycetes form symbiotic relationships with host trees or shrubs. For example, ECM species like *Laccaria laccata* and *Laccaria amethystine* produce basidiospores that are resistant to atmospheric conditions. These spores supposedly migrate in the air for longer distances before germinating in new soil (Vincenot et al., 2012).

Biocontrol microbes are beneficial because they reduce the population of pathogens and insect pests. In this context, let us consider fungi that act as antagonists and are used as biological control agents (BCA) against fungal plant pathogens. These fungi are distributed in the troposphere above the crop's canopy, as well as on leaves, fruits, stems, and branches. According to a literature survey and estimates by Thambugula et al. (2020), there are over 300 species belonging to 113 fungal genera that can be utilized to prepare biological control formulations. These formulations can be applied as aerial sprays in liquid, dust, or granule form, or even placed directly on the crop's canopy (Guetsky et al., 2001; Freeman et al., 2004; O'Brien, 2017). Currently, the most well-known plant pathogens that are targeted by biocontrol fungi include *Alternaria, Botrytis, Colletotrichum, Fusarium* (Datnoff et al., 1995), *Lasiodipolia, Phytophthora, Sclerotinia,* and *Verticillium* (Ordentlich et al., 1990). *Trichoderma* is a prominent biocontrol fungi, with 25 *Trichoderma* species that can be used in biocontrol formulations (Thambugula, 2020; Datnoff et al., 1995;). *Trichoderma* formulations are utilized worldwide in various agricultural regions. It appears that *Trichoderma* has antagonistic effects on over 100 different plant pathogens (Herman, 2006; Ommati et al., 2013). Other strong candidate fungal species that qualify to be used as biocontrol agent (BCA) formulations include *Sordariomycetes, Dothideomycetes, Eurotiomycetes, Leotiomycetes, Saccharomycetes* (Ascomycota), *Agaricomycetes, Exobasidiomycetes, Ustilaginomycetes, Microbotryomycetes, Cystobasidiomycetes, Tremellomycetes* (Basidiomycota),

Glomeromycetes (Glomeromycota), and *Mucoromycetes* (Mucoromycota) (Thambugula et al., 2020).

Knowledge about aerial distribution and dispersal traits of fungal species used as biological control agents (BCA) is indeed essential. Fungal propagules such as spores and hyphae are often found in the troposphere. These propagules can be pathogens, commensal organisms, or even BCA targeting other fungi or insects. The propagules of fungi utilized as BCA should be small in size, light, buoyant, resistant to radiation effects, and capable of withstanding a certain degree of desiccation, among other characteristics.

The above traits will allow the BCA fungus to survive in soil and on the crop's canopy. They can also survive floating or drifting in the air above agricultural fields. Additionally, lighter spores that are affected by wind may travel short or long distances. The persistence of the fungus in the atmosphere depends on its survival traits. However, we are concerned about the rate of spread of BCA in the fields, both on crops and in the atmosphere above the crop's canopy. The spores of fungi used as BCA should remain virulent in the atmosphere despite the harmful conditions of the troposphere. Other essential traits include overwintering, rapid growth, and sporulation upon the onset of favorable conditions. To ensure the effectiveness of biocontrol fungal formulations, mixed cultures of fungi are also utilized.

There are examples of fungal species that serve as biological control agents (BCAs) for insect pests. One of the most prominent examples is the use of *Metarrhizium anisopliae,* which is utilized to control locusts, grasshoppers, and a few lepidopterous insects. The fungal BCA formulations are sprayed onto locust clouds and allowed to spread among them. This way, *Metarrhizium* infects a large population of grasshoppers. As locusts travel across vast regions, the BCA fungus also spreads. Aerial sprays of *Metarrhizium* have been used on short-horned grasshoppers (locusts) in the Sahara and Sahel regions. The aerial distribution, multiplication, drifting, and floating of *Metarrhizium* require attention. It is worth noting that, like chemical pesticides, BCAs based on fungi also have non-target effects. They may affect several other insects (Brimner & Boland, 2003).

Several phyla of bacteria have members that possess biocontrol characteristics. This is mainly due to their ability to produce biochemical substances that deter plant pathogens, suppress critical biochemical pathways, or affect cell wall functions. Some biocontrol *Bacillus* species can physically exclude pathogens by covering up the area, such as on leaf lamina, thereby restricting their access to nutrient sources. Among the various genera examined by microbiologists, it is the genus *Bacillus* that appears to have several members that are effective against plant pathogens.

A few examples of *B. species* affecting plant pathogens are *B. amyloliquefaciens* FZB42, which is known to be antagonistic to several genera of fungal pathogens on plants (Shafi et al., 2017; Benitez et al., 2010; Halimi et al., 2010). *Bacillus amyloliqueficien* isolate LBM 5006 suppresses *Aspergillus* spp., *Fusarium* spp., and *Bipolaris sorokiniana* (Zhao et al., 2010). This isolate impairs spore germination and germ tube formation of fungal pathogens. *B. polymixa* is antagonistic to *Pyricularia oryzae* and *Rhizoctonia solani* (Kavitha et al., 2005). *B. vallismortis* isolate ZZ185 is utilized to suppress *Fusarium graminearum, R. solani, Colletotrichum parasitica,* and *Phytophthora capsica* (Zhao et al., 2010). *B. subtilis* 14B suppresses *Alternaria solani* (Hammami et al., 2012). *B. thuringiensis* kurstaki is an effective biological control agent against insect pests of crops. It affects cell membrane functions of the host (Kamoun et al., 2005).

Shafi et al. (2017) have listed several fungal pathogens and crops on which biocontrol Bacillus strains have been utilized. For example, *B. subtilis* HU5 is used to control *Verticillium wilt* on cotton; *B. subtilis* SQR9 to control *Fusarium oxysporuum* on cucumber; *B. pumilus* to control *F. oxysporum* on pea; *B. megaterium* to control Sclerotinia sclerotiorum on oilseed rape; *B. subtilis* QST 713 to control *Fusarium, Pythium, Rhizoctonia,* and *Septoria* on different crops, etc. We need to gather more knowledge about the aerial distribution of these BCAs.

Several biochemical and physiological mechanisms of antagonism operate through which biocontrol microbes act on pathogens/pests. However, in the present context, it is important to consider their aerial distribution, residual propagules, rate of migration within the field, drifting nature in the lower and troposphere, and jet streams. Usually, the persistence of biocontrol microbes helps in keeping the pathogen population under the threshold. The atmosphere may also support biocontrol bacterial cells/spores on dust particles. Aerosolization of biocontrol cells/spores helps them transit longer distances. The aerial migration of these biocontrol agents (BCAs) and pathogens is worthwhile.

Incidentally, *Bacillus thuringiensis* is an example of *Bacillus* spp. that is effective against several major insect pests of crops. It serves as a highly successful biological control agent (BCA) and is being adopted in many agrarian regions around the world. The aerial distribution of biocontrol bacteria, such as *B. thuringiensis*, needs to be well understood before its use in controlling insects that affect crops.

B. thuringiensis forms spores that are resistant to radiation, desiccation, and low relative humidity (RH). These spores are smaller than vegetative cells and are buoyant, allowing them to float in the air above fields that have been sprayed with *B. thuringiensis* formulations. The density of spores in the air and on foliar surfaces is important. Wind and turbulence above the soil and crop canopy can release spores into the troposphere. Spores may be carried by air currents based on their buoyancy. Precipitation events or moisture in the air can also disperse the spores.

Farmers typically use high population densities of *B. thuringiensis* cells/spores in their formulations. As a result, these cells/spores may reach jet streams and travel long distances, even between different continents, from one agrarian region to another.

There are bacteriophages that are used to control bacterial diseases. Bacteriophages can regulate the population of pathogens and keep it below the threshold. For example, several bacteriophages have been identified and utilized against epidemics caused by *Xanthomonas*, a leaf blotch causal agent on various crops. The patterns of aerial spread of bacteriophages should be known before their use in the field, especially in relation to the pathogen they are intended for.

3.5 AERIAL MIGRATION OF MICROBES WITH POTENTIAL TO CAUSE EPIDEMICS AND/OR PANDEMICS ON CROPS

Historically, microbes were discovered in the atmosphere (air) and studied to some extent by Antonie Van Leeuwenhoek (1941; Baas-Becking, 1934; Beijerinck, 1913), Louis Pasteur (Gregory, 1971, 1973), and Lister (Lidwell, 1987). The dispersal of microbes in the air was demonstrated by Louis Pasteur in 1860 (1860a, b). The aerial migration (transport) of microbes in the atmosphere was often observed in relation to the spread of diseases in crops, animals, and humans. It has been noted that studies on bacterial transport in the atmosphere have mostly focused on disease spread (Burrows et al., 2009a, b; Knudsen, 1989; Sorensen et al., 2000; Gloster et al., 1981, 1982; Addiss et al., 1989; Jones, 2021). In comparison, there are now efficient methods for detecting dust plumes through satellites and collecting air samples. Molecular methods are available for detecting, identifying, and quantifying aerial microbes. These scientific advancements have led to numerous investigations into microbial migration, particularly at different altitudes in the atmosphere.

The bioaerosol is kept levitated in the air for a longer time by convection currents, buoyancy, and strong breeze and wind. The migration itself can be for a short distance, referred to as 'micro-scale dispersal.' It occurs within a range of 10 m to 100 m. Burrows et al. (2009a, b) point out that most studies on aerial dispersal of bacteria have been conducted to understand the phenomenon at the meso-scale. Mahafee & Stoll (2016) have shown that 'meso-scale' dispersal may occur from a few hundred meters to 10,000 meters from the place of origin. Meso-scale dispersal is often aided by tornadoes, strong breeze, high-altitude horizontal wind, tropical storms, low altitude storms, chinooks, etc. Long-distance and global-scale migration occurs when bacteria (bioaerosols) are at high altitude and migrate via jet streams. They may migrate for distances of 10^7 meters and longer.

Microscale dispersal involves the spread of microbes through plant parts. Minor turbulence in the air causes them to become airborne. This is usually followed by deposition in the surrounding area. This process occurs within a range of 100 m to 10,000 m from the point of origin where the inoculum was generated. Dispersal mechanisms like these can cover a few plants, a field, or a small region of crops. Microbial dispersal can also occur across landscapes, agroecosystems, or large food grain generating regions such as the Great Plains of North America. However, if a microbe (pathogen) happens to migrate over 10,000 m, it may be affected. High altitude winds, clouds, and jet-streams play a role in this type of dispersal. On a global scale, intercontinental dispersal of microbes can reach distances of 10^7 to 10^9 m. As mentioned earlier, this is facilitated by high altitude winds and jet-streams. There are examples of disease-causing microbes that undergo micro-, meso-, or global scale dispersal.

Microbes can travel along specific routes and paths, often causing pandemics and devastating crops and food grains over large areas (e.g., *Puccinia graminis tritici*-Pgt) (Hirst, 1959; Hirst et al., 1967; Liang & Evan, 2011). Micro-scale and meso-scale dispersal of microbes can occur within seconds to a few minutes, reaching their destination of deposition. For instance, from one leaf to another in a plant, or from one plant to the next in a field. Mesoscale migration, covering distances of 100 m to 10,000 m, may take a day or less than a week. High altitude global migration of microbes on dust particles, due to dust storms, may take a fortnight to a month or even longer (Mahaffee & Stoll, 2016; Liang & Evans, 2011). Dust storms have the potential to transport vast quantities of microbes, such as rust fungal inoculum, across global wheat belts, posing a significant threat to crops. We have implemented improved methods of collection, detection, and enumeration to identify disease-causing fungi (spores). Based on this, we have adopted appropriate control measures (Mohaffee & Stoll, 2016). Furthermore, Mahaffee & Stoll (2016) emphasize the importance of understanding the scale of microbial dispersal in terms of distance. This knowledge enables farmers to implement suitable prophylaxis and disease control methods, especially when dealing with plant pathogens. The source of inoculum, extent, and mode of spread are interconnected factors that require careful consideration.

Long-distance migration of microbes pathogenic to crops is a prominent cause of global food insecurity. The intercontinental dispersal of disease-causing fungi can affect large expanses of crops across different continents. Regular monitoring for such diseases, using aerial observation of affected patches and the atmosphere, is essential. For example, each year, crops are exposed to intercontinental dispersal of diseases such as wheat black stem rust, brown leaf rust, wheat stripe virus leaf blotches, etc. These diseases spread through wind and insect vectors. Aerial dispersal of pathogens often impacts the finances of individual farms and can also have wider implications on the agrarian economics of countries dependent on crop production and export.

Fungal dispersal occurs through spores with different characteristics. Hyphal fragments, which act as efficient propagules, can also migrate on aeolian particles. Alternatively, they can be carried by the atmospheric breeze or wind. Soil fungi are often lifted due to turbulence in low altitude winds, and they can even reach high altitudes and travel via jet streams. Globally, there are several examples of plant pathogenic fungi that migrate long distances along with aeolian dust, known as bioaerosols. Fungal spores that are light, buoyant, and resistant to unfavorable conditions such as humidity, temperature, and radiation, are better suited for aerial movement, allowing them to reach distant locations. Regardless of the mode of transportation, distance, or stage of the fungus, it is the new location that applies natural selection pressure. The substrate in the new location determines the survival of the fungus. The selected fungi can be plant pathogens or saprophytic heterotrophs that thrive on organic debris. A few examples of Ascomycetous fungi that disperse over long distances via aeolian dust are fusarium head blight (FHB), whose causal agent is *Gibberellazeae/Fusarium graminarium* (Leslie & Summerell, 2006). Black Leaf streak fungus (*Mycosphaerella fujiense*) (Parnell et al., 1998), and Black sigatoka (*Mycosphaerella graminicola*) (Parnell et al., 1998; Linde et al., 2002) are other examples. Basidiomycetous examples for long distance dispersal include Loose Smut, caused by

Ustilago nuda; Leaf Rust of Coffee, caused by *Hemiliavastetrix* (Isard et al., 2005); Sugarcane Rust, caused by *Puccinia melanocephala*; Wheat Yellow Rust, caused by *Puccinia striformis* (Geaegea et al., 1999); Sugar beet rust, caused by *Uromyces betae* (*see* Anikster et al., 2005; Pammel, 2017); and Soybean Leaf Rust, caused by *Pakosporapachyriza* (Carvalho et al., 2011).

The aerial routes chosen by these fungi may often be unfavorable. Fungal propagules such as basidiospores, urediospores, and teliospores are capable of withstanding harsh atmospheric conditions. They are light, buoyant, and resistant to desiccation, allowing them to migrate over long distances. Scientists from the Space Science Institute in Bern, Switzerland, conducted simulation studies on aerial microbial dispersion. These studies have shown that dust particles, along with microbes, can travel thousands of kilometers. Dust particles are typically smaller than 0.02 mm in size and remain buoyant. To some extent, these particles provide protection to microbes that rest in their crevices.

Microbes appear to follow specific routes while in transit from one location to another, particularly at higher altitudes (Schmale & Ross, 2015). Additionally, Wilkins (2011) suggests that the size of dust particles carrying microbial loads could be significant. There are microbes that are well adapted to travel easily and over long distances. Furthermore, microbes smaller than 0.009 mm can travel across continents with ease. Bacterial clumps, amoebal clumps, and large fungal spores or sporocarps are typically smaller than 1 mm in size. It is believed that microbe-laden particles slightly larger than 20 μm but smaller than 1.0 mm can float and travel between continents effortlessly. This migration of microbe-laden dust is known to enhance biodiversity in the atmosphere, as well as in the areas where they deposit due to rain, moisture, or gravity. Wilkins (2011) and Wilkins et al. (2011) further state that the smaller the microbes, the faster, longer, and quicker their transit in the atmosphere. Therefore, smaller dust particles carrying microbes have a greater potential to cause diseases in crops and farm animals. If this holds true, we can attempt to screen and filter smaller particles in the atmosphere. This aspect has important applications in controlling the spread of crop diseases caused by pathogens such as bacteria, fungi, and viruses.

Let us consider an example pertaining to bacterial diseases of vegetables such as tomatoes and capsicums. These vegetables are grown in the Eastern USA and Plains region. They are prone to harboring *Xanthomonas campestris* on their leaf surfaces, which can cause blight and spots on the fruits. The disease reaches epidemic levels when the bacterial pathogen is spread rapidly by wind and water droplets. Sprinkler irrigation, specifically center-pivot irrigation, can also contribute to the spread of the bacterial pathogen. The value of the product, in this case, the fruits, is reduced due to the presence of spots (Gitatis & Pathak, 1988). In some cases, control measures such as spraying Cu+ mancozeb or other plant protection sprays may become ineffective when the bacterial infection has reached epidemic proportions. Therefore, it is essential to practice sanitation by adopting culling, eradication, and spraying. Wind plays a major role in disseminating bacterial species, allowing the pathogens to spread from one plant to another and from one field to another rapidly. Bacterial cells can float in the sky and get dispersed through rain and irrigation droplets, also known as aerial sprays. These floating bacterial cells can deposit on soil and crops through raindrops.

The wind-mediated migration of *Puccinia graminis tritici (Pgt)* urediospores is well-documented. *Pgt* is known to produce millions of urediospores and release them for long-range spread via wind. Within the USA, this long-range spread (1,500 km or 860 miles) allows the fungus to cover the entire length of the Great Plains from south to north. The rust fungus can transit in a single season, in just a few days. In Australia and New Zealand, the wheat rust fungus (*Pgt*) is known to travel up to 2,000 km annually on the wind. The losses due to windborne fungus can be severe if timely prophylaxis is neglected (Zadocks & Schein, 1979; Roelf, 1982; Roelf & Bushnell, 1985; Roelfs et al., 1992; Luig, 1985). Peanut rust caused by *P. arachidis* also spreads rapidly in the wind. This rust pathogen has been observed to transit over oceans and terrain in the 1970s and 1980s. Initially noticed in America and China, it later spread rapidly in the Groundnut basin of Senegal in the 1980s. It is a wind-borne fungus (Mendgen, 1997; Subramaniam et al., 1985; Savary et al., 1988).

The Saharan dust that migrates annually to the Caribbean and the Americas, in the form of plumes, storms, and via jet streams, is believed to have facilitated the spread of Sugarcane rust fungus (*P. melaniephala*). The earliest documented instances of wind-mediated intercontinental spread of sugarcane rust fungus were in 1978. The trans-Atlantic migration of the rust fungus from West Africa, which takes 2–3 days through the aeolian route, has affected the sugarcane crop in the Dominican Republic. The urediospores migrate at an altitude of 3,000 ft above the terrain and settle on susceptible cane fields. With a speed of 12 mm s^{-1} and an uninterrupted journey at an altitude of 2,000 ft–3,000 ft, it takes only 2–3 days for them to reach a sugarcane field in the Caribbean Islands (Mendgen, 1997; Purdy, 1985; Comstock, 1992). Therefore, urediospores can travel rapidly and quickly initiate an epidemic in a distant location.

Coffee leaf rust is another wind-mediated disease. Mendgen (1997) states that *Hemileia vastatrix* (rust fungus) first devastated coffee gardens in South Asia, particularly in Sri Lanka, in 1867. However, it was actually first discovered in Kenya. The rust fungus predominantly spreads via aeolian dust, with the free floatation of urediospores in the atmosphere being important. While wind dispersion is the main mode of transmission, human travel and distribution of plant material also contribute to its spread. The epidemic severely affects the productivity of coffee gardens. The windborne microbe (rust fungus) later spread rapidly into South American coffee gardens. Intercontinental spread is believed to occur via high-speed winds and jet streams.

3.6 INTERCONTINENTAL MIGRATION OF MICROBES AND GLOBAL AGRICULTURE

Global migration of microbes via atmospheric currents, long-range dispersal of aerosols (dust), bioaerosols, strong winds, storms, hurricanes, and jet streams requires attention from agricultural researchers. These phenomena may have both direct and indirect effects on global crop productivity. In the past, we have developed, modified, and refined various agronomic procedures, plant protection sprays, and cropping tactics, many of which were likely in response to the long-distance migration of microbes. Long-distance dispersal of microbes is a highly relevant natural event in farming as it impacts global-scale epidemics and pandemics of crops. Furthermore, long-distance migration may aid in the preservation of individual microbial species and complex microbial communities, serving as a mechanism to overcome unfavorable conditions in certain locations. Additionally, long-distance dispersal can facilitate the mixing of native microbes from distant locations, leading to increased genetic diversity. This inter-mixing process creates new genotypes of microbes that may possess the ability to adapt to various atmospheric conditions. Long-distance dispersal of microbes may introduce new genotypes of microbes that were previously unknown to colonize the agrarian region. It also provides a better opportunity for microbial species to reach and flourish in the new ecoregion. Perhaps, it allows them to explore and scavenge new agricultural zones for nourishment, multiplication, and population build-up. Long-distance migration of microbes may be a natural phenomenon aided by land and atmospheric weather phenomena. Vice versa, i.e., long-distance dispersal of microbes in the atmosphere (troposphere), may also affect weather phenomena such as nucleation, precipitation, induction of disease, etc.

Man-made modes of long-distance dispersal of microbes through human travel, migrating cattle, sheep, and poultry, airplanes, drones, blimps, and spacecraft may also need careful study. They have the potential to transport microbes over intercontinental distances, which can cause ecological disturbance. These modes of transport can introduce specific microbes into agroecosystems located far away on different continents. Additionally, they have the potential to initiate epidemics or pandemics if the dispersed microbe is a virulent crop pathogen. It is also worth considering the possibility of symbionts being spread through the same means of microbial transport. Furthermore, the lack of phyto-sanitation and quarantine measures could be responsible for the long-range dispersal of microbes through man-made methods.

Microbes dispersing over long distances require specific methods for detection. Firstly, to locate drifting dusts, satellite imagery, drones with lidar cameras, and radars are usually employed. Sequentially, the process generally involves: (i) collecting dust particles and studying the physicochemical characteristics of aeolian dust, with particle size distribution determined by scanning electron microscopy; (ii) visualizing microbial cells on dust particles; (iii) visualizing microcolony-forming microbial cells on dust particles (*see* Yamaguchi et al., 2012; Tang et al., 2018). Microbiologists have adopted a few specialized methods to identify and quantify aeolian microbes, particularly those that travel in the atmosphere as bioaerosols. Commonly utilized methods include visualizing microcolony-forming microbial cells on dust particles. Bacteria found in aeolian dust are cultivated, followed by DNA extraction, random cloning, and quantitative real-time PCR (Yamaguchi et al., 2012).

Long-distance migration of microbes is often aided by high-altitude winds and dust plumes that migrate to intercontinental regions. High-speed jet streams and low-altitude constant winds in the troposphere also induce microbial migration. These atmospheric phenomena allow for the rapid migration of dust above seas and oceans. Long-distance migratory birds and aerial vehicles (man-made mode) also aid in the intercontinental migration of microbes. Long-distance animal and human migration also leads to the transit of microbes from one continent to another.

The long-distance migration of microbes, whether natural or man-made, often affects agrarian regions worldwide. All cropping zones are susceptible to receiving microbial deposits from the atmosphere at any given time. These deposited microbes may be harmless commensals to crops, or they could be pathogens capable of causing severe damage. Over the past few decades, numerous studies have demonstrated the inter-continental migration of microbes and their impact on crops. Some notable plant pathogens that have been identified include Rust fungi (*Puccinia* spp.), Bacterial blotches (*Xanthomonas* spp.), Bacterial rots (*Pseudomonas syringae*), and Scabs (apple scab – *Venturia inaequalis*). These pathogens are known to travel long distances at high altitudes in the troposphere, often accompanied by aeolian dust particles. They can reach different continents after traveling distances ranging from 1,000 km to 15,000 km in the air (Yamaguchi et al., 2012; Acosta-Martinez et al., 2015; Griffin et al., 2001, 2003).

In comparison to disease-causing organisms, we have made relatively fewer attempts to understand the short, long, and extra-long-distance migration of symbionts such as *Bradyrhizobia,* AM fungi, or ECM fungi. Similarly, the migratory patterns of nitrifiers, methanogens, methane (CH_4) degrading bacteria, sulfate transformers, etc., have not been extensively studied.

According to Yamaguchi et al. (2012), dust storms may carry as much as 10^5 microbes per particle, particularly during the initial phase of intense wind turbulence in the Gobi and Taklamakan desert regions. These are sparsely vegetated areas in Central and Eastern Asia. As the dust plumes taper, we may find about 10^3 microbes per particle. The occurrence of microbes as single cells, clumps, or large clouds in dust plumes has an impact on the distance of transit and their survival in the new location. A denser microbial inoculum supposedly has a better chance of survival during transit and upon deposition on crop canopies or soil in a new location. Extreme environments like those encountered in dry, sparsely vegetated zones, deserts with sand, high-temperature regions, and locations with low humidity may affect microbes. Snowy and cold regions may deter or inhibit the establishment of migrated microbes, and snow may also induce hibernation and/or sporulation.

Let us consider a few facts about the migration of aerosol microbes between continents. Microbes adapt to high-speed winds, particularly jet streams that occur at higher latitudes on Earth. It seems that billions of tons of dust laden with microbes make the transit each year (Acosta-Martinez et al., 2015). Perhaps migration occurs several times in different seasons. We should note here that such microbes, while they are in transit, have the potential to affect the weather patterns in the location. Migrating bacteria that aid in the nucleation of ice crystals/snow may actually be influencing the seasonal precipitation patterns. So, microbes influence a large area in terms of weather. There are also indications that during long-distance transit, microbes may survive by drawing nutrition from organic dust and clouds. Furthermore, it seems that researchers from the Georgia Institute of

Technology in Atlanta, USA, have cataloged about 314 distinct species of microbes, including bacteria, fungi, and archaea. On average, at an altitude of 10 km above the terrain, they enumerated 5×10^4 microbial cells per cubic meter of air sample. They adopted a 16S rRNA-based technique to identify microbes. Obviously, certain microbial species could be more efficient in feeding in the atmosphere, allowing them to survive better and float for longer distances. About 17 species of microbes were common in most air samples. Microbial endurance in the troposphere obviously depends, to a certain extent, on nutritional sources. It seems that oxalic acid was utilized by most microbe species that survived for a longer period in the atmosphere, enabling them to transit longer distances. Several of the microbes conduct the journey in the troposphere as resting spores, dormant hyphae, or other dormant cells. These species of microbes are supposed to help in nucleation, creating rain or snow depending on the region's moisture, clouds, relative humidity, and temperature conditions.

Now, let us consider a few well-known intercontinental journeys by microbes. The Sahara and Sahelian sandy region is the largest generator of aeolian dust. The African dust plume is a conspicuous natural phenomenon. Firstly, it disperses microbes across vast desert regions. Microbes migrate from locations above Chad to Burkina Faso, and then on to Senegal. Later, the dust plumes are known to carry microbes to the Caribbean Islands. For example, the wheat rust fungus – *Puccinia graminis,* may often migrate to intercontinental locations. The rust fungus transits from locations in Kenya to West Asia, and then from Chad to Western Sahara. Later, microbes (e.g., fungal spores) float above the Atlantic to reach the North American wheat belt (Table 3.1). Wheat rust fungi from Kenyan cropping zones are also known to migrate high above the troposphere to West Asia, Pakistan, and the Western Gangetic belt. Migratory routes of certain isolates of pathogenic fungi may differ. They may not move towards the Indo-Gangetic belt. Instead, they reach Mediterranean locations. It is useful to monitor and know the migratory routes of microbes. We can then aim at proper prophylaxis at crucial points and periods in their transit route. Planting resistant genotypes in the path where fungal spores deposit is recommended.

The intercontinental migration of dust plus microbes (bioaerosol) should not be overlooked. For instance, the atmosphere in Florida receives significant amounts of dust from West Africa. African dust can also be detected in regions adjacent to Florida, such as the Caribbean and southeastern USA. The presence of bacteria in the aerosol and clouds is a crucial factor that could potentially impact crops if the microbes happen to be pathogens. Additionally, these bacteria have the ability to induce nucleation and precipitation. The bacterial load of each dust particle that reaches Florida's atmosphere from the African coast is of great importance, estimated to carry approximately 10^9 bacteria (Burrows et al., 2009a, b; Table 3.1). Observations indicate that the bacterial load in the ambient atmosphere is higher during a dust storm event reaching the Florida coast. The increased bacterial counts could be attributed to long-distance migration, multiplication of bacteria on the dust particles during the journey, or *in situ* tornadoes that stir up dust, mineral particles, and microbes (Schlesinger et al., 2006; Prospero, 1999; Prospero et al., 2005).

Dust storms that reach Amazonia from the Sahel/Sahara also carry a microbial load (Yu et al., 2015). The microbial diversity depends on the locations from which the dust is levitated, due to turbulence. Reports suggest that bacteria reach the aerial space above Brazilian forests and farming regions. The dust plumes play an important role in fertilizing the soil with diverse microbes. Microbes are known to induce nucleation. The organic dust and microbes aid in creating rain, leading to precipitation on the forest vegetation and crop land (Prenni et al., 2009; Yu et al., 2015). Therefore, the long-distance transport of bacteria in the aerosols influences both the spread of bacterial crop diseases and nucleation leading to rain in the Amazonia (*see* Table 3.1).

Dust storms originating from Syria, Saudi Arabia, and North Africa have the potential to impact the microbial diversity of the atmosphere. These dust storms travel long distances across the Mediterranean Sea to reach southern Europe (Goudie, 2009; Achilleos et al., 2020). Studies conducted by Peng (2021) have clearly demonstrated that the location, intensity of particulate matter (PM), size of PM, and organic matter composition in PM influence the microbial diversity. Their investigation revealed that dust particles originating from extremely dry conditions in Syria contained fungi that were able to tolerate such xerophilic conditions (e.g., Cladosporiales). Particles smaller than 10 μm in size harbored relatively fewer fungi. Consequently, dust storms abundant in smaller PM (< 2.5 μm) predominantly exhibited a diverse range of bacterial species. Gat et al. (2017) examined the PM of these storms using 16S rRNA sequences and discovered a wide variety of bacterial genera in fine particles. Often, these dust storms transport microbes (fungi, bacteria, archaea, metazoans) for distances of 800 km to 1,000 km, particularly during dry weather conditions. Dust storms with larger PM (>10 μm) were characterized by a predominance of diverse fungi. The population of microscopic metazoans showed a negative correlation with the concentration of PM_{10} μm.

TABLE 3.1 Long Distance Migration of Microbes on Aeolian Dust Plumes/Storms: A Few Examples

North African Dust Migrates to Caribbean and North American Agrarian Regions

Aeolian dust laden with microbes is generated in the Sahara Desert and Sahelian regions of Morocco, Mauritania, Algeria, Tunisia, and Senegal. This trans-Atlantic aeolian dust, carrying microbes with appreciable genetic diversity, reaches the Virgin Islands, Caribbean Islands, Cuba, Florida, Texas, and Kansas.

The migratory distance of microbes is approximately 9,000 to 15,000 km. Dust plumes occur during the dry seasons of summer and winter. At high altitudes of 3,000 ft (jet-streams), it takes about 7-15 days to cover the distance before the deposition of dust plumes in North America. Microbial species found in the migratory dust plumes survive the long transit period and distance (Acosta-Martinez et al., 2015; Griffin et al., 2001, 2003; Prospero, 1999; Prospero et al., 2005; Moulin et al., 1997).

Migration of Aeolian Dust and Microbes from Sahara/Sahel to Amazonia in South America

Microbe-loaded dust and sand particles from Western Sahara and Sahel (Senegal, Morocco, Mauritania) are known to transit long distances to the Brazilian east coast and Cerrados. Furthermore, aeolian dust continues its journey to reach locations in the Amazonian forests (Prenni et al., 2009; Yu et al., 2015). These microbes travel 4,000–5,000 km, utilizing high altitude jet-streams as well as rapid winds at lower heights in the atmosphere. As a result, Brazilian cropping zones and the Amazonian Forest receive both atmospheric dust and the accompanying microbes.

Dust Plumes and Microbes Migrate from North Africa and West Asia over Mediterranean Sea to Reach Southern Europe

Microbes from Morocco, Southern Algeria, Libya, Sudan, Egypt, Iran, Iraq, and Saudi Arabia are transported by dust storms and plumes, as shown in Plate 4.1 in Chapter 4. Dust storms are generated by low and high-altitude winds, such as jet streams. These storms carry microbes to various locations, including southern European farmland (North Italy), the Alp's region, Southern Spain, and even the Pyrenees (Meola et al., 2015). The migratory distance of approximately 7,000-10,000 km is covered in 15 days. Migratory dust plumes occur during two seasons: the fallow season (winter) and the dry summer. The microbial species encountered in these dust plumes are diverse (Aalismail et al., 2020; Bory et al., 2003; Diaz et al., 2017; Gat et al., 2017; Hervas et al., 2009; NASA, 2021; Rosselli et al., 2015).

Microbe-laden dust travels for over 2,000-5,000 km across the Mediterranean Sea from locations in North Africa, Israel, and Syria into Southern Europe. Migration occurs for 2,000-3,000 km in the upper troposphere. Dust particle sizes range from PM < 2.5 (2.5 μm) up to PM10 (10 μm). The diversity of microbes depends on the source/origin of dust storms. Fungi were lower in diversity if PM size is $<$ PM10. Xerophilic fungi were present in dust emanating from the arid deserts of Syria (Peng et al., 2021).

Migration of Dust and Microbes from Eastern Africa to West Asia, Mediterranean, and Southern Europe

Bioaerosols from Kenyan farming regions traverse through the atmosphere until they reach Egypt, the Mediterranean region, and then southern Europe. They float over the eastern Mediterranean Sea (see Plate 4.1 in Chapter 4). For example, the Ug 99 isolate of P. graministritici, a fungal pathogen on wheat, travels a distance of 3,500 to 4,500 km along with high altitude dust storms.

Bioaerosols also migrate from Kenyan farming zones to Pakistan and the Indo-Gangetic cereal belts. Microbes migrate for 4,500 to 5,000 km at an altitude of 3,000 m. Migratory microbes include the dreaded Brown and Black Stem Rust fungi (urediospores) (Brown & Hovmoller, 2002).

TABLE 3.1 *(Continued)*

Asian Aeolian Dust Storms from Gobi and Taklamakan Migrate to Fareast and Japanese Farming Zones
Aeolian dust, along with microbes, is generated in the Gobi and Taklamakan deserts. It then migrates to Northeast China, the Korean Peninsula, and Japan.
The migration distance is approximately 3,000 km and is completed in 7–8 days. High altitude winds carry dust containing microbes that can be commensals or crop pathogens (Hua et al., 2007; Yamaguchi et al., 2012).
Migration of Asian Dust and Microbes to the Pacific Northwest of America via Far Eastern Regions of the Korean Peninsula and Japan.
Migratory dust, along with microbes generated in Central Asia, Mongolia (Gobi, Taklamakan), and China, travels to the Korean Peninsula and Japan. Subsequently, Asian dust and microbes embark on a trans-Pacific journey to Oregon, Washington State, and Idaho (Pacific Northwest) in the USA. The migratory distance spans approximately 8,000-10,000 km. It takes about 15 days for the dust streams to reach locations in Northwest America. Dust plumes in the upper troposphere carry a diverse range of bacteria, fungi, and archaea into the pasture and grass seed production zone in the Pacific Northwest (Duce et al., 1980; Smith et al., 2012, 2013; Duan et al., 2021; Tang et al., 2018).
Australian Dust Plumes and Migration of Microbes
Aeolian dust plumes are generated from sandy terrain, lake beds, dust, and sediments from crusts of dried water bodies and other regions, and they reach the lower troposphere. These plumes can travel for short or long distances before depositing on cropping zones in southern Australia. The migratory distances range from 1,000 km to 2,000 km. It takes approximately 7–8 days for the dust plumes to be transported through this distance (Abed et al., 2012).
Global Circum-Migration of Microbes on Aeolian Dust
Microbes traverse the entire circumference of the Earth and get deposited at the place of dust generation or levitation into the atmosphere. The migratory distance is approximately 25,000 miles (42,000 km). It may take a month or more to circumnavigate the Earth at high altitudes in the upper troposphere. Spore traits such as size, structure, spore wall features, weight, buoyancy, survival period, resistance to radiation, tolerance to temperature, and desiccation are important (Uno et al., 2009).

Note: Microbial migration occurs above all continents of the world. The data above proves that long-distance aeolian transport is a global phenomenon of great significance to general weather patterns and the productivity of expanses of natural vegetation. Microbial migration has important consequences for the upkeep of agrarian regions and their productivity. In particular, the spread of diseases, epidemics, and pandemics are dependent, to a certain extent, on the aerial migration of microbes. Weather patterns, nucleation, and precipitation events are influenced by the migration of dust and microbes. Microbial migration also adds genetic diversity and improves opportunities for wide crosses of microbes. Overall, global dust/microbial migration has immense influence on atmospheric processes, nutrient dynamics, and biodiversity, including microbial diversity, above crop production zones.

Rosselli et al. (2015) state that annually, dust plumes and storms generated in the North African region comprising Tunisia, Algeria, and Libya migrate to Southern Europe. Regions in Italy and vicinity are target zones where the dust particles deposit after transit, in 2–3 days (Escudero et al., 2005; Querol et al., 2009). Furthermore, in this region of the globe, they state that 80 to 120 Tg of PM reaches Southern Europe from the Mediterranean each year. Dust mass travels a distance of over 1,000 km to 1,500 km. The wind and jet streams at altitudes of 8 km above the ground mediate the rapid migration of dust plus microbes (D'Almeida et al., 1986; Dulac et al., 1996). Dust events that reach southern Europe occur during May to November. They may also occur during other periods of the year. Therefore, microbes capable of initiating epidemics/pandemics are expected to reach Southern European farming zones in this same period (Israelevich et al., 2012; Table 3.1). The occurrence and transit routes of dust storms were reported earlier. However, the microbial composition of dust storms and their specific ability to affect crops are not available.

Microbes associated with $PM_{2.5}$ (2.5 μm) and PM_{10} (10 μm) exhibited a diverse species composition. PM could potentially transfer microbes capable of inducing epidemics in crops. Cultural methods were employed to identify microbes on aeolian dust collected in the Mediterranean midway (Sardinia). The study revealed that the microbial diversity was significant enough to noticeably alter the *in situ* microbial composition at the deposition site (Bovallius et al., 1978b; Rosselli et al., 2015). Currently, methods based on 16S rRNA are utilized to rapidly determine microbial diversity on PM 2.5 μm and PM 10 μm.

Roselli et al. (2015) have reported several microbial species occurring on dust that reaches midway (Sardinia) of the Mediterranean Sea and into Italian farming regions. These include *Alpha-* and *Gammaproteobacteria, Sphingomonas, Paracoccus, Klebsiella, Pantoea, Stenotrophomonas,* etc. Additionally, several fungi, protozoa, archaea, and viruses also migrate over this distance on dust particles. This migration actually leads to a certain level of uniformity in microbial composition across different regions on Earth. It is necessary to investigate the movement of beneficial microbes such as N-fixing *Bradyrhizobia, Azotobacter,* Ecto- and AM Mycorrhizas, as well as those microbes that aid in nutrient transformations.

Analysis of dust using 16S rRNA, culturing on selective medium if available, and other methods showed that the area of origin of the storm is an important factor (Gat et al., 2017). It determines, to a great extent, the diversity and population of microbes detected in dust plumes from North

Africa (Morocco), West Asian locations in Syria, Israel, and Turkey. Clearly, microbes (bacteria, fungi, and archaea) detected in samples from North Africa or Syria differed from those originating from Saudi Arabia. As global drying and warming increase, the effects on dust plumes and their microbial load and diversity are expected to be severe. Perhaps, only more resistant mutants of microbes may survive the journey from North Africa (Sahara) to the Eastern Mediterranean. The selection pressure is applied by the atmosphere during the approximately 3,000 km transit above the Mediterranean Sea. It may determine the microbial species that survive, deposit, and regenerate (Alpert et al., 2016; Schlesinger et al., 2006; Zittis, 2016; Krasnov, 2016). A few observations indicate that in Israel, extreme temperatures may generate more sandstorms. This, in turn, induces microbial migration from the desert region. Incidentally, the proportion of dust storms from the Sahara is higher than that from other locations. As stated earlier, the Sahara and Sahel regions are the world's largest dust generating zones, followed by the Gobi and Taklamakan in Mongolia (Table 3.1).

Asian dust migration is a significant natural phenomenon with profound implications for microbial dispersal. The dust plumes originating from the Central Asian, North Chinese, and Mongolian regions and reaching the Pacific Northwest Cropland are commonly referred to as 'Asian Dust' (Tang et al., 2018). These dust storms originate in the dry regions (deserts) and croplands of Central Asia, and are known to travel all the way to Japan and then to the Pacific Northwest. The dust deposits into forest stands and farming regions of Oregon and Washington State in the USA (Prospero et al., 2005; Griffin et al., 2001, 2003). Microbes become airborne and reach higher altitudes in the troposphere, eventually entering jet streams and being transported along with the dust particles. Microbes present in the bioaerosols can travel distances of approximately 3,000–10,000 km (Duce et al., 1980; Duan et al., 2021). Kellog & Griffin (2006) have reported that microbes can travel around 15,000 km from farming locations in the Far East to grasslands in the Northwest Pacific. It is possible that these microbes deposit in Northwest American farms, where they can survive, multiply, and establish themselves as native microflora.

Researchers have detected good phylogenetic diversity among bacteria migrating on Asian dust particles (Yamaguchi et al., 2012). Aeolian dust microbes, such as Sphingobacteria, are exposed to extreme atmospheric conditions during migration and should possess characteristics to overcome these challenges. Upon deposition, they must also adapt to survive in different terrains and ecosystems. Larger particles tend to deposit after shorter transits, while lighter particles may continue their journey in high-altitude atmospheric conditions. Aeolian dust in parts of Gobi and Taklamakan is generated in salty, dry, sandy zones. The survival of dust microbes depends on the presence of congenial terrains at the point of deposition. However, if aeolian dust originates from wheat fields in North China, these microbes may need to find vegetated and cropped fields to survive.

Smith et al. (2013) state that collecting aerosol samples from high altitudes is challenging. Several studies have shown a significant number of bacterial and archaeal genera present in the air. The researchers conducted 16S rRNA analysis to identify the microbes. Their investigation focused on *Proteobacteria, Actinobacteria,* and *Firmicutes* found at an altitude of 2.8 km above the natural vegetation and agricultural zone in the Pacific Northwest. Interestingly, these samples were collected from dust plumes that had undergone long-distance migration, traveling over the Pacific Ocean. The researchers suggest that inter-continental dust plumes originating from Asian locations generally exhibit a rich diversity of microbes.

A certain extent of microbial migration in the troposphere above the Indian subcontinent has been studied. The focus has been on deciphering the migratory patterns of fungal and bacterial pathogens that cause crop diseases. These studies were first initiated during the early 20th century (1920s, 1930s). One notable example is the discovery of the migratory route of the wheat stem/leaf rust fungus (*Pgt*) from the Southern Indian plateau and hills (Nilgiris) to the Gangetic Plains, extending up to the Terai and Shivalik hills. The fungal spores (teliospores and hyphae) are able to withstand harsh atmospheric conditions. These fungal spores transit over a distance of 2,500 km to 3,000 km every year during the fallow period. The spores germinate afresh each season when the wheat crop is initiated. It appears that collateral hosts for the *Pgt* harbor spores in the natural vegetation and sparse wheat crops of southern India (*see* Nagarajan & Joshi, 1985; Nagarajan & Singh, 1990).

3.7 MIGRATION OF MICROBES IN STRATOSPHERE AND BEYOND

According to DasSarma & DasSarma (2017), the stratosphere represents the altitudinal limits for microbial life on Earth. Microbes that are tolerant to stratospheric conditions, such as desiccation, extreme cold, and radiation, are still capable of migrating from one location to another. These microbes also undergo vertical journeys from the stratospheric heights to the ground surface, terrains, natural vegetation, and most importantly, the agrarian belt on Earth. Their downward vertical migration to the soil surface and crop canopy may induce genetic crossing with the *in situ* microflora. This phenomenon needs to be examined with great accuracy and care. It is believed that the stratosphere allows these microbes to migrate longer distances horizontally. Overall, DasSarma & DasSarma (2017) believe that the stratospheric microflora that survives the harsh conditions may have some influence on the weather patterns of Earth. They may also be the ones that escape Earth's atmosphere and reach space. Perhaps, they continue their migration as long as it is possible.

Microbial migration in the stratosphere excessively depends on their ability to endure the harsh conditions. For example, in an experiment on microbial survival in the stratosphere, only 267 bacterial spores survived out of the 40 million spores injected at high altitude. Therefore, survival is a crucial aspect for any short or long-distance migration in stratospheric conditions (Khodadad et al., 2017). The survival of microbes is the central issue concerning dispersal and long-distance migration in the stratosphere (Smith et al., 2009, 2011).

Aerial flotation is perhaps an important feature of migratory microbes. In the absence of strong winds, convection currents, and jet streams, it is possible that vertical movement becomes conspicuous. Buoyancy and tolerance to harsh environments are essential for stratospheric microbes to migrate. Deposition on agrarian soils, crop canopies, and vegetation may allow intermixing and recycling of soil microbes into the stratosphere. This process depends on breezes, winds, the extent of turbulence, and other factors that lift soil microbes to the upper layers of the atmosphere, specifically at altitudes of 50 km and above.

Microbiologists have analyzed the surfaces of spacecraft returning to Earth after a trip. The most frequently traced bacteria on the spacecraft's body surface are *Acinetobacter, Alcaligenes, Bacillus, Corynebacterium-Brevibacterium* group, *Flavobacterium, Micrococcus, Staphylococcus,* and *Streptococcus* (Nicholson et al., 2018; La Duc, 2004, 2007; Schuerger, 2004). Fungi encountered on spacecraft include *Alternaria, Aspergillus, Bipolaris, Candida, Cladosporium, Fusarium, Mucor, Penicillium,* and *Trichoderma* (Nicholson et al., 2018). The estimated microbial load on the spacecraft's surface ranged from 10^4 to 10^8 microbes (Schuerger, 2004; Nicholson et al., 2018).

We have imagined, guessed, and forecasted about space and interplanetary migration of microbes. We have tried to examine it scientifically. However, we have not studied it to any extent, even to guess accurately about space microbiology and its impact on earthly agrarian regions. Astrobiology is supposedly a recent approach to studying microbes capable of withstanding and adapting to conditions that might prevail on Earth, interplanetary space, and other planetary bodies. For the present, astrophysicists and space travel experts believe that Mars and a few moons like Europa and Titan may have conditions that could support microbes (Hall, 2001). Perhaps, we should conduct a thorough search for highly resistant versions of microbial species and attempt to cultivate them in harsh space or simulated space-like conditions. These microbes, intended for examination in space, may be selected from any ecosystem conditions prevalent on Earth, including dryland agricultural zones, totally arid deserts, or deep-water bodies. Undoubtedly, they could also be from the atmosphere above agrarian regions. Such microbes are known as 'extremophiles' – microorganisms adapted to thrive in relatively inhospitable conditions such as heat, cold, radiation, and pressure. There are chemolithotrophs that may tolerate and transit long distances in the inhospitable stratosphere and space-like conditions. Microbiologists are evaluating bacteria, fungal spores, and archaea that return after a journey to the stratosphere, testing their ability to thrive in a range of environmental conditions. The question here is how they will function in the agricultural sky.

We do not have any authentic references to quote regarding the natural dispersal or transit of microbes from one celestial body to another. While we may have several forecasts and inferences based on a few simulations, the physicochemical conditions in the stratosphere, mesosphere, and space may not support extensive monitoring or research on these microbes. We have no idea about their influence on agricultural crops or their growth and productivity upon deposition. Discussions on the interplanetary migration of microbes seem to be beyond the scope of this volume. However, there is a prerequisite. We should first devise and standardize methods that differentiate between microbes native to Earth's atmospheric layers and any potential alien forms.

In addition to the above examples of intercontinental migration of microbes, there is also intracontinental movement of dust and microbes, together. These movements also have impacts on agrarian regions. For instance, the movement of dust and microbes from the Texan plains to North Dakota and Alberta within North America, as well as the movement of atmospheric dust and microbes from the Deccan Plateau in Southern India to the Gangetic Plains and Nepal in the Indian subcontinent. Furthermore, there is the migration of dust and microbes from the Atacama Desert in Chile to farming regions in the South American continent.

KEYWORDS

- **aeolian dust**
- **aerial migration**
- **biological nitrogen fixation**
- **clouds**
- **dust plumes**
- **epidemics**
- **extreme environment**
- **gobi**
- **intercontinental migration**
- **jet streams**
- **microbial highways**
- **pandemics**
- **precipitation**
- **Sahara**
- **Sahel**
- **Taklamakan**
- **wind**

REFERENCES

Aalismail, N. A., Díaz-Rúa, R., Ngugi, D. K., Cusack, M., & Duarte, C. M., (2020). Aeolian prokaryotic communities of the global dust belt over the red Sea. *Frontiers in Microbiology, 11*, 538476. doi: 10.3389/fmicb.2020.538476/.

Abed, R. M. M., (2012). Microbial diversity of aeolian dust sources from saline lake sediments and biological soil crusts in arid Southern Australia. *FEMS Microbiology Ecology, 80*, 294–304.

Achilleos, S., Mouzourides, P., Kalivitis, N., Katra, I., Kloog, I., Kouis, P., Middleton, N., et al., (2020). Spatio-temporal variability of desert dust storms in Eastern Mediterranean (Crete, Cyprus, Israel) between 2006 and 2017, using a uniform methodology. *Science: Total Environment, 714*, 136693.

Acosta-Martinez, V., Van, P. S., Moore-Kucera, J., Baddock, M. C., & Zobeck, T. M., (2015). Microbiology of wind eroded sediments: Current knowledge and future. *Aeolian Research, 18*, 99–113.

Addiss, D., Davis, J., Laventure, M., Wand, P., Hutchinson, M., & McKinney, R., (1989). Community-acquired legionnaires' disease associated with a cooling tower: Evidence for longer-distance transport of *Legionella pneumophila*. *American Journal of Epidemiology, 130*, 557–568.

Alicai, T., Szymiszewska, A. M., Omongo, C. A., Abidrabo, P., Kko-Okuja, Baguma, Y., Ogwok, E., et al., (2019). Expansion of the cassava brown streak pandemic in Uganda revealed by annual field survey data for 2004 to 2017. *Nature Scientific Data, 6*, 327https://doi.org/10.1038/s41597-019-0334-9.

Alpert, P., Egert, S., & Uzan, L., (2016). The 7−13 September extreme dust downfall over the East Mediterranean: A two ceilometers study. *EGU General Assembly Conference Abstracts, 18*, 3788.

Amato, P., (2012). Clouds provide atmospheric oases for microbes. *Microbe.,7*(3), 119–123.

Anikster, Y., Eilam, T., Bushnell, W. R., & Kosman, E., (2005). Spore dimensions of *Puccinia* species of cereal hosts as determined by image analysis. *Mycologia, 97*(2), 474–484. doi: 10.3852/mycologia.97.2.474./.

Antignus, V., Lachman, O., & Pearlsman, M., (2007). Spread of tomato apical stunt viroid (TAVSd) in greenhouse tomato crops is associated with seed transmission and bumble activity. *Plant Disease, 91*, 47–50.

Aparcio, F., Sanchez-Pina, M. A., Sanchez-Navarro, J. A., & Pallas, V., (1999). Location of Prunus necrotic ringspot ilavirus within pollen grains of infected nectarine trees: Evidence from Rt-PCR dot-blot and *in situ* hybridization. *European Journal of Plant Pathology, 105*, 623–627. 10.1023/A:1008783705183/.

Aylor, D. E., & Lukens, R. G., (1974). Liberation of *Helminthosporium maydis* spores by wind in the field. *Phytopathology, 64*, 1136–1138.

Aylor, D. E., (1990). The role of intermittent wind in dispersal of fungal pathogens. *Annual Review of Phytopathology, 28*, 73–92.

Baas-Becking, L. G. M., (1934). *Geobiologie of Inleiding Tot de Milieukunde* (p. 147). W. P. Van Stockum and Zoon N. V., The Hague, The Netherlands. https://www.worldcat.org/title/geobiologie-of-inleiding-tot-de-milieukunde/oclc/44189670 (accessed on 19 January 2024).

Behzad, H., Mineta, K., & Gobori, T., (2018). Global ramifications of dust and sandstorm microbiota. *Genome Biology and Evolution, 10*, 1970–1987. doi: 10.1093/gbe/evy134/.

Beijerinck, M. W., (1913). *De Infusiesen de Ontdekking der Backterie¨n, Jaarboek Van de Koninklijke Akademie v* (p. 79). Wetenschappen. Mu¨ller, Amsterdam, The Netherlands.

Benitez, L. B., Velho, R. V., & Lisboa, M. P., (2000). Isolation and characterization of antifungal peptides produced by *Bacillus amyloliquefaciens* LBM5006. *Journal of Microbiology, 48*(6), 791–797.

Blackman, R., & Eastop, V., (2000). *Aphids on the World's Crops: An Identification and Information Guide* (2nd edn., p. 414). Chichester-Wiley.

Bory, A. J. M., et al., (2003). Two distinct seasonal Asian source regions for mineral dust deposited in Greenland (NorthGRIP). *Geophysical Research Letters, 30*, 1167–1171.

Bottenberg, H., & Irwin, M. E., (1991). Influence of wind speed on residence time of *Uroleuconambrosiae* alatae (*Homoptera: Aphididae*) on bean plants in bean monocultures and bean-maize mixtures. *Environmental Entomology, 20*, 1375–1380.

Bovallius, A., Bucht, B., Roffey, R., & Anas, P., (1978a). Three-year investigation of the natural airborne bacterial flora at four localities in Sweden. *Applied and Environmental Microbiology, 35*, 847–852.

Bovallius, A., Bucht, B., Roffey, R., & Anas, P., (1978b). Long-range air transmission of bacteria. *Applied and Environmental Microbiology, 35*, 1231, 1232.

Bové, J. M., (2006). Huanglongbing: A destructive, newly emerging, century-old disease of citrus. *Journal of Plant Pathology, 88*, 7–37.

Boylan-Pett, W., Ramsdell, D. C., Hoopingarner, R. A., & Hancock, J. F., (1991). Honeybee foraging behavior, in hive survival of infectious pollen borne blue berry leaf mottle virus and transmission, in highbush blue berry. *Phytopathology, 81*, 1407–1412.

Brimner, T. A., & Boland, G. J., (2003). A review of the non-target effects of fungi used to biologically control plant diseases. *Agriculture Ecosystem and Environment, 100*(1), 3–16. doi: 10.1016/S0167-8809(03)00200-7/.

Brlansky, R. H., & Rogers, M. E., (2007). *Citrus Huanglongbing: Understanding the Vector-Pathogen Interaction for Disease Management* (pp. 1–5). APS net Feature Story. doi: 10.1094/APSnetFeature-2007-1207/. https://www.apsnet.org/edcenter/apsnetfeatures/Pages/Huanglongbing.aspx (accessed on 19 January 2024).

Brown, A. D., (1953). The survival of airborne microorganisms: 1. Experiments near 0°C with psychrophilic bacteria. *Australian Journal of Biology Science, 6*(3), 463.

Brown, J. E., Dangler, J. M., Woods, F. M., Tilt, K. M., Henshaw, M. D., Griffey, W. A., & West, M. S., (1993). Delay in mosaic-virus onset and aphid vector reduction in summer squash grown on reflective mulches. *Hortscience, 28*, 895–896.

Brown, J. K. M., &, & Hovmoller, M. S., (2002). Aerial dispersal of pathogens on the global and continental scales and its impact on plant disease. *Science, 297*, 537–541.

Brown, R. M., Larson, D. H., & Bold, H. C., (1964). Airborne algae: Their abundance and heterogeneity. *Science, 143*, 583–585.

Brown, S. P., & Jumponnen, A., (2019). Microbial ecology of snow reveals taxa specific biogeographical structure. *Microbial Ecology, 77*, 946–958.

Burrows, S. M., Butler, T., Jöckel, T. P. H., Kerkweg, A., Pöschl, U., & Lawrence, M. G., (2009a). Bacteria in the global atmosphere. Part 2. Modeling of emissions and transport between different ecosystems. *Atmospheric Chemistry and Physics, 9*, 9281–9297.

Burrows, S. M., Elbert, W., Lawrence, M. G., & Posch, U., (2009b). Bacteria in the global atmosphere- Part 1. Review and synthesis of literature data for different ecosystems. *Atmospheric Chemistry and Physics, 9*, 9263–9280.

Burt, P. J. A., (2002). Airborne dispersal of *Mycosphaerella fujiense* in *Mycosphaerella* leaf spot diseases of bananas: Present status and outlook proceedings 2nd international workshop on *Mycosphaerella* leaf spot diseases, held in San Jose, Costa Rica. In: Jacome, L., (ed.), *The International Network for the Improvement of Banana and Plantain* (pp. 111–122). Montpellier, France.

Card, S. D., Pearson, M. N., & Clover, G. R. G., (2007). Plant pathogens transmitted by pollen. *Australasian Plant Pathology, 36*, 455–461.

Caroll, T. W., & Mayhew, D. E., (1976). Anther and pollen infection in relation the pollen and seed transmissibility of two strains of barley stripe mosaic virus in barley. *Canadian Journal of Botany, 54*, 1604–1621.

Carvalho, C. R., Fernandes, R. C., Carvalho, G. M. A., Barreto, R. W., & Evans, H. C., (2011). Crypto-sexuality and the genetic diversity paradox in coffee rust, *Hemileiavastatrix* (pp. 1–12). *PLoS One, 6*, e26387.

Chagnon, P., Bradley, R. L., Maherali, H., & Klironomos, J. N., (2013). A trait-based framework to understand life history of mycorrhizal fungi. *Trends in Plant Science, 18*, 484–491. doi: 10.1016/j.tplants.2013.05.001/.

Chu, J., (2017). *A Light Rain Can Spread Soil Bacteria Far and Wide, Study Finds* (pp. 1–5). Massachusetts Institute of Technology, Boston, MA, USA. https://news.mit.edu/2017/light-rain-spread-soil-bacteria-0307/ (accessed on 19 January 2024).

Compton, S., (2002). Sailing with the wind: Dispersal by small flying insects. In: Bullock, J. M., Kenward, R. E., & Hails, R. S., (eds.), *Dispersal Ecology: The 42nd Symposium of the British Ecological Society Held at the University of Reading* (pp. 113–133). Oxford: Blackwell Science.

Comstock, J. C., Shine, J. M. Jr., & Raid, R. N., (1992). Effect of rust on sugarcane growth and biomass. *Plant Disease, 76*, 175–177.

Cory, L., & Hewitt, W. B., (1968). Some grapevine viruses in pollen and seed. *Phytopathology, 58*, 1316–1319.

D'Almeida, G. A., (1986). A model for Saharan dust transport. *Journal of Applied Meteorology and Climatology, 25*, 903–916.

Da Graça, J. V., & Korsten, L., (2004). Citrus huanglongbing: Review, present status, and future strategies. In: Naqvi, S. A. M. H., (ed.), *Diseases of Fruits and Vegetables* (Vol. I, pp. 229–245). Kluwer Academic Press, Dordrecht, The Netherlands.

Da Graça, J. V., (1991). Citrus greening disease. *Annu. Rev. Phytopathol., 29*, 109–136.

DasSarma, P., & DasSarma, S., (2017). Survival of microbes in earth's stratosphere. *Current Opinion in Microbiology, 43*, 24–30. https://doi.org/10.1016/j.mib.2017.11.002/.

Datnoff, L. E., Nemec, S., & Pernezny, K., (1995). Biological control of *Fusarium* crown and root rot of tomato in Florida using *Trichoderma harzianum* and *Glomus intraradices*. *Biological Control, 5*(3), 427–431. doi: 10.1006/bcon.1995.1051/.

Delort, M., Vaitilingom, M., Amato, P., Sancelme, M., Parazols, M., Maihot, G., Laj, P., & Deguillaume, L., (2010). A short review of the microbial population in clouds: Potential roles in atmospheric chemistry and nucleation processes. *Atmospheric Research, 98*, 249–260. https://doi.org/10.1016/j.atmosres.2010.07.004/.

Department of Environment, Food and Rural Affairs, (2014). *Pest Alert: Zebra Chip Disease of Potato* (pp. 1–3). Plant health Controls, UK. https://planthealthportal.defra.gov.uk/assets/uploads/Zebra-chip-pest-alert-v8.pdf (accessed on 19 January 2024).

Desjardins, P. R., Drake, R. J., Sasaki, P. J., Atkins, E. L., & Bergh, B. O., (1984). Pollen transmission of avocado sunblotch viroid and the fate of the pollen recipient trees. *Phytopathology, 74*, 845–848.

Díaz, J., Linares, C., Carmona, R., Russo, A., Ortiz, C., & Salvador, P., (2017). Saharan dust intrusions in Spain: Health impacts and associated synoptic conditions. *Environmental Research, 156*, 455–467.

Dillon, K., & Fennel, D. E., (2021). *Bacteria and Blue Green Algae Get Rides on Clouds* (pp. 1–4). School of Environment and Biological Sciences, Rutgers University, New Brunswick, NJ, USA. https://www.rutgers.edu/news/bacteria-and-algae-get-rides-clouds (accessed on 19 January 2024).

Duan, X., Guo, C., Zhang, C., Li, H., Zhou, Y., Gao, H., Xia, X., et al., (2021). Effect of East Asian atmospheric particulate matter deposition on bacterial activity and community structure in the oligotrophic Northwest Pacific. *Environmental Pollution, 283*, 117088. https://doi.org/10.1016/j.envpol.2021.117088/.

Duce, R. A., (1980). Long-range atmospheric transport of soil dust from Asia to the tropical North Pacific: Temporal variability. *Science, 209*, 1522–1524.

Dulac, F., (1996). Quantitative remote sensing of African dust transport in the Mediterranean. In: Guerzoni, S., (ed.), *The Impact of Desert Dust Across the Mediterranean* (pp. 25–49). Kluwer Academic Publishers, Dordrecht, Netherlands.

Egan, E., li, D. W., & Klironomos, J., (2014). Detection of arbuscular mycorrhizal fungal spores in the air across different biomes and eco regions. *Fungal Ecology, 12*,1–7. doi: 10.1016/j.funeco.2014.06.004/.

Escudero, M. S., (2005). Wet and dry African dust episodes over eastern Spain. *J. Geophys. Res., 110*, D18S08. doi: 10.1029/2004JD004731/.

Everts, S., (2008). *Bacteria in Clouds* (Vol. 86, No. 15, pp. 23–28). Chemical Engineering News. https://cen.acs.org/articles/86/i15/Bacteria-Clouds.html/ (accessed on 19 January 2024).

FAO, (1981). Food and Agricultural Organization of the United Nations, Rome, Italy. *Blue Green Algae for Rice Production* (pp. 1–99).

Finlay, B. J., (2002). Global dispersal of free-living microbial eukaryote species. *Science, 296*, 1061–1063.

Freeman, S., Minz, D., Kolesnik, I., Barbul, O., Zveibil, A., Maymon, M., et al., (2004). Trichoderma biocontrol of *Colletotrichum acutatum* and *Botrytis cinerea* and survival in strawberry. *European Journal of Plant Pathology, 11*(4), 361–370. doi: 10.1023/B:EJPP.0000021057.93305.d9.

Gat, D., Mazar, Y., Cytryn, E., & Rudich, Y., (2017). Origin-dependent variations in the atmospheric microbiome community in Eastern Mediterranean dust storms. *Environmental Science and. Technology, 51*, 6709–6718. doi: 10.1021/acs.est.7b00362/.

Geagea, L., Huber, L., & Sache, I., (1999). Dry-dispersal and rain-splash of brown (*Puccinia recondita f. sp. tritici*) and yellow (*P. striiformis*) rust spores from infected wheat leaves exposed to simulated raindrops. *Plant Pathology, 48*, 472–482.

Gilbert, G. S., & Reynolds, D. R., (2005). Nocturnal fungi: Airborne spores in the canopy and understory of a tropical rain forest. *Biotropica, 37*, 462–464.

Gitatis, R., & Pathak, S., (1988). Airborne dispersal of bacteria in tomato and pepper transplant fields. *Plant Disease*, 575–579. https://www.researchgate.net/publication/256598128/ (accessed on 19 January 2024).

Gloster, J., Blackall, R., Sellers, R., & Donaldson, A., (1981). Forecasting the airborne spread of foot-and-mouth disease. *Veterinary Research, 108*, 370–374.

Gloster, J., Sellers, R., & Donaldson, A., (1982). Long distance transport of foot-and-mouth disease virus over the sea. *Veterinary Research, 110*, 47–52.

Goudie, A. S., & Middleton, N. J., (2001). Saharan dust storms: Nature and consequences. *Earth Science Reviews, 56*, 179–204.

Goudie, A. S., (2009). Dust storms: Recent developments. *Journal of Environment Management, 90*, 89–94.

Goudie, A., & Middleton, N., (2006). *Desert Dust in the Global System* (p. 287). Berlin; New York, NY: Springer. Berlin, Heidelberg, Germany.

Graham, D. C., & Harrison, M. D., (1975). Potential spread of *Erwinia spp.* in aerosols. *Phytopathology, 65*, 730–741.

Graham, D. C., Quinn, C. E., & Bradley, L. F., (1977). Quantitative studies on the generation of aerosols of *Erwinia carotovora var. atroseptica* by raindrop impaction on blackleg infected potato stems. *Journal of Applied Bacteriology, 43*, 413–424.

Gregory, P., (1971). The Leeuwenhoek Lecture-1970. Airborne microbes: Their significance and distribution. *Proceedings of Royal Society B- Biology, 177*, 469–483.

Gregory, P., (1973). *The Microbiology of the Atmosphere* (p. 377). Leonard Hill, Aylesbury, UK. doi: https://doi.org/10.5962/bhl.title.7291/.

Griffin, D. W., Garrison, V. H., Herman, J. R., & Sinn, E. A., (2001). African desert dust in the Caribbean atmosphere: Microbiology and public health. *Aerobiologia. 17*, 203–213.

Griffin, D. W., Kellog, C. A., Garrison, V. H., Lisle, J. T., Borden, T. C., & Sinn, E. A., (2003). Atmospheric microbiology in the northern Caribbean during African dust events. *Aerobiologia, 19*, 143–157.

Guetsky, R., Shtienberg, D., Elad, Y., & Dinoor, A., (2001). Combining biocontrol agents to reduce the variability of biological control. *Phytopathology, 91*, 621–627. doi: 10.1094/PHYTO.2001.91.7.621/.

Haine, E., (1955). Aphid take-off in controlled wind speeds. *Nature, 175*, 474, 475.

Halimi, B., Dortu, C., & Arguelles-Arias, A., (2010). Antilisterial activity on poultry meat of amylolysin, a bacteriocin from *Bacillus amyloliquefaciens*GA1. *Probiotics: Antimicrobial Proteins, 2*(2), 120–125.

Hall, J., (2001). *Astrobiology: The Sky's No Limit* (pp. 1–4). Science Magazine. https://www.sciencemag.org/careers/2001/11/astrobiology-skys-no-limit (accessed on 19 January 2024).

Hammami, I., Jaouadi, B., & Bacha, A. B., (2012). *Bacillus subtilis* bacteriocin Bac 14B with a broad inhibitory spectrum: Purification, amino acid sequence analysis, and physicochemical characterization. *Biotechnology and Bioprocess Engineering, 17*(1), 41–49.

Heintzen, C., & Liu, Y., (2007). The *Neurospora crassa* circadian clock. *Advances in Genetics, 58*, 25–66.

Hendrie, L. K., Irwin, M. E., Liquido, N. J., Ruesin, W. G., Mueller, E. A., Voegtlin, D. J., Achtemeier, G. L., et al., (1985). Conceptual approach to modeling aphid migration. In: MacKenzie, D. R., Barfield, C. S., Kennedy, G. C., & Smith, J. C., (eds.), *The Movement and Dispersal of Agriculturally Important Biotic Agents.* Rouge, LA: Claitor's Publishing Division USA.

Hennings, J., & Lynch, H., (2018). *Cloud Microbes* (pp. 1–4). Earth Date. https://www.earthdate.org/cloud-microbes/ (accessed on 19 January 2024).

Herman, G. E., (2006). Overview of mechanisms and uses of *Trichoderma* spp. *Phytopathology, 96*(2), 190–194. doi: 10.1094/PHYTO-96-0190/.

Hervàs, A., Camarero, L., Reche, I., Casamayor, E., Hervàs, A., & Camarero, L., (2009). Viability and potential for immigration of airborne bacteria from Africa that reach high mountain lakes in Europe. *Environmental Microbiology, 11*, 1612–1623. doi: 10.1111/j.1462-2920.2009.01926.x.

Hirst, J. M., (1959). Spore liberation and dispersal. In: Holton, C. S., Fischer, G. W., Fulton, R. W., Hart, F. A., & McCallan, S. E. A., (eds.), *Plant Pathology: Problems and Progress 1908–1959* (pp. 529–538). The University of Wisconsin, Madison, Wisconsin, USA.

Hirst, J. M., Stedman, O. J., & Hogg, W. H., (1967). Long distance spore transport; methods of measurement, vertical spore profiles and the detection of immigrant spores. *Journal of General Microbiology, 48*, 329–355.

Hua, N. P., Kobayashi, F., Iwasaka, Y., Shi, G. Y., & Naganuma, T., (2007). Detailed identification of desert-originated bacteria carried by Asian dust storms to Japan. *Aerobiologia, 23*, 291–298.

Irwin, M. E., & Thresh, J. M., (1988). Long-range aerial dispersal of cereal aphids as virus vectors in north America. *Philosophical Transactions Royal Society London B. Biological Science, 321*, 421–446.

Irwin, M. E., Kampmeier, G., & Weisser, W., (2007). Aphid movement: Process and consequences. In: Van, E. H., & Harrington, R., (eds.), *Aphids as Crop Pests* (pp. 153–186). Wallingford: CABI, U.K.

Isard, S. A., Gage, S. H., Comtois, P., & Russo, J. M., (2005). Principles of the atmospheric pathway for invasive species applied to soybean rust. *Bioscience, 55*, 851–861. http://dx.doi.org/10.1641/0006-3568(2005)055 [0851:POTAPF]2.0.CO;2. /.

Isard, S. A., Irwin, M. E., & Hollinger, S. E., (1990). Vertical distribution of aphids (*Homoptera*: *Aphididae*) in the planetary boundary layer. *Environmental Entomology, 19*, 1473–1484.

Israelevich, P., Ganor, E., Alpert, P., Kishcha, P., & Stupp, A., (2012). Predominant transport paths of Saharan dust over the Mediterranean Sea to Europe. *Journal of Geophysics and Research on Atmosphere, 117*, D02205. doi: 10.1029/2011JD016482?.

Johnson, C. G., (1954). Aphid migration in relation to weather. *Biological Reviews, 29*, 87–118.

Johnson, C. G., (1969). *Migration and Dispersal of Insects by Flight* (p. 763). London: Methuen & Co Ltd. Great Britain.

Jones, A. M., & Harrison, R. M., (2004). The effects of meteorological factors on atmospheric bioaerosol concentrations: A review. *Science: Total Environment, 326*, 151–181.

Jones, R. A. C., (2021). Global plant virus disease pandemics and epidemics. *Plants (Basel), 10*(2), 233. doi: 10.3390/plants10020233.

Kamoun, F., Mejdoub, H., & Aouissaoui, H., (2005). Purification, amino acid sequence and characterization of Bacthuricin F4, a new bacteriocin produced by *Bacillus thuringiensis. Journal of Applied Microbiology, 98*(4), 881–888.

Kavitha, S., Senthilkumar, S., & Gnanamanickam, S., (2005). Isolation and partial characterization of antifungal protein from *Bacillus polymyxa* strain VLB16. *Process Biochemistry, 40*(10), 3236–3243.

Kellogg, C. A., & Griffin, D. W., (2006). Aerobiology and the global transport of desert dust. *Trendsin Ecology and Evolution, 21*, 638–644.

Kennedy, J. S., & Booth, B. O., (1963). Free flight of aphids in the laboratory. *Journal Experimental Biology, 40*, 67–85.

Kennedy, J. S., (1990). Behavioral post-inhibitory rebound in aphids taking flight after exposure to wind. *Animal Behavior, 39*, 1078–1088.

Khodadad, C., Gregory, W. M., Leandro, J. M., Thakrar, P. J., Micheal, L. A., Catechis, J. A., & Smith, D. J., (2017). Stratosphere conditions inactivate bacterial endospores from a mars spacecraft assembly facility. *Astrobiology, 17*(4), 337–350. doi:10.1089/ast.2016.1549/.

Kim, S., Park, H., Gruszewski, H. A., Schmale, D. G., & Sunghwan, J., (2019). Vortex-induced dispersal of a plant pathogen by raindrop impact. *Proceedings of National Academy of Science, 116*, 4917–4922. www.pnas.org/cgi/doi/10.1073/pnas.1820318116/ (accessed on 19 January 2024).

Kivlin, Hawkes, C. V., & Treseder, K. K., (2011). Global diversity and distribution of arbuscular mycorrhizal fungi. *Soil Biology and Biochemistry, 43*, 2294–2303. doi:10.1016/j.soilbio.2011.07.012.

Knudsen, G., (1989). Model to predict aerial dispersal of bacteria during environmental release. *Appl. Environmental Microbiology, 55*, 2641–2647.

Kramer, C. L., & Pady, S. M., (1968). Viability of airborne spores. *Mycologia, 60*, 448–449.

Krasnov, H., Katra, I., & Friger, M., (2016). Increase in dust storm related PM10 concentrations: A time series analysis of 2001−2015. *Environmental Pollution, 213*, 36−42.

Kulkarni, P., Baron, P. A., & Willeke, K., (2011). Introduction to aerosol characterization. In: *Aerosol Measurement: Principles, Techniques, and Applications* (3rdedn.). Hoboken, NJ: John Wiley & Sons, Inc. doi: 10.1002/9781118001684.ch1/.

La Duc, M. T., (2004). Microbial monitoring of spacecraft and associated environments. *Microbial Ecology, 47*, 150–158.

La Duc, M. T., (2007). Isolation and characterization of bacteria capable of tolerating the extreme conditions of clean room environments. *Applied and Environmental Microbiology, 73*, 260–2611.

Lebenonturf (2013). *Effects of Light Exposure on Mycorrhizal Spores* (pp. 1–7). https://www.lebanonturf.com/education-center/biological-plant-treatments/effects-of-light-exposure-on-mycorrhizal-fungi (accessed on 19 January 2024).

Lee, T. F., & Eggleston, P. M., (1989). Airborne algae and cyanobacteria. *Grana, 28*, 63–66. doi: 10.1080/00173138909431014/.

Legg, J. P., Jeremiah, S. C., Obiero, H. M., Maruthi, M. N., Ndyetabula, I., Okao-Okuja, G., Boumeester, H., et al., (2011). *Comparing the Regional Epidemiology of the Cassava Mosaic and Cassava Brown Streak Virus Pandemics in Africa,159*, 161–170. doi: 10.1016/j.virusres.2011.04.018.

Legg, J. P., Shirima, R., Tajebe, L. S., Guastella, D., Boniface, S., Jeremiah, S., Nsami, E., et al., (2014). Biology and management of *Bemisia* whitefly vectors of cassava virus pandemics in Africa. *Pest Management Science, 70*(10), 1446–1453. doi: 10.1002/ps.

Leslie, J. F., & Summerell, B. A., (2006). *The Fusarium Laboratory Manual* (pp. 1–388). Blackwell, Ames, IA. https://onlinelibrary.wiley.com/doi/book/10.1002/9780470278376/.

Lewis, T., & Taylor, L. R., (1964). Diurnal periodicity of flight by insects. *Transactions of Royal Entomological Society* (Vol. 116, pp. 393–479). London, U. K.

Liang, A., & Evans, J. L., (2011). *Introduction to Tropical Meteorology*. University Corporation for Atmospheric Research. Boulder, Colorado, USA, Version 4.0 https://www.meted.ucar.edu/training_module.php?id=738/ (accessed on 19 January 2024).

Lidwell, O., (1987). Joseph lister and infection from the air. *Epidemiology and Infection, 99*, 569–578.

Linde, C. C., Zhan, J., & McDonald, B. A., (2002). Population structure of *Mycosphaerella graminicola*: From lesions to continents. *Phytopathology, 92*, 946–955. http://dx.doi.org/10.1094/PHYTO.2002.92.9.946./.

Lindemann, J., Constantinidou, H., Barcheet, W. R., & Upperi, C. D., (1982). Plant as sources of airborne bacteria including ice nucleation. *Applied and Environmental Microbiology, 44*, 1059–1063.

Loxdale, H. D., Hardie, J., Halbert, S., Foottit, R., & Kidd, N. A. C., (1993). Carter CI: The relative importance of short- and long-range movement of flying aphids. *Biological Reviews, 68*, 291–311.

Luig, N. H., (1985). Epidemiology in Australia and New Zealand. In: Roelfs, A. P., & Bushnell, W. R., (eds.), *The Cereal Rusts* (Vol. 2, pp. 301–328). Academic Press, Orlando.

Mahaffee, W. F., & Stoll, R., (2016). The Ebb and flow of airborne pathogens: Monitoring and use of disease management decisions. *Phytopathology Review, 106*, 420–431.

Mann, J. A., & Harrington, R., (1996). *Key Factors for Modeling Secondary Spread of Barley Yellow Dwarf Virus* (Vol. 129, pp. 1–24). Home Grown Cereal Authority- Project report: HGCA, U. K.

Mayol, E., Arrieta, J. M., Jimenez, M. A., Martinez-Asensio, A., Garcias-Bone, N., Dachs, J., Gonzalez-Gaya, B., et al., (2017). Long-range transport of airborne microbes over the global tropical and subtropical ocean. *Nature: Communications, 8*, 201, 202.

McInnes, T. B., Gitaitis, R. D., McCarter, S. M., Jaworski, C. A., & Pathak, S. C., (1988). Airborne dispersal of bacteria in tomato and pepper fields. *Plant Disease, 72*, 575–579.

Mendgen, K., (1997). The uridinales. In: Caroll, G. C., (ed.), *The Mycota; Plant Relations Part B* (pp. 79–94). Springer, Berlin, Germany.

Meola, M., Lazzaro, A., & Zeyer, J., (2015). Bacterial composition and survival on Sahara dust particles transported to the European Alps. *Frontiers in Microbiology, 6*, 1454. doi: 10.3389/fmicb.2015.01454/.

Meredith, D. S., Lawrence, J. S., & Firman, I. D., (1973). Ascospore release and dispersal in black leaf streak disease of bananas. *Transactions of British Mycological Society, 60*, 547–554.

Michigan State University, (2018). *MSU Biologist Sheds Light on Fog's Role in Microbial Transport*. MSU Today.

Monteil, C. L., Bardin, M., & Morris, C. E., (2014). Features of air masses associated with the deposition of *Pseudomonas syringae* and *Botrytis cinerea* by rain and snowfall. *International Society of Microbial Ecology Journal,8*, 2290–2304.

Morris, C., & Sands, D., (2017). Impacts of microbial aerosols on natural and agro-ecosystems: Immigration, invasions, and their consequences. In: Delort, A. M., & Amato, P., *Microbiology of Aerosols* (pp. 269–279). John Wiley and Sons, Inc., Hoboken, New Jersey, USA.

Moulin, C., Lambert, C. E., Dulac, F., & Dayan, U., (1997). Control of atmospheric export of dust from North Africa by the North Atlantic oscillation. *Nature, 387*, 691.

Mundt, C. C., Sacket, K. E., Wallace, L. D., Cowger, C., & Dudley, J. P., (2009). Long-distance dispersal and accelerating waves of disease: Empirical relationships. *American Naturalist, 173*, 456–466.

Nagarajan, S., & Joshi, L. M., (1985). Epidemiology in the Indian subcontinent. In: Roelfs, A. P., & Bushnell, W. R., (eds.), *The Cereal Rusts: Diseases, Distribution, Epidemiology, and Control* (Vol. 2, pp. 371–402). Academic Press, Orlando, FL, USA.

Nagarajan, S., & Singh, D. V., (1990). Long distance dispersion of rust pathogens. *Annual Review of Phytopathology, 28*, 139–153.

NASA, (2021). *Jet Streams* (pp. 1–3). National Aeronautics and Space Agency, Washington, USA. https://www.nasa.gov/mission_pages/sunearth/news/gallery/atrex-jetstream-locations.html#.Ygp0NVVBzIU/ (accessed on 19 January 2024).

Nicholson, W. L., Race, M. S., & Schuerger, A. L., (2018). Migrating microbes and planetary protection. *Trends in Microbiology* (pp. 389–392). Science and Society. doi: 10.1016/j.tim.2009.07.001. https://www.researchgate.net/publication/26785528/ (accessed on 19 January 2024).

Nottingham, S. F., & Hardie, J., (1989). Migration and targeted flight in seasonal forms of the black bean aphid. *Aphis fabae*. *Physiology Entomology, 14*, 451–458.

Nottingham, S. F., Hardie, J., & Tatchell, G. M., (1993). Flight behavior of the bird cherry aphid, *Rhopalosiphumpadi*. *Physiology Entomology, 16*, 223–229.

O'Brien, P. A., (2017). Biological control of plant diseases. *Australasia. Plant Pathology, 46*(4), 293–304. doi: 10.1007/s13313-017-0481-4/.

Ommati, F., Zaker, M., & Mohammadi, A., (2013). Biological control of *Fusarium* wilt of potato (*Fusarium oxysporum* f. sp. *tuberosi*) by *Trichoderma* isolates under field condition and their effect on yield. *Journal of Crop Protection, 2*(4), 435–442.

Oneto, D. L., Golan, J., Mazzino, A., Pringle, A., & Seminara, A., (2020). Timing of spore release dictates survival during atmospheric transport. *Proceedings of National Academy of Science,USA, 117*, 5134–5143. https://www.pnas.org/lookup/suppl/.doi: 10.1073/pnas.1913752117/-/DCSupplemental.y.

Ordentlich, A., Nachmias, A., & Chet, I., (1990). Integrated control of *Verticillium dahlia* in potato by *Trichoderma harzianum* and captan. *Crop Protection,9*(5), 363–366. doi: 10.1016/0261-2194(90)90008-U/.

Osowski, V., (2018). *Microbes Catch a Ride on Fog but Change During the Trip* (pp. 1–3). Cary Institute of Ecosystem Studies, Michigan State University, East Lansing, Michigan, USA. https://www.futurity.org/fog-microbes-1876442/ (accessed on 19 January 2024).

Pammel, L. H., (2017). *Fungus Diseases of Sugar Beet* (Vol. 2, No. 15). Iowa State University, Ames, Iowa, USA Bulletin: Article 4. http://lib.dr.iastate.edu/bulletin/vol2/iss15/4/ (accessed on 19 January 2024).

Parnell, M., Burt, P. J. A., & Wilson, K., (1998). The influence of exposure to ultraviolet radiation in simulated sunlight on ascospores causing black sigatoka disease of banana and plantain. *International Journal of Biometeorology, 42*, 22–27. http://dx.doi.org/10.1007/s004840050079/ (accessed on 19 January 2024).

Parry, H. R., (2013). Cereal aphid movement: General principles and simulation modeling. *Parry Movement Ecology, 1*, 14 http://www.movementecologyjournal.com/content/1/1/14/ (accessed on 19 January 2024).

Pasteur, L., (1860a). Experiences relatives aux générationsditesspontanées. *Comptes Rendus Hebdomadaires des Séances de L'académie des Sciences, 50*, 303–307. (English translation: Pasteur, L., (1860a). Experiences relating to so-called spontaneous generations. *Weekly Reports of the sessions of the Academy of Sciences, 50*, 303–307).

Pasteur, L., (1860b). Suite a une pr ` ec´ edente communication relative aux 'gen' erationsditesspontanees. *Comptes Rendus Hebdomadaires ´ des Seances de L'acad'emie Des Sciences, 51*, 675–678. (English translation: Following a previous communication relating to so-called spontaneous generations. *Weekly Reports of the Sessions of the Academy of Sciences, 51*, 675–678).

Peng, X., Gat, D., Paytan, A., & Rudich, Y., (2021). The response of airborne mycobiome to dust storms in the Eastern Mediterranean. *Journal of Fungi, 7*, 802.

Pepper, I. L., & Gerba, C. P., (2015). Aeromicrobiology. In: *Environmental Microbiology* (3rdedn., pp. 89–111). University of Arizona, USA. doi: http://dx.doi.org/10.1016/B978-0-12-394626-3.00005-3.

Prenni, A., Petters, M., Kreidenweis, S., Heald, C., Martin, S., Artaxo, P., Garland, R., et al., (2009). Relative roles of biogenic emissions and Saharan dust as ice nuclei in the Amazon basin. *Nature, 2*, 402–405. doi: 10.1038/NGEO517/.

Prospero, J. M., (1999). Long-range transport of mineral dust in the global atmosphere: Impact of African dust on the environment of the South-Eastern United States. *Proceedings of National Academy of Sciences, U.S.A., 96*, 3396–3403. doi: 10.1073/pnas.96.7.3396/.

Prospero, J. M., Blades, E., Mathison, G., & Naidu, R., (2005). Interhemispheric transport of viable fungi and bacteria from Africa to the Caribbean with soil dust. *Aerobiologia, 21*, 1–19. doi: 10.1007/s10453-004-5872-7/.

Purdy, L. H., (1985). Sugarcane rusts. In: Roelfs, A. P., & Bushncll, W. R., (eds.), *The Cereal Rusts* (pp. 237–256). Academic Press, Orlando.

Querol, X., Pey, J., Pandolfi, M., Alastuey, A., Cusack, M., Perez, N., Moreno, T., et al., (2009). African dust contributions to mean ambient PM10 mass-levels across the Mediterranean basin. *Atmosphere and Environment, 43*, 4266–4277. https://doi.org/10.1016/j.atmosenv.2009.06.013/.

Reche, I., D'Ortaz, G., Mladenov, N., Wingets, D. M., & Suttle, C. A., (2017). Deposition rates of viruses and bacteria above the atmosphere boundary layer. *The ISMA Journal, 12*, 1154–1162. https://doi.org/10.1038/s41396-017-0042-4/.

Robert, Y., (1987). Aphids and their environment: Dispersion and migration. In: Minks, A. K., & Harrewijn, P., (eds.), *Aphids: Their Biology, Natural Enemies, and Control; World Crop Pests* (pp. 299–313). Elsevier Publishers, Amsterdam, Holland.

Roelfs, A. P., & Bushnell, W. R., (1985). *The Cereal Rusts* (p. 592). Academic Press, Orlando, USA.

Roelfs, A. P., (1982). Effects of barberry eradication on stem rust in the United States. *Plant Disease, 66*, 177–181.

Roelfs, A. P., Singh, R. P., & Saari, E. E., (1992). *Rust Diseases of Wheat: Concepts and Methods of Disease Management* (p. 89). CIMMYT. Mexico.

Roger, P. A., & Kulasooriya, S. A., (1982). *Blue Green Algae and Rice* (p. 116). International Rice Research Institute, Manila, Philippines.

Roger, P. A., & Watanabe, I., (1982). *Research on Algae, Blue-Green Algae and Photo-autotropic Nitrogen Fixation* (pp. 1–20). The International Rice Research Institute (1963–81), Summarization, Problems and Prospects. International Rice Research Institute, Manila, Philippines.

Rosselli, R., Fiamma, M., Deligios, M., Pintus, G., Pellizrro, G., Canu, A., Duce, P., et al., (2015). Microbial immigration across the Mediterranean via airborne dust. *Scientific Reports, 5*, 16306. doi: 10.1038/srep16306.

Sahu, N., & Tangatur, A. D., (2015). Airborne algae: Overview of the current status and its implications to environment. *Aerobiology, 31*, 89–97.

Savary, S., Bosc, I. P., Noirot, M., & Zadocks, I. C., (1988). Peanut rust in West Africa: A new component in a multiple pathosystem. *Plant Disease, 72*, 1001–1009.

Schlesinger, P., Mamane, Y., & Grishkan, I., (2006). Transport of microorganisms to Israel during Saharan dust events. *Aerobiologia, 22*, 259–273.

Schmale, D. G. III., & Ross, S. D., (2015). Highways in the sky: Scales of atmospheric transport of plant pathogens. *Annual Review of Phytopathology, 53*, 591–611. doi 10.1146/annurev-phyto-080614-115942/.

Schuerger, A. C., (2004). Microbial ecology of the surface exploration of mars with human-operated vehicles. In: Cockell, C. S., (ed.), *Martian Expedition Planning* (pp. 363–386). American Astronautical Society publication AAS03-322, Univelt Publishers, Santa Barbara, CA.

Shafi, J., Tian, H., & Ji, M., (2017). Bacillus species as versatile weapons for plant pathogens: A review. *Biotechnology and Biotechnology Equipment, 3*, 446–459. http://dx.doi.org/10.1080/13102818.2017.1286950.

Shao, Y., Wyrwoll, K. H., Chappell, A., Huang, J., Lin, Z., & McTainsh, G. H., (2011). Dust cycle: An emerging core theme in Earth system science. *Aeolian Research, 2*, 181–204. doi: 10.1016/j.aeolia.2011.02.001/.

Sharma, N. K., & Singh, S., (2010). Differential aerosolization of algal and cyanobacterial particles in the atmosphere. *Indian Journal of Microbiology, 50*, 468–473.

Sharma, N. K., Rai, A. K., Singh, S., & Brown, R. M., (2007). Airborne algae: Their present status and relevance. *Journal of Phycology, 43*(4), 615–627.

Shepherd, G., (2016). *Global Assessment of Sand and Dust Storms* (pp. 1–32). United Nations Environment Program. Nairobi, Kenya. https://wesr.unep.org/media/docs/assessments/global_assessment_of_sand_and_dust_stormsx.pdf (accessed on 19 January 2024).

Slijepcevic, P., (2020). *Bacteria and Viruses are Travelling the World on Highways in the Sky.* https://theconversation.com/bacteria-and-viruses-are-travelling-the-world-on-highways-in-the-sky-142854/ (accessed on 19 January 2024).

Smith, D. J., Griffin, D. W., & Schuerger, A. C., (2009). Stratospheric microbiology at 20km over the Pacific Ocean. *Aerobiologia, 26*, 35–46. doi: 10.1007/s10453-009-9141-7/.

Smith, D. J., Griffin, D. W., McPeters, R. D., Ward, P. D., & Schuerger, A. C., (2011). Microbial survival in the stratosphere and implications for global dispersal. *Aerobiologia, 27*, 319–332.

Smith, D. J., Jaffe, D. A., Birmele, M. N., Griffin, D. W., Schuerger, A. C., & Hee, J., (2012). Free tropospheric transport of microorganisms from Asia to North America. *Microbial Ecology, 64*, 973–985.

Smith, D. J., Ravichandar, J. D., Jain, S., Griffin, D. W., Yu, H., Tan, Q., Thissen, J., et al., (2018). Airborne bacteria in earth's lower stratosphere resemble taxa detected in the troposphere: Results from a new NASA aircraft bioaerosol collector (ABC). *Frontiers in Microbiology, 9,* Article 1752. https://doi.org/10.3389/fmicb.2018.01752/.

Smith, D. J., Timonen, H. J., Jaffe, D. A., Griffin, D. W., Birmele, M. N., & Perry, K. D., (2013). Intercontinental dispersal of bacteria and archaea by transpacific winds. *Applied Environmental Microbiology, 79*, 1134–1139.

Solomon, S., Daniel, J. S., Neely, R. R., Vernier, J. P., Dutton, E. G., & Thomason, L. W., (2011). The persistently variable "background" stratospheric aerosol layer and global climate change. *Science, 333*, 866–870.doi: 10.1126/science.1206027/.

Sørensen, J. H., Mackay, D. K. J., Jensen, C., & Donaldson, A. I., (2000). An integrated model to predict the atmospheric spread of foot-and-mouth disease virus. *Epidemiology, and Infection, 124*, 577–590.

Sousa, E., Marques, A., Mimoso, C., & Cardoso, F., (2012). First occurrence of pear decline disease in Portugal. Phytoplasma in fruit trees: Multidisciplinary approach toward. *Proceedings of 22nd International Conference on Viruses and Other Graft Transmissible Diseases of Fruit Crops* (p. 11). Centro di ricera per la Patologiavegetale. Via carloGiusseppo, Bertero 22, Rome, Italy, Abstract book. https://icvg.org/data/2018proceedings.pdf (accessed on 19 January 2024).

Springer, M. L., (1993). Genetic control of fungal differentiation: The three sporulation pathways of *Neurospora crassa. Bioassays, 15*, 365–374.

Subrahmanyam, P., Reddy, L. J., Gibbons, R. W., & McDonald, D., (1985). Peanut rust: A major threat to peanut production in the semiarid tropics. *Plant Disease, 69*, 813–819.

Tang, K., Huang, Z., Huang, J., Maki, T., Zhang, S., Shimizu, A., Ma, X., et al., (2018). Characterization of atmospheric bioaerosols along the transport pathway of Asian dust during the dust-bioaerosol-2016 campaign. *Atmospheric Chemistry and Physics, 18*, 7131–7148. https://doi.org/10.5194/acp-18–7131-2018.

Taylor, L. R., (1974). Insect migration, flight periodicity and the boundary layer. *Journal of Animal Ecology, 43*, 225–238.

Teixeira, D. C., Saillard, C., Jagoueix-Eveillard, S., Danet, J. L., Ayres, A. J., & Bové, J. M., (2005). *"Candidatus Liberibacter americanus" APSnet Feature Story: Citrus Huanglongbing: Understanding the Vector-Pathogen Interaction* (pp. 1–4). https://swfrec.ifas.ufl.edu/hlb/database/pdf/00000818.pdf (accessed on 19 January 2024).

Temkiv, T. S., Finster, K., Hansen, B. M., Nielsen, N. W., & Karlson, U. G., (2012). The microbial diversity of a storm cloud as assessed by hailstones. *FEMS Microbiology Ecology, 81*, 684–695.

Thambugula, K. M., Darangama, D. A., Phillips, A. J. L., Kananangara, D., & Promputtha, I., (2020). Fungi vs fungi biocontrol: An overview of fungal antagonists applied against plant pathogens. *Frontiers in Cell, Infection and Microbiology*. https://doi.org/10.3389/fcimb.2020.604923/.

Thomas, I., & Vevai, E. J., (1940). Aphis migration. An analysis of the results of five seasons' trapping in North Wales. *Annals of Applied Biology, 27*, 393–405.

Tignat-Perrier, R., Dommergue, A., Thollot, A., Keuschnig, C., Magnad, O., Vogel, T. M., & Larose, C., (2019). Global airborne microbial communities controlled by surrounding landscapes and wind conditions. *Scientific Reports,9*, 14441 | https://doi.org/10.1038/s41598-019-51073-4/.

Uno, I., Eguchi, K., Yumimoto, K., Takemura, T., Shimizu, A., Uematsu, M., Liu, Z., et al., (2009). Asian dust transported one full circuit around the globe. *National. Geosciences, 2*, 557–560. https://doi.org/10.1038/ngeo583 (accessed on 19 January 2024).

USAID, (2018). *West African Sahel: Climate Change Risk Profile: Regional Fact Sheet* (p. 10). United States Agency for International Development, Washington D.C. USA. https://www.climatelinks.org/resources/climate-risk-profile-west-africa-sahel (accessed on 19 January 2024).

USDA-REE, (2013). *Understanding Microbes Blowing in the Wind* (pp. 1–3). Science News. https://www.sciencedaily.com/releases/2012/05/120509092415.htm (accessed on 19 January 2024).

Van, L. A., (1941). *The Collected Letters of Antoni Van Leeuwenhoek* (Vol. 2, pp. 1–7). Sets and Zeitlinger Co. Amsterdam, The Netherlands.

Vincenot, L., Nara, K., Sthultz, C., Labbé, J., Dubois, M. P., Tedersoo, L., Martin, F., & Selosse, M. A., (2012). Extensive gene flow over Europe and possible speciation over Eurasia in the ectomycorrhizal basidiomycete *Laccariaamethystina* complex. *Molecular Ecology, 21*, 281–299. https://doi.org/10.1111/j.1365-294X.2011.05392.x.

Vinette, J. R., & Kennedy, B. W., (1975). Naturally produced aerosols of *Pseudomonas glycinea. Phytopathology, 65*, 737–738.

Waggoners, P. E., (1973). The removal of *Helminthosporium maydis* spores by wind. *Phytopathology, 63*, 1252–1255.

Wallin, J. R., & Loonan, D. V., (1971). Low-level jet winds, aphid vectors, local weather, and barley yellow dwarf virus outbreaks. *Phytopathology, 61*, 1068–1070.

Wehner, M., (2018). Scientists say earth's atmosphere is raining viruses and bacteria down on the entire planet. *Journal of International Society of Microbial Ecology* (pp. 1–3). https://bgr.com/2018/02/06/viruses-in-the-sky-bacteria-atmosphere-earth/ (accessed on 19 January 2024).

Weinbauer, M. G., (2004). Ecology of prokaryotic viruses. *FEMS Microbiological Reviews, 28*, 127–181.

Wilkins, D. M., Koumoutsaris, S., Mitchell, E. A. D., & Bey, I., (2011). Modeling the effect of size on the aerial dispersal of microorganisms *Journal of Biogeography*. https://doi.org/10.1111/j.1365-2699.2011.02569.x/.

Wilkins, D., (2011). Life on the wind: Study reveals how microbes travel the Earth (phys.org). *Earth and Climate* (p. 239). Willey Blackwell Inc. https://phys.org/news/2011-08-life-reveals-microbes-earth.html (accessed on 19 January 2024).

Yamaguchi, N., Ichiijo, T., Sakoyani, A., Baba, T., & Nasu, M., (2012). Global dispersion of bacterial cells on Asian dust. *Scientific Reports, 2*(1), 525. doi: 10.1038/srep00525.

Yong, B., (2013). *Frequent-Flier Microbes Ride Hurricanes and Live in Clouds* (p. 3). National Geographic. https://www.nationalgeographic.com/science/article/frequent-flier-microbes-ride-hurricanes-and-live-in-clouds/ (accessed on 19 January 2024).

Yu, H., (2015). The fertilizing role of African dust in the Amazon rainforest: A first multiyear assessment based on CALIPSO lidar observations. *Geophysics Research Letters, 42*, 1984–1991.doi: 10.1002/2015GL063040/.

Zadocks, J. C., & Schein, R., (1979). *Epidemiology and Plant Disease Management* (p. 427). Oxford University Press, New York, USA.

Zhang, J. X., Fernando, W. G. D., & Xue, A. G., (2005). Daily and seasonal spore dispersal by *Mycosphaerellapinodes* and development of *Mycosphaerella* blight of field pea. *Canadian Journal of Botany, 83*, 302–310.

Zhao, Z., Wang, Q., & Wang, K., (2010). Study of the antifungal activity of *Bacillus vallismortis*ZZ185*in vitro* and identification of its antifungal components. *Bioresource Technology, 101*(1), 292–297.

Zittis, G., Hadjinicolaou, P., Fnais, M., & Lelieveld, J., (2016). Projected changes in heat wave characteristics in the eastern Mediterranean and the Middle East. *Regional Environmental Change, 16*, 1863−1876.

CHAPTER 4

Aeolian Dust Storms and Microbes

ABSTRACT

Aeolian dust storms are natural events in the sky that potentially affect global weather systems. They are highly relevant to natural vegetation, agricultural cropping systems, and crop productivity. This chapter focuses on the occurrence of inter-continental dust storms and their influence on aeolian microflora, particularly the microbes that thrive on particulate matter (PM) found over agrarian ecosystems. Aeolian dust storms play a role in dispersing PM laden with microbes from one farming location to another. There are various estimates of the mass transfer of PM across continents, with an annual migration of about 0.5 to 5 billion tons of particle matter across the sky. These particles transit above oceans, mountains, and agrarian areas. Oceans receive a total of 477 Tg y^{-1} of particle matter. Along with dust plumes, microbes also disperse into other regions. The dust plumes harbor and carry microbes with enormous genetic and physiological diversity. The microbial diversity and density vary depending on the location where they are generated, and they may also acquire microbial species during aerial transit. The survival of microbes in the location of deposition depends on the physicochemical characteristics of the soil, such as the substratum and ambient environmental factors.

Reports state that globally, about 1 to 3 gigatons of aerosols are generated. The aerosols are usually sorted into particles of size PM 2.5–5 μm and PM 10 μm. Denser and heavier particles stay afloat for a shorter length of time. They transit only for a shorter distance and descend onto the ground surface rapidly. Smaller PM floats for a longer time and covers intercontinental distance. It stays buoyant. Microbial diversity, population density, and survival in the PM are crucial. Microbes traced in dust storms

Microbiology of the Agricultural Sky. K. R. Krishna (Ed)

and the atmosphere that has recently experienced a dust storm belong to α and √- Proteobacteria, Firmicutes, Actinobacteria, Bacteroidetes. However, microbial assemblages noticed differ based on the location from where dust storms got generated and its migratory path. For example, Pseudomonadaceae, Bacillaceae, Rhodobacteriaceae, Cytophagaceae, Frankiales, and Chlorofexaceae were common in the Mediterranean dust storms. We may also note that a 'dusty agrarian sky could also be a 'sickly agrarian sky' by harboring plant pathogens.

Globally, the Sahara/Sahelian region is the largest generator of dust plumes/storms. *Harmattan, Haboob, Irifi, Guebli, Khamsin,* and several others that are generated in the Sahara are well-documented events. Dust storms generated in the Sahara/Sahel are predominantly natural. It is estimated that about 840 million tons of PM loaded with nutrients and microbes are generated annually from Western Sahara. Their relevance to fluctuations in atmospheric conditions and crops is understood, to a certain extent. Saharan/Sahelian dust plumes/storms transport massive amounts of particle matter laden with diverse microbes to the North American East coast. Asian dry regions in Gobi, Taklamakan, and adjoining areas also generate large dust plumes. Annually, about 202 million tons of dust plus microbes travel from Central Asia to the Far East (Japan) and on to the Pacific Northwest in North America. Such dust storms have the potential to alter earth system phenomena, such as precipitation patterns. The dispersal and occurrence of epidemics/pandemics on food crops are immensely influenced by such dust storms. The nutrient status of soils utilized for crop production could be influenced by the PM and its nutrient load.

Dust storms occur in all continents that support sizable agrarian expanses. Major agrarian regions such as the Great Plains of North America, Pampas, Cerrados, European plains, Sahelian farming zones, Indo-Gangetic plains, Northeast China, and Australian cropping zones are all exposed to dust storms periodically. Detailed discussions on each of these food grain generating regions are worthwhile. It is important to explore agricultural practices that could be adopted in each region.

Such dust storms are a major source of dust-borne nutrients and microbes. The microbial load, its diversity, and physiological manifestations are topics of relevance in farming zones. Aeromicrobiology of individual crop fields needs attention. Additionally, 'Intercontinental Dust Storms' that aid the migration of microbes may also affect farms and crop productivity. Intercontinental dust storms also disseminate detrimental microbes, particularly those that generate epidemics or pandemics.

We may realize that globally, such dust storms laden with microbes represent the largest migratory event of life forms. This event is larger than previously known migrations of birds, land animals, insects, fishes, or other life forms.

4.1 INTRODUCTION

4.1.1 AEOLIAN DUST STORMS: BACKGROUND

The aeolian dust storms are natural phenomena observed periodically in different continents. They have immediate consequences on ecological services. In this chapter, we focus our attention on the relevance of aeolian dust storms on agroecosystems. In certain regions, dust storms can be severe and even major determinants of biomass generation by vegetation. Aeolian dust storms do affect the productivity of cultivated crops. They also serve as habitats for diverse species of microbes. These microorganisms, residing on dust particles, provide several ecosystem services. An important aspect of dust storms is their ability to spread diseases, epidemics, and pandemics among major food grain crops. At the same time, there are several beneficial microbes dispersed by the same aeolian dust storms (Womack et al., 2010).

Globally, approximately 1,536 million tons of dust are emitted into the atmosphere each year. Aeolian dust storms transport vast quantities of natural material across continents, dispersing a significant amount of microbial population. The composition of the dispersed microbial population in aerosols varies depending on the microbial load of dust particles. Natural sources account for about 75% of global dust emissions, while the remaining 25% is attributed to anthropogenic activities such as farming and industries. Agricultural pursuits, particularly in North America, are major contributors to dust emissions, with 54% of emissions being anthropogenic in this region. North American and Australian farming regions are known for generating higher levels of dust emission due to anthropogenic reasons. The largest amount of dust is emitted from Western Sahara and Sahel, with each region contributing 840 million tons per year. In the Sahara, dust emissions are predominantly *natural*, accounting for 92% of the total. China and the Far East region are significant dust emitters, releasing 202 million tons per year (U.N.O., 2018). Similarly, the Pampas and Chilian dry lands are prone to desertification and severe dust storms. The Indo-Gangetic plains are also recognized as a major dust emitter in South Asia (Ginoux et al., 2012; Tozer & Leys, 2013). Numerous estimates exist regarding global dust storms and their capacity to transport dust particles.

Lack of appropriately devised land use patterns, despite knowledge about dust storms, causes economic losses. Dust storms are not just a natural phenomenon; they also result in economic losses for the farming community in terms of natural resources and the exchequer. For example, a special report by the United Nations on the environment points out that a single dust storm called 'Red Dawn' in Southern Australia, lasting two days, resulted in a loss of 262 million US dollars. In northern China, between 2010 and 2013, dust storms caused a loss of 964 million US dollars. In Mongolia, a single-day dust storm resulted in a loss of 0.5 million US dollars (UNO, 2018). The amount of dust generated may depend on natural resources, terrain, and anthropogenic pursuits adopted by humans. Agricultural fields, particularly open fields, often experience the greatest loss due to severe dust storms (Tozer & Leys, 2013; Zou & Zhai, 2004). Furthermore, unwise farming practices such as lack of cover crops, absence of mulches, and over-grazing are deteriorating soils in farm belts. Farming regions in the Central Plains of the USA, Sahel, West Asia, Indo-Gangetic Plains, Northeast China, the Far East, and Australian semi-arid regions are most prone to deterioration. The microbial load in the dust particles is an important aspect. We need to understand the microbial processes and their relevance to farming procedures under such dry conditions.

A report by the World Meteorological Organization states that approximately 40% of aerosols recorded in the troposphere are created by the erosion of terrain, including crop fields. Turbulent wind, updrafts, and convective currents scatter dust particles into the air. It is important to determine the quantity of topsoil loss as dust particles from fields.

Griffin (2009) reminds us that 34% of the terrestrial portion of the Earth can be classified as deserts. They are prone to initiating dust storms. The Sahara, Sahel, Gobi, and Taklamakan are supposed to be the most prolific dust storm generators. Each year, these deserts account for 50–75% of the dust liberated into the atmosphere. About 0.5 to 5 billion tons of dust are generated in the deserts. They affect air quality above the crop belt and human settlements. In fact, many agrarian regions constantly experience dust deposition from deserts. Regarding Saharan dust, it is said to take 3–5 days to reach the Caribbean and other North Atlantic coastal locations. The large Asian dust storms are known to transit long distances. Dust particles plus microbes may be carried high into the atmosphere, say, above 10,000 ft, and then blown to different continents (NASA, 2001). Asian dust storms take about 7–9 days to reach North American crop belts. Since the mid-1960s, a combination of climate flux and anthropogenic activity (development and agriculture) has resulted in a general increase in atmospheric dust transported from the deserts of North Africa and Asia. Perkins (2001); and Prospero & Lamb (2003) estimate that global dust and aerosols generated each year may be 0.5 to 5.0 billion tons. The majority of it (>50%) originates in the Sahara and Sahel. Regions close to these arid belts receive the most dust, but ecosystems located far off at long distances also receive the dust deposits.

As mentioned previously, dust storms are prominent natural phenomena in the atmosphere. They are widely distributed across various terrains around the world. Agrarian regions also experience dust storms, particularly those developed on dry sandy soils or silt/alluvial soils. Crops must endure dust deposits, even if only for a short period. However, seasonal and long-lasting dust storms are also common on different continents.

A dust storm consists of particles of different sizes and intensities. The intensity of a dust storm is determined by factors such as wind, turbulence, atmospheric conditions (dryness, relative humidity, season), and precipitation patterns. Dust storms can carry mineral particles containing various nutrients (P, S, Ca, Fe, Mn) that are essential for crops. These mineral-rich dust particles can contribute to improving soil fertility. The nutrient content of the particles may vary depending on the region of origin.

Dust storms also transport biological particles, including microbes from diverse genera with different functionalities. These microbes play a role in various ecosystem functions. Bacterial and fungal spores are commonly found in dust storms occurring in drylands. These airborne spores can be commensals or plant pathogens (e.g., Rust fungi). Additionally, dust storms carry pollen, which gets dispersed, facilitating pollination. Insect eggs are also encountered in dust storms. Particulate organic matter is an important component of dust storms (Kellogg & Griffin, 2006; Katra et al., 2014).

The biological components traced in dust storms are important as they regulate several ecosystem functions at the confluence of the biosphere and atmosphere. There are several bacterial genera known to cause enucleation, aiding in the formation of clouds, frost, and snowflakes (e.g., *Pseudomonas syringae*). Microbes residing on dust particles may also contribute to various biogeochemical cycles (Brodie et al., 2007; Rahav et al., 2016, 2018). The microbial diversity in dust storms plays a crucial ecological role (Mazar et al., 2016; Bowers et al., 2013; Womack et al., 2010).

Drylands currently cover approximately 46.2% (±0.8%) of the global land area. These regions are home to 3 billion people (Mirzabaev & Wu, 2019). Let's name a few examples of deserts, arid, and semi-arid belts with loose sandy soil that are prone to dust generation. They include the dusty loose mollisol regions of the Great Plains of North America, the deserts and dryland cropping zones of Peru, the Pampas with semi-arid cropping systems, the Atacama dry regions, the European soils susceptible to strong winds, the Sahelian region, the Saharan desert region, the Kalahari and adjoining semi-arid cropping zones of Southern Africa, the Namibian Desert, West Asian drylands, the Thar region in Southern Pakistan, the semiarid land in the Russian Steppes, the Mongolian deserts in Gobi and Taklamakan, the Australian deserts, and more. All of these regions, along with a few others, contribute to the presence of dust in the atmosphere. Such dust is prone to disperse to different agrarian regions on Earth. As a consequence, copious quantities of PM, sand with mineral-rich components, and microbes are added to various regions on Earth. The dust added to cropland in different continents can be immensely useful to crops and to accomplish other ecosystem functions.

Geographic regions of Earth that are prone to generating dust storms of great consequence to crop growth, productivity, and vegetation in general, as well as human health, occur in North Africa. North Africa is the prime area that generates 57.6% of the global dust recorded each year. The Arabian Peninsula generates 11.8% of the dust, Eastern China 7.8%, Coastal Asia 7.5%, Australia 5.6%, Western China 3.6%, South Africa 3.4%, and North America 0.1% (Lloyd, 2021).

The destructive consequences of dust storms on global agriculture may depend on the agricultural prowess, cropping pattern, and dependence of the human population in a particular area. Therefore, even if dust storms are feeble, if the crop that gets affected is important, then dust storms need greater attention.

Globally, approximately 1 to 3 gigatons of dusty aerosols are generated each year. Dust storms with large-sized particles descend quickly onto fields, but they only travel shorter distances from their origin. As the dust storm progresses, fine particles are retained in the troposphere and can travel longer distances as they remain afloat for a longer period of time. Dense larger particles of 10 μm may stay afloat for a brief period of a few hours, while fine dust particles smaller than 1 μm can stay afloat in the wind for 10 days or more. Microbes can be found on dust particles regardless of their size, mineral composition, or organic nature (World Meteorological Organization, 2020).

Guieu et al. (2014) have compiled evidence to show that global dust storms are important natural processes. These dust storms deposit significant amounts of dust into the oceans each year. This dust can serve as a source of nutrients such as carbon (C), nitrogen (N), phosphorus (P), aluminum

(Al), manganese (Mn), iron (Fe), and calcium (Ca) for agricultural soils. Aeolian dust also contains a substantial number of diverse microbes with various metabolic capabilities. Many of these microbes are tolerant of harsh environments.

Dust storms can also affect the oceans. For instance, the North Atlantic Ocean receives approximately 202 Tgy^{-1} of dust, the Indian Ocean receives 118 Tgy^{-1}, the North Pacific receives 72 Tgy^{-1}, the Mediterranean Sea receives 40 Tgy^{-1}, the South Pacific receives 29 Tgy^{-1}, the South Atlantic receives 17 Tgy^{-1}, and the Arctic receives 5.7 Tgy^{-1}. In total, the world's ocean surface is impacted by 477 Tgy^{-1} of dust.

Dust storms can also occur over land. Agricultural areas are not exempt from dust storm events. It is crucial to gather comparable information about storms that introduce particulate matter (PM), minerals, and microbes into major crop-producing regions. Specific information about dust mass, nutrients, and microbes added to agricultural fields is needed.

Some parts of the world frequently experience dust storms. Dust storms can be defined by a lack of visibility beyond 1 km or by qualitative descriptions, such as dust in suspension, blowing dust, dust storm, and severe dust storms. One definition of a dust storm states that it is an increase in the PM density in the air to greater than 100 $\mu g/m^3$. This information can be determined with monitors that measure PM and the size distributions based on aerodynamic diameters ≤ 2.5 μm and ≤ 10 μm. Other important factors include the path of the dust storm, the season, the geographic area of involvement, and the duration (Reed & Nugent, 2018). Mayoll et al. (2017) make it clear that the atmosphere is a habitat for microorganisms derived from both the ocean surface and terrain. Their estimations indicate that 2.2×10^{21} microbial cells could be detected in the troposphere. Atmospheric samples from above the ocean surface had 33–68% of microbes derived from oceans. Microbes derived from terrain, soils, and vegetation are also picked up by the atmosphere. Dust storms support the trans-oceanic dispersal of microbes. A dust storm with microbes actually helps in improving the genetic diversity of microbes in the soil profile. This happens after it has landed in a location. It also introduces microbes freshly from different continents.

4.1.2 BIOAEROSOLS

Bioaerosols are atmospheric particles that range in size from nanometers up to about a tenth of a millimeter. They consist of living and dead organisms such as algae, fungi, bacteria, and archaea, as well as dispersal units like bacterial spores, fungal spores, and plant pollen. Additionally, they may contain various fragments or excretions (Fröhlich-Nowoisky et al., 2016a, b). Several researchers have studied the role of dust events in the transportation of bioaerosols (Hervas et al., 2009; Prospero et al., 2005, 2014; Yamaguchi et al., 2012).

Dust storms are a worldwide phenomenon. Dust cycles refer to the movement of trillions of minute particles, consisting of sand, organic matter, and microbes, in the atmosphere above the Earth's surface. These events are often seasonal and repetitive (Middleton, 2020). They occur when loose soil, desert, or dusty terrains are subjected to intense winds. Turbulence causes soil particles to become airborne, resulting in the formation of a dust storm due to high-speed horizontal winds. Bioaerosols are extremely small particles suspended in the atmosphere, ranging in size from nanometers to 0.1 mm. Plant debris is the most commonly found material in bioaerosols. Sand particles, organic matter, and dead or live microbes are the most common mixtures of bioaerosols found in different regions. These dust storms can travel short distances of a few meters or even up to kilometers in the atmosphere. There are also long-distance dust storms that can carry live and/or dead microbes for up to 15,000 km, crossing continents. High altitude winds and jet streams that occur in the atmosphere support such long-distance movement of dust or bioaerosols. Genetically identical microbes can be traced all over the world because of large scale movement of dust storms. Jet streams are known to aid movement of microbes in the atmosphere over the entire circumference of earth (Lindsay, 2018). Viruses and bacteria, it seems, travel long distance at altitudes of 3,000 m and above in the atmospheric jet-streams. About 800 million virus particles get hurled into atmosphere per m^2 surface. Such phenomena can spread viral diseases on crops, easily. In simple terms, dust storms may also distribute virulent crop pathogenic strains and cause epidemics/pandemics, if other conditions are congenial. By adding a new strain of microbe, dust storms may even induce formation of new virulent stains of plant pathogens in agroecosystems.

Not only open deserts/plains, but agrarian regions also generate dust particles that are laden with microbes. These microbes are known to tolerate UV radiation and harsh atmospheric conditions prevalent at high altitudes (Behzad et al., 2018; Kuske et al., 1997; Makhalanyane, 2015). Microbes present in bioaerosols can overcome restricted availability of moisture and unfavorable temperatures. They play a role in organic matter degradation and aid in bio-geochemical cycles. Some species found on dust particles could have symbiotic relationships with crops, such as BNFmicrobes. Dust storms globally transfer and recycle soil microbes, introducing new genetic diversity into cropping belts. Burrows et al. (2009) conducted simulations of global microbe transport using dust particles of 1–3 μm and found that bacteria can perform diverse ecosystem functions. Another example is the yellow sandstorms originating in East Asian deserts, such as the Gobi and Taklimakan. These sandstorms impact ambient air quality by increasing fungal spore levels (Ho et al., 2005; Wu et al., 2004). Studies from South Korea also demonstrate that Asian dust affects both airborne fungal concentrations and fungal communities (Jeon et al., 2011, 2013; Yeo & Kim, 2002).

We can classify dust storms based on the microbial diversity of aerosols. Cultural and molecular methods can be used to identify the microbes present on dust particles (Acosta-Martinez et al., 2015). Acosta-Martinez et al. (2015) discuss various methods used to culture, identify, and classify dust microbes. Evaluations by Rahav et al. (2016) have shown that dust plumes originating from the Sahara and Eastern Europe are related, as they possess genetically similar microbial flora and biogeochemical activity compared to Mediterranean dusts. The microbial composition of dust storms allows us to infer the potential bio-geochemical functions that may be induced after the deposition of a particular dust storm.

Large dust storms occur when strong winds become turbulent, capable of lifting, scattering, and carrying surface soil, organic debris, microbes, and small insects over long distances. Dust particles can also acquire microbes during their journey from the source until they are deposited in distant locations. The intercontinental transport of dust particles can contribute to the microbial diversity of dust storms, as they have the opportunity to pick up microbes even during transit. The attachment of microbes to dust particles occurs through a process called adhesion, which takes place in the atmosphere (Griffin et al., 2002). For instance, Rahav et al. (2016) suggest that Eastern European and Mediterranean dust storms differ in terms of microbial diversity, population, and ecosystem functions. These characteristics of dust storms can be attributed to the microbes they acquire during intercontinental transport (Rahav et al., 2016).

Dust storms can cause soil erosion. They carry mineral-rich sand, silt, and microbes, depositing them in different locations. Dust storms have the potential to disrupt soil structure. They can either enhance soil fertility and the biotic fraction of soil layers, or deplete them. Dust storms also influence various soil microbial processes. It is important to note that dust storms can transport microbes and introduce them to new areas, whether nearby or far away.

The composition of dust particles in the storms plays a significant role in determining the microbes they carry from one location to another. Cultural methods have been used to investigate the microbial diversity on dust particles. Dust particles are collected and analyzed for different genera of microbes. In recent years, the analysis of 16S rRNA sequences has gained acceptance as it is easier to conduct molecular analysis on small samples compared to cultivating microbes on nutrient media. It is worth noting that some microbial genera may not be easily cultured using nutrient media-based methods for enumeration (Smets et al., 2016). Dust storms support a wide range of bacterial and fungal communities, and even viral particles can be found on inert organic particles. The geographic features, vegetation, and cropping systems at the origin of dust storms influence the diversity of microbes found on dust particles (Mazar et al., 2016; Barberan et al., 2015; Polymenakou, 2012). In fact, there are several ambient processes where microbes residing on dust particles can participate. For example, Gat et al. (2017) have clearly demonstrated that Mediterranean dust storms and clouds harbor various genera of microbes that play a role in several biogeochemical functions (Mazar et al., 2016).

Microbes that are carried by dust storms are clearly dispersed to any areas that these storms reach. It takes approximately 3 days for Saharan dust to reach the North American Great Plains. Consequently, the same amount of time is required for harmful crop pathogens (e.g., Rust Fungi) to spread themselves into the wheat belt in the Central Plains of Kansas (Griffin, 2009). Dust storms may also assist in the natural introduction of beneficial microbes into agrarian soils, which are involved in nutrient transformation, biological nitrogen fixation, and biological control of pathogens, among other things. In temperate regions, microbes traced in dust storms may also contribute to cloud nucleation and snow formation. They may even trigger precipitation events (Krishna, 2022).

Griffin (2007, 2009) highlights that a gram of soil from arable regions may contain nearly a million to a billion CFU (colony forming units). Microbiologists can identify a large number of microbial genera, including mostly fungi (~1 million per gram), viruses (virus particles ~100 million per gram), and protozoa (~10,000 per gram). Furthermore, even desert soils contain a high population of microbial cells, reaching up to 1 billion bacterial cells per gram. Considering the size of dust storms and the high density of microbes, it truly represents an exceptionally large migratory event of microbes. The migration of microbes facilitated by dust storms has an impact on ecosystem services, atmospheric processes, and crop productivity. Undoubtedly, dust storm events introduce significant amounts of particulate matter laden with minerals and microbes into agroecosystems. In a way, dust storms distribute nutrients and microbes to different crop belts.

4.1.3 AGRARIAN REGIONS, DUST STORMS, AND MICROBES

Globally, there are approximately 8–10 large food grain producing systems that span across vast terrains (see Pearson, 1992; Krishna, 2003, 2015). We require data on the amount of dust received by each crop belt, as the minerals in dust particles contribute to soil enrichment. The agrarian regions in these areas receive a diverse range of microbes, which could have significant versatility. Dust microbes can offer various benefits to the agroecosystem, including their potential involvement in nutrient transformation. These resident dust microbes are capable of performing several biochemical transformations. One crucial aspect to consider is the identification of microbes involved in biological nitrogen fixation (BNF). Extrapolations suggest that dust, sand, and BNF organisms may contribute noticeable amounts of atmospheric nitrogen to vast agrarian belts in different continents. In nature, the replenishment of microbial inoculum through dust storms plays a crucial role in sustaining agrarian regions. However, it is important to note that data on dust storms, dust quantity, and microbial presence may also reveal the deposition of a large inoculum of crop pathogens each year. This information should guide us in implementing appropriate preventive measures. Several reports indicate that dust storms and wind, in general, carry small-sized cyanobacteria, bacteria, fungi, archaea, and viruses. These microbes could have ecological implications in agrarian regions (Griffin, 2007; Lang-Yona et al., 2014; Sharoni, 2015; Smith, 2013; Tegen et al., 2004).

Agrarian regions experience a loss of topsoil, resulting in a decline in soil fertility and microbial assemblages present in dust particles. The annual loss of soil fertility and subsequent decrease in crop productivity is significant (Acosta-Martinez, 2015). The biogeochemical cycling of nutrients, facilitated by aeolian microbes, could have implications for agricultural crop production. Additionally, Acosta-Martinez et al. (2015) suggests that while we acknowledge the long-range dispersal of dust plumes, the understanding of short-distance disturbance within smaller agro-zones is limited. The loss of microbes appears to depend on the size of dust particles and soil aggregates that are lost due to wind erosion. It is important to consider the organic substrates that support microbes on dust particles. The viability of dust microbes is crucial.

Periodic droughts in agrarian regions are a significant factor contributing to dust storms. In recent years, these droughts have often been attributed to climate change. Other factors include the lack of cover crops and mulches. Pasture overgrazing and deforestation are also considered by climate change scientists to be causes of these devastating droughts.

Annually, about 500 to 3,300 Tg of dust particles are eroded from the terrain. This includes vast stretches of agrarian regions (Shao et al., 2011; McTainish & Strong, 2007). Dust storms are induced by natural causes such as desertification, sandy arid terrains, loose soils, and excessive tillage operations. Agrarian regions contribute to approximately 10% of the total dust emissions, estimated at around 50 million MT per year (Prospero et al., 2005, 2014). The dust that is carried by wind erosion from organic soils contains a rich microflora. It is estimated that about 105 to 107 microbes per gram of agricultural soil could be lost with the dust. Based on 16S rRNA sequences, it is estimated that the aeolian dust generated in fertile farming zones may harbor around 8.3 million species of microbes.

Acosta-Martinez (2015) lists several properties of agricultural soil that initiate dust storms within agroecosystems. Dry sandy soils are prone to the loss of soil aggregates, resulting in dust. Farm operations such as deep and surface tillage make soils susceptible to erosion and dust. This loss of soil also leads to a decrease in microbial diversity and population. The absence of mulching, crop residue application, or cover crops further contributes to soil loss as dust. Agrarian regions that experience dust plumes may also experience a decline in microbial diversity due to harsh atmospheric conditions. Unfavorable weather conditions, humidity, radiation, temperature, wind, and particle size can all impact the survival of microbes in dust storms. It has been hypothesized that dust particle size, clay, silt, sand complexes, and organic matter help microbes residing on dust particles to survive until the particles are deposited. Dust storms that deposit in different locations can contribute to microbial diversity, along with the nutrients attached to the particles. The loss of microbial diversity and/or population is essentially the difference between erosion and deposition in a given location.

Agrarian regions exhibit soil loss due to wind erosion. The loss of topsoil can occur even at wind speeds as low as 3 m/s. However, when wind speeds reach 6 m/s, it has enough force to detach soil particles, sand grains, and break aggregates, resulting in the emission of dust. Drylands are particularly susceptible to wind-aided topsoil erosion, often leading to localized dust events that do not travel as far as trans-continental dust storms that create large dust plumes.

Dust storms of varying intensity and travel ability occur in specific types of terrain found in different continents. Globally, certain areas are known for their significant food production. For instance, Krishna (2003, 2014, 2015, 2022) identifies major food grain-producing regions. These regions may themselves generate dust storms of varying sizes. Some dust storms may travel longer distances. Food grain-producing regions also experience the deposition of dust particles from storms that pass through the atmosphere. This means that agroecosystems can both lose and gain microbes due to dust storms, which can have devastating effects on food crops.

Microbial communities identified in the atmosphere so far consist of commensals, pathogens, as well as beneficial organisms such as symbionts and organic substrate fermenters. According to McGrath (2009), commensals were found more frequently and belonged to various genera and families. However, it is worth noting that certain well-known plant pathogens, like rust fungi, can be airborne during certain stages of their life cycle. They are able to withstand harsh conditions in the ambient atmosphere while being transported on dust particles or when their urediospores float freely in the air. Viruses that cause diseases in farm animals and farmers can also be dispersed through dust storms. Airborne viruses, such as the foot and mouth disease virus, can spread among cattle (Maki, 2012; McGrath, 2009).

Atmospheric microbes may possess unique and beneficial functions for humans, which should be further explored. For example, airborne fungi and bacteria found on dust particles at an altitude of 3,000 ft have been cultured to produce flavorful fermented foods, like 'Sky Natto' (Maki, 2012). In conclusion, McGrath (2009) suggests that we need to improve our sampling methods at different altitudes in the atmosphere. Cultural techniques for various microbes found at different altitudes should be expedited, and molecular methods should be refined and accelerated to enable rapid identification of dust microbes (Wu et al., 2021). Considering the context of this volume, it is pertinent to propose that we should explore the atmosphere specifically for microorganisms that are beneficial to crops. The adoption of microbial species found in the atmosphere above agricultural land should be emphasized.

4.2 AEOLIAN DUST PLUMES IN AMERICAS: MICROBIAL ASPECTS

Intercontinental dust storms reach North American cropland from Asian deserts (Fischer et al., 2009; Smith et al., 2013). The dust plumes reach the Pacific Northwest from Asian deserts such as Mongolia, Gobi, and Taklamakan. These plumes are rich in aerial microbes that drift freely with the wind, as well as those trapped in the dust particles. Let's consider an example. The dust plumes were sampled at Mt. Bachelor Observatory in Oregon. The microbial load was examined using 16S rRNA sequences to identify specific bacteria on the dust particles. Using micro-array molecular methods, several genera and species were identified (Smith et al., 2013). Thousands of bacterial species were recorded in the dust plume samples. Bacteria, fungi, and archaea originating from the high seas were traced during and after the Asian dust reached North American locations. Clearly, the microbes that completed the transpacific journey were tolerant to harsh atmospheric conditions. About 15 families of bacteria were found in the dust samples, with Proteobacteria, Actinobacteria, and Firmicutes being frequently observed. Furthermore, Smith et al. (2013) suggest that microbes with diverse metabolic abilities are introduced into North American farmland. Fertile soils may play a specific role in hosting alien microbes from the Asian desert. Therefore, this aspect requires further investigation, particularly regarding their role in biogeography and nutrient transformation. The alien microbes in American farms may have certain useful functions, but it is also important to understand any potential detrimental effects. For instance, dust plumes could serve as a source of plant pathogens. Among the negative effects of microbes found in dust particles, the presence of crop pathogens requires attention. The movement of dust plumes across short or long distances could determine the aerial dispersal routes of plant pathogens (Gonzalez et al., 2014). Clearly, monitoring dust plumes and microbes is crucial.

Wind erosion primarily occurs in dryland ecosystems. It is estimated that annually, 363 teragrams (Tg, 1×10^9kg) of soil are eroded (Ginoux et al., 2012). A significant portion of this soil forms dust plumes. The practice of wheat-fallow in the Pacific Northwest makes the topsoil vulnerable to erosion and dust storm formation, especially during the fallow period. Periodic droughts in the Columbia Plateau (Oregon, USA) contribute to the dust formation process. Dust events generated from agrarian fields in the Pacific Northwest may contain diverse microbes capable of various bio-geochemical transformations, including N-fixation and potential pathogenicity. Pathogens that affect wheat, legume, pasture seeds, and other crops may be present. Additionally, we may also encounter microbes that are pathogenic to farm animals and farmers. Crops, farm animals, and farmers all suffer due to sandstorms that carry microbes with diverse potentials (Sharratt & Feng, 2009; Sharratt & Lauer, 2006; Sharratt et al., 2012; Arimoto et al., 1996). Acosta-Martinez et al. (2015) state that dust particles from soils that travel in the atmosphere provide a microhabitat for microbes. Microbial propagules are protected within the crevices, folds, and surface features of the particles. The organic fraction of dust particles provides a suitable substrate for heterotrophic microbes to survive. Finer particles smaller than 10 μm can travel longer distances to other regions. Fine dust disseminates a large number of microbial genera.

Fungal genera traced in the atmosphere above soils exposed to different tillage practices were *Fusarium, Ulocladium, Chitomiaceae,* and several unidentified Ascomycota. About 2,832 bacterial colonies were identifiable into 90–148 genera. Prominent genera included *Nocardiaeaceae, Kaistobacter, Amycolatopsis,* and *Bradyrhizobium.* These findings were consistent regardless of tillage or fertilizer practices. It is important to note that dust storms can transport these microbes for short distances, up to a few hundred kilometers from their origin. Therefore, these dust plumes do not disperse pathogens over long distances. However, they may also spread symbionts like Bradyrhizobia, as well as spores of symbiotic fungi that require further identification (Schlatter et al., 2018). The significance of natural inoculation of agroecosystems with symbionts needs to be properly understood and highlighted.

Wind-mediated topsoil erosion can have devastating effects on crops. It reduces soil fertility and poses a threat to the productivity of agricultural soils. Additionally, wind erosion leads to the formation of dust. In the inland region of the Pacific Northwest and Great Plains, farmers utilize winter wheat (*Triticum aestivum* L.)-fallow rotations and adopt conservation tillage practices to control wind erosion. Moreover, in this dry region, farmers apply organic matter to the soil. Schlatter et al. (2018) have emphasized the importance of understanding the effects of soil management practices, in addition to wind erosion. Specifically, we need to comprehend the generation of dust plumes in response to large-scale soil management practices (Tegen et al., 2004). The influence of soil management practices on the emissions of dust microbiota into the atmosphere remains understudied. For instance, wind-blown dust from plots with different tillage practices (conventional and conservation) exhibited differences in microbial load and diversity. Ascomycota was found to be more abundant in dust from conventional tillage fields, while dust from conservation tillage plots had a higher number of unidentified phyla.

West Texas experiences dust storms each year between March and June. The dust storms originate from Albuquerque, New Mexico, and travel to the El Paso region in West Texas. These dust storms have negative impacts on crops and human health. The high concentration of dust particles leads to haze and reduced visibility. Additionally, the dust particles carry microbes, which can pose health risks if inhaled without precautions. There are also dust storms that reach the Rolling Plains of Texas, near Lubbock (Lee et al., 1993, 1994). The microbial content of the dust storms that reach the agricultural areas of Texas greatly affects major crops such as maize, sorghum, wheat, cotton, and pulses.

'Haboobs' are dust storms that occur in the southwest USA (Roberts & Knippertz, 2012). In 2011, a massive haboob transported large quantities of dust particles larger than PM_{10} and PM_{10}. The peak wind speeds reached 29 m/s (Raman et al., 2014). Turbulence and wind cause dust particles to become airborne, allowing them to reach heights of 2–4 km above the agricultural terrain. Hourly measurements near Phoenix, Arizona, showed that between 907 g/m^3 and 1974 g/m^3 of dust could be deposited. Consequently, a significant number of microbes are also transported and deposited on the soil surface.

Eager et al. (2017) reported that between 2005 and 2014, approximately 3 to 20 haboob events occurred in the southwest USA (Arizona, Nevada, California). During these events, an average deposition of 259 kg ha^{-1} to 2,950 kg ha^{-1} of sandy dust particles was observed (Eager et al., 2017). The quantity of larger dust particles deposited was relatively higher. These particles contained a significant amount of mineral-rich dust particles carrying a variety of microbes. Further investigations on the microbiological aspects are necessary.

According to a report by Creamean et al. (2013), dust and aerosol from Sahel may have an impact on the east coast of the USA and the Great Plains. Additionally, dust plumes from Asia may influence weather patterns in the western region of the USA. The consequences of such aerosol deposits on the microbial diversity of soil require investigation. It is also important to consider the dust deposition rates on cropping zones and their potential role in mediating various biogeochemical cycles. These aspects have already been discussed in Chapter 3.

Now let us consider regions in South America that are prone to dust storms. The Argentine Pampas is known for its vast expanse of agricultural crops. Farmers cultivate various crops such as wheat, maize, sorghum, millets, sunflower, soybean, and cotton on loess and sandy alfisols. These soils, with their fine sediments and silt, are susceptible to creating dust plumes in the air, especially during fallow periods. Schmaltz (2019) provides evidence from satellite imagery. The Moderate Resolution Imaging Spectroradiometer on NASA's AQUA satellite was used to observe a dust storm.

Satellite observations reveal that the Pampas region experiences dust drift above the crops. The dust storm moves from east to the Andean mountainous regions. Undoubtedly, such dust drifts above the Pampas aid in the dispersal of microbes. This dust may settle on the soil along with the microbes. The nutrient and microbial content of dust particles require greater attention. The loss of topsoil fertility is a concern, as fertilizer practices and productivity depend on soil characteristics. If we encounter a loss of soil fertility, it means that the nutrients lost from a field location are higher than those replenished via dust storms.

The Atacama region in Chile generates dust storms of varying magnitudes. It is one of the driest deserts known in the world. Similar to other dust bowl regions, it releases dust plumes into the atmosphere (Rech et al., 2006). Currently, it is being utilized by space scientists as a region that resembles the Martian environment. Atacama has a hyper-arid core desert region where scientific experiments have been conducted to study the dust, its mineral and organic composition, and microbial diversity (Azua-Bustos, 2015, 2019). Azua-Bustos (2019) and several others have discovered bacteria and fungi that can tolerate the harsh conditions prevalent in Atacama. These microbes can withstand extremely dry conditions, near-zero relative humidity, UV radiation, cold temperatures, lack of precipitation, and moisture (Houston & Hartley, 2006; McKay, 2003; Navarro-Gonzalez, 2003). Viable bacteria have been observed on dust particles despite such harsh conditions.

Dust samples that pass over Atacama have been analyzed for their mineral content. X-ray diffraction analysis has revealed the presence of silicate quartz (SiO_2), (Ca) $NaAlSi_3O_8$, $Mg_5Al(AlSiO_3)$, and Microperthite (K, Na) $AlSiO_3$. Microbes survive in the crevices and on the surface of dust particles with such mineral composition. The wind speeds recorded in the hyper-arid region fluctuate around 20 kmph. Therefore, it has been hypothesized that dust particles may travel only up to 100 km in the hyper-arid desert. This transit occurs within a time span of 5–6 hours (Hartley et al., 2005; Azua-Bustos et al., 2015, 2019). The survival and perpetuation of these microbes depend on the characteristics of the soil substrate.

Azua-Bustos et al. (2019) conducted direct DNA extraction and 16S rRNA analysis on dust borne microbes. They found *Oceanobacillus oncorhynchi*, *Paenibacillus* sp., *Terribacillus saccharophilus*, *Arthrobacter* sp., *Staphylococcus equorum*, and *Bacilluus firmus*. There were several bacteria that did not lend to cultural techniques. They were traced only via specific DNA sequences. Several species of archaea were also detected in the dust samples from hyper arid region of desert.

Sporulation is an important characteristic of microbes detected in the aeolian dust sampled from the hyper-arid core regions of the Atacama Desert in Chile. Approximately 71% of the identified bacterial species were of the sporulating type. The presence of thick spore walls and the ability to hibernate in unfavorable conditions may be crucial. Sporulation provides resistance against the harsh Atacama environment. Bacterial species like *Microbacter* sp. appear to be transported by wind into the core of the desert region from peripheral areas (Azua-Bustos et al., 2019).

4.3 SOUTHERN EUROPEAN AND MEDITERRANEAN DUST STORMS: MICROBIOLOGICAL ASPECTS

Dust storms of varying intensity are generated in the cropping zones of European plains. These dust storms can contribute to particulate matter (PM) in the air. The bioaerosols present above crops contain a wide range of microbes, including several important plant pathogens. Some fungal species found in coarse particles (>3 μm) can cause health issues for farm animals and humans. Many of these fungi are natural inhabitants of the atmosphere and can be found at different altitudes. Frohlich-Nowoisky et al. (2009, 2016a, b) conducted a detailed study on aerosols in both rural cropping areas and urban regions. They specifically examined the diversity and population trends of microbes that reside on aeolian dust particles. The relative abundance of fungal classes was as follows: Basidiomycetes 64%, Ascomycetes 34%, Fungi imperfecti or Deuteromycetes 1%, and others 0.5%. Fungal classes, genera, and species were identified using 15S rRNA molecular methods (Frohlich-Novoisky, 2009; Womiloju et al., 2003). Airborne Basidiomycetes have been extensively studied (Levitin, 1990). Among Basidiomycetes, *Agaricomyces* was the most prevalent in dust samples, accounting for 87% of the basidiomycetes found in the location. Pucciniomycete spores were also commonly observed. This class of fungi includes the notorious rust fungi that cause epidemics in major cereals. *Tremellomyces, Ustillagomycetes* (smut fungi), *Exobasidiomycetes, Wallemomycetes, Mirbasidiomyctes, Cystobasidiomycteas,* and a few others were also detected in the dust samples.

The families of Ascomycetes traced in dust were *Sordarimycetes, Dothiomycetes, Eurotiomycetes, Leotiomyctes,* and *Saccharomyces.* Their relative abundance ranged from 10-24% of the total ascomycetes. Other ascomycetes traced were *Lecanoromycetes* and *Orbilliomycetes. Orbillomycetes* ranged from 0.5 to 2% of fungal flora. Fungal diversity in dust samples was similar to many other locations in the European plains. Seasonal changes in fungal spores were prominent, with the autumn season supporting a greater population of fungal spores in the bioaerosols. Fungal genera that are potential pathogens were prominent in cropping zones. Frohlich-Nowoisky et al. (2009, 2016a) reported that fungal species encountered also depended on the particle size of dust storms. Fine PM had more basidiomycetes. Species richness was also affected by the size of dust particles. Dust storms originating from the African continent transport copious quantities of nutrient-rich PM, which could harbor microbes. Several fungal pathogens of plants have been traced in the dust storms reaching the Mediterranean region (e.g., rust fungi).

The Pyrenees in Northeast Spain are said to receive dust plumes. A large population of microbes is attached to dust from the Western Sahara/Sahel, reaching the Spanish region. Dust storms are also generated locally in Western Europe. Barberan et al. (2014, 2015) attempted to identify the source of microbes in the dust that settled in the Southwest European region. They conducted molecular typing using 16S rRNA sequences. They found that immediately after a dust storm, the wind brought particles and microbes that were more related to Saharan assemblies. The microbial population in the dust was related to those found in the dry soils of Mauritania (Western Sahara). A high microbial diversity was noticed in the aerosol that deposited in the Pyrenees. Microbial assemblages in the soil surface, surface waters, and dust exhibited certain similarities.

In the following paragraphs, let us consider details about the dust storms related to the Mediterranean region. Studies depicting the diversity and relative abundance of microbes in the dust storms generated across the Mediterranean are limited. Knowledge about microbes in dust storms is essential, especially considering their importance to natural ecosystem functions. The particulate matter (PM) encountered in the Mediterranean, Arabian, and North African dust storms was less than 53 $\mu g/m^3$. The dust storms may harbor several genera of bacteria. The nature of the dust storm, its origin, and the microbial population in the origin are important factors to consider. During the transit of microbes on PM, several physiological aspects are crucial to their survival, particularly their ability to tolerate the harsh conditions experienced in the atmosphere. The nature of dust particles is also crucial for the survival and movement of microbes in the atmosphere. Let us consider an example from the dust-prone locations of the Mediterranean. The agrarian regions in North Africa, Syria, Israel, Egypt (Nile Delta), Sinai, and Saudi Arabia were analyzed for bacterial composition in the atmosphere, particularly during dust storm events. The microbes were compared with a control when the agrarian sky experienced a tranquil (no dust, clear sky) condition.

Gat et al. (2017) have reported interesting facts about dust storms and microbial loads. North African dust storms showed predominance of *Deinococcus thermus* compared to dust storm originating in Syria or Saudi Arabia. They suspect that differences in the composition of dust and trajectories might contribute to differences in the composition of microbial species. For example, dust from Saharan locations in the east was exposed to only shorter distances in the air. Cyanobacteria and Chloroflexi were predominant (Hanada, 2014). Additionally, the period of the year when samples were drawn from dust storms of different origins might have influenced the bacterial composition. Microbial diversity could be influenced by the distance the dust storm has moved when samples were drawn. For example, in the case of samples from Syria, the population of *Firmicutes* sp. was consistently higher (These were named Cluster I samples in the study).

The dust storm in Saudi Arabia was composed of a higher population of Firmicutes. Firmicutes constituted 29% of the total recorded bacteria. *Acinetobacter, Proteobacteria,* and *Bacteroides* were also dominant. In a few samples collected in Saudi Arabia, the population of Acinetobacter reached 30% of the total bacteria. Other bacteria observed in the dust samples from Saudi Arabia belonged to *Chloroflexaceae.* These results confirm that the microbial density in a dust storm depends on its origin, distance traveled, and the nature of particles. We should expect dust storms from different origins to support various types of plant pathogens and symbionts. The North African dust samples (cluster II samples) contained Frankiales (Actinobacteria), which were ubiquitous in soils. Some species were able to fix atmospheric nitrogen (Normand, 2006). Rubrobacter (Actinobacteria) was found to be radio-tolerant, and those isolated from desert soils were tolerant to desiccation (Ferreira et al., 1999). Pontibacter (Bacteroidetes) were found in both desert and forest soils (Zhou et al., 2007). Truepera (Deinococcus– Thermus) were desiccation-resistant and radio-resistant bacteria (Ho et al., 2016). Kineosporiaceae (Actinobacteria) were commonly isolated from soils and plant matter (Ho et al., 2016; Lee et al., 2016). Cellulomonadaceae (Actinobacteria), which are cellulose-degrading soil bacteria, were also identified (Stackebrandt et al., 2006). It is possible that a few microbial species may be lost during transit. This aspect also requires attention from microbiologists.

The dust storm samples collected during clear days in Cluster III contained *Sphingomonas* (α proteobacteria), which were found in aqueous environments, rhizosphere, foliage, and other parts of plants (White et al., 2017). Acinetobacter (γ proteobacteria), traced in the samples, are known human pathogens (Teixeira & Merquior, 2014). Dermabacteraceae (Actinobacteria) were associated with human skin, poultry litter, and cheese surfaces (Stackebrandt, 2014). *Escherichia, Shigella* (Proteobacteria), along with other human pathogens and commensals, were also present (Janda, 2006). Pathogenic Clostridia (Firmicutes) were isolated from various environments, including ice, desert soil, and dust (Wiegel et al., 2006). *Streptococcus* (Firmicutes) were frequently observed as commensal colonizers, but were also found on plant surfaces or in association with rich carbon sources (Lory, 2014). It appears that dust storms were not a major source of antibiotic resistance gene sources (ARGS), as these genes were derived from anthropogenic activity (Gat et al., 2017; Mazar et al., 2016; D'Costa et al., 2006; Martinez, 2008; WHO, 2016). Clearly, there is a need to focus on microbes capable of causing diseases in crops.

Dust storms emanating from crop production zones in Syria, Egypt, Israel, and other Arabian locations contained *Sphingomonas, Arthrobacteria, Microcoleus, Blastococcus,* and *Solirubrobacterales* (Belnap & Gardner, 1993). A compilation by Gat et al. (2017) lists the following microbes found in desert sand, which differed significantly from those found in agricultural soils (Sghaier et al., 2016). Some examples of microbes traced in the dust storms include Frankiales (Actinobacteria, soil species) (Normand, 2006), *Rubrobacter* (Actinobacteria, radio-tolerant bacteria), *Pontibacter* (Bacteroidetes, found in desert and forest environments), *Deinococcus* (desiccation tolerant), *Acinetobacter* (soil species found in desert and forest soils), Cellulomonadeceae (cellulose degrading soil species), Dermabacteria (human pathogen) (Stackebrandt, 2014; Stackebrandt et al., 2006), *Clostridium, Escherichia, Streptococcus* (commensals or pathogens found in human inhabited areas) (Wiegel et al., 2006), and *Shigella* (commensal or human pathogen) (Janda, 2006), among others. Many of these microbes were already present as commensals on crop surfaces before dust formation. As mentioned earlier, the abundance of microbial species depended on the specific dust cluster from which the sampling was conducted. For example, *Sphingomonas* was found in soil, slush, rhizosphere, and leaf surfaces (White et al., 2017).

Leski et al. (2011) have traced diverse microbes in the dust plumes created by deserts and cropping zones of the Arabian region (Plate 4.1). Detailed sequencing and comparison have led them to conclude that pathogens affecting crops were present in those dust storms. It is worth noting that several studies have shown rust fungi infect cereals in Asia, Africa, and North America, residing in the dust particles. They hitchhike a ride on dust storms and jet streams across the Sahara Desert, eventually reaching the North American wheat belt.

In the West Asia region, a few nations like Iran, Iraq, and even Saudi Arabia have practiced mulching of their agricultural crops. Mulching has reduced the formation of dust storms and the loss of precious topsoil. However, Nourmoradi et al. (2015) point out that such programs have been discontinued in the recent past, resulting in more frequent dust storms. For example, Draxler et al. (2001) reported the accumulation of dust in the atmosphere with PM of <10 μm size. In parts of the Mediterranean and Arabia, they traced dust storms that caused the accumulation of 3,000 μm m^{-3} dust in the air. Such dust is rich in microbes that could cause crop diseases. Dust storms also reduce the air quality above cropping zones.

Furthermore, Nourmoradi et al. (2015) have reported that on a clear day without a dust storm, the air contained the following microbes: *Bacillus* sp., *Cladosporium, Trychophyla, Mycobacterium,* and *Aureabasidium.* In the dust storm region, the air contained the following bacteria and fungi: *Bacillus* sp., *Mycosporium, Cladosporium, Chryosporium,* and *Pseudomonas aeruginosa.* It is clear that the microbial load and diversity differ due to dust storms compared to clear days.

The Saharan dust is dispersed over long distances within the African continent. Mineral dust is transported from the Sahara in the form of strong pulses into the Mediterranean region (Loye-Pilot et al., 1986; Bergametti et al., 1989a, b; Moulin et al., 1998). Saharan dust storms also reach locations across Southern Europe. Reports indicate that North African dust from the Sahara even reaches locations in the British Isles. Let's consider an example of Saharan dust originating from the North African regions of Morocco, Algeria, Tunisia, Libya, and Egypt. Simulations based on trajectory have also revealed that dust particles reaching the Mediterranean could have originated from regions in North Africa (MeteoTrentino, 2017; Scheuvens et al., 2013; Stockdale et al., 2016; *see* Plate 4.1).

There are studies about microbes in North African dust that reach the snow fields and Alps. Weil et al. (2017) state that massive amounts of sand and organic particulate matter reach the Italian Alps each year. Several genera of bacteria and fungi from African dust storms have been found in the snow deposits of the Italian Alps. The Italian Alps receive diverse microbes along with nutrient-rich dust particles. The following are a few families of bacteria and fungi recorded by Weil et al. (2017). They are *Geodermatiphillaceae, Bacillaeae, Micrococcaceae, Methylobacteriaceae, Rhadobacteriaceae, Rubrobaceriaceae, Psueudonocardiaeae, Kineosporaceae, Nocardiaceae, Planococcaceae, Gemmatimonadiaceae,* and *Micromonosporaceae.* In addition, soil microbes beneficial to vegetation like *Beijerinkia* were also traced in the Saharan dust. A few species of *Beijerinckia* fix atmospheric-N.

Microbes that are transported on dust storms from the Sahara have been found to be radiation tolerant. They can withstand desiccation, limited nutrients, and varying moisture levels. According to Weil et al. (2017), the presence of fungi in air samples collected by Charles Darwin in 1838 from the Sahara provides undeniable evidence of the survival capabilities of dust microbes. Bacteria found on dust particles often form spores to endure the harsh conditions during their journey through the Sahara. Additionally, there are specific microbial species that can tolerate the freezing conditions and snowflakes in the Southern Alps (Bonasoni et al., 2014; Chuvochina et al., 2011; Darwin, 1845a, b; Furman, 2003; Gorbushina et al., 2007; Smith, 2013; Smith et al., 2011; Meola et al., 2015).

Rosselli et al. (2015) have also conducted a study on the dust storms originating from North Africa and reaching Southern Europe, specifically Italy. These dust storms travel across the Mediterranean Sea, suspended in the atmosphere. It is evident that the dust storms that settle in European regions, including agricultural areas, carry a significant population of microbes. The researchers detected microbes in the dust particles captured in the Sardinian atmosphere during their transit. Through 16S RNA analysis, they identified diverse genera of microbes involved in various biogeochemical and nutrient transformations that are beneficial for crop growth. Additionally, they also identified certain microbes capable of causing plant diseases. Rosselli et al. (2015) reported the presence of several bacterial genera in the dust sample, including *Alphaproteobacteria, Gammaproteobacteria, Paracoccus, Sphingomonas, Methylobacter, Corinebacteria, Caulobacter, Brevundimonas, Citrobacter, Cronobacter, Klebsiella,* and *Stenotrophomonas.*

As mentioned previously, agrarian regions in southern Europe are susceptible to dust, various harmful microbes, and pollutants, if any. These particles travel long distances across the Mediterranean Sea. Some of the dust microbes are pathogens that can have severe consequences for crops such as cereals, legumes, vegetables, and fruits. It is evident that Saharan dust from North Africa can impact crop fields in the Mediterranean and southern European regions. Rusts, blights, blotches, mildews, and cankers are among the diseases spread by airborne microbes. These plant pathogens are transported from the African continent to Southern Europe through dust storms.

Weil et al. (2017) conducted a study on microbes in the Dolomite mountains of Italy. This region consistently experiences dust storms originating from the Sahara. In soil layers containing Saharan sand/dust, *Acinetobacter* and *Dothideomyces* were found to be abundant. Italian locations that were free from Saharan dust had the following microbes: *Acinetobacter,*

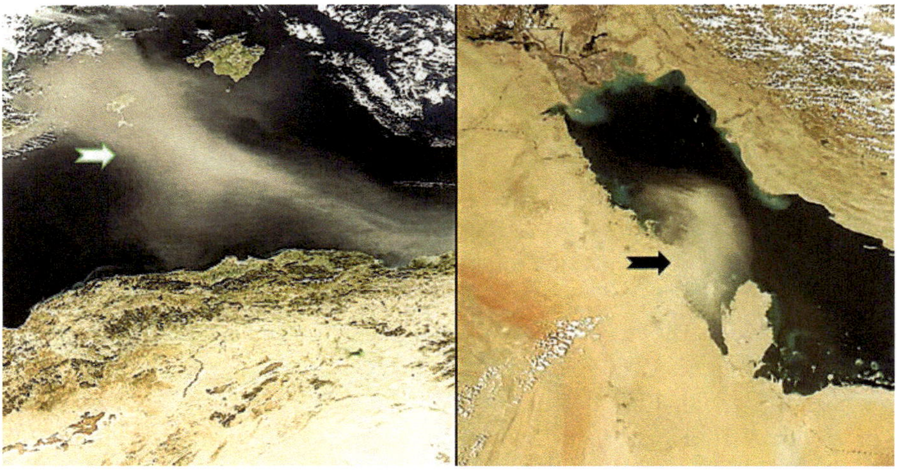

PLATE 4.1 Dust storms transit intercontinental distances and disperse microbes. [Left: Dust storm above the Mediterranean Sea. It originates in North Africa and reaches the Spanish and Italian coasts. Right: Dust storm over the Persian Gulf. Such storms originate from Saudi Arabia and reach the Middle Eastern region, affecting crops in Iran, Syria, and Israel].

Note: Such dust storms carry microbes. Firstly, they disperse microbes into different regions. Dust microbes could be commensals, crop pathogens that cause diseases, or symbionts. Dust storms are perhaps the most important natural processes globally that aid the migration of microbes. Dust facilitates the intermixing of the microbial population. Dust storms have the potential to generate microbial recombinants. Perhaps inter-continental dust storms laden with microbes represent the largest migratory event of life forms on Earth.

Source: Schmaltz, J. Goddard Space Observatory, NASA, Washington, USA, also see NASA (2001).

Caulobacter, Acquabacterium, Beijerinckia, Novosphingobium, Granullicella, Terriglobus, Pelomonas, Ralstonia, and *Phenylobacterium.*

Bacteria encountered in Saharan dust above the Italian region also possessed enucleation properties. Such dust-driven snowfalls are becoming common in the Mediterranean region. The bacterial genera related to snow formation were *Pseudomonas* and *Acinetobacter*. Spores of a few molds were also traced. The genus *Penicillium* was predominant in most soil samples taken from snow-clad regions of the Dolomite mountains. Undoubtedly, the Mediterranean agrarian region is exposed to periodic or routine incursions of Saharan dust, organic PM, and microbes. These could have massive consequences for the ecosystem. The Italian Dolomite mountains received asymbiotic-N fixers like *Beijerinkia.* This indicates that the agrarian sky aids in the mixing of microbial components, dust, minerals, and organic particulates. Crops do respond to such changes in the soil caused by aerial deposits.

Gemmatimonadetes are normally not present in the Southern European air, but they have been noticed in recent years. Clearly, they are transported from a different region. In due course, they may become endemic to the region. This proves that new microbial species are added by dust storms, providing an opportunity for crossing with native microbes and resulting in the creation of new genetic diversity.

Aeolian dust in the Mediterranean region may harbor nutrients that are beneficial to crops. When a dust storm occurs, it contributes to the fertility of the soil by adding nutrients such as PO_4, SO_2, Fe, and Mn (Rahav et al., 2016, 2018). Additionally, aeolian dust storms can transport a large population of biological N fixers, known as diazotrophs. If the soil lacks native diazotrophs, this dust can stimulate higher levels of N fixation. Experiments conducted along the Red Sea coast and adjacent terrain have revealed that aeolian dust storms actually provide the nutrients required by algae. Therefore, these dust storms could be beneficial for farming in sandy arid soils. However, it is important to note that dust storms deplete the surface soil, reduce rooting depth, and decrease the population of surface microbial organisms that are beneficial to crops. The physicochemical properties of the soils where sandstorms originate are also significant.

Recently, Krupnik et al. (2021) made an interesting observation regarding the interaction between algal epibionts and dust microbes. They demonstrated that dust plumes contain microbes that are beneficial for restoring optimal growth of algae. The presence of dust nutrients and microbes positively influenced the epibionts on *Ulva* sp., resulting in the optimum growth of marine algae. Similarly, positive interactions between dust microbes and epiphytes on the phyllosphere of crops may occur. The phyllosphere, which is another area in cropped fields that receives significant amounts of dust from storms, is inhabited by vital microbes that serve various ecosystem functions.

The cropping expanses in Israel are frequently exposed to dust storms from different directions. Dust storms from the Mediterranean, Syria, Western Sahara, and Saudi Arabia in the east reach Israel. In fact, Israel receives dust from all directions (Weizmann Institute of Science, 2017). Crop damage due to severe sandstorms is possible. The dust microbiome consists of various microbes such as fungi, bacteria, archaea, viruses, etc. Fungal pathogens can have devastating effects on crops. Microbes with antibiotic resistance genes (ARG) have also been detected (Rudich, 2017). There is a certain level of specificity regarding the desert dust, microbes, and types of antibiotic resistance. Bacterial species serve as 'signatures' of the dust storms, allowing us to determine the origin of the dust storm samples. This can be achieved by studying the assemblages of microbes. Researchers at the Volcani Centre in Israel state that dust may also carry biological nitrogen fixers, which have the potential to improve soil nitrogen status (Rudich, 2017).

Microbes detected using 16S rRNA analysis belonged to the five major phyla: Acidobacteria, Actinobacteria, Bacteroidetes, Proteobacteria, and Firmicutes. Furthermore, Rahav et al. (2016) reported that the heterotrophic microbial community detected in the Mediterranean dust storms contained members of Pseudomonadaceae (3–11%), Bacillaceae (4–34%), Rhodobacteriaceae (4–13%), Cytophagaceae (3–16%), Frankiales (2–8%), and Chloroflexaceae (2–8%).

4.4 AEOLIAN DUST STORMS AND MICROBIAL ACTIVITY IN SAHELIAN FARMING REGION AND SAHARA

The West African Sahara and Sahelian region is the world's largest generator of dust storms. Such dust storms are composed of mineral-rich sand particles, organic dust, and microbes. Satellite images show the occurrence of dust storms, depicting the movement of massive quantities of dust in Sahelian West Africa. These dust storms and heat waves generated in the Sahel are important natural phenomena (e.g., harmattan; haboobs; see Plate 4.2). The consequences to natural vegetation, crops, and their productivity could vary depending on the geographic location. The Sahelian dust is, in reality, an important repository of diverse microbes and other microscopic biotic constituents. The bioaerosol generated in the Sahel needs greater attention (Sokolik & Toon, 1999). Specifically, we need to understand its consequences on sparse natural vegetation and dryland crops of the Sahel, as well as its influence on the maintenance of microbial diversity in the soil profile. Fluctuations in microbial assemblages need attention.

Seasons influence the formation and movement of dust storms across Sahelian farm regions. Surface wind velocities are regulated by seasonal patterns. However, dust generation is higher during winter-spring. The duration of a dust storm and its transport process can last for 1–6 days, depending on local conditions. Dust storms carry dust plumes laden with bioaerosols, microbes, and organic/mineral constituents. These dust plumes also contain diverse microbes. Dust emission during the dry season is variable. The microbial assemblages on dust particles may also vary based on the season.

Studies conducted by Marticorena et al. (2010) involved three stations located in the Sahel region, namely Mbora in Senegal, Banizoumba in Niger, and Cinzana in Mali. Measurements in agricultural stations in the

Sahel have shown that monitoring meteorological parameters such as wind direction, speed, altitude, temperature, and relative humidity is important as they affect dust and its biotic composition. The daily concentration of dust material generated can range from 10 $\mu g\ m^{-1}$ to 1,000 $\mu g\ m^{-1}$, depending on the season. Local wind patterns can also influence the dust pattern near the ground surface and the natural movement of dust storms at higher altitudes. Modeling the emanation, intensity, size, transport, nutrients, and microbial load of dust storms could be useful (Marticorena & Bergametti, 1995, 1996; Laurent et al., 2008; Mborou et al., 1997; Rajot, 2010).

The Sahelian agricultural belt is a region where the atmosphere is characterized by dust storms. These dust storms are known to harbor the highest quantities of suspended particles. According to Marticorene et al. (2010), the alternation of harmattan dust storms and the wet season is an important meteorological aspect of the Sahel. Meteorological parameters such as temperature, precipitation, relative humidity, and wind direction are key aspects of Sahelian agriculture. Atmospheric dust concentrations follow a seasonal cycle, and dust storm patterns can be characterized on a regional scale. This seasonal cycle of dust concentrations is an important aspect of farming (Marticorena et al., 2010; Jones & Harrison, 2004; Rajot, 2010).

The concentration of dust can vary seasonally, and this variability may be influenced by the frequency of intense dust storms. The seasonality of these storms also impacts the presence of dust microbes. It is now evident that Sahelian dust storms annually remove and transport significant amounts of sand particles, nutrients, and microbes (Acosta-Martinez, 2015). These storms are notable migratory events for the biotic components of the atmosphere. Imbler (2020) highlights that approximately 182 million tons of dust are emitted and transported from the western Sahara each year. This migration of dust, mineral nutrients, organic PM, and microbes is one of the largest migratory events on Earth, occurring consistently year after year. While animal or insect migrations are generally recognized as similar events, they are smaller in scale.

In certain locations, sudden changes in wind velocity are known to affect convective currents and dust generation. Each dust plume event removes microbes from agricultural fields, displacing them to different sites in Sahel or even farther. Studying the microbial diversity, intensity, and total population shifted by such dust storms could be useful. It is important to better understand the overall dynamics of dust and microbial shifts caused in Sahel. There is some information on plant pathogens dispersed in dust storm traits. Additionally, these dust storms may also contain beneficial microbes (Imbler, 2020).

PLATE 4.2 Harmattan dust storm in the Sahelian West Africa.

Note: Harmattan dust carries and distributes particulate matter laden with nutrients and microbes. The dust hangs and sometimes moves slowly over agrarian regions.

Source: Dr. K. R. Krishna, International Crops Research Institute for the Semi-arid Tropics, Hyderabad, India, Location: South of Niamey near Bernin-n-Konni, Niger, West Africa (1985).

It is not enough if the dust cycles occurring in various locations of the Sahel are known. The microbes carried by dust, particularly their potential to cause diseases (e.g., rust, downy mildews, blotches, streaks/mosaics caused by viruses, etc.), are equally important (Imbler, 2020). Dust may also contain beneficial microbes such as nitrogen fixers and mycorrhizal spores. Microbial commensals are also present. We need detailed knowledge about the spatial and temporal changes in microbial load during Saharan dust storms. We should avoid agricultural practices that may lead to the widespread dispersal of crop disease-causing microbes. It is important to protect the limited natural vegetation that endures harsh conditions in the Sahel. If possible, we should confine disease-causing microbes to a single location.

The dust storms, such as Harmattan, Haboob, Irifi, Guebli, Khamsin, and several others, that are generated in the Sahara have been well documented. They have been observed for several millennia. These storms can be devastating to crop plants in the Sahel region. Sometimes, we even encounter their effects in North America (Roberts & Knippertz, 2012). Hoppe (2004) explains the process of African dust storm generation and their impact on crops along the Atlantic coast and the Great Plains. It appears that Sahelian dust is rich in mineral nutrients. In fact, the sand particles that travel across the ocean to North America are rich in silicates, iron, zinc, phosphorus, and other elements. Dust particles also contain microbes such as algae, bacteria, fungi, archaea, and minute insects and their eggs. Due to the harsh conditions in the ambient atmosphere, we often encounter resting stages such as bacterial spores, fungal spores, and fungal hyphae with the potential to regenerate. It is said that silicate/iron-rich dust and sand particles that land on water bodies can induce red tides, which are red algal mats.

The quantity of dust storms generated in the Sahara and Sahel is extremely high, ranging in teragrams (Tg) each year (Acosta-Martinez et al., 2015). In recent years, North American crops have experienced 2-3 African dust storms, which are carried by high-altitude jet streams. These sandstorms deposit dust along with nutrient-rich particles, as well as carry deadly crop pathogens. For instance, rust fungi that attack wheat, maize, and sugarcane can travel long distances from crop belts in North Africa (Egypt, Morocco, Algeria, etc.). Drought years typically generate more dust from fields, especially those without crops, mulches, or cover crops. Cover crops are rarely practiced in the Sahel during winter, leaving fields vulnerable until the next rainy season. Consequently, large volumes of dust are transported across the Atlantic during winter months. It is important to note that a single dust storm can contain up to 65 species of microbes, with approximately 25% of the diversity being contributed by crop-pathogenic microbes. Rust fungi (urediospores and telia) are known to disperse over long distances, reaching as far as the Caribbean and Great Plains. These fungi reside on dust particles generated in the Sahel, Kenya, North Africa (Egypt), and the Mediterranean.

Dust storms allow microbes to reside and survive on them. They also aid microbes in migrating. An interesting phenazomenon related to the migration of microbes from West Africa to the U.S. Virgin Islands and the Caribbean region was observed by Griffin et al. (2001, 2003, 2006). Earlier studies by Prospero et al. (1970, 2005) and Prospero & Nees (1986) had already shown that dust particles above Caribbean cropping zones were associated with African Dust events. African dust was even traced in the Amazonia (Swap et al., 1992; Schmidt, 2001; Griffin et al., 2006; Tulet et al., 2008). These studies demonstrated that microbial counts in the atmosphere over the US Virgin Islands were linked to dust storm events in Africa. Dust and microbes migrate long distances through jet streams, clouds, and winds into the

Caribbean. Researchers utilized the auto-fluorescence of bacteria and fungi to count them under a fluorescence microscope. They identified bacteria-like and fungi-like organisms in air samples collected from both 'dusty sky' and 'clear sky' conditions. Microbes cultivable on nutrient medium were 2–3 times more abundant in dust storm samples compared to 'clear sky' samples. The microbial population in the Caribbean atmosphere increased, coinciding with dust storm events in Africa. Furthermore, Griffin et al. (2001) were only able to detect autotrophs (bacteria-like) organisms in samples from storms originating in Africa, and not in other samples. Undoubtedly, the ability of microbes to survive, metabolize, and synthesize energy is essential. Energy is required for the normal activity of microbes residing on dust particles during transit. Energy is also needed in the soil (terrain) after microbes have landed in a remote location. The nature of the substrate and its ability to provide an energy source determine whether microbes will survive or perish. These are crucial aspects of the long-distance dispersal of aeolian microbes. Such African dust storm particles may also induce pollution and crop diseases, which can be hazardous to human health if consistently inhaled (Schmidt, 2001; Griffin et al., 2003).

Dust storms may contain up to 15 bacteria per liter of air. Most of them complete the long-distance transit into the Caribbean or even to the Great Plains of the USA. They survive the harsh transit. Kellogg & Griffin (2006), Prospero (1999), and Prospero et al. (2005) found that dust storms enhance the bacterial population in the air above the US Virgin Islands by several folds. Surprisingly, such dust storms may be generated in the Eastern Sahara, particularly in regions like Sudan and Egypt. They also pick up microbes as they transit across the Sahara. Several dust storms may also be generated *in situ* in the West Sahel, for example in Senegal, Mali, etc. The dust contains well-shrouded and protected microbes. Several genera of bacteria, fungi, archaea, and viral particles have also been traced in the dust.

Kellogg & Griffin (2003, 2006) have found that several microbial species complete trans-Atlantic journeys with organic PM as a substrate. The bright pigmentation is known to protect the microbes against harmful radiation in the atmosphere. The microbial load of a dust particle is an important aspect. They state that about 10,000 to 15,000 microbes per gram of dust is possible. A conservative estimate indicates that 1 million tons of dust are transported per dust event, landing in locations in North America. Proportionately, approximately a quadrillion microbes from Sahelian cropping zones in Mali, Burkina Faso, and Senegal could land in the Caribbean isles (Griffin et al., 2003, 2006).

It is important to note that African dust storms also contain several plant pathogenic species. Dust storms undoubtedly disperse crop diseases into North American crop belts. The rust fungi that affect wheat, maize, and other grasses are good examples of transatlantic dispersal of crop diseases via dust storms and wind (jet streams). Fungal, bacterial, and viral attacks on rice and beans have been traced to dust movement into the Caribbean from the Western Sahel. Sugarcane rust is also transported to the Caribbean via dust storms generated in eastern Africa. Livestock diseases are also spread via dust storms on the African continent. Dust particles may harbor pathogens that attack livestock. For example, Foot and Mouth disease virus particles resting on dust particles may find their way into the Caribbean (Kellogg & Griffin, 2003; Griffin et al., 2006). Often, a crop disease may reach epidemic proportions if the dust is richer in pathogenic microbes. The expression of coral reef diseases in the Caribbean strictly coincided with peaks of Sahelian dust storms that occurred in 1974, 1985, and 1994. At the same time, we should note that Sahelian storms also carry beneficial microbes. We have to evaluate the population of Biological N-fixers (BNF), mycorrhizal fungal spores, and phyllosphere microbes in the dust storms. The periodic infusion of beneficial microbes from the Sahel is a topic that needs greater attention. A 'dust storm' naturally inoculates the world's cropping zones with both beneficial and pathogenic microbes throughout its path.

Dust storms that settle into Florida have induced algal blooms. Nutrient-rich dust storms supplement oceanic water, leading to the growth of algal blooms. Similar effects on agrarian soils cannot be ruled out. It may be worth investigating the quantity of nutrient-rich dust added to the spodosols of Central Florida. The dust storms that reach the soils could be promoting vegetation and crop growth in Florida and the Caribbean. This suggestion requires further investigation (Griffin, 2009). Additionally, it should be noted that dust from the Sahara may contribute to the population of *Vibrio cholera* in the water and marshes of the Florida Keys (Bucher, 2017).

Sandstorms also occur in the semi-arid regions of Southern Africa, including the Kalahari region. These sandstorms could have an impact on cropping systems in the vicinity and even in farther locations on the continent.

4.5 ASIAN DUST STORMS-MICROBIAL DIVERSITY AND THEIR ROLE IN NATURE

Asian drylands and deserts are prone to inducing dust storms. According to Tang et al. (2018), Asian dust storms, particularly those seen in the Chinese mainland, are known to travel long distances. They carry bioaerosols as they transit from one location to another, including organic dust particles, pollen, and microbes. Dust storms originating in West and Central China often reach Japan, resulting in the presence of dust particles and microbes from Central Asia, Mongolia, and the Chinese Plains in the atmosphere of the Far East and Japan (Iwasaka et al., 1983, 2009; Huang et al., 2008a, b, 2010). Such dust storm events in Asia, in general, are known to contain a high density of microbial debris, bacterial cells, fungi, hyphae, spores, viral particles, and organic dusts (Hua et al., 2007; Maki et al., 2015, 2017, 2018; Yeo & Kim, 2002; Wang et al., 2010, 2014, 2015).

Recent reports by Chen et al. (2017) and others suggest that the Gobi Desert contributes dust to the cropping belts of East Asia. It also plays a significant role in disseminating microbes in this region. Approximately 35% of the dust generated in the Gobi region reaches Far Eastern Asia. Another source of dust storms is the Taklimakan. Asian dust storms are known to carry dust particles and microbes across trans-oceanic locations in North America and the Arctic. These storms also transport microbes to locations in the French Alps and Greenland (Bory et al., 2003; Grousset et al., 2003; Wang, 2018; Huang et al., 2015). It appears that Asian storms are capable of completing one full circuit of the Earth's surface in 13 days (Uno et al., 2009).

There are reports that deal with the detection, identification, and even enumeration of microbes on the dust storms generated in the Gobi region of Mongolia. For example, a survey of field soils, lakebeds, cropped areas, and aeolian dust particles was conducted by Hagiwara et al. (2020). They utilized 16S rRNA analysis to detect specific microbes. Soil bacterial flora was among the most dominant microbes detected in the dust particles. They were able to identify 2 phyla and 7 families of bacteria. These bacteria were frequently encountered in the dust storms. In fact, Hagiwara (2021) comments that the terrain and its characteristics affect the composition of airborne bacteria.

Dust storms cause a large-scale disturbance in the aeolian microbial component. The aeolian microbial population can be detected using 16S rRNA methods. Long-term survival of microbes is necessary during inter-continental transit. The microbial population on dust particles will have to withstand harsh conditions. To perpetuate, microbes should find a suitable substrate before the microbe loses viability (Huang et al., 2010, 2017a, b). Upon deposition, dust storm microbes may affect the soil microbial flora.

Dust storms are important natural phenomena observed in agrarian regions (Rao et al., 2020). The Chinese cropping belts are no exception. Dust storms lead to atmospheric contamination (Rao et al., 2020). According to a study by Rao et al. (2020) in the agrarian region of eastern China, dust microbes respond to changes in atmospheric conditions. It is reported that several genera of microbes reside on PM. Microbes in the aerosols have been observed to travel over 1,000 km in Eastern China (Rao et al., 2020; Hara & Zhang, 2012; Huang et al., 2008a, b). These microbes must survive for over a week or two during transit before settling in a different location (Sun et al., 2003, 2004; Wang et al., 2015).

Liao et al. (2013) observed that microbial aerosols were influenced by immediate meteorological parameters. The physicochemical constituents of dust particles are also important. Dust particles may contain SO_2, SO_4^{2-}, NO_3^-, NH_4^+, and Ca_2^+. These constituents have an impact on microbial diversity. The bacterial population generally decreases on sunny days. However, a dusty atmosphere supports a higher bacterial load during a storm. Several reports indicate that the microbial population during a dust storm in Eastern China was 10 times higher compared to non-storm tranquil conditions in the air (Jeon, 2011; Maki et al., 2015). The bacterial population is higher near the ground surface. *Aquabacterium* sp., *Flavobacteria, Prevotellaceae, Propionibacterium* sp., *Bacillus* sp., and *Acinetobacter* sp. were commonly detected during dust storms in the atmosphere (Lee et al., 2009; Maki et al., 2015). Rao et al. (2020) has detected Proteobacteria, Actinomycetes, and Firmicutes in dust storms of eastern China. Others were *Rubellimicrobium, Paracoccus, Chroococcidiopsis, Deinococcus, Clostridium. Deinococcus* were noticed during three stages of progression of dust storm. Microbes detrimental to human health too could be traced in the PM (Griffin, 2007). Further, circulation of aerosol microbes due to dust storms could have important impact on ecological manifestation of cropping regions.

Dust storms of various intensities occur in China. These storms can be transported over long distances. As mentioned earlier, dust storms have the potential to carry microbes during their journey. Such storms can have an impact on both human and crop health. Understanding the pathway of Asian Dust storms originating from mainland China, Mongolia, and adjacent regions of Fareast is valuable (Tang et al., 2018; Fu et al., 2016; Huang et al., 2008a, b, 2011). Detailed investigations are required to study the dynamics of bioaerosols in order to develop preventive measures. Tang et al. (2018) conducted a study on Asian dust storms originating from the Gobi Desert, Taklimakan, and other regions. They monitored the microbial load and diversity of bioaerosols, analyzing bacteria through microscopic analysis and staining with 4,6-diamidino-2-phenylindole (DAPI) and 16S rRNA analysis. The relative abundance of bacteria varied among bioaerosol samples collected from different locations. Microbial populations typically increased significantly immediately after a dust event. For instance, in most locations, the population of Firmicutes did not exceed 5% of the total bacteria counts in the dust samples. Interestingly, previous studies have indicated that the nature of terrain, soil, vegetation, crops, and even infrastructure at the place of origin of dust storms were crucial factors. Such factors determine bacterial and fungal diversity in dust samples. For example, the relative abundance of microbes and their diversity depended on the location and dust storm clusters (Gat et al., 2017). However, Tang et al. (2018) conducted a detailed examination of dust samples and non-dust clear aerosols. They concluded that the bacterial and fungal population were specific to dust storms, their routes, altitude, and movement. The microbial population and diversity traced on dust or clear bioaerosol samples from the Gobi Desert differed. The nature of soil, sand, or vegetation had a negligible effect on microbial diversity in dust samples.

Regarding Archaea, Tang et al. (2018) report that dust storm samples contained two phyla of archaea kingdom. They were identified as Crenarchaeote and Euryarchaeotic. Crenarchaeote includes Thaumarchaeota. While Euryarchaeotic includes the more easily traced *Methanobacteria*, *Thermoplasmata*, *Methanomicrobia*, and *Halobacteria*. Yet, the relative abundance of archaea was much lower than bacteria or fungi encountered in the dust storm samples. Archaea constituted less than 2% of the total microbes in the dust samples and were not commonly found in the dust storms. Tang et al. (2018) and Grantham et al. (2015) further state that the bacteria and fungal spores detected in dust storm samples could be used to identify the movement of air masses across long distances, even continents. Certain microbial species could serve as markers to determine the origin of dust storms and the routes they have traveled. Therefore, it is important to conduct sampling carefully and investigate microbial diversity thoroughly. Microbes present in the dust can also be used as indicators of wind/dust storms and their potential to cause diseases or epidemics in crops.

Tang et al. (2018); and Maki et al. (2015, 2017) have noticed diversity of bacterial species in the dust sample drawn from different locations in the Gobi Desert (*see* Table 4.1). Sample from farming regions near Ulaanbaatar and Erenhot has revealed predominance of *Acidobacter*, *Actinobacter*, *Bacteroides*, *Chloroflexi*, *Fermicutes*, *Gemmatimonades*, and *Protobacteria*. A few of the bacterial species mentioned above were also found in the clear air sample. Generally, the population of bacterial species was much higher in the dust sample compared to the clear-air samples.

Tang et al. (2018) mention that Asian deserts are known to impact the atmosphere above cropping belts by introducing microbes. Dust from western China carries not only bacteria but also harmful fungal species. Asian dust storms transport fungal phyla over long distances, including the yellow dust storm that reaches Taiwan. Airborne fungal genera have been detected in samples from these dust events, even in far-off South Korea. Dust particles with organic matter provide a favorable surface for resting spores of various fungi, enhancing their survival. Several fungal phyla encountered in aerosol samples above the Mongolian desert and cropping zones are listed in Table 4.1. Periodic Asian dust storms can disseminate fungal spores and contribute to epidemics. Agricultural experts often express concerns about the rapid spread of fungal diseases such as rust, blotches, smuts, fusarial disease, and more.

Airborne dust samples collected from the Gobi Desert and along the storm's pathway have revealed a diverse range of fungal propagules (Tang et al., 2018). This diversity may be attributed to periodic additions to the fungal flora as the dust storms progress. The predominant fungi identified in the Gobi Desert dust samples were Ascomycota, specifically Dothideomycetes and Sordarimycetes. Several strains of Agaricomycetes belonging to Basidiomycota were also detected, albeit with much lower diversity compared to ascomycetes. Chytridiomycota were also found in the dust samples. Additionally, there was an abundance of *Microbotryomycetes, Pucciniomyces* (rust fungi), *Eurotiomyces,* and *Tremellomyces.* Some of these fungi may have originated from the storms. Notably, a significantly high population of Eurotiomycetes was observed in the dust samples from Ulaanbaatar, Mongolia. However, it appears that a few genera of fungi are present in both dust and non-dust clear air samples, regardless of the geographic origin of the dust storms. There was no definitive pattern of fungal genera identified in the 'dust' and 'non-dust clear air' samples. The dispersal route and the adaptation of fungal spores to substrates are important factors that determine the population and diversity of fungi in dust storms.

Dust storms generated in the Central Asian deserts of Gobi and Taklimakan are known to travel long distances. Typically, they reach agricultural areas in Northeast China, the Korean Peninsula, and Japan. This phenomenon is commonly referred to as 'Yellow Dust' due to the color of the sand and organic matter in the particles. Yellow dust storms occur from spring to summer each year, depositing nutrient-rich particulate matter, microbes, and pollutants. These storms pick up dust and microbes from various locations along their route (Maki et al., 2017). Traveling on jet streams from Mongolia to Japan, the yellow dust can transit over 4,000 km eastwards at altitudes

TABLE 4.1 Relative Abundance of Bacterial and Fungal Phyla Traced in the 'Dust Storm' and 'Clear Air' (Non-Dust Storm) Samples Drawn from Above Agrarian Regions Near Ulaanbaatar, in Mongolia, During 2017

Bacterial Phyla	
Dust storm samples	*Saprosporia* (24.0%), *Actinobacteria* (16.1%), *Alpha Proterobacteria* (16.1%), *Bacillus* (3.0%), *Beta Proterobacteria* (3.0%), *Cytophaga* (3.5%), *Gamma Proterobacteria* (12.1%), *Chloroflexi* (NA), *Gemmatinomoides* (2.8%), *Thermoliophila* (4.8%), *Sphingobacilli* (NA), *Solibacteria* (NA), *Delta Proterobacteria* (NA), Others (15.1%).
Clear air (non-dust)	*Saprosporia* (21.4%), *Actinobacteria* (20%), *Alpha Proterobacteria* (13.8%), *Bacillus* (2.9%), *Beta Proterobacteria* (4.2%), *Cytophaga* (5.6%), *Gamma Proteobacteria* (2.3), *Thermoliophila* (3.6%), *Sphingobacilli* (3.3%) Solibacteria (3.0%), *Delta Proterobacteria* (4.5%) and others (17.1%).
Fungal Phyla	
Dust storm samples	*Agaricomycetes* (38.6%), *Dothideomycetes* (16.6%), *Eurotiomycetes* (16.8%), *Pezzizomycetes* (2.3%), *Sordariomycetes* (25.1%), *Pucciniomycetes* (NA), *Microbotryomycetes* (NA), *Tremellomyces* (NA), Others (0.6%).
Clean air (non-dust)	*Agaricomycetes* (66.4%), *Dothideomycetes* (12.3%), *Eurotiomycetes* (NA), *Pezzizomycetes* (NA), *Sordariomycetes* (21.0%), *Pucciniomycetes* (13.3%), *Microbotryomycetes* (1.1%), *Tremellomyces* (3.5%), and others (0.3%).

Note: Percentage predominance relative to the total microbial diversity detected in the samples has been shown in parentheses. 'Dust' and 'non-dust clear air' samples were drawn from similar altitudes. The bacteria were detected using a 16S rRNA-based method, along with staining using 4,6-diamidino-2-phenylindole (DAPI). Fungal genera were detected and identified based on morphological features of spores and hyphae.

Source: Tang et al. (2018); Rao et al. (2020); Maki et al. (2015, 2017); and Hanada (2014).

ranging from 1,000 to 4,000 m in the atmosphere. The yellow dust is known to contain abundant bacteria and other microbes. Aeolian specialists from different countries have extensively studied the dispersal of bacteria through yellow dust. They have traced the yellow dust along its entire path, including in Japan. Based on their observations of microbial load and diversity, they have concluded that dust storms facilitate the dispersal of microbes. The microbial composition may vary depending on the height of the dust storm, the characteristics of the terrain where it originates, the distance traveled, and the microbial richness along the dispersal route. Bacteria, fungi, and archaea were easily identified by their fluorescence characteristics. Temporal changes in dust and microbial diversity were also observed. During spring, dust particles ranging from 0.2 to 30 μm were most prevalent.

Maki et al. (2017) reported that microbial particles fluctuated from 3.8 $\times 10^4$ to 2.0 $\times 10^6$ cells m^{-3} in aerosols. During dusty days, 5- to 10-fold more microorganisms were detected in the atmosphere. Bacterial and fungal species were detected using 16S rRNA gene sequences. Samples collected in Korea (Yongin and Yonago) and Japan after the dust storm had almost completed its transit showed over 285 and 247 taxonomic units belonging to 21 and 27 phyla, respectively. Approximately 190 and 179 families of bacterial species were easily recognizable in the two locations. It is important to note that dust storms in Asia, known as 'yellow dust,' serve as a repository of genetically diverse species of bacteria, including plant pathogens. Some of these bacteria are commensals. Bacteria from the samples collected in Yongin and Yonago could be categorized into 282 and 245 operational taxonomic units, respectively, consisting of 21 and 27 phyla as well as 194 and 170 families. The following three key points were recognized: 1) 'yellow dust' may carry bacterial plant pathogens, 2) efforts should be made to identify and take precautionary measures against them, and 3) this will benefit the cereal belts of Northeast China and the Far East.

Dust storms generated in the Arabian Peninsula are known to rapidly transit into the western shores and inland regions of India. This area is abundant in vegetation and crops, making it susceptible to intense microbial activity. This phenomenon occurs despite the semi-arid agroclimate. In 2015, an Arabian dust storm carrying a diverse range of microbes reached India after traveling over 2,000 km. These microbes eventually settled in Western India (Ghosh, 2016). The microbial composition of the dust particles was captured on Petri plates containing nutrient-rich medium. Selective media for bacteria and fungi were utilized. The fine fraction of the dust storm exhibited a significant presence of microbes. The dust storm indeed had an impact on the microbial diversity and population in the soil where it settled. A total of 92 microbial species were detected, including 32 bacterial genera and four fungal species. Archaea were also found in the dust samples. These microbes had traveled on dust particles for 2,000 km. They had overcome harsh dry climate over the Arabian desert. Then, transited through humid conditions generated by Arabian sea. The dust microbes also withstood strong radiation and desiccation if any, during the transit. Microbial species belonging to *Bacillus*, *Arthrobacter*, *Staphylococcus*, *Kocuria*, *Streptomyces*, and *Lysinibacillus* were frequently traced, in the dust samples. Osmotolerant and halotolerant (salt-tolerant) bacterial genera, such as *Psychrobacter,* and *Exiguobacterium*, were recorded, specifically, during the dust storm. A small proportion of microbes were opportunistic pathogens on crops. The versatile ability of microbes found in Arabian dust samples was similar to those reported from locations in Iran, Egypt, and Israel. Leski et al. (2011) adopted molecular techniques (15S rRNA) to detect microbes on dust floating above the atmosphere of deserts in Iraq and Kuwait.

4.6 AUSTRALIAN DESERT EXPANSES AND MICROBES

A large dust storm, known as 'The Red Haze,' swept across southern Australia in 2009. This event followed a decade-long drought. The dust carried diseases that affected crops, animals, and humans, in addition to causing environmental pollution. Approximately 5,000 tons of dust material, containing microbes, were carried over New South Wales. In comparison, dust storms in Northeast China can carry up to 30,000 tons of dust material (Kelly, 2009). These dust storms reduce the amount of light available for photosynthesis in crops, but they also contribute minerals to the soil. Furthermore, they have the potential to spread crop diseases such as rusts and blotches. Dust storms in this region can also be triggered by forest fires and crop burning.

KEYWORDS

- **aeolian dust**
- **Atacama**
- **bioaerosol**
- **dust bowl**
- **dust plumes**
- **Geubli**
- **gobi**
- **great plains**
- **haboob**
- **harmattan**
- **Irifi**
- **Mediterranean**
- **Sahara**
- **Taklamakan**

REFERENCES

Acosta-Martinez, V., Van, P. S., Moore-Kucera, J., Baddock, M. C., & Zobeck, T. M., (2015). Microbiology of wind eroded sediments: Current knowledge and future. *Aeolian Research, 18*, 99–113.

Arimoto, R., Duce, R. A., Savoie, D. L., Prospereo, J. M., Talbot, R., Cullen, J. D., Tomza, U., et al., (1996). Relationship among aerosol constituents from Asia and North Pacific during PEM- West A. *Journal of Geophysics Research, 101*, 2011–2024. doi: 10.1029/95jd01071.

Azua-Bustos, A., Caro-Lara, L., & Vicuña, R., (2015). Discovery and microbial content of the driest site of the hyper arid Atacama Desert, Chile. *Environmental Microbiology Reporter, 7*, 388–394.

Azua-Bustos, A., Gonzalez-Silva, C., Fernandez-Martinez, M. A., Arenas-Fajerdo, C., Fonseca, R., Martin, T., Fernandez-Sampedro, M., et al., (2019). Aeolian transport of viable microbial life across the Atacama Desert, Chile: Implications for Mars. *Nature Research Scientific Reports,9*, 11024 https://doi.org/10.1038/s41598-019-47394-z/.

Barberan, A., Henley, J., Fierer, N., & Casamayor, E. O., (2014). *Structure, Inter-Annual Recurrence, and Global-Scale Connectivity of Airborne Microbial Communities* (pp. 1–14). https://doi.org/10.1016/j.scitotenv.2014.04.030.

Barberan, A., Ladau, J., Leff, J. W., Pollard, K. S., Menninger, H. L., Dunn, R. R., & Fierer, N., (2015). Continental-scale distributions of dust associated bacteria and fungi. *Proceedings of National Academy of Science U. S. A., 112*, 5756–5761.

Behzad, H., Mineta, K., & Gojobori, T., (2018). Global ramifications of dust and sandstorm microbiota. *Genome and Biological Evolution, 10*, 1970–1987. doi: 10.1093/gbe/evy13.

Belnap, J., & Gardner, J. S., (1993). Soil microstructure in soils of the Colorado plateau: The role of the *Cyanobacterium Microcoleusvaginatus*. *Great Basin Naturalist, 53*, 40–47.

Bergametti, G., Dutot, A. L., Buat-Ménard, P., Losno, R., & Remoudaki, E., (1989b). Seasonal variability of the elemental composition of atmospheric aerosol particles over the north-western Mediterranean. *Tellus B., 41*, 353–361.

Bergametti, G., Gomes, L., Remoudaki, E., Desbois, M., Martin, D., & Buat-Ménard, P., (1989a). Present, transport and deposition patterns of African dusts to the North-Western Mediterranean. In: Leinen, M., & Sarnthein, M., (eds.), *Paleoclimatology and Paleometeorology: Modern and Past Patterns of Global Atmospheric Transport* (pp. 227–252). Kluwer Academic Inc., Dordrecht, Netherlands.

Bonasoni, P., Cristofanelli, P., Marinoni, A., Duchi, R., Calzolari, F., & Roccato, F., (2014). *Italian Climate Observatory "O. Vittori" at Mt. Cimone.* GAW-WMO Global Station-WINTER 2013/2014 Report [Internet]. Available from: http://www.isac.cnr.it/cimone/sites/default/files/ICOVCimoneReport_3.pdf (accessed on 19 January 2024).

Bory, A. J. M., Biscaye, P. E., & Grousset, F. E., (2003). Two distinct seasonal Asian source regions for mineral dust deposited in Greenland (NorthGRIP). *Geophysics Research Letters, 30*, 149–153. https://doi.org/10.1029/2002GL016446,2003.

Bowers, R. M., Clements, N., Emerson, J. B., Wiedinmyer, C., Hannigan, M. P., & Fierer, N., (2013). Seasonal variability in bacterial and fungal diversity of the near-surface atmosphere. *Environmental Science Technology, 47*, 12097–12106.

Brodie, E. L., DeSantis, T. Z., Parker, J. P. M., Zubietta, I. X., Piceno, Y. M., & Andersen, G. L., (2007). Urban aerosols harbor diverse and dynamic bacterial populations. *Proceedings of National Academy of Sciences. U. S. A., 104*, 299–304.

Bucher, G. B., (2017). *Could Dust Storms from Africa be Damaging Ecosystems in the Florida Keys* (pp. 1–4). Pacific Standard. https://psmag.com/environment/its-getting-dusty-down-in-the-florida-keys/ (accessed on 19 January 2024).

Burrows, S. M., Butler, T., J¨ockel, P., Tost, H., Kerkweg, P¨oschl, U., & Lawrence, M. G., (2009). Bacteria in the global atmosphere – Part 2: Modeling of emissions and transport between different ecosystems. *Atmospheric Chemistry and Physics, 9*, 9281–9297. www.atmos-chem-phys.net/9/9281/2009/ (accessed on 19 January 2024).

Chen, S., Huang, J., Li, J., Jia, R., Jiang, N., Kang, L., Ma, X., & Xie, T., (2017). Comparison of dust emissions, transport, and deposition between the Taklimakan Desert and Gobi Desert from 2007 to 2011. *China Earth Science, 60*, 1338–1355. https://doi.org/10.1007/s11430-016-9051-0,2017.

Chuvochina, M. S., Marie, D., Chevallier, S., Petit, J. R., Normand, P., & Alekhina, I. A., (2011). Community variability of bacteria in Alpine snow (Mont Blanc) containing Saharan dust deposition and their snow colonization potential. *Microbes and Environment, 26*, 237–247.

Creamean, J. M., Suski, K. J., Rosenfeld, D., Cazoria, A., DeMott, Sullivan, R. C., White, A. B., et al., (2013). Dust and biological aerosols from the Sahara and Asia influence precipitation in the Western U.S. *Science, 339*, 1572–1578. doi: 10.1126/science.1227279.

D'Costa, V. M., McGrann, K. M., Hughes, D. W., & Wright, G. D., (2006). Sampling the antibiotic resistome. *Science, 311*, 374−377.

Darwin, C. R., (1845a). An account of the fine dust which often falls on vessels in the Atlantic Ocean. *Quarterly Journal of Geological Society of London, 2*, 26–30.

Darwin, C. R., (1845b). Journal of researches into natural history and geology of the countries visited during the voyage of HMS beagle round the world. Under the command of captain Fitzroy, R. N. John Murray, London. United Kingdom. Quoted in: Acosta-Martinez, V., Van, P. S., Moore-Kucera, J., Baddock, M. C., & Zobeck, T. M., (2015). Microbiology of wind eroded sediments: Current knowledge and future. *Aeolian Research, 18*, 99–113.

Draxler, R. R., Gillette, D. A., Kirkpatrick, J. S., & Heller, J., (2001). Estimating PM10air concentrations from dust storms in Iraq, Kuwait, and Saudi Arabia. *Atmospheric Environment, 35*(25), 4315–4330. http://dx.doi.org/10.1016/S1352-2310(01)00159-5.

Eager, J. D., Pierre, H., & Hartnet, H. E., (2017). *The Characterization of Haboobs and the Deposition in Tempe, AZ from 2005 to 2014* (pp. 1–4). National Aeronautics and Space Agency, Washington, USA. https://www.science.gov/topicpages/h/haboob+dust+storms (accessed on 19 January 2024).

Ferreira, C. A., Nobre, F. M., Moore, E., Rainey, A. F., Battista, R. J., & Da Costa, M. S., (2019). Characterization and radiation resistance of new isolates of *Rubrobacter radiotolerans* and *Rubrobacter xylanophilus*. *Extremophiles,3*, 235−238.

Fischer, E. V., Hsu, N. C., Jaffe, D. A., Jeong, M. J., & Gong, S. L., (2009). A decade of dust: Asian dust and spring time aerosol load in the U. S. Pacific Northwest. *Geophysics Research Letters, 36*, L03821. doi: 10.1029/2008GL036467.

Fröhlich-Nowoisky, J., Hill, T. C. J., Pummer, B. G., Yordanova, P., Franc, G. D., & Pöschl, U., (2015). Ice nucleation activity in the widespread soil fungus *Mortierellaalpina. Biogeoscience, 12*, 1057−1071.

Frohlich-Nowoisky, J., Kampf, C. J., Weber, B., Huffman, J. A., Pöhlker, C., Andreae, M. O., Lang-Yona, N., et al., (2016a). Bioaerosols in the earth system: Climate, health, and ecosystem interactions. *Atmospheric Research, 182*, 346–376. https://doi.org/10.1016/j.atmosres.2016.07.018.

Frohlich-Nowoisky, J., Pickersgilla, D. A., Despre, V., & Poschla, U., (2009). High diversity of fungi in air particulate matter. *Proceedings of National Academy of Science, 106*, 12814–12819. www.pnas.org.cgi (accessed on 19 January 2024).
Frohlich-Nowoisky, J., Schmüser, L., Mauri, S., Scheel, J. F., Knopf, D. A., Pöschl, U., Bonn, M., & Weidner, T., (2016b). Ice-nucleating bacteria control the order and dynamics of interfacial water. *Scientific Advances*. 2.e150163010.1126/ sciadv.1501630.
Fu, P. Q., Zhuang, G. S., Sun, Y. L., Wang, Q. Z., Chen, J., Ren, L. J., Yang, F., et al., (2016). Molecular markers of biomass burning, fungal spores and biogenic SOA in the Taklimakan desert aerosols. *Atmosphere and Environment, 130*, 64–73. https://doi.org/10.1016/j.atmosenv.2015.10.087,2016.
Furman, H. K. H., (2003). Dust storms in the middle east: Sources of origin and their temporal characteristics. *Indoor Built Environment, 12*, 419–426.
Gat, D., Mazar, Y., Cytry, E., & Rudich, Y., (2017). Origin-dependent variations in the atmospheric microbiome community in Eastern Mediterranean dust storms. *Environmental Science and Technology, 51*, 6709–6718.
Ghosh, S., (2016). *Dust Storms Could Spread Microbes, Finds Study* (pp. 1–10). Prevention Web. https://www.preventionweb.net/news/view/69767.
Ginoux, P., Prospero, J. M., Gill, T. E., Hsu, N. C., & Zhao, M., (2012). Global-scale attribution of anthropogenic and natural dust sources and their emission rates based on MODIS deep blue aerosol products. *Reviews in Geophysics, 50*. http://dx.doi.org/10.1029/2012RG000388. RG3005.
Gonzalez, C., Teigell, M. N., Valaldares, P. B., & Griffin, D. W., (2014). *The Global Dispersion of Pathogenic Microorganisms by Dust Storms and its Relevance to Agriculture*. https://www.sciencedirect.com/science/article/pii/B9780128001318000017?via%3Dihub; https://doi.org/10.1016/B978-0-12-800131-8.00001-7(accessed on 19 January 2024).
Gorbushina, A. A., Kort, R., Schulte, A., Lazarus, D., Schnetger, B., & Brumsack, H., (2007). Life in Darwin's dust: Intercontinental transport and survival of microbes in the nineteenth century. *Environmental Microbiology, 9*, 2911–2922.
Grantham, N. S., Reich, B. J., Pacifici, K., Laber, E. B., Menninger, H. L., Henley, J. B., Barberán, A., et al., (2015). Fungi identify the geographic origin of dust samples. *PLoS One, 10*. e0122605.
Griffin, D. W., (2007). Atmospheric movement of microorganisms in clouds of desert dust and implications for human health. *Clinical Microbiology Review, 20*(3), 459–477. doi: 10.1128/CMR.00039-06.
Griffin, D. W., (2009). *Desert Dust-Storm Microbiology* (pp. 1–7). U.S. Geological Survey, Tallahassee, Florida. https://www.accessscience.com/content/desert-dust-storm-microbiology/YB090025 (accessed on 19 January 2024).
Griffin, D. W., Garrison, V. H., Herman, J. R., & Shinn, E. A., (2001). African desert dust in the Caribbean atmosphere: Microbiology and public health. *Aerobiologia, 17*, 203–213. https://doi.org/10.1023/a:1011868218901.
Griffin, D. W., Kellogg, C. A., Garrison, V. H., & Shinn, E. A., (2002). The global transport of dust. *Atmospheric Science, 90*, 228–235.
Griffin, D. W., Kellogg, C. A., Garrison, V. H., Lisle, J. T., Borden, T. C., & Shinn, E. A., (2003). Atmospheric microbiology in the northern Caribbean during African dust events. *Aerobiologia, 19*, 143–157. https://doi.org/10.1023/B:AERO.0000006530.32845.8d.

Griffin, D. W., Westphal, D. L., & Gray, M. A., (2006). Airborne microorganisms in the African desert dust corridor over the mid-Atlantic ridge, ocean drilling program, leg 209. *Aerobiologia, 22*, 211–226. https://doi.org/10.1007/s 10453-006-9033-z.

Grousset, F. E., Ginoux, P., Bory, A., & Biscaye, P. E., (2003). Case study of a Chinese dust plume reaching the French alps. *Geophysics Research Letters, 30*, 23–26. https://doi.org/10.1029/2002GL016833.

Guieu, C., Dulac, F., Ridame, C., & Pondaven, P., (2014). Introduction to project DUNE, a DUST experiment in a low nutrient, low chlorophyll ecosystem. *Biogeoscience, 1*, 425–442. www.biogeosciences.net/11/425/2014/ doi: 10.5194/bg-11-425-2014.

Hagiwara, K., (2021). *Viable Bacteria in Dust Generating Area.* https://encyclopedia.pub/2600 (accessed on 19 January 2024).

Hagiwara, K., Matsumoto, T., Tsedendamba, P., Baba, K., & Hoshino, B., (2020). Distribution of viable bacteria in the dust-generating natural source area of the Gobi Region, Mongolia. *Atmosphere, 11*, 893 https://doi.org/10.3390/atmos11090893.

Hanada, S., (2014). *The Phylum Chloroflexi, the Family Chloroflexaceae, and the Related Phototrophic Families Oscillochloridaceae and Roseiflexaceae: The Prokaryotes* (pp. 515–532). Springer, Berlin Heidelberg, Germany.

Hara, K., & Zhang, D., (2012). Bacterial abundance and viability in long-range transported dust. *Atmosphere and Environment, 47*, 20–25, https://doi.org/10.1016/j.atmosenv.2011.11.050,2012.

Hartley, A., Chong, G., Houston, J., & Mather, A., (2005). 150 million years of climatic stability: Evidence from the Atacama Desert, Northern Chile. *Journal of Geological Society. London, 162*, 421–424.

Hervas, A., Camarero, L., Reche, I., & Casamayor, E. O., (2009). Viability and potential for immigration of airborne bacteria from Africa that reach high mountain lakes in Europe. *Environmental Microbiology, 11*, 1612–1623. https://doi.org/10.1111/j.14622920.2009.01926.x,2009.

Ho, H. M., Rao, C. Y., Hsu, H. H., Chiu, Y. H., Liu, C. M., & Chao, H. J., (2005). Characteristics and determinants of ambient fungal spores in Hualien, Taiwan. *Atmospheric Environment, 39*, 5839–5850. https://doi.org/10.1016/j.atmosenv.2005.06.034,2005.

Ho, J., Adeolu, M., Khadka, B., & Gupta, R. S., (2016). Identification of distinctive molecular traits that are characteristic of the phylum "Deinococcus-Thermus" and distinguish its main constituent groups. *Systematic and Applied Microbiology, 39*, 453–463.

Hoppe, K. K., (2004). *Out of Africa: Hitching a Ride on African Dust Plumes* (pp. 1–3). http://baysoundings.com/legacy-archives/fall04/dust.html (accessed on 19 January 2024).

Houston, J., & Hartley, A., (2003). The central Andean west-slope rain shadow and its potential contribution to the origin of hyper-aridity in the Atacama Desert. *Desert International Journal Climatology, 23*, 1453–1464.

Hua, N. P., Kobayashi, F., Iwasaka, Y., Shi, G. Y., & Naganuma, T., (2007). Detailed identification of desert-originated bacteria carried by Asian dust storms to Japan. *Aerobiologia (Bologna), 23*, 291–298. https://doi.org/10.1007/s10453-007-9076-9.

Huang, J., Fu, Q., Zhang, W., Wang, X., Zhang, R., Ye, H., & Warren, S. G., (2011). Dust and black carbon in seasonal snow across Northern China. *Bulletin of American Meteorology Society, 92,* 175–181, https://doi.org/10.1175/2010BAMS3064.1,2011.

Huang, J., Li, Y., Fu, C., Chen, F., Fu, Q., Dai, A., Shinoda, M., et al., (2017a). Dryland climate change: Recent progress and challenges. *Reviews in Geophysics, 55*, 1–60. https://doi.org/10.1002/2016RG000550.

Huang, J., Minnis, P., Chen, B., Huang, Z., Liu, Z., Zhao, Q., Yi, Y., & Ayers, J. K., (2008a). Long-range transport and vertical structure of Asian dust from CALIPSO and surface measurements during PACDEX. *Journal of Geophysics Research-Atmosphere, 113*, 1–13. https://doi.org/10.1029/2008JD010620,2008a.

Huang, J., Yu, H., Dai, A., Wei, Y., & Kang, L., (2017b). Drylands face potential threat under 2°C global warming target. *Nature Climate Change, 7*, 417–422. https://doi.org/10.1038/nclimate3275.

Huang, J., Yu, H., Guan, X., Wang, G., & Guo, R., (2015). Accelerated dryland expansion under climate change. *Nature Climate Change,6*, 166–171. https://doi.org/10.1038/nclimate2837.

Huang, J., Zhang, W., Zuo, J., Bi, J., Shi, J., Wang, X., Chang, Z., et al., (2008b). An overview of the semi-arid climate and environment research observatory over the loess plateau. *Advances in Atmospheric Science, 25*, 906–921, https://doi.org/10.1007/s00376-008-0906-7,2008b.

Huang, K., Zhang, G., Li, J., Wang, Q., Sun, Y., Lin, Y., & Fu, J. S., (2010). Mixing of Asian dust with pollution aerosol and the transformation of aerosol components during the dust storm over China in spring 2007. *Journal Geophysics Research, 15*, 1307–1314. https://doi.org/10.1029/2009JD013145.

Huang, Z., Huang, J., Bi, J., Wang, G., Wang, W., Fu, Q., Li, Z., Tsay, S. C., & Shi, J., (2010). Dust aerosol vertical structure measurements using three MPL lidars during 2008 China-US joint dust field experiment. *Journal Geophysics Research, 115*, D00K15. https://doi.org/10.1029/2009JD013273,2010.

Huang, Z., Huang, J., Hayasaka, T., Wang, S., Zhou, T., & Jin, H., (2015). Short-cut transport path for Asian dust directly to the Arctic: A case study. *Environment Research Letters, 10*, 114018 https://doi.org/10.1088/1748-9326/10/11/114018,2015.

Imbler, S., (2020). *A Giant Dust Storm is Heading Across the Atlantic* (pp. 1–4). https://www.theatlantic.com/science/archive/2020/06/saharan-dust-storms-giving-earth-life/613441/ (accessed on 19 January 2024).

Iwasaka, Y., Hiroaki, M., & Nagaya, K., (1983). The transport and spatial scale of Asian dust storm clouds: A case study of the dust storm event of April 1979. *Tellus B., 35*, 189–196. https://doi.org/10.3402/tellusb.v35i3.14594,1983.

Iwasaka, Y., Shi, G. Y., Yamada, M., Kobayashi, F., Kakikawa, M., Maki, T., Naganuma, T., et al., (2009). Mixture of Kosa (Asian dust) and bioaerosols detected in the atmosphere over the Kosa particles source regions with balloon borne measurements: Possibility of long-range transport. *Air Quality and Atmospheric Health, 2*, 29–38. https://doi.org/10.1007/s11869009-0031-5,2009.

Janda, J. M., (2006). New members of the family *Enterobacteriaceae*. In: Dworkin, M., Falkow, S., Rosenberg, E., Schleifer, K., &Stackebrandt, E., (eds.), *The Prokaryotes: Proteobacteria: Gamma Subclass* (Vol. 6, pp. 5–40). Springer: New York, NY, USA.

Jeon, E. M., Kim, H. J., Jung, K., Kim, J. H., Kim, M. Y., Kim, Y. P., & Ka, J. O., (2011). Impact of Asian dust events on airborne bacterial community assessed by molecular analyses, atmospheric. *Environment, 45*, 4313–4321. https://doi.org/10.1016/j.atmosenv.2010.11.054.

Jeon, E. M., Kim, Y. P., Jeong, K., Kim, I. S., Eom, S. W., Choi, Y. Z., & Ka, J. O., (2013). Impacts of Asian dust events on atmospheric fungal communities. *Atmospheric Environment, 81*, 39–50. https://doi.org/10.1016/j.atmosenv.2013.08.039,2013.

Jones, A. M., & Harrison, R. M., (2004). The effects of meteorological factors on atmospheric bioaerosol concentrations: A review. *Science and Total Environment, 326*, 151–180. https://doi.org/10.1016/j.scitotenv.2003.11.021.

Katra, I., Arotsker, L., Krasnov, H., Zaritsky, A., Kushmaro, A., & Ben-Dov, E., (2014). Richness and diversity in dust storm borne biomes at the Southeast Mediterranean. *Science Reporter, 4*, 5265.

Kellogg, C. A., & Griffin, D. W., (2003). *African Dust Carries Microbes Across the Ocean: Are They Affecting Human and Ecosystem Health?* United States Geological Survey Washington US. http://coastal.er.usgs.gov/african_dust/ (accessed on 19 January 2024).

Kellogg, C. A., & Griffin, D. W., (2006). Aerobiology and the global transport of desert dust. *Trendsin Ecology and Evolution, 21*, 638–644.

Kelly, J., (2009). *Mega Dust Storms Spread Diseases Worldwide But Can Help Mitigate Global Warming* (p. 104). Green Car Congress. https://www.greencarcongress.com/2009/09/mega-dust-20090928.html (accessed on 19 January 2024).

Krishna, K. R., (2003). *Agrosphere: Nutrient Dynamics, Ecology and Productivity* (p. 356). Science Publishers Inc. Enfield, New Hampshire, USA.

Krishna, K. R., (2014). *Agroecosystems: Soils, Climate, Crops, Nutrient Dynamics, and Productivity* (p. 576). Apple Academic Press Inc. Waretown, New Jersey, USA.

Krishna, K. R., (2015). *Agricultural Prairies: Natural Resources and Crop Productivity* (p. 499). Apple Academic Press Inc. Waretown, New Jersey, USA.

Krishna, K. R., (2022). *Agricultural Sky: A Concept to Revolutionize Farming* (p. 435). Apple Academic Press Inc. Burlington, Ontario, Canada.

Krupnik, N., Asis, D. T., Belkin, N., Rubin-Blum, M., Israel, A., Paytan, A., Meiri, D., Herut, B., & Rahav, (2021). Dust-borne microbes affect *Ulva ohnoi*'s growth and physiological state. *FEMS Microbial Ecology, 97*, fiab020. https://doi.org/10.1093/femsec/fiab020/.

Kuske, C. R., Barns, S. M., & Busch, J. D., (1997). Diverse uncultivated bacterial groups from soils of the arid southwestern united states that are present in many geographic regions. *Applied and Environment Microbiology, 63*(9), 3614–3621.

Lang-Yona, N., Lehahn, Y., Herut, B., Burshtein, N., & Rudich, Y., (2014). Marine aerosol as a source for endotoxins in coastal areas. *Science and Total Environment, 499*, 311–318. doi: 10.1016/j.scitotenv.2014.08.054.

Laurent, B., Marticorena, B., Bergametti, G., L´eon, J. F., & Mahowald, N. M., (2008). Modeling mineral dust emissions from the Sahara Desert using new surface properties and soil database. *Journal of Geophysics Research, 113*, D14218. doi: 10.1029/2007JD009484.

Lee, D., Jang, J. H., Cha, S., & Seo, T., (2016). *Thalassiellaazotovora gen. Nov., sp. Nov.*, a new member of the family Kineosporiaceae isolated from Sea water in South Korea. *Current Microbiology, 73*, 676–683.

Lee, J. F., Allen, B. L., Peterson, R. E., Gregory, J. M., & Moffett, K. E., (1994). Environmental controls on morning dust direction in Lubbock, Texas, USA. *Earth Surface Processes and Landforms, 19*, 437–449.

Lee, J. F., Wigner, K. A., & Gregory, J. M., (1993). Drought, wind, and blowing dust on the southern high plains of the United States. *Physical Geography, 14*, 56–67.

Lee, S., Choi, B., Yi, S. M., & Ko, G., (2009). Characterization of microbial community during Asian dust events in Korea. *Science and Total Environment, 407*, 5308–5314. https://doi.org/10.1016/j.scitotenv.2009.06.052.

Leski, T. A., Malanoski, A. P., Gregory, M. J., Lin, B., & Stenger, D. A., (2011). Application of a broad-range resequencing array for detection of pathogens in desert dust samples

from Kuwait and Iraq. *Applied Environment Microbiology, 77*, 4285–4292. https://doi.org/10.1128/AEM. 00021-11.
Levitin, E., (1990). Studies on airborne basidiospores. *Aerobiologia, 6*, 177–180.
Liao, X., Hu, A. Y., Yang, X. Y., Chen, J. S., Yu, C. P., & Lin, J. Q., (2013). Community composition and environmental sources of bacterial and eukaryotic microbiomes in fine air particles (PM2.5) of Xiamen in winter. *Ecology and Environmental Science, 22*, 1395.
Lindsay, B., (2018). *Dust Storms and Ocean Spray Lift Viruses into Atmosphere by the Millions, Study Says* (pp. 1–8). CBC News. https://www.cbc.ca/news/canada/british-columbia/dust-storms-virus-ocean-spray-atmosphere-study-1.4523728/ (accessed on 19 January 2024).
Lloyd, G., (2021). *Our Environment Victoria University of Environmental Science* (pp. 1–8). Wellington, New Zealand. https://ourenvironment.ac.nz/2021/05/22/dust-with-a-side-of-disease/ (accessed on 19 January 2024).
Lory, S., (2014). The family *Streptococcaceae*. In: Rosenberg, E., DeLong, E. F., Lory, S., Stackebrandt, E., & Thompson, F., (eds.), *The Prokaryotes: Firmicutes and Tenericutes* (pp. 367–370). Springer: Berlin, Heidelberg, Germany.
Loye-Pilot, M. D., Martin, J. M., & Morelli, J., (1986). Influence of Saharan dust on the rain acidity and atmospheric input to the Mediterranean. *Nature, 321*, 427–428. doi: 10.1038/321427a0.
Makhalanyane, T. P., (2015). Microbial ecology of hot desert edaphic systems. *FEMS Microbiological Reviews, 39*(2), 203–221.
Maki, T., (2012). Outbreak of foot-and-mouth disease in Miyazaki from March to July 2010—Effect of yellow sand and local surface wind. *Journal of Arid Land Studies, 22*, 167–170.
Maki, T., Hara, K., Kobayashi, F., Kurosaki, Y., Kakikawa, M., Matsuki, A., Chen, B., et al., (2015). Vertical distribution of airborne bacterial communities in an Asian-dust downwind area, Noto peninsula. *Atmospheric Environment, 119*, 282–293. https://doi.org/10.1016/j.atmosenv.2015.08.052.
Maki, T., Kurosaki, Y., Onishi, K., Lee, K. C., & Pointing, S. B., Jugder, D., Yamanaka, N., et al., (2017). Variations in the structure of airborne bacterial communities in Tsogt-Ovoo of Gobi Desert area during dust events. *Air Quality and Atmospheric Health, 10*, 249–260. doi 10.1007/s11869-016-0430-3.
Maki, T., Lee, M. C., Kawai, K., Onishi, K., Hong, C. S., Kurosaki, Y., Shinoda, M., et al., (2018). Aeolian dispersal of bacteria associated with desert dust and anthropogenic particles over continental and oceanic surfaces. *Journal of Geophysical Research: Atmospheres, 124*, 5579–5588. doi: 10.1029/2018JD029597.
Marticorena, B., & Bergametti, G., (1995). Modeling the atmospheric dust cycle: 1-Design of a soil derived dust production scheme. *Journal of Geophysics Research, 100*, 16415–16430.
Marticorena, B., & Bergametti, G., (1996). Two-year simulations of seasonal and interannual changes of the Saharan dust emissions. *Geophysics Research Letters, 23*, 1921–1924.
Marticorena, B., Chatenet, B., Rajot, J. L., Traor, S., Coulibaly, M., Diallo, A., Kon, I., et al., (2010). Temporal variability of mineral dust concentrations over West Africa: Analyses of a pluriannual monitoring from the AMMA Sahelian dust transect. *Atmospheric Chemistry and Physics, 10*, 8899–8915. www.atmos-chem-phys.net/10/8899/2010/; doi: 10.5194/acp-10-8899-2010 (accessed on 19 January 2024).
Martínez, J. L., (2008). Antibiotics and antibiotic resistance genes in natural environments. *Science, 321*, 365–367.
Mayoll, E., Arrieta, J. M., Jimenez, M. A., Martinez-Asensio, A., Garcia-Bonnet, N., Dachso, J., Gonzalez-Gaya, B., et al., (2017). Long-range transport of airborne microbes over

the global tropical and subtropical ocean. *Nature Communications, 8*, 201. doi: 10.1038/s41467-017-00110-9. www.nature.com/naturecommunications/ (accessed on 19 January 2024).

Mazar, Y., Cytryn, E., Erel, Y., & Rudich, Y., (2016). Effect of dust storms on the atmospheric microbiome in the Eastern Mediterranean. *Environmental Science and Technology, 50*, 4194–4202.

Mbourou, G., Bertrand, J., & Nicholson, S. E., (1997). The diurnal and seasonal cycles of wind-borne dust over Africa north of the equator. *Journal of Applied Meteorology, 36*, 868–882.

McGrath, C., (2009). Up in the air: The emerging science of dust and sandstorm microbes. *Genome Biology Evolution, 10*(8), 2008–2009. doi: 10.1093/gbe/evy160.

McKay, C. P., (2003). Temperature and moisture conditions for life in the extreme arid region of the Atacama Desert: Four years of observations including the El Niño of 1997–1998. *Astrobiology, 3*, 393–406.

McTainish, G., & Strong, C., (2007). The role of aeolian dust in ecosystems. *Geomorphology, 89*, 39–54.

Meola, M., Lazzaro, A., & Zeyer, J., (2015). Bacterial composition and survival on Sahara dust particles transported to the European alps. *Frontiers in Microbiology, 6*, 1454.

METEOTRENTINO, (2017). *Meteo Trentino Snow Event Records, Station Data and Season Summary Bulletins (in Italian)* [Internet]. Available from: http://www.meteotrentino.it/neve-ghiacci/neve/calendario.aspx?id=216 (accessed on 19 January 2024).

Middleton, N., (2020). *Impacts of Sand and Dust Storms on Oceans.: A Scientific Environmental Assessment for Policy Makers* (p. 395). United Nations Environment Program, Paris, https://www.unep.org/resources/report/impacts-sand-and-dust-storms-oceans/ (accessed on 19 January 2024).

Mirzabaev, A., & Wu, J., (2019). Desertification. In: *Climate Change and Land: An IPCC Special Report on Climate Change, Desertification, Land Degradation, Sustainable Land Management, Food Security, and Greenhouse Gas Fluxes in Terrestrial Ecosystems* (pp. 249–308). Intergovernmental Panel on Climate Change.

Moulin, C., Lambert, C. E., Dayan, U., Masson, V., Ramonet, M., Bousquet, P., Legrand, M., et al., (1998). Satellite climatology of African dust transport in the Mediterranean atmosphere. *Journal of Geophysics Research, 103*, 13137–13144.

NASA, (2001). *All the World's a Stage for Dust* (pp. 1–4). National Aeronautics and Space Agency, Washington, D.C. USA. https://science.nasa.gov/science-news/science-at-nasa/2001/ast26jun_1/. (accessed on 19 January 2024).

Navarro-González, R., (2003). Mars-like soils in the Atacama Desert, Chile, and the dry limit of microbial life. *Science, 302*, 1018–1021.

Normand, P., (2006). The families frankiaceae, geodermatophilaceae, acidothermaceae and sporichthyaceae. In: Dworkin, M., Falkow, S., Rosenberg, E., Schleifer, K., &Stackebrandt, E., (eds.), *The Prokaryotes: Archaea. Bacteria: Firmicutes, Actinomycetes* (Vol. 3, pp. 669–681). Springer: New York, USA.

Nourmoradi, H., Moradnejadi, K., Mohammadi, F., Khosravi, B., Hemati, L., Koshniyat, R., & Kaembeigi, (2015). The effect of dust storm on the microbial quality of ambient air in sanandaj: A city located in the West of Iran. *Global Journal of Health Science, 7*, 114–119.

Pearson, C. J., (1992). Field crops ecosystems. In: Bhattarai, E. S., (ed.), *Ecosystems of the World* (p. 576). Elsevier Science Publications Ltd. Netherlands.

Perkins, S., (2001). *Dust, the Thermostat* (Vol. 160, No. 13, pp. 200–202). Science News. doi: 10.2307/4012776.

Polymenakou, P. N., (2012). Atmosphere: A source of pathogenic or beneficial microbes? *Atmosphere, 3*, 87–102.

Prospero, J. M., & Lamb, P. J., (2003). African droughts and dust transport to the Caribbean: Climate change implications. *Science*. doi: 10.1126/science.1089915.

Prospero, J. M., & Nees, R. T., (1986). Impact of North African drought and El Nino on mineral dust in Barbados trade winds. *Nature, 320*, 735–738.

Prospero, J. M., (1999). Long term measurements of the African mineral dust to south-eastern United States. Implications of the regional air quality. *Journal of Geophysical Research, 104*, 15917–15927.

Prospero, J. M., Blades, E., Mathison, G., & Naidu, R., (2005). Interhemispheric transport of viable fungi and bacteria from Africa to the Caribbean with soil dust. *Aerobiologia, 21*, 1–19.

Prospero, J. M., Collard, F. X., Molinie, J., & Jeanot, A., (2014). Characterizing the annual cycle of African dust transport to the Caribbean Basin and South America and its impact on the environment and air quality. *Global Geochemical Cycles, 28*, 757–773.

Prospero, J., Bonatti, E., Schubert, C., & Carlton, T. N., (1970). Dust in Caribbean atmosphere traced to an African dust storm. *Earth Planetary Science Letters, 9*, 287.

Rahav, E., Ovadia, G., Paytan, A., & Herut, B., (2016). Contribution of airborne microbes to bacterial production and N2 fixation in seawater upon aerosol deposition. *Geophysics Research Letters, 43,* 719–727. doi: 10.1002/2015GL066898.

Rahav, E., Pavtan, Mescioglu, E., Galletti, Rosenfield, S., Raveh, O., Santinelli, C., et al., (2018). Airborne microbes contribute to N2 fixation in surface water of the Northern Red Sea. *Geophysical Research Letter, 45*, 6186–6194. https://doi.org/10.1029/2018GL077132.

Rajot, J. L., (2010). *Wind-Blown Sediment Mass Budget of Sahelian Village Land Units in Niger, Bulletin of Society of Geology, France, 172*(5), 523–531.

Raman, A., Arellano, A. F., & Brost, J. J., (2014). *Revisiting Haboobs in the Southwestern United States: An Observational Case Study of the 5th July 2011 Phoenix Dust Storm* (p. 1). NASA Astrophysics Data System. National Aeronautics and Space Agency, Washington, USA. https://www.science.gov/topicpages/h/haboob+dust+storms (accessed on 19 January 2024).

Rao, Y., Li, H., Chen, M., Fu, Q., Zhuang, G., & Huang, K., (2020). Characterization of airborne microbial aerosols during a long-range transported dust event in eastern China: Bacterial community, influencing factors, and potential health effects. *Aerosol and Air Quality Research, 20*, 2834–2845. Online Publisher: Taiwan Association for Aerosol Research https://doi.org/10.4209/aaqr.2020.01.0030.

Rech, J. A., Currie, B. S., Michalski, G., & Cowan, A. M., (2006). Neogene climate change and uplift in the Atacama Desert, Chile. *Geology, 142*, 761–764.

Reed, L., & Nugent, K., (2018). The health effects of dust storms in the Southwest United States. *The Southwest Respiratory and Critical Care Chronicles, 6*(22), 42–46.

Roberts, A., & Knippertz, P., (2012). Haboobs: Convectively generated dust storms in West Africa. *Weather, 67*(12), 311–316.

Rosselli, R., Fiamma, M., Deligios, M., Pintus, G., Pellizrro, G., Canu, A., Duce, P., et al., (2015). Microbial immigration across the Mediterranean via airborne dust. *Scientific Reports, 5*, 16306. doi: 10.1038/srep16306.

Rudich, Y., (2017). Wind from the desert brings more than the dust. *The Dust Storm Microbiome* (pp. 1–12). Weizmann's Earth and Planetary Sciences, The Weizmann Institute, Israel. https://wis-wander.weizmann.ac.il/earth-sciences/dust-storm-microbiome (accessed on 19 January 2024).

Scheuvens, D., Schütz, L., Kandler, K., Ebert, M., & Weinbruch, S., (2013). Bulk composition of Northern African dust and its source sediments: a compilation. *Earth Science Reviews, 116*, 170–194.

Schlatter, D. C., Schillinger, W. F., Bary, A. L., Sharratt, B., & Paulitz, T. C., (2018). Dust associated microbiomes from dryland wheat fields differ with tillage practices. *Atmospheric Environment, 185*, 29–40.

Schmaltz, J., (2019). *Dust Storm in Argentina* (pp. 1–4). NASA GSFC Source Information https://earthobservatory.nasa.gov/images/51467/dust-storm-in-argentina/(accessed on 19 January 2024).

Schmidt, L. J., (2001). *When the Dust Settles* (pp. 1–18). Goddard Space Flight Centre, EROS Data Centre. National Aeronautics and Space Agency, Washington, USA. https://earthdata.nasa.gov/learn/sensing-our-planet/when-the-dust-settles/ (accessed on 19 January 2024).

Sghaier, H., Hezbri, K., Ghodhbane-Gtari, F., Pujic, P., Sen, A., Daffonchio, D., Boudabous, A., et al., (2016). Stone-dwelling actinobacteria *Blastococcussaxobsidens, Modestobacter marinus and Geodermatophilus obscurus proteogenomes*. *ISME Journal, 10*, 21–29.

Shao, Y., Wyrwoll, K. H., Chappel, A., Huang, J., Li, Z., McTainish, S., Tanaka, T. Y., et al., (2011). Dust cycle: An emerging core theme in Earth Science. *Aeolian Research, 2*, 181–204.

Sharoni, S., (2015). Infection of phytoplankton by aerosolized marine viruses. *Proceedings of NationalAcademy of Sciences, U.S.A., 112*, 6643–6647. doi: 10.1073/pnas.1423667112 (October, 20th, 2022).

Sharratt, B. S., & Feng, G., (2009). Windblown dust influenced by conventional and undercutter tillage within the Columbia Plateau, USA. *Earth Surf. Process. Landf., 34*, 1323–1332. https://doi.org/10.1002/esp.1812.

Sharratt, B. S., & Lauer, D., (2006). Particulate matter concentration and air quality affected by windblown dust in the Columbia Plateau. *J. Environ. Qual., 35*, 2011. https:// doi.org/10.2134/jeq2006.0212.

Sharratt, B. S., Wendling, L., & Feng, G., (2012). Surface characteristics of a windblown soil altered by tillage intensity during summer fallow. *Aeolian Research, 5*, 1–7. https://doi.org/10.1016/j.aeolia.2012.02.002.

Smets, W., Moretti, S., Denys, S., & Lebeer, S., (2016). Airborne bacteria in the atmosphere: Presence, purpose, and potential. *Atmospheric Environment, 139*, 214–221.

Smith, D. J., (2013). Aeroplankton and the need for a global monitoring network. *Bioscience, 63*(7), 515–516. doi: 10.1525/bio.2013.63.7.

Smith, D. J., Griffin, D. W., McPheters, R. D., Ward, P. D., & Schuerger, A. C., (2011). Microbial survival in the stratosphere and implications for global dispersal. *Aerobiologia (Bologna), 27*, 319–322.

Smith, D. J., Timonen, H. J., Jaffe, D. A., Griffin, D. W., Birmele, M. N., Perry, K. D., Ward, P. D., & Robertse, M. A., (2013). Intercontinental dispersal of bacteria and archaea by transpacific winds. *Applied and Environmental Microbiology, 79*, 1134–1139.

Sokolik, I., & Toon, O., (1999). Incorporation of mineralogical composition into models of the radiative properties of mineral aerosol from UV to IR wavelengths. *Journal of Geophysics Research, 104*, 9423–9444.

Stackebrandt, E., (2014). The family *Dermabacteraceae*. In: Rosenberg, E., DeLong, E. F., Lory, S., Stackebrandt, E., & Thompson, F., (eds.), *The Prokaryotes: Actinobacteria* (pp. 289–299). Springer: Berlin, Heidelberg.

Stackebrandt, E., Schumann, P., & Prauser, H., (2006). The family *Cellulomonadaceae*. In: Dworkin, M., Falkow, S., Rosenberg, E., Schleifer, K., &Stackebrandt, E., (eds.), *The Prokaryotes: Archaea; Bacteria: Firmicutes, Actinomycetes* (Vol. 3, pp. 983–1001). Springer: New York, NY, USA.

Stockdale, A., Shi, Z., Berman-Frank, I., Meador, T., Tanaka, T., Violaki, K., & Paraskevi, P., (2016). The potential impact of Saharan dust and polluted aerosols on microbial populations in the East Mediterranean Sea, an overview of a mesocosm experimental approach. *Frontiers in Marine Science, 3*, 226–230.

Sun, Y., Zhuang, G., Yuan, H., Zhang, X., & Guo, J., (2004). Characteristics and sources of 2002 super dust storm in Beijing. *Chinese Science Bulletin, 49*, 698–705. https://doi.org/10.1007/BF03184268.

Sun, Z. H., Lu, J. C., Tong, Y. C., Wang, L., Chen, M. L., & Shun, R. Q., (2003). Study on laws of ground layer meteorological element variation and bacteria in air concentration distribution in Beijing. *Environment Monitor China,4*, 11–17. https://doi.org/10.19316/j.issn.1002-600 2.2003.04.006.

Swap, R., Garstang, M., Greco, S., Talbot, R., & Kallberg, P., (1992). Saharan dust in the Amazon basin. *Tellus, 44*, 133–149.

Tang, K., Huang, Z., Huang, J., Maki, T., Zhang, S., Shimizu, A., Ma, X., et al., (2018). Characterization of atmospheric bioaerosols along the transport pathway of Asian dust during the dust-bioaerosol-2016 campaign. *Atmospheric Chemistry and Physics, 18*, 7131–7148. https://doi.org/10.5194/acp-18-7131-2018.

Tegen, I., Werner, M., Harrison, S. P., & Kohfeld, K. E., (2004). Relative importance of climate and land use in determining present and future global soil dust emission. *Geophysical Research Letters, 31*. http://dx.doi.org/10.1029/2003GL019216.

Teixeira, L. M., & Merquior, V. L. C., (2014). The family *Moraxellaceae*. In: Rosenberg, E., DeLong, E. F., Lory, S., Stackebrandt, E., & Thompson, F., (eds.), *The Prokaryotes: Gammaproteobacteria* (pp. 443–476). Springer: Berlin, Heidelberg, Germany.

Tozer, P., & Leys, J., (2013). Dust storms – what do they really cost? *The Rangeland Journal, 35*, 131–142. http://www.publish.csiro.au/rj/pdf/ RJ12085/ (accessed on 19 January 2024).

Tulet, P., Mallet, M., Pont, V., Pelon, J., & Boone, A., (2008). The 7– 13 March 2006 dust storm over West Africa: Generation, transport, and vertical stratification. *Journal of Geophysics Research, 113*, D00C08. doi: 10.1029/2008JD009871.

U.N.O, (2018). *Sand and Dust Storms: Subduing a Global Phenomenon* (pp. 46–56). United Nations Environment Program's Frontiers Report.

Uno, I., Eguchi, K., Yumimoto, K., Takemura, T., Shimizu, A., Uematsu, M., Liu, Z., et al., (2009). Asian dust transported one full circuit around the globe, National. *Geoscience, 2*, 557–560. https://doi.org/10.1038/ngeo583.

Wang, G., Huang, J., Guo, W., Zuo, J., Wang, J., Bi, J., Huang, Z., & Shi, J., (2010). Observation analysis of land-atmosphere interactions over the loess plateau of Northwest China. *Journal Geophysics Research-Atmosphere, 115*, 1–15. https://doi.org/10.1029/2009JD013372.

Wang, X., Pu, W., Shi, J., Bi, J., Zhou, T., Zhang, X., & Ren, Y., (2015). A comparison of the physical and optical properties of anthropogenic air pollutants and mineral dust over Northwest China. *Journal of Meteorological Research, 29*, 180–200. https://doi.org/10.1007/s13351-0154092–0,2015.

Wang, X., Wen, H., Shi, J., Bi, J., Huang, Z., Zhang, B., Zhou, T., et al., (2018). Optical and microphysical properties of natural mineral dust and anthropogenic soil dust near dust source regions over north-western China. *Atmospheric Chemistry and Physics, 18*, 2119–2138. https://doi.org/10.5194/acp18-2119-2018,2018.

Wang, X., Xu, B., & Ming, J., (2014). An overview of the studies on black carbon and mineral dust deposition in snow and ice cores in East Asia. *Journal of Meteorological Research, 28*, 354–370. https://doi.org/10.1007/s13351-014-4005-7,2014.

Weil, T., Filippo, C., Albanese, D., Donati, C., Pindo, M., Pavarini, L., Carotuneto, F., et al., (2017). Legal immigrants: Invasion of alien microbial communities during winter occurring desert dust storms. *Microbiome, 5*, 32. doi 10.1186/s40168-017-0249-7.

Weizmann Institute of Science, (2017). *The Dust Storm Microbiome* (pp. 1–7). Science Daily. www.sciencedaily.com/releases/2017/06/170627105407.htm (accessed on 19 January 2024).

White, D. C., Sutton, S. D., & Ringelberg, D. B., (2017). The genus *Sphingomonas*: Physiology and ecology. *Current Opinions in Biotechnology, 7*, 301−306.

WHO, (2016). *Antibiotic Resistance*. http://www.who.int/mediacentre/factsheets/antibiotic-resistance/en/ (accessed on 19 January 2024).

Wiegel, J., Tanner, R., & Rainey, F. A., (2006). An introduction to the family *Clostridiaceae*. In: Dworkin, M., Falkow, S., Rosenberg, E., Schleifer, K., & Stackebrandt, E., (eds.), *The Prokaryotes: Bacteria: Firmicutes, Cyanobacteria* (Vol. 4, pp. 654−678). Springer: New York, NY, USA.

Womack, A. M., Bohannan, B. J. M., & Green, J. L., (2010). Biodiversity and biogeography of the atmosphere. *Philosophical Transactions of Royal Society. B., 365*, 3645−3653.

Womiloju, T. O., Miller, J. D., Mayer, P. M., & Brook, J. R., (2003). Methods to determine the biological composition of particulate matter collected from outdoor air. *Atmospheric Environment, 37*, 4335–4344.

World Meteorological Organization, (2020). *Sand and Dust Storms* (p. 107). World Meteorological Organization, Geneva. https://public.wmo.int/en/our-mandate/focus/areas/environment/SDS (accessed on 19 January 2024).

Wu, P. C., Tsai, J. C., Li, F. C., Lung, S. C., & Su, H. J., (2004). Increased levels of ambient fungal spores in Taiwan are associated with dust events from China. *Atmospheric Environment, 38*, 4879–4886. https://doi.org/10.1016/j.atmosenv.2004.05.039,2004.

Wu, Y., Wen, B., Li, Y., & Guo, Y., (2021). *Sand and Dust Storms in Asia: A Call for Global Cooperation on Climate Change: Comment* (pp. 1–2). Elsevier Publishers. https://doi.org/10.1016/ S2542-5196(21)00082-6.

Yamaguchi, N., Ichiijo, T., Sakoyani, A., Baba, T., & Nasu, M., (2012). Global dispersion of bacterial cells on Asian dust. *Scientific Reports, 2*(1–6), 525. doi: 10.1038/srep00525.

Yeo, H. G., & Kim, J. H., (2002). SPM and fungal spores in the ambient air of west Korea during Asian dust (Yellow sand) period. *Atmospheric Environment, 36*, 5437–5442. doi: 10.1016/S1352–2310(02)00672–6/.

Zhou, Y., Wang, X., Liu, H., Zhang, K., Zhang, Y., Lai, R., & Li, W., (2007). *Pontibactera kesuensis* sp. Nov., isolated from a desert soil in China. *International Journal of Systematics and Evolutionary Microbiology, 57*, 321−325.

Zou, X. K., & Zhai, P. M., (2004). Relationship between vegetation coverage and spring dust storms over northern China. *Journal of Geophysical Research: Atmosphere*. 109.110.1029/2003JD003913.

CHAPTER 5

Aerial Microbes and Nutrient Dynamics in Agricultural Cropping Zones

ABSTRACT

The agrarian sky is home to aerosols, which contain nutrients in a dissolved state as mist or moisture droplets, or as particles suspended in the atmosphere. Essential elements for crops and a wide range of biochemical constituents exist in dissolved form within the moisture of the troposphere. The agrarian sky also harbors genetically and physiologically diverse microbes. These atmospheric microbes are often versatile in their ability to transform nutrients, playing a crucial role in maintaining various ecosystem functions related to nutrient dynamics in agroecosystems. Bioaerosols are not only relevant to nutrient dynamics in the troposphere, but also to the entire agroecosystem, including the soil, crops, biotic components, and water resources. This is because aerosols in the troposphere are interconnected with the soil and crops.

Natural emissions from biogenic PM range from 50 to 100 Tgy^{-1}, while anthropogenic PM emissions are estimated at 3.3 Tgy^{-1}. PM that carries microbes is a significant repository of various nutrients. PM in the troposphere serves as a source of carbonaceous and inorganic substances for microbes that thrive on them. PM generated from natural vegetation, crop canopies, and organically rich soils is predominantly carbonaceous. Cellulose, sugars such as xylose, arabinose, fructose, sucrose, shikimic acid, and other anhydrous saccharides indicate the carbonaceous nature of PM. Cellulose, polyols, and sugars are often used as tracers of PM and also indicate the presence of microbial components. Therefore, it is possible to deduce likely nutrient transformations in the troposphere. PM laden with microbes

Microbiology of the Agricultural Sky. K. R. Krishna (Ed)

in the troposphere regulates nutrient dynamics in the troposphere, soil phase, and crop canopies/biomass to some extent. Aeolian dust often contains C, N, P, S, and several micronutrients such as Fe, Mn, Zn, Pb, Ni, Co, Cu, and silicate minerals. The source of clay, silt, and sand particles observed in the agrarian troposphere is often the soil. The dynamics of minerals complexed with organic substances found on PM are greatly influenced by microbial assemblages present on them. Environmental parameters such as RH, temperature, moisture, and acidification, if any, affect the stoichiometry of nutrient conversions in the air.

Globally, major dust storms like the Harmattan in the Sahara/Sahel region transport enormous quantities of carbonaceous and mineral matter from the sandy soils. Aerial microbes also play a role in this process. It is important to note the nutrient load, nutrient ratios, and microbial load associated with these dust storms. The microbial diversity found on the dust particles has significant consequences on nutrient dynamics.

Microbes in the soil, phyllosphere, crop canopy, and troposphere have an impact on gaseous emissions into the troposphere. These microbes contribute to the release of CO_2, CH_4, nitrous oxide (N_2O), and volatile organic compounds (VOCs) in the agricultural atmosphere. Repeated tillage practices create oxidative conditions in the soil, resulting in a higher degree of microbial respiration and CO_2 emission. Free-living diatoms, green algae, and blue-green algae can be found in the troposphere, where they play a role in fixing CO_2 from the air. Methanogenic microbes (such as *Methanobacterium, Methanosarcina, Methanosaeta*) and methanotrophic microbes (such as *Bacillus methanicum, Methyloacidiphylum* sp.) can regulate the CH_4 content in the airspace above crops, although to a limited extent. Methanogens found in submerged paddy fields are major contributors to CH_4 emissions in the global agricultural atmosphere. There are also aerobic methanogens, as well as methanotrophs that can survive in extreme acidophilic and high-temperature conditions. Undoubtedly, microbes play a crucial role in regulating the carbon cycle in agroecosystems, not just in the troposphere.

Atmosphere is the largest reservoir of N_2 gas. The natural vegetation, agrarian belts, and other biotic aspects of the terrain form the second largest source of N_2 and nitrogenous compounds. Aerial microbes have a say over the transformations of N that occur in the troposphere. Also, to a certain extent, N in water bodies and soil phase. Nitrogen dynamics in the troposphere are intricately linked to microbial transformations. Organic matter degradation, emission of NH_3, N_2O, NO_2, N_2 are a few steps mediated by aerial microbes. Nitrification mediated by microbes is actually a process that occurs both in the air and in the soil phase. Nitrifiers like Nitrobacter and Nitrosomonas in the troposphere mediate N conversions. Atmospheric N fixation is a process that converts N_2 in the gaseous phase into organic compounds. The biological N fixers (BNF) play a vital role in regulating the N dynamics of agrarian regions.

Information on the evolution and distribution of genes for the nitrogenase enzyme and its loss in the microbial lineages is interesting. The BNF requires micro-aerophilic or anaerobic conditions. There are various adaptations that help microbes to overcome oxygen-related ill-effects on biological N-fixation.

Microbes also play a vital role in regulating the dynamics of other elements such as P, S, and micronutrients that are essential for crops. The role of aerial microbes in P transformations requires attention. Microbes may release PO_4 from PM through enzymatic action. Intercontinental dust storms transport large quantities of organic-P. Upon deposition, they undoubtedly enrich agrarian soil with P compounds. For example, diatoms that drift from the oceans to land masses carry and deposit organic P onto soils. Aerial microbes, particularly those found in smog, affect the sulfur cycle. They accelerate S transformation in the atmosphere. Microbes that potentially detoxify sulfurous compounds in the troposphere are relevant to farming regions. Sulfur-oxidizing and reducing bacteria have a significant impact on the S transformations that occur in the agrarian sky. Microbes also influence P and S transformations that contribute to acid rain formation. The distribution of chemolithotrophs like *Beggitao alba, Thiobacillus thioxidense,* and related species involved in S transformation in the air needs evaluation.

Aerial microbes are present even in extreme climates, particularly those encountered above agrarian regions. Firstly, microbes adapt to the harsh ambient climate. They also induce several biogeochemical transformations that are relevant to crops.

5.1 INTRODUCTION

5.1.1 AERIAL PARTICULATE MATTER (PM) AND NUTRIENTS

The mineral and organic composition of aerosols above crops can be complex. It depends on the characteristics of the location, sources of organic matter, natural vegetation, crops, cropping seasons, soil, and its characteristics. In addition, weather patterns such as wind, temperature, convection currents, dryness of soil, and a few other factors may determine the extent of particulate matter (PM) and its biochemical nature. The nutrient dynamics that occur in the atmosphere undoubtedly depend on the physicochemical nature of PM of different sizes. The PM could be predominantly carbonaceous (cellulosic, lignocellulose, polysaccharides, glucose, etc.). Elemental carbon and minerals are also present in PM. These components undergo various biochemical transformations. In most agricultural areas, only 20–30% of PM can be thoroughly characterized in terms of biochemical composition. Often, we encounter complex secondary metabolites such as lignin, phenolic compounds, pigments, and more. Our current understanding of the biochemical composition and mineral content of PM is still limited. Especially when it comes to seasonal changes and transformations. Additionally, there may be several unknown biochemical fractions, making it difficult to track their transformations (Karagulian et al., 2015; Wagenbrenner et al., 2017). Understanding their dynamics in relation to seasons, weather conditions, crops, and agronomic procedures is even more challenging to assess (Yttri et al., 2007; Putaud et al., 2004a, b).

We can study the emission of PM into the troposphere above crops by using indicator biochemicals. For example, the presence of cellulose or polyols suggests the occurrence of predominantly OC-rich PM. There are several reports about the production of aerosols (PM) that possess Organic Matter (OM) and volatile organic compounds (VOCs) (Brighty et al., 2021). Such aerosols originate from biogenic material like crops or vegetation. Solid airborne material, biological material, microorganisms, fragments of plant material including cellulose and polyols can be found in the aerosol above crops (Samake et al., 2020, 2021; Alves, 2017). In addition to plant debris, PM (> 100 μm) may contain fungal spores, bacterial cells, viral particles, pollen fragments, and remnants of dead insects. They may also reach the atmosphere from the soil or crop's canopy. The nutrient transformations that ensue in the atmosphere due to microbial degradation need attention. Upon deposition, such microbes may modify the chemical composition of the soil. Clouds containing the PM may hasten biochemical degradation. Such partially or completely degraded PM gets deposited on the soil surface whenever precipitation occurs.

Let us consider an example where cellulose estimation is conducted to assess the contribution of plant debris-derived particulate matter (PM). Brighty et al. (2021) and Geibi et al. (2007) investigated cellulose as an indicator for plant debris, utilizing HPLC and molecular marker methods to analyze organic compounds in the PM. The concentration of carbonaceous compounds varied depending on the location, with rural vegetated areas showing a higher fraction of plant debris in the PM compared to urban areas. The measured cellulose levels also varied with season and weather patterns. It is evident that in farming regions, plant debris contributes to the PM in the atmosphere. The average annual mean cellulose concentration in some rural agro-zones was 284 ng m^{-3}, with occasional spikes reaching 2,200 ng m^{-3}. To gain a better understanding of carbon dynamics, it may be necessary to identify cellulose degraders in the atmosphere, particularly on the PM. Additionally, xylose, arabinose, fructose, sucrose, shikimic acid, and anhydrous saccharides derived from plant parts and pollen are also used as indicators of organic compounds in the PM (Marynowski et al., 2021).

Natural emissions of biogenic PM may range from 50 to 100 $Tgyr^{-1}$ (Brighty et al., 2021; Klimont et al., 2017). Together with anthropogenic emissions of organic PM (3.3 $Tgyr^{-1}$), these provide a substrate for atmospheric microbes to act upon. The nutrient dynamics that ensue could be complex and need to be deciphered as accurately as possible. This is mainly because it helps us gauge the net nutrient input into agrarian soils, wherever they get deposited. Characterization of PM above agrarian regions using chemical tracers could be efficient (Brighty et al., 2021; Weber et al., 2001, 2019). For example, cellulose is used as a molecular marker to estimate and quantify ambient concentrations of PM and biogenic aerosol (Sanchez-Ochoa et al., 2007; Martin et al., 2010; Bozetti et al., 2016; Andrae & Merlet, 2001; Poschl, 2005).

Generally, cellulose-derived carbon, cellulose-bound lignin, and hemicellulose contribute to atmospheric carbon. Free cellulosic material is also found in the air. Therefore, cellulose-degrading microbes are active in the airspace above crops.

In the air above the crop's canopy, the cellulose content often peaks during certain seasons. The emission of cellulosic-C follows seasonal cycles. Peaks in cellulose-containing particulate matter (PM) are observed during spring and summer. We can expect the population of microbes that utilize cellulosic material as a substrate to peak around the same time. According to Samake et al. (2020, 2021), the populations of fungi, archaea, and bacteria also peaked around the same time as cellulosic PM emissions. However, they found only a few specific microbial species to be accentuated, despite the large diversity of microbes in the ambient atmosphere. Furthermore, there were moderate to strong correlations ($r = 0.28 - 0.78$, $p < 0.0001$)

between cellulose and typical biogenic chemical tracers such as polyols and glucose across the nine sites. Seasonality was strongest at sites with stronger biogenic correlations, suggesting that cellulose arises from biogenic origins. The other compounds found in PM were secondary metabolites such as complex lignin and pigments.

5.1.2 AEOLIAN DUST, MICROBES, AND NUTRIENT TRANSFORMATIONS

Wind and air currents above agricultural expanses induce a large-scale shift of atmospheric gases. Particulate matter (PM) and dissolved biochemicals transit in the troposphere, happening through clouds and precipitation. Wind and air turbulence just above the crops can physically alter the dynamics of PM, resulting in a shift in the physical location of nutrients and microbes above the crop's canopy. Inorganic and organic compounds found in the dust particles are prone to transformations by the microbes during transit and upon deposition. Microbes in the soil and those resident on the crop's canopy may act upon the deposited dust. On a larger scale, inter-continental dust storms that transit thousands of kilometers also affect the nutrient dynamics. Nutrients in the atmosphere are physically shifted, and microbes also migrate through the dust storms, altering microbial load, diversity, and activity. Aerial microbe-mediated transformations also affect the nutrient dynamics in the atmosphere. Additionally, microbial mass also shifts in the atmosphere due to intercontinental dust storms/plumes. For example, Acosta Martinez et al. (2015) state that over 50 Tg of particulate matter laden with microbes is shifted each year. This has a large-scale influence on the nutrient dynamics in the atmosphere. Moreover, dust storms harbor microbes with diverse biochemical transformation abilities. The natural vegetation and crops that receive these depositions experience the net effects of nutrient shifts caused by storms and microbial transformations. Aeolian dust storms, in fact, facilitate the long-distance and large-scale inoculation of microbes originating from distant locations. This contributes to the microbial diversity of soils at the deposition sites. It may also help in achieving a certain level of uniformity in microbial diversity across different distant locations, both in the atmosphere and soil profile (due to deposition). To authenticate this aspect, well-designed sampling, detection, identification, and enumeration procedures for aeolian microbes need to be conducted.

Africa is responsible for approximately half of the global aeolian dust emissions. This means that nutrient dynamics in the atmosphere are proportionately affected by events in Africa. World Meteorological Reports suggest that Saharan dust plumes fertilize Amazonian forests and large cropping expanses of Cerrados, particularly with organic matter, mineral nutrients like P, S, and Fe. Microbes are also inoculated in the region of dust deposition (WMO, 2016, 2020; Ginoux et al., 2012; Kaufman et al., 2005; Ridley et al., 2012).

In the atmosphere, abiotic factors such as wind, jet stream, clouds, snow, fog, rain, and particulate matter (PM) play a vital role in transporting significant amounts of nutrients (FAO of the United Nations, 2020; Plates 5.1–5.4). Microbes are among the crucial biotic factors that influence the nutrient dynamics in the atmosphere. They facilitate various transformation reactions. Microbes have a significant impact on the availability of essential nutrients for crop productivity. Additionally, they convert compounds present in the atmosphere that may otherwise be toxic or harmful to crops and other organisms. As mentioned earlier in previous chapters, microbes play a crucial role in nutrient transformation not only in the atmosphere but also in soil and water bodies. In this context, we are specifically concerned with the utilization and conversion of nutrients present in atmospheric dust, smog, and dust plumes, and their effects on crop productivity in soil. The microbial load, referring to the population and diversity of microorganisms in different atmospheric layers, has been extensively discussed in Chapters 2 and 4 (*see* Krishna, 2023).

Storms are known to induce destruction of forest stands, natural vegetation, and crops. At the same time, they bring in microbes and certain nutrients. Microbes aid in the rapid degradation of wood logs, branches, foliage, and dust matter (FAO of the United Nations, 2020). It is the combination of substrate and microbial diversity that determines the direction of biological transformations and end products. The stoichiometry of biochemical transformations is also dependent on similar factors. Additionally, abiotic conditions in the atmosphere play a role.

Aeolian dust is known to impact the dynamics of several nutrients, including C, N, P, S, micronutrients, and various other minerals present in sand and organic particles. For instance, the aeolian dust from the Sahel region is believed to contribute different nutrients to agricultural areas, along with the introduction of microbes. These microbes facilitate transformations that can enhance soil fertility (FAO of the United Nations, 2020; Acosta-Martinez et al., 2015; Griffin et al., 2001, 2003). One observation regarding

PLATE 5.1 A large mass of looming clouds above the Sahelian cropping zone.

Note: Clouds shift large quantities of microbes and nutrients as they traverse through the troposphere and deposit wherever they shower on the fields. They enhance the microbial load and soil fertility status in the place where they shower. Clouds enmeshed with dust/sand particles may travel a distance of 50 to 100 km before dissipating. [Location: Near Dosso, South of Niamey, Niger in West Africa.]

Source: Dr. K. R. Krishna, International Crops Research Institute for the Semi-arid Tropics, Hyderabad, India.

the movement of aeolian dust in the troposphere suggests that each year, migratory aeolian dust adds valuable nutrients and microbes to the vegetation (forests) and cultivated areas of the Amazon (FAO of the United Nations, 2020; WMO, 2020).

PM is an important factor affecting nutrient dynamics in the agrarian sky. Aeolian dust affects nutrients at both lower and upper altitudes of the troposphere. The chemical composition of dust is crucial. The stoichiometry of nutrient transformation is determined by prevailing atmospheric conditions in the surroundings. The population of microbes capable of degrading and transforming troposphere dust needs to be deciphered. The release of nutrients from dust/PM is influenced by the physicochemical features of PM. For example, dust storms in West Asia contained 55% clay, 18% silt, and 18% sand. The dust matter can be primarily classified as sandy-silty-clay, while other fractions can be termed sandy-clay-silt.

Minerals are also present in dust/PM that settles on agrarian soils. It is common to find heavy minerals like pyroxene, hornblende, zircon, chlorite, epidote, and garnet. Light minerals like quartz, feldspar, mica, calcite, gypsum, and dolomite are also frequently found in PM of different sizes. X-ray diffraction studies have shown that clay minerals encountered in PM can be montmorillonite, illite, kaolinite, chlorite, etc. (Shepherd, 2016).

PLATE 5.2 A dust storm in progress in Sahel [Location: Near Niamey, Niger, West African Sahel (1985)].

Note: Usually, a dust storm with lots of particulate matter containing mineral nutrients (sandy particles with Si, Al, Fe, P, and K) and organic compounds gathers just prior to precipitation. A storm shifts large quantities of nutrients. Simultaneously, it transports microbes of diverse species and physiological abilities. The storm may distribute disease-causing propagules of plant pathogens as well as symbionts (e.g., BNF organisms, mycorrhizal fungal spores, etc.).

Source: Dr. K. R. Krishna, International Crops Research Institute for the semi-arid tropics, Hyderabad, India.

The pH of the troposphere, PM, and microbe complex determine, to a certain extent, the nutrient availability for crops in the soil upon deposition. Usually, dust particles may contain a significant amount of N in the organic fraction. Additionally, P, K, and microcontinents occur in trace amounts. Micronutrients such as Fe, Pb, Zn, Ni, Co, Cd, and Cu have been found in dusts originating from dry (arid) zones in West Asia (Shepherd, 2016). Microbial activity is the main mediator of nutrient release from dust particles. Physicochemical processes such as lightning, dissolution, adhesion, pH, and acidification affect the nature of the substrate, thus influencing the overall nutrient dynamics of the troposphere.

Microbial analysis has indicated that water extract and washings from dust matter may show 10^3 to 10^5 colony forming units (CFU) per ml. Reports about dust storms in West Asia have indicated that gram-positive bacteria could account for 40–43% of the microbial population, Aspergillus for 18.6%, gram-negative rods like *Escherichia coli* for 9.5%, Gram + *Streptococci* for 8.7%, and *Enterobacter* species. Chapter 2 provides several examples of microbial load and diversity. The terrain, vegetation, and nutrient content of the region where the dust was created essentially affect the microbial population and diversity. As the dust storm moves, it may contribute to the microbial diversity (Shepherd, 2016). Additionally, some microbial species may be lost during transit in the sky.

The source of organic matter for the PM could be related to the soil. No doubt, it is related to the organic fraction, plant species in the natural vegetation, or crops found at the point of formation of dust storms or convection currents. Sometimes, dust above the crops may not show rapid movement or disturbance. They may stay *in situ* in the troposphere, just floating in a small area above crops. Yet, the hanging dust particles in the troposphere above crops affect the nutrient dynamics in the ecosystem.

Sandstorms are integral to Earth's agroecosystems, at least in a few continents. They have impacts on soil structure, nutrients, and crop productivity. This is in addition to their direct influence on atmospheric conditions (Goudie & Middleton, 2006; Shao et al., 2011; UNEP, 2016). Aeolian dust storms influence climate processes, troposphere chemistry, soil characteristics, water quality, and nutrient dynamics. They affect biogeochemical cycling in the environment. Sandstorms also result in damage to crops. Sandstorms remove fertile topsoil, thus adversely affecting food production (Plate 5.4). Dust plumes emanating from Asian deserts have been evaluated for their ability to add plant nutrients and adverse effects on crops, if any (Aciego et al., 2017; Jaffe et al., 2001). No doubt, they do influence nutrient dynamics in the farming zones where dust deposits after transit. Upon deposition, microbes from dust particles also modify the microbial biodiversity of the soil. Microbe-mediated nutrient transformations follow suit. Several new biogeochemical transformations could be mediated by the microbes found in the dust particles. They may

PLATE 5.3 A dust and moisture-laden storm occurred in the troposphere above agrarian fields and the Niger River near Niamey, Niger, West Africa in 1985.

Note: Such dust and moisture-laden storms add microbes and nutrients to the troposphere, soil, and water bodies. For example, in the case mentioned above, the River Niger collects enormous quantities of particulate matter. As a result, the river water and sediments are rich in nutrients and microbes. The river distributes them throughout its course as it flows. These small-scale dust storms occur periodically during the cropping season. They actually enhance the microbial diversity and nutrient load of the entire ecosystem.

Source: Dr. K. R. Krishna, International Crops Research Institute for the semi-arid tropics, Hyderabad, India.

take place in the atmosphere and in the soil, i.e., at the place of deposition. Locations touched by the dust plumes during transit may also experience alterations in nutrient dynamics. It is important to note that sandstorms can cause erosion of topsoil, resulting in the loss of nutrients and the entire microbial flora in the affected area. Consequently, locations experiencing erosion may also experience a decrease in microbial population (Plate 5.4). As a result, various microbe-mediated nutrient transformations may occur at lower rates in those locations.

The particulate matter (PM) that is deposited onto agrarian terrain varies in terms of physical size, shape, weight, density, buoyancy, etc. Dust particles potentially improve soil fertility upon deposition. Undoubtedly, dust particles contain complex organic compounds and inorganic minerals.

PLATE 5.4 A calm and still atmosphere after the dust storm and heavy precipitation. A field near Dosso, Niamey, Niger.

Note: The above fields have suffered from 'sheet erosion' due to strong turbulent winds. There is a severe loss of topsoil, and the dust storm has depleted the field of precious mineral nutrients and organic matter. The loss of topsoil also reduces seedling establishment, especially when cereals or cowpea are sown in such fields. Therefore, a dust storm and precipitation event common to the Sahel is not just a weather phenomenon. It has an immense influence on the microbial load and nutrient dynamics of the soil, as well as the agrarian sky. Such storms induce the migration of microbes for a distance of 50 km to 100 km, and they remain confined to altitudes within the lower troposphere. These storms mix up microbes and nutrients often during the cropping season (May/June till October). During June/July, storms can remove approximately 76 kg C ha^{-1}, 18 kg N ha^{-1}, 6 kg P ha^{-1}, and 57 kg K ha^{-1} through eroded sand particles and dust. Harmattan, for example, results in massive loss of topsoil fertility. Nutrient ratios on sandy PM ranged from 1.18 to 4.15 for C and from 1.12 to 2.31 for N and 1.1 to 1.4 for P (Sterk et al., 1998; Michels et al., 1995).

Source: Dr. K. R. Krishna, International Crops Research Institute for the Semi-arid Tropics, Hyderabad, India.

The chemical constituents primarily depend on the source of PM. Mixing of PM from different regions may occur in the atmosphere, facilitated by wind and convection currents. Some of these factors also affect the microbial component present in the PM. Microbes that are able to utilize the nutrients found in PM find a niche and adopt suitable biochemical processes to derive energy. PM may deposit on agrarian soils, water bodies, crop canopies, and leaf laminae. Let us consider an example. The main inorganic nitrogen compounds in the water-soluble fraction of aerosol particles were nitrate (60%) and ammonium (38%), with mean concentrations of 35 and 22 nmol m^{-3} of air, respectively. Soluble phosphate ranged between 0.09 and 2 nmol m^{-3} of air, with a mean value of 0.4 nmol m^{-3} of air. The soluble inorganic nitrogen and soluble phosphate accounted for approximately 86% and 69% of the total soluble nitrogen and total soluble phosphorus, respectively. The mean concentration of soluble iron was 0.3 nmol m^{-3} of air.

Upon deposition, aeolian dust particles proportionately alter the nutrient dynamics in the soil profile or water bodies such as rivers, ponds, lakes, or even the crop's canopy. Microbial species with diverse biochemical abilities are introduced into soils. Often, large quantities of aeolian dust deposits occur in a specific location (Plate 5.5). These significant amounts of microbes are capable of altering the nutrient dynamics in farming regions located far away (Aalismail et al., 2020). *In situ* microbes also transform the nutrients received through such dust storms (Chen et al., 2007).

Let us consider an example from the Sahelian agricultural zone. Dust storms such as Harmattan, Haboob, or others have a significant impact on nutrient dynamics, both at the place of origin and deposition (*see* Plate 4.2 in Chapter 4). These storms also affect the atmosphere through which they pass. Generally, there is a depletion of mineral and organic nutrients in the topsoil at the place of origin. Even intermediate locations may experience nutrient depletion if they contribute to the particulate matter (PM) of the dust plumes. During transit, dust plumes may lose a certain quantity of PM through deposition. As a result, the nutrient dynamics of soils and the microbial component in intermediate locations are affected. The mineral nutrients and organic matter at the destination become enriched depending on the quantity of PM deposited, as well as the nutrient carrying capacity of the PM. It is worth noting that, in addition to nutrients, we should also consider the microbial diversity and population introduced by the Harmattan. Knowledge about the possible biochemical transformations facilitated by microbial inoculation via dust could be valuable.

In Sahelian West Africa, the dust and rainstorms that occur early in the cropping season bring in particulate matter (PM) that is rich in minerals (Plates 5.1–5.4). The inorganic sand particles add nutrients that are beneficial to natural vegetation and crops. The nutrients in PM have an impact on the dynamics upon deposition. Crops are able to utilize the nutrients that are added to the soil through aeolian dust plumes (Bielders et al., 2002a, b; Krishna, 2008; Sterk et al., 1996, 1998; Romheld et al., 2000; UNEP, 2016).

PLATE 5.5 A large sized dust storm in progress in Sahelian West Africa.

Note: Large quantities of nutrients, both organic and inorganic, and microbial populations shift in the troposphere. This occurs due to the occurrence of massive dust storms. Smaller storms that travel shorter distances also cause shifts in nutrients and microbes, although to a lesser extent. As a result, the overall nutrient dynamics of the atmosphere, soil, and other bodies of water are affected. If a dust storm originates in a particular location, fields in that area may experience a loss of nutrients and microbes. Under natural conditions, the fertility of the soil and the microbial load in a field are determined by the difference between the net inputs and outputs via large dust storms. These intense dust storms have the potential to carry microbes and dust particles to higher altitudes, possibly placing them in the jet streams of wind. Consequently, microbes and nutrient-rich dust particles can migrate over long distances and potentially impact both microbial diversity and nutrient levels in remote regions, even on other continents.

Source: The World Meteorological Organization Sand and Dust Storm Project (2004); WMO (2020) Sand and Dust Storms. World Meteorological Organization, Zurich, Switzerland (pp. 1–12); http://public.wmo.int/our-mandate/focus-areas/environment/SDS.

Infusing dust storms with selective microbes at the point of origin of aeolian dust storms could be a useful approach. For instance, infusing Sahelian dust with Bradyrhizobia (BNF bacteria) that can survive the transit conditions could be a beneficial method, especially for inoculating soybean fields in Latin America. Initially, the inoculum of BNF bacteria at the destination fields may be small, but over time, the symbionts may accumulate to sufficient population levels in the soil and affect nitrogen dynamics.

There is also a possibility that dust storms may carry plant pathogens at the point of origin. In such cases, crops at the deposition sites may be susceptible to diseases, resulting in reduced carbon fixation and biomass accumulation by the crops.

5.1.3 THE GREENHOUSE GAS (GHG) EMISSIONS ABOVE FARMING ZONES

Globally, about 79–81% of GHG emissions are constituted by carbon dioxide (CO_2), 11% by methane (CH_4), 5% by nitrous oxide (N_2O), and 2% by hydrofluorocarbons (HFCs) (European Parliament, 2018).

Agriculture, forestry, and related land use seem to have generated 22% of global GHG emissions between the years 2010 to 2019. Agriculture and forestry are among the major options to reduce GHG emissions. They can significantly reduce emissions by 2050. CH_4 emissions need to be reduced by 34% by 2030 and by 44% by the year 2044. The question now is how aeolian microbes can alter the situation so that GHGs are utilized, leading to their lower concentration. Or how can microbes reduce emissions by converting them in the atmosphere (Marston, 2022).

It appears that around 60% of greenhouse gas (GHG) emissions are contributed by a small number, approximately 10–12, of developed nations. Agricultural crop production is one of the major sources of GHG emissions in this region. Of course, there are other sectors such as energy generation, transportation, and industry that also contribute. Clearly, this means that we are causing harmful effects in the atmosphere above the crops. We can consider various measures to reduce these GHG emissions from crops into the troposphere. Additionally, we should also explore remedies that utilize atmospheric microbes, which are already widely distributed. Alternatively, we can introduce species that can transform GHG and other harmful gases (e.g., SO_2, H_2S) in the agricultural sky.

Reports from the World Resources Institute in Washington D.C., USA state that among greenhouse gases (GHGs), CO_2 emissions are prominent. They could constitute up to 74% of the total GHG in some agrarian locations. The majority of CO_2 emissions are from cropping expanses. Methanogenic microbes found in rice cultivation zones liberate CH_4. There are also methanotrophs that could utilize CH_4 emissions. Together, CH_4 (17%) and NO_2 (6%) emissions can be detrimental (Ge & Friedrich, 2020). They enhance ambient temperatures. There are soil-related remedies to lessen these emissions. At the same time, we should consider modifying GHGs using microbes if possible. Troposphere microbes may alleviate the situation if they possess metabolic processes that consume these GHGs to a certain extent. For example, photoautotrophs can consume CO_2 or methanotrophs can consume CH_4.

Let us consider an interesting example about nitrogen (N) emissions from agricultural fields. Around 2020–2021, Argentina began cultivating genetically engineered, drought-tolerant wheat varieties (e.g., HB_4). These wheat varieties possess traits for better water exploration and use efficiency, as well as relatively higher nitrogen efficiency. This helps in reducing greenhouse gas (GHG) emissions, as farmers are able to use less fertilizer-N. The efficient wheat varieties also result in less soil-N accumulation, leading to lower N-emissions.

Reports suggest that an average of 2.82 t/ha wheat grain yield was derived from 70 kg N/ha input, resulting in GHG emissions of 148.5 kg CO_2 eq/t wheat. By cultivating such low-emission varieties in 33% of the wheat area in Argentina, it seems possible to reduce GHG emissions by 0.86 million metric tons of CO_2 equivalent per year (Kovack & Blustein-Rejitos, 2022). However, there are other ways to reduce N emissions. The N dynamics in the wheat field and in the atmosphere above can also be maintained at optimum levels through the use of inhibitors of nitrification.

Heavy metals are disseminated through the atmosphere and water bodies into the agricultural environment. They are primarily released into the atmosphere as particulate matter (PM). The application of fertilizers through sprinkler sprays, pesticide dusting, and irrigation introduces heavy metals such as Pb, Cd, Hg, As, Zn, and Cu into the atmosphere (Mosbaek et al., 1989). The presence of heavy metals in the atmosphere affects the activity of atmospheric microbes. In agroecosystems, heavy metals are mostly depleted through soil leaching and irrigation water. However, certain heavy metals like Zn and Cu can be deposited from the atmosphere through smog, dust particles, and precipitation. These metals are essential minerals for crops. Currently, the role of atmospheric microbes in heavy metal dynamics in the atmosphere is not well understood, except for Zn and Cu as identified by Hou et al. (2013). These metals may serve as effective co-factors for enzymes found in atmospheric microbes.

5.2 AERIAL MICROBES AND NUTRIENT CYCLES RELEVANT TO GLOBAL FARMING

5.2.1 MICROBES AND CARBON CYCLE

Carbon dynamics on Earth are an intricate process. It extends into the atmosphere, hydrosphere, lithosphere, natural flora, agricultural crops, fauna, and microbes. Here, we are concerned with microbes related to carbon stocks found in the atmosphere, in general. Microbes mediate several processes directly affecting the carbon dynamics. However, specifically, we are concerned with carbon that is cycled in the troposphere of agrarian regions. As such, it is difficult to treat carbon stocks in the atmosphere exclusively. Most often, both microbes and carbon stocks are highly mobile. They transgress into soil, water, natural vegetation, crops, aerial dust, etc. Microbes may migrate from soil to the atmosphere and reside for a period. During this period, they conduct several biogeochemical transformations crucial to the carbon cycle on Earth. Atmospheric carbon stocks are primarily found as carbonaceous gases like CO_2, CO, CH_4, and VOC. It also occurs in sufficiently large quantities in intercontinental PM that floats through the atmosphere. Microbes do have clear access to atmospheric carbon found on PM. In fact, dust supports microbes with a niche that has nutrients, including carbon. Yet, the atmosphere is relatively a feeble storehouse for carbon compared to soil, vegetation, or agricultural crops.

Let us explore the distribution of carbon stocks in various compartments of the Earth, such as the atmosphere, lithosphere (including land and plants), hydrosphere (including oceans, freshwater lakes, ponds, rivers, and sediments), and atmosphere (including gases, particulate matter, smog/dust storms, etc.) (*see* Reerburgh, 1997; Prentice et al., 2001; Falkowski, 1997; Falkowski et al., 2008; Canfield et al., 2010). Microbes can inhabit many of these regions, exhibiting active physiological processes as they utilize carbon stocks for growth and reproduction. Additionally, they play a crucial role in various biogeochemical transformations. These microbes can contribute to the emissions of greenhouse gases such as CO_2, CO, and CH_4. Gaseous CO_2 is utilized during the process of photosynthesis, also known as carbon fixation. It sequesters carbon stocks into microbial cells (e.g., Bluegreen Algae, Green Alga, diatoms). Natural vegetation, forest stands, plantations, and field crops also fix carbon found in the atmosphere as CO_2. Several atmospheric microbes are capable of utilizing carbon stock in organic form heterotrophically and emit CO_2. Reerburgh (1997) and Prentice et al. (2001) have reported on the rate of increase of carbon stocks, i.e., accumulation in the atmosphere and other regions of the earth. The following is the list: Atmospheric CO_2 (750 Pg) accumulated at a rate of 3.2 Pg C per year with a residence time of 3-5 years. The lithosphere has 15×10^5 PgC, the deep ocean contains 38,000 Pg of dissolved inorganic carbon (DIC) and 700 Pg of dissolved organic carbon (DOC). Lakes and river sediments possess 150 Pg C, microbial biomass may possess 15–30 Pg C, inorganic carbonates 1,220

Pg C, CH_4 clathrates 11 Pg C, soil peat 360 Pg C, and limestone 60 Pg C. The residence time and rates of accumulation of carbon stocks vary with each compartment. Carbon stock accumulation may vary enormously even within a compartment. It is based on location. For example, carbon stocks build up in the atmosphere vary with region and altitude. Similarly, carbon stock in microbes may vary depending on microbial population, diversity, and activity in the atmosphere.

Carbon sequestration in agrarian terrain can be achieved by adopting farming tactics that enhance carbon stocks in the soil. This can be aided by crops themselves and/or the cultivation of photoautotrophic microbes such as blue-green algae (BGA) on azolla, as well as free-living phototrophs like green algae and diatoms. Microbes that are present in the atmosphere or in particulate matter (PM) can also contribute to the increase in carbon sequestration trends. Additionally, by practicing no-tillage and minimizing soil disturbance, we can reduce soil microbial respiration and prevent unnecessary loss of soil carbon through biological and chemical oxidative processes (Lal, 2008). It is important to manage microbes in atmospheric dust and PM. The population of photoautotrophs versus heterotrophs in aeolian PM may hold the key to adding organic carbon into the atmosphere.

The global carbon cycle has been studied in greater detail than any other climate change factor. Currently, global carbon dynamics are out of balance. This is mainly due to the burning of fossil fuels and changes in land use, which result in the conversion of carbon-rich forests, marshes, and grasslands into agricultural crops with lower carbon-fixing capacity. It is worth noting that the activity of sequestering carbon, as well as the accumulation and conversion of atmospheric carbon, is carried out by photoautotrophic aeolian microbes. These microbes reside on dust and particulate matter and require attention. Blue-green algae (BGA) and diatoms, which are found freely floating in the troposphere, also contribute to carbon stocks. Upon deposition, they add carbon to agricultural soils. Additionally, aeolian dust deposits provide nutrients such as nitrogen, phosphorus, sulfur, iron, and silicon, which are essential for crops.

5.2.1.1 CARBON EMISSIONS INTO ATMOSPHERE

Rice production zones are sources of greenhouse gases (GHGs) such as CO_2, CH_4, and carbon monoxide (CO). The extent of GHG emissions depends on various factors, including the crops grown in rotation with rice, the organisms living in the soil, the chemical and physical properties of the soil, and the climate. The management of paddy fields plays a role in the amount of CH_4 and N_2O emitted. When fields are flooded, anaerobic soil conditions lead to the production of CH_4. At the same time, N_2O emissions can be produced by soil microbes in aerobic, microaerophilic, and anaerobic soils. Nitrogen emissions are influenced by nitrogen inputs, such as fertilizer. Typically, N_2O emissions are low in permanently flooded fields, while CH_4 emissions are high. There is a trade-off between them, depending on water-management practices in paddy fields (Lagomarsino et al., 2016).

Microbes in soil definitely influence C and N emissions. It has been observed that soil kept flooded and anaerobic conditions emit more CO_2 and NO_2. The composition of bacteria and archaea determines the extent of GHG emissions. In the case of C dynamics, it is said that reducing methanogenesis by CH_4-emitting microbes is important. To reduce the population of methanogens, it is preferable to alternate the soils between flooded and aerobic conditions. Not only soils, but the methanogen population in the troposphere can also be effectively managed (Lagomarsino et al., 2016).

CH_4 emissions form an important fraction of carbon dynamics in agrarian regions. Methanogenic microbes account for substantial quantities of CH_4 emissions. Global farm reports indicate a marginal decrease in greenhouse gas (GHG) emissions from agricultural belts. This decrease has been attributed to reduced use of fossil fuels and a shift from coal to natural gas for various farm-related activities. In 2017, the agricultural sector accounted for 5.8 billion metric tons of CO_2 equivalent, amounting to 9% of all GHG emissions. Over the past decade, CH_4 emissions have shown a decrease and stability (Newton, 2019). CH_4 emissions hover at 10% compared to 80% for CO_2 from cropping belts. However, it is important to note that CH_4 has a high global warming potential. CH_4 emissions are particularly large from ruminant cattle and sheep production centers. Wetland crops like rice also emit CH_4 in large quantities. Farmers have only reduced CH_4 emissions per unit of milk produced by 10% over the past decade. Farmers have affected carbon dynamics by sequestering atmospheric carbon into crops, forest stands, and by adopting agronomic measures that reduce emissions. For example, by adopting reduced or no-tillage practices, incorporating crop residue, etc. (Newton, 2019).

Lal (2008) points out that soil can be maintained as a net carbon emitter, i.e., a source for CO_2 or a sink, depending on the land use pattern and management strategies adopted. Agricultural fields could be disturbed and frequently plowed, for example, after each crop season. Such fields will be rich in heterotrophic aerobic microbial population. It is these fields that emit a large quantity of CO_2 through rapid respiration. The CO_2 emission occurs due to soil oxidative reactions as well as microbial respiration. The atmosphere receives large quantities of CO_2 due to reasons like the conversion of native stabilized vegetation to open field cropping systems. Burning of crop residues releases carbonaceous material from agricultural fields. Additionally, it may cause dust/smoke clouds containing S (SO_2, H_2S), CH_4, CO, and CO_2. Microbes on the surface and crop debris also get lofted into the atmosphere during crop burning. Generally, there is an assortment of microorganisms resident in the atmosphere. Therefore, we can expect a series of biological/chemical transformations commensurate with the biodiversity of atmospheric microbes.

Several biochemical transformations in the dust/clouds are mediated by atmospheric microbes. These are mainly identified as methanogenic, methanotrophic, CO_2-emitting heterotrophs, N-fixers, ammonia (NH_3) oxidizers, nitrifiers, S reducers, sulfur oxidizers, etc. Crop burning results in complex chemical combinations with carbon (C) as the major component in the atmosphere. Additionally, a variety of microbes are found in smog that results from crop burning. It may be necessary to identify and select suitable detoxifying microbes. It is said that continuous crop burning can even reduce the carbon stock of organically rich soils. It appears that approximately 36% of soil carbon has been lost as CO_2, PM, and CH_4 over a few decades. During the same period, soil nitrogen (N) has also been lost, reducing it by 16% of the original soil N status (Schaufler et al., 2010).

Loss of carbon (C) from topsoil, where rooting is intense, can be detrimental to crop productivity. No-till practices help conserve C and nitrogen (N) in the topsoil, preventing excessive microbial activity that would otherwise result in the loss of CO_2 through respiration (Schaufler et al., 2010).

It is widely acknowledged that prior to the Industrial Revolution, the concentration of CO_2 in the atmosphere was approximately 280 ppm. However, this concentration has significantly increased due to carbon emissions from various sources, including agriculture, industry, and transportation. In 2005, the average measured CO_2 concentration in the atmosphere was 380 ppm (Denman et al., 2007; Forster & Ramaswamy, 2007). Clearly, our actions have had a negative impact on C dynamics, leading to greater emissions than sequestrations in the soil. Climatologists believe that the increase in atmospheric CO_2 is largely attributed to agricultural practices, such as excessive plowing of the soil. This practice allows soil microbes to respire the C that is sequestered in the soil organic matter (SOM). Additionally, the burning of crop stubble instead of incorporating it into the soil after harvest has contributed to pollution through the release of carbonaceous carbon particles. Atmospheric microbes, known as heterotrophs, may utilize a significant portion of these particles and subsequently emit CO_2 through respiration.

Agricultural practices, such as submerged wet cropping systems, cattle dairying, and fossil fuel burning, are major causes of CH_4 emissions in the agrarian region. Methanogens in the soil, air, and cattle sheds release CH_4. It appears that during the preglacial era, the concentration of CH_4 in the atmosphere was about 700 ppb. Currently, it is 1,774 ppb (Denman et al., 2007; Forster & Ramaswamy, 2007; Le Treut et al., 2007; IPCC, 2007). Therefore, during the intermission, we have released excessive CH_4 into the airspace above crops. Methanogens found in microaerophilic soil conditions in paddy fields are among the important CH_4 emitters. Of course, there are also methanotrophs that utilize CH_4 as a carbon source. In fact, CH_4 emissions are higher than what is utilized by microbes. Incidentally, CH_4 has a greater potential to cause global warming, about 27 times greater than CO_2.

5.2.1.2 METHANOGENIC MICROBES AND CARBON DYNAMICS

Among carbonaceous emissions into the troposphere, CO_2 has shown a marked increase due to reasons such as a higher human population that utilizes fossil fuels, industries, and farming operations like crop burning. Reports indicate that since 1970 when ambient CO_2 was 288 ppm, it has increased to 400 ppm. This is a matter of concern, as CO_2 is a prominent greenhouse gas (GHG) that could trigger higher temperatures in agricultural zones. Next, CH_4 is another carbonaceous emission commonly encountered in paddy fields, cattle sheds, marshes, and biogas generating units. The total volume of CH_4 emitted from wetlands is of a lower order, yet it is still detrimental. This is mainly because the global warming potential of CH_4 is 25-27 times higher than CO_2 (Intergovernmental Panel on Climate Change, 2007). The half-life of CH_4 in the atmosphere is higher. It lasts in the ambient aerospace above crops for a longer duration, up to 25–27 years. It seems CH_4 concentration in atmosphere has increased by 150% (i.e., 1060 ppb), if 1750 A.D. is considered as the base year (Pramanik & Kim, 2017, 2018).

As stated earlier, rice production zones are the major emitters of CH_4 into the atmosphere. Dairying also generates a high degree of CH_4. Out of the 20 to 40 Tg CH_4 generated each year, anthropogenic sources like rice production could account for 11% (Yan et al., 2009; Anastasic et al., 1999).

Since we revise the global rice production trend each year, it proportionately increases CH_4 emissions. Therefore, it contributes to the burden of enhanced global warming.

The generation of CH_4 in rice fields is intricately linked to soil chemistry and microbial activity. Submerged paddy fields create anaerobic conditions with a negative redox potential. The redox potential (Eh) value in submerged soil starts decreasing after 72 hours of flooding the fields with water. Throughout the duration of rice cultivation, the Eh values in submerged anaerobic soils typically vary around –200 eV (Haque et al., 2013). Under such conditions, NH_4^+, H_2S, Mn_2^+, Fe_2^+, and CH_4 are dominant. In contrast, in aerobic soil that supports arable or dryland crops, NO_3^-, SO_4^{2-}, Mn_4^+, Fe_3^+ and CO_2 are predominant. The reduced aeration (low oxygen tension) in paddy fields immensely affects nutrient dynamics. The diffusion of gases into the soil in submerged fields is significantly slower. The water layer effectively separates the atmosphere and soil phase, creating microaerophilic or anaerobic conditions. As flood water is withdrawn, aeration is restored, leading to an increase in redox potential and a drastic change in the microflora of the soil phase to aerobic heterotrophic microbes (Greenwood & Goodman, 1961; Saree et al., 2012).

Now, how is the alteration of soil chemistry connected with methanogenic microbes and net CH_4 emission? CH_4 is generated in soil under anaerobic and chemically reduced conditions (Garcia et al., 2000). Initially, CH_4 emission may be feeble. However, CH_4 evolution increases as anaerobic conditions set in and the rice crop grows. Anaerobic conditions in flooded rice production zones are conducive to CH_4 generation (Ali et al., 2009; Kern et al., 2012). CH_4 emanates intensely when the soil redox potential reaches –200 eV (Ali et al., 2008).

Methanogens utilize organic carbon as an energy source. They are heterotrophic (Kumar & Sharma, 2014). CH_4 is the end product of the metabolic process with methanogenic microbes. Methanogens utilize simple carbon compounds like carboxylic acids, acetate, and formate. This is accomplished through a multistep process. However, substrate requirements could be stringent for a few species of methanogens (Kumar & Sharma, 2014; Welte & Deppenmeier, 2011). Hydrophilic methanogens may utilize CO_2 as an energy source (Wraith & Sharma, 1998).

In flooded rice fields, cellulolytic bacteria act upon carbonaceous polymers to release simple, low molecular weight organic compounds. Methanogenic bacteria utilize the resultant organic carbon sources to produce CH_4 (Pramanik & Kim, 2014a). The physicochemical nature of organic matter added to submerged paddy fields affects methanogenesis (Pramanik & Kim, 2014b). The population and diversity of methanogenic bacteria and archaea found in the soil are also important. Archaea are relatively less tolerant to oxygen tension in the soil medium (Wolfe & Metcalf, 2010). Sometimes, it is better to isolate specific efficient methanogens, culture them in the laboratory, and subsequently increase the inoculum size and add them to paddy fields for experimentation (Khelaifa et al., 2013).

As stated earlier, archaea found in the flooded paddy fields also generate CH_4 (Lee et al., 2010). This is in addition to the usual methanogenic bacterial load. The population of methanogenic bacteria is relatively higher. Isolating methanogenic Archaea requires stringent cultural conditions. They are slow to grow and extremely intolerant to oxygen (Wolfe & Metcalf, 2010; Khelaifa et al., 2013). It appears that the *mcrA* gene, which codes for the 'methyl co-enzyme M reductase' enzyme, is directly involved in the evolution of CH_4 (Lehmchar & Klenk, 1994). Overall, the dynamics of C, particularly CH_4, in submerged paddy fields are immensely influenced by methanogenic bacteria and archaea. Optimum expression of relevant genes, substrate, and its chemical constitution, as well as environmental conditions, require attention. We may alter any of these factors or all of them to regulate CH_4 emissions.

CH_4 emissions in agricultural zones depend primarily on the production, transport, and oxidation of CH_4 in rice fields (Conrad, 1989, 1993, 2002). The production of CH_4 is a prerequisite for any flux. Factors that affect CH_4 fluxes include soil type, rice variety, temperature, soil redox potential, water management, fertilizer inputs, and organic carbon content of the soil. The process of CH_4 production in paddy fields can be characterized into three stages: a lag phase, followed by an accelerated production phase of CH_4, and finally a tapering of CH_4 generation. These stages appear to be related to the activity of methanogenic microbes in the paddy fields (Conrad et al., 1989, 1993, 2002; Yao et al., 1999).

There are methods to mitigate methanogenesis. Rice cultivation under flooded conditions is considered an important source of CH_4 emission. Lee et al. (2010) observed that the active form of Fe (ionic form) decreased CH_4 emission. CH_4 emission from submerged soil may be reduced by approximately 15% through the application of industrial sludge and silicate fertilizers. Pramanik & Kim (2014a, b); and Pramanik et al. (2014) found that EDTA application, at smaller doses (up to 5.0 ppm), reduced CH_4 emission from submerged soils without affecting crop productivity. Methanogenic activity of microbes subsided in EDTA-treated soil. However, the activity of all microorganisms (microbial respiration) in the soil initially decreased due to EDTA application during rice cultivation.

Rice cultivation under submerged conditions involves generous applications of chemical fertilizers (NPK, micronutrients). In addition, soils are incorporated with larger quantities of organic matter to improve soil carbon status. Rice straw is recycled into the soil, and then the fields are flooded. Several reports depict the dynamics of population and diversity of methanogens. Rice straw incorporation affects both bacteria and archaea involved in the decomposition of soil organic matter and the emission of CH_4 gas (Adachi et al., 1996; Asakawa et al., 1998; Cahyani et al., 2004; Joulian et al., 1996; Kaku et al., 2000; Sugano et al., 2005a; Ueki et al., 1997). Sugano et al. (2005b) have adopted molecular sequencing of 16S DNA to identify the microbes and monitor changes in the population and community structure of methanogens in Japanese rice fields. In the case of methanogenic archaea, they found that *Methanosarcina* and *Methanobacterium* were dominant in paddy soils (Weber, 2001; Watanabe et al., 2004). Major bacteria traced belonged to Proteobacteria and Spirochaetes. Methanogenic bacteria and archaea are said to stratify themselves at different soil depths in paddy fields. Vaksmaa et al. (2017) have reported the occurrence of bacteria like *Acidobacteria, Chloroflexi, Proteobacteria, Planctomycetes,* and *Actinobacteria.* Archaea like *Euryarchaeota* and *Bathyarchaeota* were prominent compared to *Thaumarchaeota* in the upper 30 cm layer of soil. About 45% of archaea were traced in the top 5 cm layer of submerged soil. *Methanosaeta* occurred in a higher number than *Methanosarcina* in some fields. The measured rates of CH_4 evolution also varied. It depended on the population size of methanogens. The highest rate of CH_4 generation recorded was 442 nmol gdw^{-1} ha^{-1}. Air sampling above rice fields and cattle ranches to monitor methanogens should be possible.

In Asia, rice is often grown in sequence with arable crops. This procedure actually results in alternating submerged (negative redox) conditions with dry aerobic (oxidized soil conditions). Sometimes, the soil may be left fallow and dry. Rice fields may harbor about 10^5MPN methanogens per gram of dry soil. Investigations have shown that the population of methanogenic bacteria reduces by a quarter if submerged soils are drained. A few species of methanogens can withstand higher temperatures in the soil for a longer duration. The diversity, survival, and activity of methanogens depend upon soil types, oxygen tension, temperature, and substrate quality.

At this juncture, we should note that wind, air turbulence, and convection currents can introduce these methanogens into the atmosphere. The survival of methanogenic bacteria and archaea in the atmosphere is to be investigated. We cannot rule out the quick transfer of methanogenic microbes from one field to another through wind, precipitation, or dust storms. Migration of methanogenic microbes may occur despite aerobic conditions in the atmosphere. It depends on the extent of tolerance of methanogenic microbes to atmospheric conditions.

5.2.1.3 METHANE (CH_4) UTILIZATION BY METHANOTROPHS AND METHYLOTROPHS

Methanotrophy is encountered among prokaryotes found in many extreme environments, including the atmosphere. Methanotrophs show a wide range of abilities to tolerate temperature extremes from 4°C to 52°C. Methanotrophs were first identified and isolated in 1906. *Bacillus methanicum* is the earliest species of methanotroph identified. However, at present, over 100 bacterial species with the ability to oxidize CH_4 have been isolated (Jiang et al., 2011; Bowman, 2005). Aerobic methanotrophs belong to 14 genera. They are grouped into 'type 1,' 'type 2,' and 'type X' based on morphogenetics, cytoplasmic membrane structure, and biochemical pathways. There are filamentous methanotrophs like *Crenothrix polyspora* (Stoeker et al., 2006) and *Clonothrix fusca* (Vigliotta et al., 2007). There are also methanotrophs that tolerate extreme acidophilous conditions, such as *Methylacidiphilum infernorum* V4T (Dunfield et al., 2007), *Methylacidiphilum fumariolicum* SolV (Dunfield et al., 2007), and *Methylacidiphilum kamchatkense* Kam1 (Kruse et al., 2019). No doubt, methanotrophs are potentially useful in reducing the accumulation of CH_4.

Aerobic methanotrophs encountered in soil or the atmosphere metabolize emitted CH_4. This ability is found in bacteria belonging to the phyla Proteobacteria, mainly in the classes Alpha and Gamma Proteobacteria, and Verrucomicrobia (Dedysh & Dunfield, 2011). Methanotrophs oxidize C_1 compounds such as CH_4 to methanol. The enzyme responsible for this conversion is CH_4 monooxygenase (Crombie & Murrell, 2011). Methylotrophs have the ability to further convert the produced methanol to formaldehyde. Methylotrophs utilize specific biochemical pathways that capture the cytotoxic formaldehyde with cofactors like tetrahydrofolate, glutathione, or mycothiol before it is oxidized to CO_2 and emitted (Giri et al., 2021; Crombie & Murrell, 2011). Methylotrophs like *Methylomonas*, *Methylosphaera*, *Methylomicrobium*, *Methanocladium*, and *Methylobacter* utilize formaldehyde via methanol as intermediate. Clearly, the population dynamics of methanotrophs and methylotrophs in the ecosystem, including the troposphere, influence the rate of oxidation of emitted CH_4. In an agroecosystem, we may encounter methylotrophs as obligate converters of CH_4, such as *Methylomonas aminofaciens*. There are also a few others that are facultative. Thermotolerant isolates of methylotrophs operate at higher temperatures in the ecosystem, for example, *Bacillus brevis* (Yurimoto et al., 2002).

As mentioned previously, CH_4 is a greenhouse gas emitted from agricultural regions, particularly those involved in submerged paddy cultivation and dairying. CH_4 has low solubility in water, making it less accessible to airborne bacteria and other prokaryotes. Methylotrophic bacteria are known to utilize CH_4 as their sole source of carbon. They help detoxify CH_4 in the atmosphere and in soils. Methylotrophic bacteria are effective converters of atmospheric CH_4. Some fungi, such as *Fusarium solani*, found in the atmosphere and dispersed through airborne spores, are also capable of degrading CH_4. They transform CH_4 emitted from farms into the atmosphere. Laboratory trials have shown that *F. solani* can convert approximately 300 mg m^{-3} h^{-1}, similar to levels achieved by common methylotrophic bacteria like *Methylomicrobium album* and *Methylocystis sp.* (Vergara-Fernandez et al., 2019). In practical farming, we encounter an assortment of species of methylotrophic microbes capable of efficiently utilizing CH_4. Hence, it seems worthwhile to inoculate a mixture of methylotrophic microbes into farming regions. For example, inoculating methylotrophs into cattle ranches, sheds, pastures, and paddy fields may help reduce CH_4 in the atmosphere.

There are links between the N and C cycles in nature (Wang et al., 2017, 2019, 2022). Bacterial groups like Candidatus *Methylomirabilis oxyfera* and the archaeal group *Candidatus Methanoperedens nitroreducens,* as well as denitrifying anaerobic methane oxidation (DAMO) archaea, convert nitrate to nitrite using electrons derived from CH_4. The DAMO bacteria then reduce nitrite to nitric oxide, which is further converted to nitrogen and oxygen through the inter-aerobic denitrification pathway.

5.2.2 MICROBES AND NITROGEN CYCLE IN AGROECOSYSTEM

The atmosphere is the largest repository of nitrogen (3.9×10^{21} g N). The second largest nitrogen reservoir is the biosphere. The global nitrogen content is 4,100 Tg in vegetation and 5,000 Tg in organic matter. Approximately 95,000 Tg is present in soils. Oceans possess $2{,}000 \times 10^2$ Tg of organic-N (Holland & Weitz, 2003).

Nitrogen occurs in different physicochemical forms and compounds in different ecosystems of Earth. Galloway & Cowling (2002) point out that molecular N_2 found in the air is inert. About 79% of atmospheric gases are constituted by such inert N_2 gas. Often, this N_2 is called non-reactive, i.e., 'inert N gas.' However, it has exceptions. The biological N-fixing microorganisms resident in the troposphere or those in the soil can utilize gaseous N_2. They convert gaseous N to NH_4 compounds. There are asymbiotic and symbiotic microbes capable of converting atmospheric-N to organic-N. Apart from this, N compounds other than N_2 gas are collectively termed 'reactive-N (Nr).' Organic and inorganic N compounds, i.e., Nr compounds, are amenable to biological and chemical transformation in the atmosphere, soil, and biotic systems (Galloway et al., 2004; Canfield et al., 2010). Furthermore, Galloway et al. (2003) suggest that there are three main reasons for the enhanced rate of Nr creation on the globe. They are excessive use of legumes and other crops that convert atmospheric N_2 via BNF, combustion of fossil fuels and crop burning that increases Nr PM in the atmosphere, and the Haber-Bosch process used in the manufacture of fertilizer-N that creates Nr pool.

Nitrogen dynamics in the atmosphere are affected by both advection and gaseous diffusion. Microbial movement in the atmosphere also affects the extent of utilization of N compounds and their distribution in the troposphere. Furthermore, in the atmosphere, the conversion of gaseous N compounds to particulate form and *vice versa* may also affect the N dynamics, particularly through microbe-mediated conversions.

Nitrogen deposits from the atmosphere that reach the soil surface and water bodies need to be considered. Microbial conversions in soil and water depend on the rate of nitrogen deposits. Reactive nitrogen (Nr) compounds that reach the soil are important in terms of the reactions that occur. There are actually two types of nitrogen deposits: wet deposits and dry deposits. Wet deposits involve clouds, precipitation, fog, and snow during the transit of nitrogen from the atmosphere to the soil surface. Dry deposition is the direct transport of nitrogen gas into the soil profile. These physical processes do have an impact on microbe-mediated transformation.

Major forms of nitrogen (N) encountered in agrarian regions, including soil, water, and the troposphere, are as follows. Our specific focus is on atmospheric N. Commonly traced inorganic gases above crops include ammonia (NH_3), nitrogen dioxide (NO_2), nitrogen oxide (NO), nitric acid (HNO_3), and nitrous acid (HNO_2) (Yan et al., 2003; Aneja et al., 2001). Particulate airborne N consists of ammonium (pNH_4) and nitrate (pNO_3).

NH_3 and ammonium ion (pNH_4) are the major reduced forms in the atmosphere, while the rest of the N compounds are oxidized forms. N_2O and nitric oxide (NO) are collectively referred to as NO_X. Nitrogen dioxide (N_2O), an important greenhouse gas (GHG), is excluded from airborne N as it is unlikely to contribute to atmospheric deposition. Organic N also exists in the atmosphere, such as organic nitrates like peroxy-acetyl nitrate (PAN) that are traced in the air. Other examples of organic-N in the agrarian sky include atmospheric bacteria, archaea, fungal propagules, and their organic debris (Neff et al., 2002; Zhang & Liu, 2018). Organic particulate matter (PM) is one of the major sources of organic-N in the atmosphere and is susceptible to transformations and degradation by aeolian microbes.

Nitrogen dynamics in the lower troposphere, soil profile, and water bodies that feed the farming regions are always interconnected (Plate 5.6). Microbes that metabolize nitrogen-containing compounds are also widely distributed and interconnected. Together, they have a significant influence on the cropping procedures adopted, nitrogen absorption and assimilation by natural vegetation, weeds, and main crops in a given agrarian region. In this context, we are concerned with the nitrogen dynamics mediated by microbes that often reside in the troposphere, whether floating freely, inside particulate matter, on clouds, or in moisture, among other locations. These troposphere microbes may have originated in the soil phase of the agroecosystem. They reach the troposphere through aerosolization, levitating dust masses, or from water bodies due to turbulence. Microbes that are traced on crop canopies and the phyllosphere also find their way into the lower troposphere. Regardless of the source of troposphere microbes, their role in mediating nitrogen transformation in the air is of concern. They may contribute small to large amounts of nitrogen to the soil phase of the agroecosystem.

5.2.2.1 BIOLOGICAL NITROGEN FIXATION ($2N_2 + 8H = NH_3 + 8H_2$)

Biological nitrogen fixation (BNF) in the atmosphere occurs via aeolian microbes, such as BGA (blue-green algae), specifically cyanobacteria like *Anabaena* and *Nostoc,* as well as free-living bacteria like *Azotobacter* and *Azospirillum,* among others. It also occurs in the soil profile through both asymbiotic and symbiotic microbes, such as *Bradyrhizobia* and *Frankia.* Nitrogen fixation requires microaerophilic or anaerobic conditions to facilitate the conversion of atmospheric N_2 into organic forms. For instance, this can happen inside nodules on roots or cells of *Azotobacter* kept in microaerophilic conditions, or by producing excessive amounts of polysaccharides. Free-living BGA produce specialized cells known as 'heterocysts,' which provide anaerobic conditions for atmospheric N_2 to be fixed into ammoniacal compounds.

Major genera that mediate N-fixation in the asymbiotic state, in the atmosphere under free-living aerobic conditions, include *Azotobacter, Azospirillum, Beijerinckia, Klebsiella,* and BGA. Additionally, Archaea such as *Euryarcheota* (methanogens), *Crenacheota, Thaumarchaeota,* and *Nanoarchaeota* (Boyd & Peters, 2013) are involved. Asymbiotic N-fixers that operate under anaerobic conditions are *Clostridium* and *Desulfovibriode sulfuricans* (Purple sulfur bacteria).

Nitrogen gas (inert) is abundantly available in the troposphere. However, its distribution is not uniform in terms of geographic location and altitude above cropping zones. It also varies temporally depending on the season of the year. It is important to consider the conversion of non-reactive N_2 gas in the atmosphere to reactive oxidized forms like N_2O, nitric oxide (NO), and/or reduced forms like NH_3 (Falkowski et al., 2008).

Nitrogen transformation in the troposphere has a significant impact on soil nitrogen fertility and crop growth. In nature, N_2 gas is a stable component, and breaking the strong triple bonds between N atoms is challenging. However, there are physicochemical reactions, such as those caused by lightning, that can lead to the formation of NH_4. In this section, we are primarily concerned with microbe-mediated nitrogen dynamics, specifically biological nitrogen fixation. The conversion of atmospheric N into organic N is crucial for the perpetuation of crops and other biotic forms on Earth. The key enzyme responsible for mediating the formation of NH_4 from inert N_2 gas is nitrogenase, which functions under anaerobic conditions. The evolution of nitrogen fixation genes (nif) is crucial, particularly for microbes in the atmosphere to convert the large quantities of available N_2 gas. A microaerophilic or anaerobic environment and an energy source, such as ATP, are necessary for this process. Only then is it possible for N dynamics to result in NH_4 inputs into the troposphere or soils. The nitrogenase enzyme, found among certain microbes, plays a vital role in the nitrogen dynamics of agroecosystems.

According to Boyd & Peters (2013), the evolution of *nif* genes in microbes is restricted to a few genera and species. There could be a few that have lost the ability for biological N-fixation during the course of evolution. For example, they state that a few bacterial species often found in the atmosphere, such as *Actinobacteria, Proteobacteria, Firmicutes, Chloroflexi, Chlorobi,* and *Bacteroidetes,* might have had their *nif* genes deleted during the course of their evolution. Alternatively, it is possible that nif genes never evolved in these microbes. In nature, there are several microbial species that affect N dynamics via *nif* genes (i.e., BNF). Of course, necessary modifications to create partially or completely anaerobic conditions are needed. For example, the production of heterocysts, polysaccharides, or nodules (core), etc. Microbial species including BGA, bacteria, actinomycetes, and archaea are involved in the conversion of tropospheric N_2 to NH_4^+.

The evolution of nitrogenase, i.e., the ability to fix atmospheric N_2 in microbes, seems to be related to N dynamics in the agroecosystem. The conversion of tropospheric N_2 to NH_4 compounds has been occurring for millions of years (i.e., since the Archaean geologic period). Boyd & Peters (2013) and several others have suggested that, in nature, a scarcity of NH_4 compounds might have occurred due to insufficient conversion of gaseous N_2 to NH_4^+ via abiotic sources, such as lightning and other means. Hence, it could have necessitated and induced several microbial species to evolve into species with nitrogenase (*nif* genes). Microbes supporting the conversion of N_2 have evolved several methods by which they seclude nitrogenase from O_2. These adaptations seclude nitrogenase into partial or totally anaerobic conditions (Schoonen & Xu, 2001; Navarro-González et al., 2001; Kasting & Siefert, 2001; Fani et al., 2000). In the context of N_2 conversion in the troposphere, the rate-limiting factors are the presence and distribution of aerial microbes, the activity of the nitrogenase enzyme (*nif* genes), and favorable conditions. The presence of Mo-dependent nitrogenase genes in aerial microbes is crucial for biological nitrogen fixation (BNF). However, there are alternative nitrogenase enzymes that utilize different metal ligands. These alternative enzymes may become necessary in environments with low Mo availability. Given the availability of energy, microbial population, and organic substrates, the troposphere has the potential for large-scale transformation of N_2 into NH_4^+ and organic-N.

5.2.2.2 *NITROGENASE ENZYME*

Nitrogenase enzyme and its activity determine, to some extent, the dynamics of N_2 conversion to NH_4^+ compounds. Therefore, we aim to identify the microbes that carry genes encoding nitrogenase (*nif*). Several species of archaea possess *nif* genes and are capable of converting tropospheric N into NH_4^+ compounds. For instance, Boyd & Peters (2013) state that nitrogenase is found in archaea classified as *Methanococcales, Methanobacteriales,* and *Methanosarcinales.* A few other archaea that possess N-fixation genes include Crenarchaeote, Thaumarchaeota, and Nanoarchaeota. Additionally, a few thermophilic bacteria also have the ability to convert atmospheric N, influencing the nitrogen dynamics in the environment. For example, *Thermocrinis albus* and *Hydrogenobacter thermophilus* (Zeytun et al., 2011) are thermophilic N fixers.

Nitrogen fixation is associated with microorganisms that display a wide variety of other physiological abilities. The BNF organisms operate at a wide range of O_2 tensions. They could be obligate aerobes, micro-aerophilic, or totally anaerobes (Raymond et al., 2004; Boyd et al., 2011; Dos Santos et al., 2012). Microbes in the troposphere or elsewhere in the soil profile may adopt a few different mechanisms to segregate the N-fixation apparatus from aerobic or oxic conditions (Fay, 1992; Berman-Frank et al., 2003; Boyd & Peters, 2013). In the case of nodule bacteria (rhizobium), it is achieved via leghemoglobin, which provides a chemically reduced condition with low oxygen tension. In cyanobacteria (BGA), the N fixation proceeds in 'heterocysts.' Heterocysts are specialized cells in algae. Heterocysts exclude aerobic/oxic conditions. These cells provide anaerobic conditions for nitrogenase enzyme activity. Several other microbes utilize cytochrome oxidase and respire off O_2. It creates low oxygen tension or microaerophilic conditions (Poole & Hill, 1997). Whatever is microbial species or method of supporting nitrogenase activity, the bottom line is that *nif* genes (nitrogenase) and anaerobic conditions are needed for N fixation, even though the microbe may be aerobic in nature.

Farmers often apply excessive amounts of chemical nitrogen (N) to their fields, resulting in its accumulation in the soil. The presence of excessive nitrogen in the soil makes it susceptible to transformation by microbes, ultimately leading to greenhouse gas (GHG) emissions. However, Stewart et al. (2005) have demonstrated that cultivating crops without the use of commercial chemical nitrogen fertilizers in the Great Plains region can lead to decreased yields. Cereal crops such as corn, wheat, rice, barley, and sorghum experience a decrease in grain yield ranging from 16–41%. Cotton grown without chemical fertilizers experiences a 31% loss in yield. However, in soils that are rich in nitrogen-fixing (BNF) microbes, legumes that are not provided with chemical fertilizers do not suffer a decrease in yield.

The BNF also occurs in the atmosphere. Particularly, in regions endowed with organic dust (PM), energy-providing carbon compounds, and BNF organisms. Microbes in the troposphere may also contribute to nitrogen via BNF if they happen to possess nitrogenase activity. We may need to explore methods of enhancing the microbial population of the troposphere at the appropriate time above large cropping zones. Wind drifts and/or precipitation may affect the deposition of atmospheric nitrogen added via aeolian microbes.

5.2.2.3 THE NITROGEN (N) LOAD AND FLUXES IN ATMOSPHERE

There is a term known as 'nitrogen load (N load)' that depicts the concentration of nitrogen in the agricultural atmosphere. Essentially, it deals with the amount of nitrogen present in the troposphere above crops and the subsequent nitrogen dynamics. This includes the accumulation and transformations of nitrogen, as well as the limits to the accumulation of nitrogen compounds in the atmosphere (lower troposphere). Major emissions such as NH_3, NO_2, N_2O, NO, and N_2 are often observed in agricultural regions. In the case of the Far East and Japan, it is said that excessive use of fertilizers has led to nitrogen accumulation, frequently reaching critical nitrogen load levels. In China, the nitrogen load in the atmosphere above agricultural areas has reached 3.56 Tg N $year^{-1}$. Japan also adopts intensive farming practices and high nitrogen inputs, resulting in an air nitrogen load of 0.59–0.69 Tg N $year^{-1}$. Accurate nitrogen inputs, at appropriate intervals, followed by removal by crops and soil microbes, and proper cycling of nitrogen through residue incorporation, are essential. Reducing the number of inputs of nitrogen to the soil is necessary to avoid greenhouse gas emissions (Hayashi & Yan, 2009). A higher nitrogen load in the troposphere can lead to increased deposits in agricultural soils. In fact, in some locations in Japan, deposits of around 60 kg of nitrogen per year have been recorded. It is important to understand the role of aerial microbes in nitrogen conversions, as they have the potential to reduce nitrogen accumulation in agroecosystems. If emissions are reduced, organic nitrogen in the soil could be stabilized. Additionally, we need to study the relationship between nitrogen loads and the microbes involved in nitrogen transformations in the atmosphere, soil, and water bodies (Plate 5.6). Microbes that fix atmospheric nitrogen or those that cause emissions of soil nitrogen may play a vital role in regulating tropospheric nitrogen loads.

Generally, achieving altered N load in an agricultural field or large cropping expanse can be done by adopting a few suitable agronomic procedures. Some examples include direct leaching of N from the soil, reduction in atmospheric deposition of N via PM, reduction of fertilizer-N inputs, and reduction in airborne-N to maintain air quality. High N inputs can lead to pollution and overflow of N from agroecosystems into other regions. These measures may not require microbial intervention within the agroecosystem. However, by adopting atmospheric-N fixation through microbes and reducing emissions of N_2O, NO_2, NH_3, N_2, the N load can be regulated. Nitrification inhibitors combined with fertilizers are also used in several agrarian regions to regulate the loss of N through emissions.

On a larger scale, agrarian regions worldwide have experienced an increase in N inputs for crop production. Fertilizer N, BNF via symbiotic and free-living soil bacteria, crop burning, and aeolian deposits have caused an enhanced turnover of N in cropping belts. The microbial component in soil, atmosphere (lower troposphere), and water bodies such as lakes, ponds, rivers, and canals are intricately linked to N dynamics (Plate 5.6).

The industrial conversion of free-N (inert N_2 gas) through the Haber-Bosch process has resulted in significant inputs of nitrogen into the soil ecosystem of agrarian regions. As a result, the nitrogen cycle in the soil environment is impacted by the introduction of N through fertilizers. Fertilizer-N inputs have exponentially increased in most agrarian regions in the USA since the 1960s, particularly in areas where intensive cropping is practiced. Consequently, unutilized N is susceptible to chemical and bacterial transformations, leading to the emissions of NO_2, N_2O, NH_3, and N_2. The concentration of atmospheric N_2O is observed to be increasing by 0.8 parts per trillion annually. The adoption of chemical fertilizer inputs has caused an increase in atmospheric N_2O concentration from 270 ppt to 314 ppt (Prather et al., 2001). However, current farming trends have reduced the supply of fertilizer inputs, resulting in a plateau in atmospheric N_2O levels.

Biological nitrogen fixation (BNF) supported by legumes is estimated to contribute up to 40 Tg N per year. Blue-green algae and free-living bacteria also contribute to soil-N through BNF. Extensive areas of soybean and other legumes efficiently convert atmospheric N into organic N in the soil. BNF can even occur in atmospheric particulate matter if it harbors N-fixing microbes.

Global nitrogen fluxes have changed due to excessive use of chemical fertilizer. It has increased from 90-140 g year^{-1} since the pre-chemical fertilizer era. About 210 Tg N year^{-1} has been added to these fluxes by the use of chemical fertilizer-N (urea, KNO_3, NH_4NO_3), expansion of legumes (i.e., BNF), deposition of N-containing pollutants, aeolian dust deposits, etc.

Denitrification in soil removes accumulated nitrogen, to a certain extent. It is a natural way of avoiding eutrophication. Denitrification of fertilizer nitrogen inputs is estimated to emit 6.9 Tg of nitrogen into the atmosphere. Denitrification is mediated by several species of microbes. Overall, about

PLATE 5.6 Atmospheric wind, mild breeze, turbulence due to farm installations, clouds, mist, may all distribute microbes present in the ambient troposphere. [Left: Troposphere breeze and/or turbulence due to farm installation (windmill), clouds, mist, and canals distribute/mix nutrients and microbes in the ecosystem. Location: Kinder Dijk, near Rotterdam, Holland. Right: A tulip field at seedling stage, natural forest stand, fertile chernozemic soil, and irrigation channel under a misty microbe and nutrient loaded atmosphere. Location: Near Keukenhof, Lisse, South Holland, Netherlands].

Note: Left: Windmills can create winds/turbulence that transport particulate matter and microbes over short distances. This mixing affects the gaseous composition of the troposphere. It is true that the microbial load in the agrarian sky is interconnected with crops, soil, water bodies, and other abiotic and biotic factors in nature. The atmosphere above the field supports diverse microbes capable of countless biochemical transformations. These microbes are closely connected with nutrient dynamics, especially nitrogen load, which occurs in the field. This includes the air, natural vegetation, crops, soil, and water bodies (such as irrigation channels).

Source: Mr. Sharath Kowligi and Roopashree, B. N., Eindhoven, Noord Brabant, Netherlands.

20–30 Tg of gaseous nitrogen is removed into the atmosphere each year, by crop/stubble burning.

Ammonification (i.e., R-NH_2NH_3 + H_2O > NH_4^+ + OH-) is the cause for ammonium (NH_4^+) and nitrification is the cause for nitrate (NO_3^-) emissions. Both are biologically available nitrogen forms. Nitrogen-bearing compounds traced in microflora, plants, fauna, soil organic matter, or crop residues are degraded and recycled into the atmosphere as NH_3. Ammonification is accomplished by diverse microbial species resident in the troposphere, soils, and water surface. Primarily, ammonification is an important process mediated by microbes (bacteria, fungi, and other prokaryotes) in the ecosystem. It releases large quantities of NH_3 into the troposphere.

5.2.2.4 AMMONIA (NH_3) OXIDATION BY BACTERIA AND ARCHAEA

Ammonia (NH_3) emissions into the atmosphere are attributable to several factors, such as excessive use of ammoniacal fertilizers, crop burning, submerged paddy cultivation, and industrial emissions. Fertilizer-N (e.g., urea, NH_4NO_3) supplied to crops is a major source of NH_3. NH_3 is captured by clouds, rain, fog, surface water, and PM. It is then recycled to vegetation, crops, and terrain, which is an important aspect of the N cycle. Atmospheric NH_3 also catalyzes transformations like sulfur dioxide to sulfur trioxide (Renard et al., 2004; Pai, 2021).

Gao et al. (2016) further state that NH_3 emissions could be detrimental to the environment that is already polluted. It generates toxic smog in the troposphere above farm belts. Under these conditions, the role of ammonia oxidizing bacteria (AOB) needs to be given due consideration. Investigation regarding the diversity, abundance, and activity of AOB in the PM is useful. Nitrification, of course, is the first step in NH_3 oxidation. It results in N_2O. Lehtovirta-Morley (2018) believes that NH_3 oxidizing microbes play a vital role in the geochemical cycling of N. Microbe-mediated nitrification is the first step in NH_3 conversion. It is also the rate-limiting step in nature for the conversion of NH_3.

Examination of smog and PM in Asian farming regions has shown that *Nitrosopumilus multiformis,* and *N. aesturii* were common on atmospheric PM. *Candidatus Nitrospira inopinata* was dominant in finer $PM_{2.5}$ (< 2.5 μm). Reports from Chinese locations have shown that NH_3 oxidizing archaea (AOA) and AOB population fluctuated at 10^3 to 10^4 cells per m^3. Hence, they could mediate nitrification to significant levels. Good enough to mitigate NH_3 accumulation in air. The maximum rates of nitrification due to AOA and AOB populations in finer PM were estimated at 0.14 μg (NH_4–N) m^{-3} air.hr^{-1}. Depending on the location, *Nitrosphera* species and *Nitrosomonas eutropa* were commonly found in PM (Gao et al., 2016).

NH_3 is the most abundant reactive N (Nr) emitted into the troposphere. It is highly mobile and disturbs the normal geochemical cycling of N in nature. Atmospheric PM may also contain inorganic N forms such as NH_4NO_3, $(NH_4)_2SO_4$, and $NH_4.HSO_4$. As stated above, these are derived from fertilizer inputs in agrarian regions and are abundant in finer PM (< 2.5 μm). Examination of the atmosphere above farms in Northeast China showed diverse species of AOA and AOB. Most AOB belonged to ß- and √-Proteobacteria, including *Nitrosococcus, Nitrosomonas,* and *Nitrosospira.* Surveys have also shown the common occurrence of *Nitrosopumilus, Nitrosotalea, Nitrosocaldus,* and *Nitrososphaera.* Archaea belonging to *Thaumarchaeota* were also found. Gao et al. (2016) state that, overall, microbial oxidation of NH_3 is perhaps a central step in regulating N cycling in agrarian regions.

5.2.2.5 NITRIFICATION

The nitrification process, mediated by microbes, is an important aspect of the nitrogen cycle in soil and the atmosphere (PM). Soil NO_3 is largely contributed by fertilizer inputs. Nitrification involves two steps. The first step involves the conversion of NH_3 to nitrite, followed by the conversion of nitrite to nitrate (Step 1: $NH_4^+ + 3O_2 > 2NO_2 + E + 4H$; followed by Step 2: $2NO_2 + O_2 > 2NO_3 + E + 4H$). Nitrification that occurs in nature is mediated by a few microbes found in soil, the atmosphere, or on the phyllosphere.

Nitrifiers are aerobic microbes. They oxidize NH_3 while utilizing CO_2 to synthesize organic compounds. Nitrification occurs at optimum rates under slightly acidic to neutral pH, in the atmosphere/soil or any medium. Nitrification is mediated by at least two groups of microbes. Nitrosofication is mediated by *Nitrosomonas, Nitrospira, Nitrosolobus, Nitrosovibrio,* etc. The second step is mediated by nitrite oxidizers. For example, nitrifiers like *Nitrobacter, Nitrococcus, Nitrospira.* Nitrification actually links oxidation of NH_3 that is produced by the degradation of organic compounds to gaseous N_2 emission. NH_3 oxidation, as stated earlier, is accomplished by AOA and AOB. Most of these microbial species are aerobic and a few are chemoautotrophs.

The concentration of N_2O in the atmosphere increased from 270 ppb during the pre-industrial revolution period to about 319 ppb in 2005 (Le Treut et al., 2007; IPCC, 2007; Forster & Ramaswamy, 2007). This increase has been attributed to the use of fertilizer-N. The excess nitrogen in the soil, which accumulates each year, is susceptible to conversion by nitrifying bacteria. Failure to use nitrification inhibitors when there is a higher use of fertilizer-N exacerbates N_2O emissions from agrarian regions. Additionally, crop burning releases N_2O into the atmosphere. The particulate matter released into the atmosphere may contain microbes that convert organic nitrogen compounds. Denitrification is also a possibility in the atmosphere. It is often suggested that nitrification and denitrification transformations mediated by atmospheric microbes may be equal, thus canceling each other out. This provides a better nitrogen balance in the atmosphere compared to the excessive release of N_2O (Signor et al., 2013; Knowles, 1982).

5.2.2.6 DENITRIFICATION (NO_3>NO_2> NO + N_2O >N_2)

Denitrification involves the conversion of nitrate to gaseous intermediates like N_2O, resulting in nitrogen (N_2) emission. Denitrifying bacteria utilize nitrate as a terminal electron acceptor for respiration instead of O_2. Microbes that are facultative anaerobes carry out denitrification. This microbial process depletes soil nitrogen (N). Some examples of denitrifying microbes include *Thiobacillus denitrificans, Serratia, Pseudomonas,* and *Achromobacter.* Several of these denitrifiers can be found in the particulate matter (PM) of aeolian dust. They occur at population levels that can impact nitrogen dynamics in agricultural areas. The primary source of denitrifiers in the troposphere is surface soil.

Denitrification occurs in the absence of oxygen and releases N_2 gas back into the troposphere. By definition, denitrification is the conversion of fixed nitrogen. In the soil and atmosphere, it is actually the decomposition of organic matter. This process takes place under microaerophilic conditions and is mediated by facultative anaerobes. Non-biological chemo-denitrification also occurs under certain atmospheric conditions, resulting in the release of N_2 gas into the air. In agrarian regions, denitrification of nitrogen is an important process that leads to the loss of nitrogen from the soil. Soil nitrogen is released as N_2O or N_2. Tiedje (1988) has stated that denitrification completes the nitrogen cycle in agrarian soil and sky, almost balancing the total biological nitrogen fixation that contributes to nitrogen in agrarian soils. The sequence of nitrogen conversions in the agroecosystem during denitrification begins with NO_3, which is transformed to NO_2, followed by NO, N_2O, and finally N_2. Factors such as rainfall events, soil drainage, and oxygenation regulate denitrification. The oxygen tension in the region determines the conversion of NO_3 to N_2. Denitrifying bacteria utilize carbon compounds as electron donors (Tiedje et al., 1983).

5.2.3 MICROBES AND PHOSPHORUS CYCLE IN AGRARIAN REGIONS

The atmosphere is not the primary source or reservoir of phosphorus (P) on Earth. Unlike nitrogen (N_2), which makes up 79% of the gaseous component of the atmosphere, phosphorus is not frequently cycled into the atmosphere but rather exchanged between the lithosphere and hydrosphere. The quantity of phosphorus in the atmosphere is not significant compared to nitrogen. In the troposphere, particulate matter (PM) contains relatively smaller amounts of phosphorus. Phosphorus is present in inorganic or sand/mineral PM in a chemically bound form. Organic-P can be found in organic dust, minute organic debris, plant residue, and other sources. The abundance of P compounds in the troposphere is proportional to the microbes involved in the biochemical transformations of phosphorus.

Atmospheric phosphorus (P) is derived from both natural and anthropogenic activities. The main source of P is mineral dust, which is generated through turbulence on the soil surface (Weinberger, 2016; Shi et al., 2019; Tipping et al., 2014). Organic-P in the atmosphere is contributed by microbes (bioaerosols) and organic compounds (Violaki et al., 2021). The solubility of P depends on the source of atmospheric particulate matter (PM). When PM is sourced from natural mineral particles, the solubility of P is low. However, if dust particles are derived from anthropogenic sources, the solubility is higher. The acidification process is responsible for this solubility. Additionally, the solubility of P in PM varies depending on the season when the dust particles were introduced into the atmosphere. Samples collected in winter were more acidified and had higher solubility compared to those collected in summer. Moisture and relative humidity (RH) are also believed to influence the solubility of P in PM (He et al., 2011; Herbert et al., 2018). Furthermore, microbes residing on PM may contribute to its solubilization, as many microbial species are known to secrete organic acids into the substrate (i.e., particulate matter (PM)).

Tian et al. (2021) have pointed out that the phosphorus (P) cycle differs from the nitrogen (N) and carbon (C) biogeochemical cycles. Phosphorus compounds are scarce in the atmosphere, with only small amounts of phosphoric acid (H3PO4) entering the ambient troposphere. This can contribute to acid rain. Okin et al. (2004) note that gaseous inputs to the soil P pool are negligible. Therefore, terrestrial ecosystems and agricultural crop production are said to primarily rely on inherent soil P pools in the long term, with additional contributions from particulate deposits received from the atmosphere. However, reports on soil-P vs crop production often overlook the accumulation of atmospheric dust/deposits in the soil. It is important to consider the significant role of P-rich aeolian dust deposits in various crop regions across different continents (Artaxo et al., 2002). Additionally, we should also acknowledge the contribution of diverse microbial communities, particularly those capable of transforming organic-P sources. Each dust particle may introduce microbes that can transform P sources found in particulate matter (PM) upon deposition into the soil profile.

Miller (2019) highlights the importance of aeolian dust inputs to agrarian regions, including their water resources. Phosphorus is, in fact, traced in multiple forms in soil, aeolian dust, and water resources within an agrarian ecosystem. Aeolian dust P that deposits on P-fixing soil may convert it into Fe and Al-P. These forms of P are not easily available to crops immediately. In soils where it forms Ca-P, it easily dissolves and is available to crops.

Aeolian dust inputs play a vital role in supporting the biotic component. The nature of aeolian dust, plus the microbes that act upon the substrate, should be noted. They may also have an impact on the P nutrition of crops that utilize water sources.

5.2.3.1 MICROBES AND PHOSPHORUS AVAILABILITY TO CROPS

Phosphorus is found in the lithosphere, mostly as phosphatic rocks (ores) or in soil matrix in chemically bound or, to a certain extent, in exchangeable form. Most of the P traced in the biosphere or hydrosphere arises from the lithosphere. Atmospheric P is also derived from the lithosphere. The atmosphere (lower troposphere) is a weak source of P and P-containing organic compounds. The general consensus is that P in crop production farms does not cycle into the atmosphere in large quantities. Phosphorus transits from agricultural soils or rock-P deposit regions to water bodies and vice versa through runoff (erosion).

Phosphorus traced in the lithosphere is not easily available for biotic activity. It is poorly soluble in water and often held in insoluble form in soils. It is chemically fixed to the soil matrix or strongly held by soil particles. Phosphorus is not easily released into the soil solution, making it difficult for crops to scavenge soil P. This is unlike NO_3 or NH_4 compounds that rapidly transit through the soil solution in a dissolved state across the soil profile.

In practical agriculture, soils may contain optimum levels of total phosphorus (P). However, its availability to the roots of crops is low due to chemical fixation. Insoluble inorganic P forms, such as $Al.Mg.PO_4$ and $Ca_3(PO_4)_2$, are also unavailable to crops. In agricultural soils, P that is found in a soluble form is accessible to plant roots. There are also exchangeable or sparingly soluble forms, as well as an unavailable form that is chemically bound, in soils. Phosphorus held in soluble and sparingly soluble forms is easily exchangeable. These P pools reach equilibrium in the soil, which is dependent on the physicochemical nature of the soil profile. Phosphorus found in the soil matrix as a chemically bound form is released very slowly and is not easily exchangeable. Exchangeable P is released into soluble P pools based on the rate of exhaustion of the soluble-P pool and equilibrium levels. Microbes and plant roots can only utilize the freely soluble form.

Phosphate solubilizing microbes (PSM) release organic acids, resulting in an increase in exchangeable-P. Some PSMs enhance available-P through the release of organic acids. Mycorrhizal fungi exploit the P found in the available P pool of the soil. However, mycorrhizal fungi are not known to actively solubilize P in the soil. Plant roots also recover P only from the soluble-P pool, unless there is a tendency to release organic acid and solubilize the sparingly soluble fraction of P. The efficiency of absorption of applied soluble fertilizer-P by plant roots is very low, often just 20% of the fertilizer-P supplied to the soil. This is due to soil chemical fixation and the conversion of fertilizer P to organic-P by soil microbes.

Phosphorus solubilizing microbes (PSM) are widely distributed in agricultural soil. Dust plumes generated from such soils may contain PSM in the particulate matter (PM). These PSMs are known to solubilize inorganic phosphorus (P) through the secretion and release of organic acids such as gluconic, oxalic, and propionic acids. Additionally, organic-P found in the PM can be solubilized through the action of phosphatase enzymes.

There are numerous reports that identify families of microbes with P solubilizing ability. For instance, Wang et al. (2021) have identified members of Micrococcaceae, Streptomycetaceae, Methylobacteriaceae, Phyllobacteriaceae, Rhizobiaceae, Pseudomonadaceae, Oxalobacteriaceae, and Burkholderiaceae. They utilized insoluble-P as the source of P and employed 16S rRNA sequencing to identify individual species of phosphate solubilizing bacteria (PSB) within each family. Phosphate solubilizing microbes belonging to bacteria, actinomycetes, archaea, and fungi have been identified, cultivated, and evaluated. Commonly found PSB are *Bacillus* sp., *B. megaterium strains*, *Pseudomonas* sp., *Pseudomonas putida* (P13), *Pantoea agglomerans* (P5), *Microbacterium laevaniformans* (P7), *Escherichia* sp., *Actinobacter* sp., etc. Several isolates of *Burkholdrea gladioli*, *B. cenocepacia* and *B.* sp. possess phosphate solubilizing ability. Fungal species identified as phosphate solubilizers are *Penicillium* sp., *Aspergillus* sps., *Mucor* sp., and *Rhizopus irregularis.* Several actinomycetes also possess phosphate solubilizing ability (Tian et al., 2021; Zhang et al., 2010). Several of these microbes' resident in the surface soil do escape to troposphere. It happens due to turbulence and low altitude wind. They may also reach upper troposphere and migrate to farther locations.

Phosphorus moves from the soil or P-rich deposits (Rock PO_4) into water sources via leaching. Phosphorus is often made soluble industrially by treating it with phosphoric acid or sulfuric acid. The most common soluble fertilizer sources supplied to crops worldwide are single and triple superphosphate. These are highly soluble and available to the crop's roots. However, only up to 20% of the fertilizer-P applied to the soil is utilized by crops. The rest either gets fixed in the soil profile or gets leached into water bodies such as rivers, lakes, ponds, and dams. Leaching of the soil is the main avenue of transport of P from the lithosphere to the hydrosphere. PM with organic-P/inorganic-P may reach the troposphere due to wind and/or turbulence. It serves as a relatively feeble source of P upon deposition to the soil.

Additions of P into the soil elicit a response from the biological component in soils. However, excess and unused P accumulates. Such accumulation of P is often the cause of eutrophication in water bodies, marshes, and stagnated soils (Reeburgh, 1997).

The dynamics of phosphite compounds under chemically reduced conditions in farms is an important aspect of phosphorus cycling. However, it is often overlooked (Ewens et al., 2021). Such chemically reduced soils occur under submerged paddy cultivation as well as in marshes with stagnating water. Microbial dissimilatory phosphite oxidation (DPO) was discovered in 2000, initially with only two microbial species identified to have DPO activity in phosphite-rich zones. However, with the help of metagenomic sequencing, 21 more microbial species with DPO ability have been identified. These DPO microbes belong to six classes: Desulfotomaculia, Synergista, Negativicutes, Syntrophia, Desulfobacteria, and Desulfomonilia. Microbes with DPO ability are found globally in areas with anoxic conditions or chemically reduced sources of substrate. It is possible that the distribution of DPO microbes in the atmosphere is due to atmospheric turbulence. Therefore, it is important to survey the atmosphere over fields that harbor DPO microbes.

5.2.3.2 INTERCONTINENTAL MOVEMENT OF PHOSPHORUS AND MICROBES LADEN PARTICULATE MATTER (PM)

The above discussion is about phosphorus (P) dynamics in soil, specifically focusing on the physicochemical transformations and dynamics found in the soil phase of agroecosystems. It is important to note that these dynamics may differ from those occurring in the atmosphere. In this chapter, our primary concern is with the presence of P in the atmosphere, particularly in the lower and upper troposphere. Phosphorus can be found in various forms such as aeolian dust, clouds, fog, snow, and rainfall. The main mode of P transport is through dust plumes, wind, and jet streams. Significant aeolian dust migrations from locations like the Sahara, Gobi, or other desert/arid sandy areas contain substantial amounts of P. The organic-P present in tropospheric dust is susceptible to the activity of aerial microbes. Microbes in particulate matter (PM) utilize the small quantities of P found on the surface and in organic compounds. The organic-P can also be traced back to microbial biomass.

Transfer of phosphorus (P) from the atmosphere to the soil phase of agricultural fields can be significant. Phosphorus-laden atmospheric particulate matter (PM) may descend onto the soil surface due to gravitational settling, low wind, precipitation, and convection currents. As Saharan dust plumes travel in the troposphere and deposit, they actually contribute PM containing P to the soil. It is important to note that sandy PM can deplete P from the source (location) of the dust storm, while the point of deposition receives additions of P-laden particles. Drifting sandstorms like Harmattan and Haboob transport large quantities of P held in inorganic sand particles and organic dust. This phenomenon has been observed in studies by Sterke et al. (1996, 1998), Stoorvogel (2000), and Romheld et al. (2000). Similarly, PM-mediated P transport via dust plumes can occur in other regions of the atmosphere.

There are reports regarding short- and long-term phosphorus (P) cycles operative in an ecosystem. The time scales of the P cycle are maintained via aeolian dust inputs. It is important to note the periodicity, nutrient load, and quantity of aeolian dust shifted through the atmosphere. For example, Swap et al. (1992) state that P inputs to Amazonia via particulate matter (PM) from North Africa control P levels in Amazonian Forest stands. The dust storms seem to maintain P cycles for time scales of 500–2,000 years. Okin et al. (2004) suggest that the shortest dust-borne P turnovers occur in the area of dust emanation itself. Often, strongly dust-generating locations are also the ones that receive dust (P-rich) from the close vicinity (atmosphere) (Tegen & Fung, 1995). Usually, the coarsest material with P and microbes descends and deposits quickly in the same location, allowing for rapid P cycling. This replenishes P in the soil profile for crops to extract. Okin et al. (2001a, b) state that in arid regions, plant material is degraded rapidly, which may generate lighter dust. Such regions also support relatively longer distance movement of aeolian dust. Therefore, P turnover will be slower. In Sahel and East Asia, P turnover rates of 102 to 104 years are encountered.

Dust-borne transport of phosphorus (P) from Africa to the North American plains is considered to be significant (Talbot et al., 1986). Similar dust-borne P deposits are also found in Amazonia. The amount of soil P relative to that found in atmospheric particles varies depending on soil type, location, and the "P carrying capacity of PM." Various locations on Earth receive P deposits that can be traced back to atmospheric dust. It is important to consider the role of dust microbes in the ecosystem when making these inferences. Additionally, the turnover of P due to snow, fog, ice nucleation, and other atmospheric events is also significant. Whenever relevant, the role of microbes in cycling P through such atmospheric events should be taken into account.

The P transported and deposited in the troposphere can be utilized by soil microbes, crop roots, and other organisms in the soil profile. Aeolian dust plumes carry microbes that are capable of solubilizing P. This includes spores/hyphae of mycorrhizal fungi (Egan et al., 2014; Kivlin et al., 2011; Vincenot et al., 2012; *see* Plate 3.3 in Chapter 3). Dust particles may also contain microbes that can solubilize organic-P through the secretion of extracellular phosphatases. Therefore, upon deposition, dust particles may contribute to the soil P pool, although to a limited extent. Aeolian dust can also introduce microbes into the soil that are capable of transforming P.

Smog with sulfur (S) and phosphorus (P) containing particulate matter (PM) may also contribute to the presence of P in the atmosphere. They could also transport P and S compounds in significant quantities. Studies on the population of P compounds in the atmosphere (lower troposphere), their diversity, distribution in relation to latitude, migratory trends, and effects on agricultural soils are necessary. Industrial ash, sludge, dust particles from mine dumps, and organic wastes with fine dust particles can all contribute to the presence of P-containing compounds and inorganic particles.

Surface water from the ocean can be released into the atmosphere due to high waves, turbulence, and hurricanes. It may carry microbes that metabolize P compounds. Seawater typically contains 0.1 ppm of phosphate and 0.016 ppm of organic-P. Therefore, ocean water is a significant source of P that reaches the atmosphere (Cambell, 2016). Wind and turbulence over the sea surface can release moisture laden with P into the atmosphere.

Phosphatic fertilizer supplied to agrarian soil is a major input to the agro-ecosystem. Phosphorus is immediately fixed or held in sparingly soluble and easily available forms for microbes/plants. Excess P or unused P is vulnerable to conversion into organic-P, chemical fixation into the soil matrix, seepage, percolation, and leaching into irrigation channels and lakes, resulting in reduced fertilizer-P efficiency. The surface soil that forms dust plumes carries a considerable amount of P in bound and soluble form. Organic dust particles may reach the upper troposphere and travel long distances via jet streams. These P-rich dust particles may also contain microbes that act on organic-P. The nutrient carrying capacity (i.e., P) and microbial diversity of PM need to be ascertained.

As mentioned previously, atmospheric processes such as dust plumes are known to transport significant amounts of phosphorus (P) from one location to another. Biomass burning, including fossil fuels, biofuels, and crop/forest burning, is a major source of P emissions into the atmosphere. Industrial emissions and volcanic ash also contribute to atmospheric P levels. O'Daya et al. (2021) suggest that the chemical nature, various forms, and quantity of P found in airborne particulate matter (PM) are still not fully understood and require improved extraction methods. The researchers have utilized X-ray absorption and nuclear magnetic resonance methods to assess the P content in aeolian dust particles. It was observed that the distribution of P-containing aeolian particles varied depending on altitude. Approximately 60 to 65% of PM from higher altitudes consisted of organic-P compounds, with mono and diesters of P being the most prevalent. This pattern was consistent regardless of particle size (PM 1-10 μm). Aeolian dust particles from lower altitudes in the atmosphere contained both organic (20–60%) and inorganic particles (30-40%). Given the variations in particle size distribution, chemical nature of P, and quantity based on altitude, it is expected that the microbial communities thriving in or on the PM will also differ (Kanakidou et al., 2012; Wang et al., 2015).

5.2.3.3 SULFUR CYCLE, TRANSFORMATIONS, AND TROPOSPHERE MICROBES

In nature, the sulfur cycle involves emissions from volcanoes, crop burning, stubble burning (smog), coal and other fossil fuel burning, industrial emissions, fumes, and smoke from transport vehicles, among others. Large-scale sulfur emissions into the troposphere occur due to volcanic eruptions, smelting of sulfur-containing ores, and large farming enterprises that release sulfur-bearing particles and smog. In recent years, the extensive burning of cereal stubble in vast plains has been contributing sulfur particles, sulfur-containing dust, and smog (Krishna, 2023). Smog has the potential to induce acid rain. There are microbes that facilitate the transformation of sulfur compounds. Normally, the troposphere contains a certain amount of sulfur-bearing products. In certain areas, sulfur emissions may lead to the formation of clouds that contribute to acid rain formation. Sulfur oxidation and reduction are important natural processes that aid sulfur cycling in nature, particularly in the lower troposphere. Microbes play a role in transforming different sulfurous compounds. We should be able to selectively utilize them to mitigate the effects of sulfur dioxide (SO_2). Sulfur dioxide is a toxic gas that affects crops, fauna, and humans alike. Hydrogen sulfide (H_2S) and mercaptans are also released into the troposphere through various abiotic and biotic transformations (Lavelle et al., 2005).

The global sulfur cycle, including its pools and fluxes, has been extensively discussed in various reports. Some of these reports highlight the role of tropospheric microbes in the sulfur cycle (Loka Bharathi, 2008; Munn et al., 2011). Lavelle et al. (2005) have provided a comprehensive summary of the pools, deposits, and fluxes of dust and sulfur compounds. Brasseur et al. (2003) and Reeburgh (1997) have reported that fossil fuel and crop burning contribute 70 to 100 Tg of sulfur particles to the atmosphere annually. Urban fires add 3 Tg of sulfur particles per year. Dust from the soil surface contributes 8 Tg of sulfur per year. Volcanoes contribute 10 Tg of sulfur per year. Additionally, rivers and oceans also release significant amounts of sulfur into the atmosphere. Atmospheric deposits are the primary sources of sulfur, along with terrestrial and agricultural systems. Microbe-mediated conversion of sulfur particulate matter could be significant in certain locations.

Acid rain contains sulfuric acid, as well as nitric, carbonic, and a few other organic acids. The acidic particles in acid rain are neutralized by the release of ash, which has an alkaline reaction. Microbes that utilize sulfurous compounds also contribute to the neutralization process. In some cases, particles containing SO_4 and microbes can induce nucleation, condensation, and ultimately result in acidic precipitation with a low pH. In certain tropical agrarian regions, various sources of sulfur compounds contribute to tropospheric sulfur. Sulfate dissolves in rainwater, forming dilute sulfuric acid, which can be deposited on the soil surface, vegetation, and crops. This can have detrimental effects on crops (Smil, 1997; Ramaswamy et al., 2001; Reeburgh, 1997; Brasseur et al., 2003).

The deposition of sulfur from the atmosphere is estimated to be 22–47 Tg per year globally. The SO_4 compounds in the aerosol are crystalline and act as nucleation agents. They have the potential to reduce greenhouse gas emissions and the warming effects. However, it is not sufficient to correct the excessive emissions of SO_2, sulfur, and acidification.

Babich & Stotzky (1978) stated that excessive release of SO_2 gas and S-containing PM prompts us to study the S cycle in nature, particularly in agrarian regions, in greater detail. Since then, several groups have attempted to decipher various aspects such as S-caused pollution, distribution of S in the atmosphere (particulate matter + gaseous matter), and transformations mediated both chemically and via atmospheric microbes (Bremner & Steele, 1978; Pfenning, 1975). Incidentally, microbes are both active sinks and sources of S-containing gases. Microbes are potential sinks of atmospheric S-containing gases and volatiles. Aeolian microbes also depend on the atmospheric S that exists freely as gas. Otherwise, S is traced on the surface of PM or found organically in a combined state within the particles. Babich & Stotzky (1978) point out that microbes have a role in S-emissions. They are also capable of assimilating atmospheric S and volatile S compounds, which may lead to the mitigation of toxic effects of S-containing gases.

In addition to the natural transformation of sulfur in different habitats, sulfur-reducing bacteria could be utilized to generate H_2S gas through industrial procedures. They also aid in the precipitation of heavy metals from natural substrates (Rabus et al., 2006; Bridge et al., 1999; Barton & Fauque, 2009; Wu et al., 2021).

There are a few sulfur transformations that need consideration. There are chemical reactions important to the progress of the sulfur cycle in nature. The oxidation states for sulfur range from +6 to -2. This permits several sulfur-containing compounds to serve as electron acceptors and electron donors during microbial transformations. The major sulfur transformations mediated by microorganisms in different habitats are:

- Sulfate reduction ($SO_4^{-2} + 4H_2 \rightarrow H_2S + 2H_2O + 2OH^-$(e.g., *Desulfovibrio* and *Desulfobacter*).
- Sulfide oxidation ($H_2S \rightarrow S^0SO_4^{2-}$ (e.g., *Beggiato alba; Thiobacillus,* and purple and green sulfur bacteria).
- Sulfur oxidation ($S^0 + 2O_2 \rightarrow H_2SO_4$ (e.g., *Thiobacillus* and *Acidothiobacillus*); and
- Sulfur reduction ($S^0 + H_2 \rightarrow H_2S$ (e.g., *Desulfuromonas acetoxidans*).

There are several reports that describe the isolation, characterization, and biochemical reactions mediated by the enzymes relevant to S dynamics (Bridge et al., 1999; Gavel et al., 1998; Barton & Fauque, 2009; Fauque & Barton, 2012; Wu, 2021).

5.2.3.4 SULFUR AND SULFATE REDUCING MICROBES

Microorganisms mediate oxidation-reduction reactions of the sulfur cycle in nature. We have accumulated knowledge about the forms of sulfur in natural systems. Furthermore, there is literature on microbially mediated transformations of sulfur within the pedosphere and hydrosphere. However, we have not deciphered much about the volatile sulfur released to the atmosphere through microbial activity. Additionally, the role of these microorganisms in the atmospheric sulfur cycle (Pfennig, 1975; Wiggington, 2014) remains unclear.

Sulfur that is utilized by microbes can be traced in hot springs, smog, polluted fog-filled skies, marshes, and other environments. Microbes are closely linked to S-related transformations and the S-cycle. Sulfur-reducing bacteria can be found in environments that support chemo-lithotrophic conditions. They utilize inorganic sulfur compounds as electron acceptors during respiration, in place of oxygen. As a result, they are found in low-oxygen or anaerobic environments, such as marshes. Low-oxygen tension troposphere regions like smog/dust plumes, areas with toxic sulfur emissions, sulfur dioxide-contaminated or polluted regions also support S-reducing microbes. Sulfur-rich dust/smog in the troposphere may harbor S-reducing bacteria. Non-methanogenic archaea also possess the ability to reduce S to H_2S in the troposphere.

A pure culture of sulfur-reducing bacteria, such as *Desulfuricans acetoxidans,* was isolated and cultivated in an artificial medium in 1976 (Glass et al., 2017). In natural environments, such as sulfur-polluted agricultural areas, it is common to encounter sulfur-reducing bacteria (SRB). These bacteria utilize both elemental sulfur and other sulfur compounds. Sulfur reduction is often coupled with the oxidation of acetate, succinate, or other organic compounds in nature. Sulfate-reducing bacteria are known to utilize both inorganic sulfur and compounds containing sulfate groups. A few microbes, such as *Proteus vulgaris, Pseudomonas,* and *Campylobacter*, are also capable of sulfate reduction (Wikipedia, 2022).

Sulfur and sulfate-reducing bacteria can be found in various habitats. They are often found in anaerobic or chemolithotrophic environments. Examples include sulfur springs, natural hydrothermal vents, sulfur-emitting industrial regions, methanogenic conditions, and more (Munn, 2011; Florentino et al., 2016). Agrarian regions that utilize sulfur compounds may also harbor sulfur-reducing microbes. These bacteria are often facultative anaerobes and can thrive under anaerobic, chemolithotrophic, or aerobic conditions.

In the atmosphere, sulfur-reducing bacteria may occur transiently until they are deposited onto the soil. They can be introduced into the troposphere through wind, turbulence, volcanic eruptions, vents, and industrial chimneys. According to the National Centre for Biotechnology Information, USA, most sulfur and/or sulfate-reducing bacteria are classified under the phylum Proteobacteria.

Major classes and genera of SRBs are as follows: δ-Proteobacteria (*Desulfuromonas*, *Desulfurella*, *Geobacter*, and *Pelobacter*,); Class Gamaproteobacteria and Epsilonproteobacteria (also known as Campylobacterota). A few other phyla that possess S-reducing microbes are Firmicutes (e.g., *Desulfitobacterium*, *Carboxythermus*), Aquifecae (Desulfurobacterium, and Aquifes), Synergistetes (*Detiosulfovibrio*), Deferribacteres (*Geovibrio*), Thermodesulfobacteria, Spirochaetes, and Chrysiogenetes (Hedderich et al., 1998; Waite et al., 2017, 2018; Jansen & Morgan, 1992). So far, there are at least 74 genera of bacteria and archaea that are identified as S-reducing organisms. They are traced in variety of habitats, including agrarian regions (Wikipedia, 2022).

Sulfur and sulfate-transforming bacteria that require anaerobic conditions may find suitable niches in the crevices and folds of atmospheric particles. The microaerophilic conditions present in particulate matter (PM) enable sulfur reduction. This aspect requires attention in agrarian regions that release PM, CO_2, and sulfur compounds through crop residue burning. Additionally, we should monitor sulfur-oxidizing microbes in areas affected by acid rain.

In nature, sulfate-reducing prokaryotes (SRP) (bacteria, archaea) play a crucial role in regulating nutrient dynamics. This is particularly true in predominantly agrarian regions that contribute to the formation of carbonaceous PM smog. According to Peng et al. (2018), SRP are ubiquitous in anoxic environments. They regulate both sulfur and carbon cycles and are essential for the completion of several interconnected biogeochemical cycles. This holds true even in agrarian habitats where smog, dust storms, and dusty clouds are common in the troposphere. Moreover, there are a few agronomic procedures that involve dusting or spraying sulfur compounds onto crops or fields.

5.2.3.5 SULFUR OXIDIZING BACTERIA (SOB)

Sulfur-oxidizing bacteria (SOB) oxidize reduced sulfur compounds such as hydrogen sulfide (H_2S), elemental sulfur (S^0), sulfite (SO_3^{2-}), thiosulfate ($S_2O_3^{2-}$), and various polythionates (SnO_6^{2-} or $-SnO_6^{-}$) into sulfate (SO_4^{2-}).

On the contrary, SO_4^{2-} can serve as an electron acceptor under anaerobic conditions. They reduce SO_4^{2-} and other oxidized sulfur compounds ($S_2O_3^{2-}$, SO_3^{2-}, S0) into H_2S. In natural systems, sulfate-reducing bacteria (SRB) reduce SO_4^{2-} through two different reduction processes: dissimilatory and assimilatory reactions. The dynamics and activity of SRB and SOB in nature require careful consideration.

There are several examples of effective removal of H_2S by sulfur-oxidizing bacteria (Giardano et al., 2015). Acidophilus sulfur bacteria are useful in metal leaching. Although it is not a highly efficient procedure, microbes like *Thiobacillus ferrooxidans, T. thiooxidans,* and *T. acidophilus* may be used to extract metals like copper. The sulfur-oxidizing bacteria that deposit onto soil may have important use in farming, particularly in transforming sulfur compounds (Giardano et al., 2015; Dopson & Johnson, 2012; Heydharzadeh et al., 2014; Simon & Kroneck, 2013). Swamps, water bodies with sulfur contamination, H_2S emission, eutrophication, and wastewater treatment around farming regions could be ameliorated using purple sulfur bacteria. On the other hand, sulfur reducers like *Desulfovibrio desulfuricans* leach and concentrate metals from their ores (Roy & Roy, 2015; Sand et al., 2001).

It is interesting to note that atmospheric pollutants that are deposited on the phylloplane, soil surface, farm equipment, and water bodies can support the activity of sulfur-oxidizing bacteria. Experiments with such deposits on the leaves of sycamore trees have clearly shown that carbon (C), nitrogen (N), and sulfur (S) present in atmospheric deposits support the multiplication of sulfur-oxidizing bacteria (Killham & Wainwright, 1981). The sulfur-oxidizing bacteria resulted in the release of S_2O_3 and SO_4.

5.3 AERIAL MICROBES AND NUTRIENT DYNAMICS IN EXTREME AGRO CLIMATE

Agricultural crop production in wet tropics, drylands, or temperate farming zones depends on certain basic levels of nutrient dynamics. The nutrient availability in the agro environment should allow normal growth rates of crops, microbes, and other biotic factors in the ecosystem. Let us consider the conditions prevalent in the arid and dryland regions where crop production occurs. Crop production occurs despite harsh soil/atmospheric conditions. Both crops and microbes negotiate the harsh temperature (< 0.0°C night temperatures to 40 to 50°C in the afternoon), desiccating conditions, low water availability, strong solar radiation (840 GJ km^{-2} $year^{-1}$), heat waves, short rainy season if any, etc. (Goudie, 2009, 2010; Ramawat, 2010; Greenville, 2018; Mohammadipanah & Wink, 2016). Microbes that thrive in such harsh atmospheric conditions are known as 'extremophiles.' These extremophiles are adapted to diverse substrates available feebly in the atmosphere or on the crop's canopy (phyllosphere), etc. No doubt, microbes mediate nutrient dynamics in the arid agricultural fields despite harsh conditions. Farmers in arid belts may also encounter salt-affected patches. Microbes in such fields may have to tolerate unfavorable osmotic conditions. However, they still thrive metabolically in extreme conditions. Nutrient transformations differ from those observed in fields with favorable atmospheric conditions. Microbes often overwinter by adopting sporulation as a means of survival. They resume physiological activity and nutrient transformation processes once favorable moisture and temperature conditions occur.

Alsharif et al. (2020) identify a few of the world's harsh desert/arid regions where vegetation is generally scanty, such as the Sahara, Sahel, Arabian desert, Atacama, Namibian deserts, Kalahari, Great Indian desert (Thar), the Gobi and Taklamakan, Australian desert region, etc. In such extreme conditions, nutrient dynamics are regulated by microbes that thrive at a commensurate growth rate. Furthermore, Alsharif et al. (2020) believe that these microbes, which thrive in harsh atmospheric conditions, have specific adaptations that may be useful in farming. These microbes are capable of conducting normal physiological activity, including nutrient transformation, under harsh farming conditions. There are also other species that exhibit modified enzyme 3D structures/activity and operate at subdued levels. Therefore, we could select and introduce appropriate microbes. In addition to the troposphere, microbes may also localize and find a niche on the crop's canopy, leaf whorls, leaf lamina, branches, twigs, flowers, and fruits. They operate in a relatively better environment and transform nutrients. It is necessary to accumulate detailed knowledge about microbes that thrive in the troposphere above the crop's canopy in the arid/dryland agro zones. We need to first enumerate and characterize them. This can be done by using both culture-based and molecular sequencing methods. After that, we should study the diverse nutrient and biochemical transformations. It is possible that we may discover some extremely useful microbial species that can be utilized in drylands or even in other regions. It is not just the atmospheric or surface microflora on crops that we should focus on, but also the endophytes (microbes) found in desert vegetation and crops. These endophytes may possess some useful traits that are relevant to nutrient dynamics in both congenial and harsh agro-climates (Goudjal et al., 2013). We should pay attention to the mechanisms that aid microbes in adapting to the extreme conditions prevalent in crop fields. Microbes that are adapted to the harsher troposphere in arid belts may possess traits that are useful for farming, especially in optimum farming conditions.

KEYWORDS

- **aeolian dust**
- **aerial microbes**
- **ammonia**
- **bio-geochemical transformations**
- **biological nitrogen fixation**
- **carbon emissions**
- **greenhouse gas**
- **methanogens**
- **methanotrophs**
- **methylotrophs**
- **nitrification**
- **nitrogenase**
- **nitrous oxide**
- **nutrient dynamics**
- **phosphorus cycle**
- **sulfur cycle**

REFERENCES

Aalismail, N. A., Díaz-Rúa, R., Ngugi, D. K., Cusack, M., & Duarte, C. M., (2020). Aeolian prokaryotic communities of the global dust belt over the Red Sea. *Frontiers in Microbiology, 11*, 538476. doi: 10.3389/fmicb.2020.538476/.

Aciego, S. M., Riebe, C. S., Hart, S. C., Blakowski, M. A., Carey, C. J., Aarons, S. M., Dove, N. C., et al., (2017). Dust outpaces bedrock in nutrient supply to montane forest ecosystems. *National Communications, 8*, 14800.

Acosta-Martinez, V., Van, P. S., Moore-Kucera, J., Baddock, M. C., & Zobeck, T. M., (2015). Microbiology of wind eroded sediments: Current knowledge and future. *Aeolian Research, 18*, 99–113.

Adachi, K., Oyediran, G., & Senboku, T., (1996). Effect of application of rice straw and cellulose on methane emission and biological nitrogen fixation in a subtropical paddy field. II. Enumeration of populations of methanogenic bacteria by most probable number method and roll tube method. *Soil Science and Plant Nutrition, 42*, 713–723.

Ali, M. A., Lee, C. H., & Kim, P. J., (2008). Effects of silicate fertilizer on reducing methane emission during rice cultivation. *Biology and Fertility of Soils, 44*, 597–604.

Ali, M. A., Lee, C. H., Lee, Y. B., & Kim, P. J., (2009). Silicate fertilization in no-tillage rice farming for mitigation of methane emission and increasing rice productivity. *Agriculture, Ecosystems and Environment, 132*, 16–22.

Alsharif, W., Saad, M. M., & Hirt, H., (2020). Desert microbes for boosting sustainable agriculture in extreme environments. *Frontiers in Microbiology, 11*, 1666. doi: 10.3389/fmicb.2020.01666.

Alves, C. A., (2017). A short review on atmospheric cellulose. *Air Quality of Atmosphere and Health,10*, 669–678.

Anastasic, C., Dowding, M., & Simpson, V. J., (1999). Future CH_4 emission from rice production. *Journal of Geophysical Research, 27*, 7521–7525.

Andrea, M. O., & Merlet, P., (2001). Emission of trace gases and aerosols from biomass burning. *Global Biogeochemical Cycles, 15*, 955–966.

Aneja, V. P., Roelle, P. A., & Murray, G. C., (2001). Atmospheric nitrogen compounds II: Emissions, transport, transformation, deposition, and assessment. *Atmospheric Environment, 35*, 1903–1911.

Artaxo, P., Martins, J. V., Yamasoe, M. A., Procopio, A. S., Pauliquevis, T. M., Andreae, M. O., Guyon, P., et al., (2002). Physical and chemical properties of aerosols in the wet and dry seasons in Rondonia, Amazonia. *Journal of Geophysical Research, 107*(D20), 8081. doi: 10.1029/ 2001JD000666/.

Asakawa, S., Akagawa-Matsushita, M., Koga, Y., & Hayano, K., (1998). Communities of methanogenic bacteria in paddy field soils with long-term application of organic matter. *Soil Biology and Biochemistry, 30*, 299–303.

Babich, H., & Stotzky, G., (1978). Atmospheric S compounds and microbes. *Environmental Research, 15*(3), 513–531. https://doi.org/10.1016/0013-9351(78)90130-5/.

Barton, L. L., & Fauque, G. D., (2009). Biochemistry, physiology, and biotechnology of sulfate-reducing bacteria. *Advances in Applied Microbiology, 68*, 41–98.

Berman-Frank, I., Lundgren, P., & Falkowski, P., (2003). Nitrogen fixation and photosynthetic oxygen evolution in cyanobacteria. *Research in Microbiology, 154*, 157–164. doi: 10.1016/S0923-2508(03)00029-9/.

Bielders, C. L., Rajot, J. L., & Koala, S., (2002a). *Wind Erosion Research in Niger: The Experience of ICRISAT and Advanced Research Organizations*. 'ICRISAT Sahelian Centre, BP 12404, Niamey, Niger 'ORSTOM, BP 11416, Niamey, Niger. http://oar.icrisat.org/6219/ (accessed on 19 January 2024).

Bielders, C., Rajot, J., Amadou, M., & Skidmore, E., (2002b). *On Farm Evaluation of Wind Erosion Under Traditional Management Practices* (p. 6). ICRISAT.Org. CCER.htm. doi:10.13031/2013.4585.

Bowman, J. P., (2005). Family V. *Methylocystaceae* fam. Nov. In: Brenner, D. J., Krieg, N. R., Staley, J. T., & Garrity, G. M., (eds.), *Bergey's Manual of Systematic Bacteriology, (The Proteobacteria) Part C (The Alpha-, Beta-, Delta-, and Epsilonproteobacteria)* (2nd edn., Vol. 2, pp. 411–413). Springer, New York.

Boyd, E. S., & Peters, J. W., (2013). New insights into the evolutionary history of biological nitrogen fixation. *Frontiers in Microbiology*. https://doi.org/10.3389/fmicb.2013.00201/.

Boyd, E. S., Anbar, A. D., Miller, S., Hamilton, T. L., Lavin, M., & Peters, J. W., (2011). A late methanogen origin for molybdenum-dependent nitrogenase. *Geobiology, 9*, 221–232. doi: 10.1111/j.1472-4669.2011.00278.x/.

Bozetti, C., Daellenbach, K. R., Heuglin, C., Fermo, P., Sciare, J., Kasper-Giebl, A., Mazar, Y., et al., (2016). Size-resolved identification, characterization, and quantification of primary biological organic aerosol at a European rural site. *Environmental Science and Technology, 50*, 3425–3434.

Brasseur, G. P., Artaxo, P., Barrie, L. A., Delmas, R. J., Galbally, I., et al., (2003). An integrated view of the causes and impacts of atmospheric changes. In: Brasseur, G. P., Prinn, R. G., & Pszenny, A. A. P., (eds.), *Atmospheric Chemistry in a Changing World* (pp. 207–230). IGBP Series, Springer: Berlin.

Bremner, J. M., & Steele, C. G., (1978). Role of microorganisms in the atmospheric sulfur cycle. *Advances in Microbial Ecology* (Vol. 2, pp. 155–201). https://doi.org/10.1007/978-1-4615-8222-9_4/.

Bridge, T. A., White, C., & Gadd, G. M., (1999). Extracellular metal-binding activity of the sulphate-reducing bacterium *Desulfococcus multivorans*. *Microbiology, 145*(10), 2987–2995.

Brighty, A., Jacob, V., Uzu, G., Borlaza, L., Conil, S., Hueglin, C., Stuart, K., et al., (2021). Cellulose in atmospheric particulate matter. *Atmospheric Physics and Chemistry Discussions*. https://doi.org/10.5194/acp-2021-930/.

Cahyani, V. R., Matsuya, K., Asakawa, S., & Kimura, M., (2004). Succession and phylogenetic profile of methanogenic archaeal communities during the composting process of rice straw estimated by PCR-DGGE analysis. *Soil Science and Plant Nutrition, 50*, 555–563.

Cambell, K., (2022). *Over 40 Minerals and Metals Contained in Seawater, Their Extraction Likely to Increase in the Future* (pp. 1–3). Mining Weekly. https://www.miningweekly.com/article/over-40-minerals-and-metals-contained-in-seawater-their-extraction-likely-to-increase-in-the-future-2016-04-01 (accessed on 19 January 2024).

Canfield, D. E., Glazer, A. N., & Falkowski, P. G., (2010). The evolution and future of Earth's nitrogen cycle. *Science, 330*, 192–194.

Chen, Y., Mills, S., Street, J., Golan, D., Post, A., Jacobson, M., & Paytan, A., (2007). Estimates of atmospheric dry deposition and associated input of nutrients to Gulf of Aqaba seawater. *Journal of Geophysics Research, 112*, D04309. doi: 10.1029/2006JD007858/.

Conrad, R., (1989). Control of methane production in terrestrial ecosystems. In: Andreae, M. O., & Schimel, D. S., (eds.), *Exchange of Trace Gases Between Terrestrial Ecosystems and the Atmosphere* (pp. 39–58). Dahlem Konferenzen. Wiley, Chichester.

Conrad, R., (1993). Mechanisms controlling methane emission from wetland rice fields. In: Oremland, R. S., (ed.), *The Biogeochemistry of Global Change: Radiative Trace Gases* (pp. 317–335). Chapman & Hall, New York.

Conrad, R., (2002). Control of microbial methane production in wetland rice fields. *Nutrient Cyclingin Agroecosystems, 64*, 59–64.

Crombie, A., & Murrell, J. C., (2011). Methods in methane metabolism: Methanotrophy. *Methods in Enzymology, 495*, 119–133.

Dedysh, S. N., & Dunfield, P. F., (2011). Methods in methane metabolism part B: Methanotrophs. *Methods in Enzymology, 495*, 31–44.

Denman, K. L., Brasseur, G., Chidthaisong, A., Ciais, P., Cox, P. M., Dickinson, R. E., Hauglustaine, D., et al., (2007). Couplings between changes in the climate system and biogeochemistry. In: Solomon, S., Qin, D., Manning, M., Chen, Z., Marquis, M., Averyt, K. B., Tignor, M., & Miller, H. L., (eds.), *Climate Change 2007: The Physical Science Basis* (pp. 499–588). Contribution of Working Group I to the Fourth Assessment Report of the Intergovernmental Panel on Climate Change. Cambridge University Press, Cambridge, United Kingdom and New York, NY, USA.

Dopson, M., & Johnson, D. B., (2012). Biodiversity, metabolism, and applications of acidophilic sulfur-metabolizing microorganisms. *Environmental Microbiology, 14*, 2620–2631.

Dos, S. P., Fang, Z., Mason, S., Setubal, J., & Dixon, R., (2012). Distribution of nitrogen fixation and nitrogenase-like sequences amongst microbial genomes. *BMC Genomics, 13*, 162. doi: 10.1186/1471-2164-13-162/.

Dunfield, P., Yuryev, A., Senin, P., & Smirnova, A. V., (2008). Methane oxidation by an extremely acidophilic bacterium of the phylum Verrucomicrobia. *Nature, 450*, 879–882. doi: 10.1038/nature06411.

Egan, E., Li, D. W., & Klironomos, J., (2014). Detection of arbuscular mycorrhizal fungal spores in the air across different biomes and ecoregions. *Fungal Ecology, 12*, pp. 1–7. doi: 10.1016/j.funeco.2014.06.004/.

European Parliament (2018). *Amount of Greenhouse Gas Emission Per Year in Europe* (pp. 1–14). Greenhouse gas emissions by country and sector (infographic) | News | European Parliament (europa.eu). https://www.europarl.europa.eu/news/en/headlines/society/20180301STO98928/greenhouse-gas-emissions-by-country-and-sector-infographic/ (accessed on 19 January 2024).

Ewens, S. D., Barnum, P., Borton, M. A., Carlson, M. A., Wrighton, K. C., & Coates, J. D., (2021). The diversity and evolution of microbial dissimilatory phosphite oxidation. *Proceedings of National Academy of Sciences, 118*(11), e2020024118. https://doi.org/10.1073/pnas.2020024118?.

Falkowski, P. G., (1997). Evolution of the nitrogen cycle and its influence on the biological sequestration of CO_2 in the ocean. *Nature, 387*, 272–275. doi: 10.1038/387272a0 /.

Falkowski, P. G., Fenchel, T., & Delong, E. F., (2008). Microbial engines that drive earth's biogeochemical cycles. *Science, 320*, 1034–1038.

Fani, R., Gallo, R., & Liò, P., (2000). Molecular evolution of nitrogen fixation: The evolutionary history of the *nifD*, *nifK*, *nifE*, and *nifN* genes. *Journal of Molecular Evolution, 51*, 1–11.

FAO of the United Nations, (2020). *Abiotic Disturbances- Storms* (pp. 1–18). Food and Agricultural Organization of the United Nations. http://www.fao.org/forestry/abiotics/71634/en/ (accessed on 19 January 2024).

Fauque, G. D., & Barton, L. L., (2012). Hemoproteins in dissimilatory sulfate- and sulfur-reducing prokaryotes. *Advances in Microbial Physiology, 60*, 1–90.

Fay, P., (1992). Oxygen relations of nitrogen fixation in cyanobacteria. *Microbiology Review, 56*, 340–373.

Florentino, A. P., Weijma, J., Stams, A. J., & Sánchez-Andrea, I., (2016). Ecophysiology and application of acidophilic sulfur-reducing microorganisms. In: Rampelotto, P. H., (ed.), *Biotechnology of Extremophiles: Advances and Challenges; Grand Challenges in Biology and Biotechnology* (Vol. 1, pp. 141–175). Cham: Springer International Publishing. doi: 10.1007/978–3-319-13521-2_5/.

Forster, P., & Ramaswamy, V., (2007). Changes in atmospheric constituents and in radiative forcing. In: Solomon, S., et al., (eds.), *Climate Change 2007: The Physical Science Basis* (pp. 3–53). Cambridge. https://www.researchgate.net/publication/224988383/ (accessed on 19 January 2024).

Galloway, J. N., & Cowling, E. B., (2002). Reactive nitrogen and the world: 200 years of change. *Ambio, 31*, 64–71.

Galloway, J. N., Aber, J. D., Erisman, J. W., Seitzinger, S. P., Howarth, R. W., Cowling, E. B., & Cosby, J., (2003). The nitrogen cascade. *Bioscience, 53*, 341–356.

Galloway, J. N., Dentener, F. J., & Capone, D. G., (2004). Nitrogen cycles: Past, present, and future. *Biogeochemistry, 70*, 153–226.

Gao, J., Fan, X., Pan, K. L., Li, H. Y., & Sun, L. X., (2016). Diversity, abundance, and activity of ammonia-oxidizing microorganisms in fine particulate matter. *Science Reports, 6*, Article No. 38785. doi: 10.1038/srep38785/.

Garica, J. L., Patel, B. K. C., & Ollivier, O., (2000). Taxonomic, phylogenetic, and ecological diversity of methanogenic archaea. *Anaerobe, 6*, 205–226.

Gavel, O. Y., Bursakov, S. A.,Calvete, J. J.,George, G. N.,Moura, J. J. G., & Moura, I., (1998). ATP sulfurylases from sulfate-reducing bacteria of the genus *Desulfovibrio*. A novel metalloprotein containing cobalt and zinc. *Biochemistry, 37*, 16225–16232.

Ge, M., & Friedrich, J., (2020). *Gas Emissions by Countries and Sectors* (pp. 1–4). World Resources Institute, Washington, D.C. USA. https://www.wri.org/insights/4-charts-explain-greenhouse-gas-emissions-countries-and-sectors/ (accessed on 19 January 2024).

Geibi, N., Gelenscer, A., Puxbaum, H., & Adrio-Pio, C., (2007). Concentration of atmospheric cellulose: A proxy for plant debris across a west east transect over. *Europe Journal of Geophysical Research Atmospheres, 112*, D23S08. doi: 10.1029/2006JD008180/.

Giardano, P. C., Spennati, F., Melone, A., Petroni, G., Verni, F., Munz, G., Mori, G., & Vannini, C., (2015). Biological sulfur oxidizing potential of primary and biological sludge in a tannery wastewater treatment. *Water Air Soil Pollution, 226*, 39–42.

Ginoux, P., Prospero, J. M., Gill, T. E., Hsu, N. C., & Zhao, M., (2012). Global-scale attribution of anthropogenic and natural dust sources and their emission rates based on MODIS deep blue aerosol products. *Reviews in Geophysics, 50*, RG3005. doi: 10.1029/2012RG000388/.

Giri, D. D., Singh, S. K., Giri, A., Dviwedi, H., & Ajay, K., (2021). Bioremediation potential of methylotrophic bacteria. In: Kumar, A., (ed.), *Microbe Mediated Remediation of Environmental Contaminants* (pp. 199–207). Woodhead Publishing.

Glass, J. B., Chen, S. I., Dawson, K. S., Horton, D. R., Vogt, S., & Ingall, E. D., (2017). Trace metal imaging of sulfate-reducing bacteria and methanogenic archaea at single-cell resolution by synchrotron x-ray fluorescence imaging. *Geomicrobiology Journal, 35*, 81–89. doi: 10.1080/01490451.2017.1321068.

Goudie, A. S., & Middleton, N. J., (2006). *Desert Dust in the Global System* (p. 287). Springer: Heidelberg, Germany.

Goudie, A. S., (2009). Global deserts and their geomorphological diversity. In: Parsons, A. J., & Abrahams, A. D., (eds.), *Geomorphology of Desert Environments* (pp. 9–20). Dordrecht: Springer Netherlands. doi: 10.1007/978-1-4020-5719-9_2/.

Goudie, A., (2010). Holocene book review: Deserts and desert environments Julie laity, oxford, and Chichester. *Holocene, 20*, 147–148. doi: 10.1177/ 0959683609345799/.

Goudjal, Y., Toumatia, O., Nasserdine, S., Mustapha, B., Florence, M., & Zitouni, A., (2013). Endophytic actinomycetes from spontaneous plants of Algerian Sahara: Indole-3-acetic acid production and tomato plants growth promoting activity. *World Journal of Microbiology and Biotechnology, 29*, 1821–1829. doi: 10.1007/s11274-013-1344-y/.

Greenville, A., (2018). The biology of deserts David Ward (2nd edn.). Oxford University Press, Oxford, 2016. xv + 370 pp. Price AUD $66 (paperback, also available as hardback and eBook). ISBN: 9780198732761. *Australian Ecology, 43*, e20. doi: 10.1111/aec.12523/.

Greenwood, D. J., & Goodman, D., (1961). Direct measurement of the distribution of oxygen in soil aggregates and in columns of fine soil crumbs. *Journal of Soil Science, 78*, 182–196.

Griffin, D. W., Garrison, V. H., Herman, J. R., & Sinn, E. A., (2001). African desert dust in the Caribbean atmosphere: Microbiology and public health. *Aerobiological, 17*, 203–213.

Griffin, D. W., Kellog, C. A., Garrison, V. H., Lisle, J. T., Borden, T. C., & Sinn, E. A., (2003). Atmospheric microbiology in the Northern Caribbean during African dust events. *Aerobiology, 19*, 143–157.

Haque, M. M., Kim, S. Y., Pramanik, P., Kim, G. Y., & Kim, P. J., (2013). Optimum application level of winter cover crop biomass as green manure under considering methane emission and rice productivity in paddy soil. *Biology and Fertility of Soils, 49*, 487–493.

Hayashi, K., & Yan, X., (2009). *Airborne Nitrogen Load in Japanese and Chinese Agroecosystems* (pp. 1–32). https://doi.org/10.1111/j.1747-0765.2009.00423.x/.

He, J., Balasubramanian, R., Burger, D. F., Hicks, K., Kuylenstierna, J. C., & Palani, S., (2011). Dry and wet atmospheric deposition of nitrogen and phosphorus in Singapore. *Atmospheric Environment, 45*, 2760–2768.

Hedderich, R., Klimmek, O., Kröger, A., Dirmeier, R., Keller, M., & Stetter, K. O., (1998). Anaerobic respiration with elemental sulfur and with disulfides. *FEMS Microbiology Reviews, 22*(5), 353–381. doi: 10.1111/j.1574-6976.1998.tb00376.x./.

Herbert, R. J., Krom, M. D., Carslaw, K. S., Stockdale, A., Mortimer, R. J. G., Benning, L. G., Pringle, K., & Browse, J., (2018). The effect of atmospheric acid processing on the global deposition of bioavailable phosphorus from dust. *Global Biogeochemical Cycles, 32*, 1–19. https://doi.org/10.1029/2018GB005880/.

Heydharzadeh, H., Najafpour, G., Ghoreyshi, A., & Younesi, H., (2014). *Hydrogen Sulfide Elimination from Natural Gas by Native Isolated Bacteria from Hot-Spring* (Vol. 14, pp. 1–6). Eco-Tech. https://doi.org/10.15626/Eco-Tech.2014.005/.

Holland, E. A., & Weitz, A. M., (2003). Nitrogen cycle. *Encyclopedia of Physical Sciences and Technology* (3rd edn., Vol. 1–18, p. 15453). Academic Press Inc. San Diego, California, USA.

Hou, Q., Yang, Z., Ji, J., Yu, T., Chen, G., Li, J., Xia, X., Zhang, M., & Yuan, X., (2013). Annual net input fluxes of heavy metals of the agro-ecosystem in the Yangtze River Valley, China. *Journal of Geochemical Exploration, 139*, 68–84. https://doi.org/10.1016/j.gexplo.2013.08.007 /.

Intergovernmental Panel on Climate Change, (2007). Climate change 2007: Mitigation of climate change—Contribution of working group III to the fourth assessment report of the intergovernmental panel on climate. *Methanogens Harboring in Rice Rhizosphere Reduce Labile Organic Carbon Compounds*. Cambridge, United Kingdom and New York, NY, USA: Cambridge University Press. http://dx.doi.org/10.5772/intechopen.73299/.

IPCC 2007, *Climate Change-2007: Synthesis Report*. Intergovernmental Panel on Climate Change Geneva: IPCC. https://www.ipcc.ch/report/ar4/syr/ (accessed on 19 January 2024).

Jaffe, D., Snow, J., & Cooper, O., (2003). The 2001 Asian dust events: Transport and impact on surface aerosol concentrations in the US. *Eos Transactions of American Geophysics Union, 84*, 501–507.

Janssen, P. H., & Morgan, H. W., (1992). "Heterotrophic sulfur reduction by *Thermotoga sp.* strain FjSS3.B1. *FEMS Microbiology Letters, 75*(2, 3), 213–217. doi:10.1111/j.1574-6968.1992.tb05419.x. PMID 1398039./.

Jiang, H., Chen, J. C., Murrell, P., Jiang, C., Zhang, X., Xing, T., & Smith, T. J., (2011). A brief review of methanotrophs. *Comprehensive Biotechnology* (2nd edn., Vol. 6, pp. 249–262). https://www.sciencedirect.com/topics/biochemistry-genetics-and-molecular-biology/methanotroph/ (accessed on 19 January 2024).

Joulian, C., Ollivier, B., Neue, H. U., & Roger, B., (1996). Microbiological aspects of methane emission by a rice field soil from the Camargue (France): 1. Methanogenesis and related microflora. *European Journal of Soil Biology, 32*, 61–70.

Kaku, N., Ueki, A., Fujii, H., & Ueki, K., (2000). Methanogenic activities on rice roots and plant residue and their contributions to methanogenesis in wetland rice field soil. *Soil Biology and Biochemistry, 32*, 2001–2010.

Kanakidou, M., Duce, R. A., Prospero, J. M., Baker, A. R., Benitez-Nelson, C., Dentener, F. J., Hunter, K. A., et al., (2012). Atmospheric fluxes of organic N and P to the global ocean. *Global Biogeochemical Cycle, 26*, GB3086. doi: 10.1029/2011GB004277/.

Karagulian, F., Belis, C. A., Dora, C. F. C., Prüss-Ustün, A. M., Bonjour, S., Adair-Rohani, H., & Amann, M., (2015). Contributions to cities' ambient particulate matter (PM): A systematic review of local source contributions at global level. *Atmosphere and Environment, 120*, 475–483.

Kasting, J. F., & Siefert, J. L., (2001). Biogeochemistry: The nitrogen fix. *Nature, 412*, 26, 27. doi: 10.1038/35083660/.

Kaufman, Y. J., Koren, I., Remer, L. A., Tanre, D., Ginoux, P., & Fan, S., (2005). Dust transport and deposition observed from the terra moderate resolution imaging spectroradiometer (MODIS) spacecraft over the Atlantic Ocean. *Journal of Geophysical Research-Atmosphere, 110*, D10S12. doi: 10.1029/2003JD004436/.

Kern, J., Hellebrand, H. J., Gömmel, M., Ammon, & Berg, W., (2012). Effects of climatic factors and soil management on the methane flux in soils from annual and perennial energy crops. *Biology and Fertility of Soils, 8*, 1–8.

Khelaifa, S., Raoult, D., & Drancourt, M. A., (2013). Versatile medium for cultivating methanogenic archaea. *PLoS One, 8*, e61563.

Killham, K., & Wainwright, M., (1981). Microbial release of Sulphur from atmospheric pollution deposits. *Journal of Applied Ecology, 18*, 889–896.

Kivlin, Hawkes, C. V., & Treseder, K. K., (2011). Global diversity and distribution of arbuscular mycorrhizal fungi. *Soil Biology and Biochemistry, 43*, 2294–2303. doi:10.1016/j.soilbio.2011.07.012.

Klimont, Z., Kupiainen, K., Heyes, C., Purohit, P., Cofala, J., Rafaj, P., Borken-Kleefeld, J., & Schöpp, W., (2017). Global anthropogenic emissions of particulate matter including black carbon. *Atmospheric Chemistry and Physics, 17*, 8681–8723.

Knowles, R., (1982). Denitrification. *Microbiological Reviews, 46*, 43–70.

Kovack, E., & Blaustein-Rejitos, D., (2022). *The World's First Genetically Engineered Wheat is Here* (pp. 1–4). The Breakthrough Institute Reports. https://thebreakthrough.org/issues/food/the-worlds-first-genetically-engineered-wheat-is-here?goal=0_8e101ace96-7672cdf36a-99726801/ (accessed on 19 January 2024).

Krishna, K. R., (2008). *Peanut Agroecosystem: Nutrient Dynamics and Productivity* (pp. 75–127). Alpha Science Publishers, Oxford, England.

Krishna, K. R., (2023). *The Agricultural Sky: A Concept to Revolutionize Farming* (p. 588). Apple Academic Press Inc. Palm Bay, Florida, USA.

Kruse, T., Ratnadevi, C. M., Erikstad, H., & Birkeland, N., (2019). Complete genome sequence analysis of the thermoacidophilicverruco microbial methanotroph "*Candidatus* Methylacidiphilum kamchatkense" strain Kam1 and comparison with its closest relatives. *BMC Genomics*, *20*(1), 642. doi: 10.1186/s12864-019-5995-4. https://doi.org/10.1186/s12864-019-5995-4/.

Kumar, A., & Sharma, M. P., (2014). Review of methodology for estimation of labile organic carbon in reservoirs and lakes for GHG emission. *Journal of Materials and Environmental Sciences, 5*, 653–660.

Lagomarsino, A., Agnelli, A. E., Pastorelli, R., Pallara, G., Rasse, D. P., & Silvennoinen, H., (2016). Past water management affected GHG production and microbial community pattern in Italian rice paddy soils. *Soil Biology & Biochemistry, 93*, 17–27. doi: 10.1016/j.soilbio.2015.10.016.

Lal, R. C., (2008). Sequestration of atmospheric CO_2 in global organic-C pools. *Energy and Environmental Science, 1*, 86–100.

Lavelle, P., Dugdale, R., & Scholes, R., (2005). Nutrient cycling. In: *Ecosystems and Human Well-Being: Current State and Trends* (pp. 333–351). https://www.millenniumassessment.org/documents/document.281.aspx.pdf/ (accessed on 19 January 2024).

Le Treut, H., Somerville, R., Cubasch, U., Ding, Y., Mauritzen, C., Mokssit, A., Peterson, T., & Prather, M., (2007). Historical overview of climate change. In: Solomon, S., Qin, D., Manning, M., Chen, Z., Marquis, M., Averyt, K. B., Tignor, M., & Miller, H. L., (eds.), *Climate Change 2007: The Physical Science Basis* (pp. 93–127). Contribution of Working Group I to the Fourth Assessment Report of the Intergovernmental Panel on Climate Change. Cambridge University Press, Cambridge, United Kingdom and New York, NY, USA.

Lee, C. H., Park, K. D., Jung, K. Y., Ali, M. A., Lee, D., & Gutierrez, J., (2010). Effect of Chinese milk vetch (*Astragalus sinicus* L.) as a green manure on rice productivity and methane emission in paddy soil. *Agriculture, Ecosystems and Environment, 138*, 343–347.

Lehmacher, A., & Klenk, H. P., (1994). Characterization and phylogeny of MCRII, a gene-cluster encoding an isozyme of methyl coenzyme-M reductase from hyperthermophilic *Methanothermus fervidus*. *Molecular and General Genetics, 243*, 198–206.

Lehtovirta-Morley, L. E., (2018). Ammonia oxidation: Ecology, physiology biochemistry, and why they must all come together. *FEMS Microbiology Letters, 365*, fny058. doi: 10.1093/femsle/fny058/.

Loka, B. P. A., (2008). Sulfur cycle. *Encyclopedia of Ecology* (pp. 3424–3431). Elsevier. doi: 10.1016/b978-008045405-4.00761-8./.

Marston, J., (2022). *Brief: New IPCC Report Says Time's Running Out to Meet Emissions Targets-But Food and Ag Can Help* (pp. 1–3). AFN Food Tech/Agtech. https://agfundernews.com/ipcc-time-running-out-to-meet-emissions-targets-but-food-ag-can-help?goal=0_8e101ace96-7672cdf36a-99726801 (accessed on 19 January 2024).

Martin, S. T., Andrea, O. M., Artaxo, P., Baumgardner, D., Chen, Q., Goldenstein, A. H., Guenther, A., et al., (2010). Sources and properties of Amazonian aerosol particles. *Review of Geophysics, 48*(2). https://doi.org/10.1029/2008RG000280/.

Marynowski, L., Brnard, R. T., & Simoneit, S., (2021). Saccharides in atmospheric particulate and sedimentary organic matter: Status overview and future perspectives. *Chemosphere, 288*, 132376. https://doi.org/10.1016/j.chemosphere.2021.132376/.

Michels, K., Shivakumar, M, V. K., & Allison, B. E., (1995). Wind erosion control using crop residue I. Effects of soil flux and soil particles. *Field Crops Research, 40*, 101–110.

Miller, S., (2019). *Aeolian Dust in Puts are an Important Source of Phosphorus to Ecosystems*. College of Agriculture and Natural Resources, Ag News, -Ecosystems https://uwagnews.com/2019/09/23/aeolian-dust-inputs-are-an-important-source-of-phosphorus-to-ecosystems/ (accessed on 19 January 2024).

Mohammadipanah, F., & Wink, J., (2016). Actinobacteria from arid and desert habitats: Diversity and biological activity. *Frontiers in Microbiology, 6*, 1541. doi: 10.3389/fmicb.2015.01541/.

Mosbaek, P., Tuell, J. C., & Hoymand, M. E., (1989). Atmospheric lead input to agricultural crops in Denmark. *Chemosphere, 19*, 1787–199. https://doi.org/10.1016/0045-6535(89)90523-7/.

Munn, C. B., (2011). *Marine Microbiology: Ecology and Applications* (2nd edn., p. 364). New York: Garland Science. https://doi.org/10.1201/9781136667527/.

Navarro-González, R., McKay, C. P., & Mvondo, D. N., (2001). A possible nitrogen crisis for Archaean life due to reduced nitrogen fixation by lightning. *Nature, 412*, 61–64. doi: 10.1038/35083537.

Neff, J. C., Holland, E. A., Dentener, F. J., McDowell, W. H., & Russel, K. M., (2002). The origin, composition, and rates of organic nitrogen deposition: A missing piece of the nitrogen cycle? *Biogeochemistry, 57, 58*, 99–136.

Newton, J., (2019). *Agriculture and Greenhouse Gas Emissions* (pp. 1–8). Farm Bureau, US Department of Agriculture, Washington, D.C. USA.

O'Daya, P. A., Nwosua, U. G., Bamesb, M., Harta, S. C., Berhea, A. A., Christiansen, J. B., & William, K. H., (2021). Phosphorus speciation in atmospherically deposited particulate matter and implications for ecosystem productivity. *Environmental Science and Technology, 54*(8), 4984–4994/ https://doi.org/10.1021/acs.est.9b06150/.

Okin, G. S., Mahowald, N., Chadwick, O. A., & Artaxo, P., (2004). Impact of desert dust on the biogeochemistry of phosphorus in terrestrial ecosystems, global biogeochemical. *Cycles, 18*, GB2005. doi: 10.1029/2003GB002145/.

Okin, G. S., Murray, B., & Schlesinger, W. H., (2001a). Degradation of sandy arid shrubland environments: Observations, process modeling and management implications. *Journal of Arid Environment, 47*(2), 123–144.

Okin, G. S., Murray, B., & Schlesinger, W. H., (2001b). In: Conacher, A., (ed.), *Desertification in an Arid Shrubland in the Southwestern United States: Process Modeling and Validation* (pp. 53–70). In Land Degradation: Papers Selected from Contributions to the Sixth Meeting of the International Geographical Union's Commission on Land Degradation and Desertification, Perth, Western Australia.

Pai, S. J., Heald, C. L., & Murphy, J. G., (2021). Exploring the global importance of atmospheric ammonia oxidation. *ACS Earth and Space Science, 7*, 1674–1685. https://doi.org/10.1021/acsearthspacechem.1c00021/.

Peng, B., Li, G., X., Yao, G. X., CXu, Y. Y., Meharg, A. A., & Zhu, Y. G., (2018). The role of sulphate reducing prokaryotes in the coupling of element biogeochemical cycling. *Science of the Total Environment (V), 613, 614*, 398–408.

Pfennig, N., (1975). The phototrophic bacteria and their role in the sulfur cycle. *Plant and Soil, 43*(1–3), 1–16. https://doi.org/10.1007/BF01928472.

Poole, R. K., & Hill, S., (1997). Respiratory protection of nitrogenase activity in *Azotobacter vinelandii*—roles of the terminal oxidases. *Bioscience Reports, 17*, 303–317. doi: 10.1023/A:1027336712748 (April, 12th, 2022).

Poschl, U., (2005). Atmospheric aerosols: Composition, transformation, climate, and health effects. *Angewandte Chemie, (A Journal of German Chemical Society, International Edition, 44*(46), 7520–7540. https://onlinelibrary.wiley.com/doi/abs/10.1002/anie.200501122/ (accessed on 19 January 2024).

Pramanik, P., & Kim, P. J., (2014a). Evaluating changes in cellulolytic bacterial population to explain methane emissions from air-dried and composted manure treated rice paddy soils. *The Science of the Total Environment, 470, 471*, 1307-1312.

Pramanik, P., & Kim, P. J., (2014b). Mitigate CH_4 emission by suppressing methanogen activity in rice paddy soils using ethylene diamine tetra acetic acid (EDTA). *Geoderma, 219, 220*, 58–62.

Pramanik, P., & Kim, P. J., (2017). Contrasting effects of EDTA applications on the fluxes of methane and nitrous oxide emissions from straw-treated rice paddy soils. *Journal of the Science of Food and Agriculture, 97*, 278–283.

Pramanik, P., & Kim, P. J., (2018). *Methanogens Harboring in Rice Rhizosphere Reduce Labile Organic Carbon Compounds to Produce Methane Gas*. Intechopen http://dx.doi.org/10.5772/intechopen.73299/.

Pramanik, P., Haque, M. M., Kim, S. Y., & Kim, P. J., (2014). C and N accumulations in soil aggregates determine nitrous oxide emissions from cover crop treated rice paddy soils during fallow season. *The Science of the Total Environment, 490*, 622–628.

Prather, M., Ehhalt, D., Dentener, F., Derwent, R., Dlugokencky, E., Holland, E., Isaksen, I., et al., (2001). Atmospheric chemistry and greenhouse gases. In: Houghton, J. T., (ed.), *Climate Change 2001: The Scientific Basis* (pp. 239–287). Contribution of Working Group I to the Third Assessment Report of the Intergovernmental Panel on Climate Change, IPCC, Geneva, Switzerland, Cambridge University Press, Cambridge, UK.

Prentice, I. C., Farquhar, G. D., Fasham, M. J. R., Goulden, G. L., Heimann, M., et al., (2001). The C cycle and atmospheric CO_2. In: *Climate Change 2001: The Scientific Basis; Volume 1 of the Third Assessment Report of the Intergovernmental Panel on Climate Change* (pp. 183–237). Intergovernmental Panel on Climate Change Cambridge University Press.

Putaud, J. P., Raes, F., Van, D. R., Brüggemann, E., Facchini, M. C., Decesari, S., Fuzzi, S., et al., (2004a). A European aerosol phenomenology 2: Chemical characteristics of particulate matter at kerbside, urban, rural and background sites in Europe. *Atmospheric Environment, 38*(16), 2579–2595.

Putaud, J. P., Van, D. R., Alastuey, A., Bauer, H., Birmili, W., Cyrys, J., Flentje, H., et al., (2004b). A European aerosol phenomenology – 3: Physical and chemical characteristics of particulate matter from 60 rural, urban, and kerbside sites across Europe. *Atmospheric Environment, 44*, 1308–1330.

Rabus, R., Hansen, T. A., & Widdel, F., (2006). Dissimilatory sulfate- and sulfur-reducing prokaryotes. In: Dworkin, M., Falkow, S., Rosenberg, E., Schleifer, K. H., & Stackebrandt, E., (eds.). *The Prokaryotes* (pp. 659–768). New York, NY: Springer New York. doi:10.1007/0-387-30742-7_22.

Ramaswamy, V., Boucher, O., Haigh, J., Hauglustaine, D., Haywood, J., et al., (2001). Radiative forcing of climate change. In: *Climate Change 2001: The Scientific Basis; Volume 1 of the Third Assessment Report of the Intergovernmental Panel on Climate Change* (pp. 349–416). Cambridge University Press.

Ramawat, K. G., (2010). *Desert Plants Biology and Biotechnology*. Berlin: Springer.

Raymond, J., Siefert, J. L., Staples, C. R., & Blankenship, R. E., (2004). The natural history of nitrogen fixation. *Molecular Biology and Evolution, 21*, 541–554. doi: 10.1093/molbev/msh047/.

Reeburgh, W. S., (1997). Figures summarizing the global cycles of biogeochemically important elements. *Bulletin of Ecological Society of America, 78*(4), 260–267.

Renard, J. J., Calidonna, S. E., & Henley, M. V., (2004). Fate of ammonia in the atmosphere: A review for applicability to hazardous releases. *Journal of Hazard Material, 108*, 29–60. doi: 10.1016/j.jhazmat.2004.01.015/.

Ridley, D. A., Heald, C. L., & Ford, B., (2012). North African dust export and deposition: A satellite and model perspective. *Journal of Geophysics. Research, 117*, D02202. doi: 10.1029/2011JD016794/.

Romheld, D., Marschner, H., Becker, K., Schecht, E., Hulsebusch, C., & Buerkert, A., (2000). *Effects of Site Factors on the Efficiency of Soil Amendments and on Nutrient Fluxes in Subsistence-Oriented Cropping of the West African Sahel* (pp. 1–24). http://www.troz.uni-hohenheim.de/research/sfb308/Ebcont/bcheck (accessed on 19 January 2024).

Roy, S., & Roy, M., (2015). Bioleaching of heavy metals by sulfur oxidizing bacteria: A review. *International Research Journal of Environment Sciences, 4*(9), 1–5.

Samake, A., Bonin, A., Jaffrezo, J. L., Taberlet, P., Uzu, G., Jacob, V., Conil, S., & Martins, J. M. F., (2020). High levels of primary biogenic organic aerosols in the atmosphere in summer are driven by only a few microorganisms from the leaves of surrounding plants. *Atmospheric Chemistry and Physics*. https://doi.org/10.5194/acp-20-5609-2020/.

Samake, A., Martins, J. M., Bonin, A., Uzu, G., Taberlet, P., Conil, S., Favez, O., et al., (2021). Variability of the atmospheric PM10 microbiome in three climatic regions of France. *Frontiers in Microbiology*. https://doi.org1040 10.3389/fmicb.2020.576750/.

Sánchez-Ochoa, A., Kasper-Giebl, A., Puxbaum, H., Gelencsér, A., Legrand, M., & Pio, C., (2007). Concentration of atmospheric cellulose: A proxy for plant debris across a west-east transect over Europe. *Journal of Geophysics Research, 112*. https://doi.org/10.1029/2006JD008180/ (May, 14th, 2022).

Sand, W., Gehrke, T., Jozsa, P. G., & Schippers, A., (2001). Biochemistry of bacterial leaching-direct vs. indirect bioleaching. *Hydrometallurgy, 59*, 159–175.

Saree, S., Ponphang-nga, P., Limtong, P., & Chidthaisong, A., (2012). Soil carbon sequestration affected by cropping changes from upland maize to flooded rice cultivation. *Journal o fSustainable Energy and Environment, 23*, 147–152.

Schaufler, G., Kitzler, B., Schindbacher, A., Skiba, U., Sutton, M. A., & Zechmeister-Boltenstem, S., (2010). Greenhouse gas emissions from European soils under different land use: Effects of soil moisture and temperature. *European Journal of Soil Science, 61*(5). doi: 10.1111/j.1365-2389.2010.01277.x. https://www.researchgate.net/publication/249430206_Greenhouse_gas_emissions_from_European_soils_under_different_land_use_Effects_of_soil_moisture_and_temperature. (accessed on 19 January 2024).

Schoonen, M. A. A., & Xu, Y., (2001). Nitrogen reduction under hydrothermal vent conditions: Implications for the prebiotic synthesis of C-H-O-N compounds. *Astrobiology, 1*, 133–142. doi: 10.1089/153110701753198909/.

Shao, Y., Wyrwoll, K. H., Chappell, A., Huang, J., Lin, Z., McTainsh, G. H., & Yoon, S., (2011). Dust cycle: An emerging core theme in Earth system science. *Aeolian Research, 2*, 181–204.

Shepherd, G., (2016). *Global Assessment of Sand and Dust Storms* (pp. 1–38). United Nations Environment Program, Nairobi, Kenya. https://wesr.unep.org/media/docs/assessments/global_assessment_of_sand_and_dust_stormsx.pdf (accessed on 19 January 2024).

Shi, J., Wang, N., Gao, H., Baker, A. R., Yao, X., & Zhanga, D., (2019). Phosphorus solubility in aerosol particles related to sources and atmospheric acidification in Asian continental outflow. *Atmospheric Chemistry and Physics, 19*, 847–860. https://doi.org/10.5194/acp-19-847-2019.

Signor, D., Eduardo, C., & Cerri, P., (2013). Nitrous oxide emissions in agricultural soils: A review. *Pesquisos Agropecuaria Tropicale., Goiânia, 43*(3), 322–338.

Simon, J., & Kroneck, P. M. H., (2013). Microbial sulphite respiration. *Advances in Microbial Physiology, 62*, 45–117.

Smil, V., (1997). *Cycles of Life: Civilization and the Biosphere* (p. 221). Scientific American Library, New York.

Sterk, G., Herman, L., & Bationo, A., (1996). Windblown nutrient transport and soil productivity changes in southwest Niger. *Land Degradation and Development, 7*, 325–335.

Sterk, G., Stroosnijder, I., & Raats, P. A. C., (1998). *Wind Erosion and Control Techniques in Sahelian one Niger* (pp. 1–4). https://www.researchgate.net/publication/40138628_Wind_Erosion_Processes_and_Control_Techniques_in_the_Sahelian_Zone_of_Niger (accessed on 19 January 2024).

Stewart, W. M., Dibb, D. W., Johnston, A. E., &. Smyth, T. J., (2005). The contribution of commercial fertilizer nutrients to food production. *Agronomy Journal, 97*, 1–6.

Stoecker, L., Bernd, B., Bjorn, S., Per, H. N., Jeppe, L. N., Christian, B., Elena, R. T., et al., (2006). Cohn's crenothrix is a filamentous methane oxidizer with an unusual methane monooxygenase. *Proceedings of National Academy of Sciences, USA, 103*, 2363–2367. doi: 10.1073/pnas.0506361103; https://www.ncbi.nlm.nih.gov/pmc/articles/PMC1413686/ (accessed on 19 January 2024).

Stoorvogel, J. J., & Smalling, E. M. A., (2000). *Assessment of Soil Nutrient Depletion in sub-Saharan Africa* (Vol. 1, p. 342). 1983–2000 Main Report. The Win and Staring Centre, Wageningen, The Netherlands.

Sugano, A., Tsuchimoto, H., Cho, T. C., Kimura, M., & Asakawa, S., (2005b). Succession of methanogenic archaea in rice straw incorporated into a Japanese rice field: Estimation by PCR-DGGE and sequence analysis. *Arachaea, 1*, 391–397.

Sugano, A., Tsuchimoto, H., Tun, C. C., Asakawa, S., & Kimura, M., (2005a). Succession and phylogenetic profile of eubacterial communities in rice straw incorporated into a rice field: Estimation by PCR-DGGE analysis. *Soil Science and Plant Nutrition, 51*, 51–60.

Swap, R., Garstang, M., Greco, S., Talbot, R., & Kallberg, P., (1992). Saharan dust in the Amazon basin. *Tellus, Series. B, 44*(2), 133–149.

Talbot, R. W., Harriss, R. C., Browell, E. V., Gregory, G. L., Sebacher, D. I., & Beck, S. M., (1986). Distribution and geochemistry of aerosols in the tropical North Atlantic troposphere: Relationship to Saharan dust. *Journal of Geophysics Research, 91*(D4), 5173–5182.

Tegen, I., & Fung, I., (1995). Contribution to the atmospheric mineral aerosol load from land surface modification. *Journal of Geophysical Research, 100*(D9), 18,707–18,726.

Tian, J., Ge, F., Zhang, D., Deng, S., & Liu, X., (2021). Roles of phosphate solubilizing microorganisms from managing soil phosphorus deficiency in mediating biogeochemical P cycle. *Biology, 10*, 158. 10.3390/biology10020158/.

Tiedje, J. M., (1988). Ecology of denitrification and dissimilatory nitrate reduction to ammonium. Scientific research – An academic publisher. *Biology of Anaerobic Microorganisms, 717*, 179–244.

Tiedje, J. M., Sexstone, A. J., Myrold, D. D., & Robinson, J. A., (1983). Denitrification: Ecological niches, competition, and survival. *Antonie Van Leeuwenhoek, 48*, 569–583.

Tipping, E., Benham, S., Boyle, J. F., Crow, P., Davies, J., Fischer, U., Guyatt, H., et al., (2014). Atmospheric deposition of phosphorus to land and freshwater. *Environmental Science, 16*, 1608–1617.

Ueki, A., Ono, K., Tsuchiya, A., & Ueki, K., (1997). Survival of methanogens in air-dried paddy field soil and their heat tolerance. *Water Science Technology, 36*(6, 7), 517–522. https://doi.org/10.2166/wst.1997.0631/.

UNEP, (2016). *Global Assessment of Sand and Dust Storms* (pp. 1–139). United Nations Environment program, World Meteorological Organization and United nations Convention to Combat Desertification, Paris.

Vaksmaa, A., Van, A. T. A., Ettwig, K. F., Lupotto, E., Valè, G., Jetten, M. S. M., & Lüke, C., (2017). Stratification of diversity and activity of methanogenic and methanotrophic microorganisms in a nitrogen-fertilized Italian paddy soil. *Frontiers in Microbiology, 8*, 2127. doi: 10.3389/fmicb.2017.02127/.

Vergara-Fernandez, A., Morales, P., Scot, F., & Guerrero, S., (2019). Methane biodegradation and enhanced methane solubilization by the filamentous *Fusarium solani. Chemosphere, 226*, 24–35. doi:10.1016/j.chemosphere.2019.03.116/.

Vigliotta, G., Nutricati, E., Carata, E., & Tredici, S. M., (2007). *Clonothrix fusca* Roza 1896, a filamentous, sheathed, methanotrophic proteobacterium. *Applied and Environmental Microbiology, 73*, 3556–3565. doi:10.1128/AEM.02678-06.

Vincenot, L., Nara, K., Sthultz, C., Labbé, J., Dubois, M. P., Tedersoo, L., Martin, F., & Selosse, M. A., (2012). Extensive gene flow over Europe and possible speciation over Eurasia in the ectomycorrhizal basidiomycete *Laccariaamethystina* complex. *Molecular Ecology, 21*, 281–299. https://doi.org/10.1111/j.1365-294X.2011.05392.x.

Violaki, K., Nenes, A., Tsagaraki, M., Paglioni, M., Jacquet, S., Semere, R., & Panagiotopoulos, C., (2021). Bioaerosols and dust are the dominant sources of organic P in atmospheric particles. *Climate and Atmospheric Science, 4*, 63. https://doi.org/10.1038/s41612-021-00215-5/.

Wagenbrenner, N. S., Chung, S. H., & Lamb, B. K., (2017). A large source of dust missing in Particulate Matter emission inventories? Wind erosion of post-fire landscapes. *Elem. Sci. Anth., 5*(2), https://doi.org/10.1525/elementa.185/.

Waite, D. W., Vanwonterghem, I., Rinke, C., Parks, D. H., Zhang, Y., Takai, K., et al., (2017). "*Epsilonproteobacteria* and Proposed reclassification to Epsilonbacteraeota (Phyl. Nov.)." *Frontiers in Microbiology, 8*, 682. doi: 10.3389/fmicb.2017.00682/.

Waite, D. W., Vanwonterghem, I., Rinke, C., Parks, D. H., Zhang, Y., Takai, K., et al., (2018). "*Epsilonproteobacteria* and proposed reclassification to Epsilonbacteraeota (phyl. nov.)." *Frontiers in Microbiology, 9*, 772. doi: 10.3389/fmicb.2018.00772/.

Wang, J., Ci, C., Li, Y., Hua, M., Wang, J., Yang, W. H., Zheng, P., & Hu, B., (2019). Denitrifying Anaerobic methane oxidation: A previously overlooked methane sink in intertidal zone. *Environmental Science and Technology, 53*, 203–212. doi: 10.1021/acs.est.8b05742/.

Wang, J., Yao, X., Jia, Z., Zheng, P., Kartal, B., & Hu, B., (2022). Nitrogen input promotes denitrifying methanotrophs' abundance and contribution to methane emission reduction in coastal wetland and paddy soil. *Environmental Pollution,302,* 119090. https://doi.org/10.1016/j.envpol.2022.119090/.

Wang, R., Balkanski, Y., Boucher, O., Ciais, P., Penuelas, J., & Tao, S., (2015). Significant contribution of combustion-related emissions to the atmospheric phosphorus budget. *Nature Geoscience, 8*(1), 48–54.

Wang, S., Walker, R., Schicklberger, M., Nico, P. S., Fox, P. M., Karaoz, U., Chakraborty, R., & Brodie, E. L., (2021). Microbial phosphorus mobilizing strategies across nutrient limitation gradient and evidence for linkage with iron solubilization. *Frontiers in Microbiology* https://doi.org/10.3389/fmicb.2021.572212/.

Wang, Y., Wang, D., Yang, Q., Zeng, G., & Li, X., (2017). Wastewater opportunities for denitrifying anaerobic methane oxidation. *Trends in Biotechnology, 35*, 799–802. doi: 10.1016/j.tibtech.2017.02.010/.

Watanabe, T., Asakawa, S., Nakamura, A., Nagaoka, K., & Kimura, M., (2004). DGGE method for analyzing 16S rDNA of methanogenic archaeal community in paddy field soil. *FEMS Microbiology Letters, 232*, 153–163.

Weber, S., Lueders, T., Friedrich, M. W., & Conrad, R., (2001). Methanogenic populations involved in the degradation of rice straw in anoxic paddy soil. *FEMS Microbiology and Ecology, 38*, 11–20.

Weber, S., Salameh, D., Albinet, A., Alleman, L. Y., Waked, A., Besombes, J. L., Jacob, V., et al., (2019). Comparison of PM10 Sources Profiles at 15 French sites using a harmonized constrained positive matrix factorization approach. *Atmosphere, 10*(6), 310. https://doi.org/10.3390/atmos10060310.

Weinberger, R., Weiner, T., & Angert, A., (2016). Isotopic signature of atmospheric phosphate emitted from coal combustion. *Atmosphere and Environment, 136*, 22–30.

Welte, C., & Deppenmeire, U., (2011). Proton translocation in methanogens; Chapter 13. In: Rosenzweig, A. C., & Ragsdale, S. W., (eds.), *Methods in Methane Metabolism: Methanogenesis* (pp. 257–280). Cambridge: Academic Press, Elsevier.

Wiggington, N. S., (2014). Dissecting ancient microbial sulfur cycling. *Science, 346*, 712, 713. doi: 10.1126/science.346.6210.712-h.

Wikipedia, (2022). *Sulfur-Reducing Bacteria.* https://en.wikipedia.org/wiki/Sulfur-reducing_bacteria (accessed on 19 January 2024).

WMO, (2016). *Interannual Variability and Decadal Trends in Mineral Dust Aerosols* (pp. 1–4). World Meteorological Organization, Zurich, Switzerland. https://sds-was.aemet.es/materials/technical-reports/SDSWASNAMEE2016001INTERANNUAL.pdf (accessed on 19 January 2024).

WMO, (2020). *Sand and Dust Storms* (pp. 1–12). World meteorological Organization, Zurich, Switzerland. https://community.wmo.int/en/activity-areas/gaw/science-for-services/sds-was (accessed on 19 January 2024).

Wolfe, R. S., & Metcalf, W. W., (2010). A vacuum-vortex technique for preparation of anoxic solutions or liquid culture media in small volumes for culturing methanogens or other strict anaerobes. *Anaerobe, 16*, 216–219.

Wraith, M. A., & Sharma, R., (1998). Technical review of methods to enhance biological degradation in sanitary landfills. *Water Quality Research Journal of Canada, 33*, 417–437.

Wu, B., Liu, F., Fang, W., Yang, T., Chen, G. H., He, Z., & Wang, S., (2021). Microbial sulfur metabolism and environmental implications. *Science of the Total Environment, 778*, 146085 Review.

Yan, X., Akimoto, H., & Ohara, T., (2003). Estimation of nitrous oxide, nitric oxide, and ammonia emissions from croplands in East, Southeast and South Asia. *Global Change Biology, 9*, 1080–1096.

Yan, X., Akiyama, H., & Yagi, K., (2009). Global estimations of the inventory and mitigation potential of methane emissions from rice cultivation conducted using the intergovernmental panel on climate change guidelines. *Global Biogeochemical Cycles, 23*, GB2002.

Yao, H., Conrad, R., Wassmann, R., & Neue, H. U., (1999). Effect of soil characteristics on sequential reduction and methane production in sixteen rice paddy soils from China, the Philippines, and Italy. *Biogeochemistry, 47*, 269–295.

Yttri, K. E., Aas, W., Bjerke, A., Cape, J. N., Cavalli, F., Ceburnis, D., Dye, C., et al., (2007). Elemental and organic carbon in PM_{10}: A one-year measurement campaign within the European monitoring and evaluation program EMEP. *Atmospheric Chemistry and Physics, 7*, 5711–5725. https://doi.org/10.5194/acp-7-5711-2007?.

Yurimoto, H., Hisashi, R. H., Yasueda, Y. R., Sakai, Y., & Kato, Y., (2002). The ribulose monophosphate pathway operon encoding formaldehyde fixation in a Thermotolerant methylotroph, *Bacillus brevis*S1. *FEMS Microbiology Letters.* https://doi.org/10.1111/j.1574-6968.2002.tb11345.x/.

Zeytun, A., Sikorski, J., Nolan, M., Lapidus, A., Lucas, S., Han, J., et al., (2011). Complete genome sequence of *Hydrogenobacter thermophilus* type strain (TK-6 T). *Stand. Genomic Sci., 4*, 131–143. doi: 10.4056/sigs.1463589/.

Zhang, L., & Liu, X., (2018). *Nitrogen Transformation Associated with N_2O Emissions in Agricultural Soils.* Intechopen http://dx.doi.org/10.5772/intechopen.71922/.

Zhang, T., Engling, G., Chan, C. Y., Zhang, Y. N., Zhang, Z. S., Lin, M., Sang, X. F., et al., (2010). Contribution of fungal spores to particulate matter in a tropical rainforest. *Environmental Research Letters, 5*(2), 24010.

CHAPTER 6

Microbes on Crop Canopies: Their Significance to Crop Productivity

ABSTRACT

The crop's canopy that extends into the atmosphere is the focus of this chapter. It is commonly referred to as the 'Phyllosphere.' The Phyllosphere is a vast area, supposedly twice the size of the land surface. It encompasses the surface features of the crop's foliage (phyllosphere), including the leaf lamina (phylloplane), branches, twigs (caulosphere), flowers (anthosphere), pollen, and fruits (carposphere). The Phyllosphere is home to microbial species with diverse genetic constitution, morphological traits, and physiological abilities. These microorganisms play a crucial role in providing various ecosystem services that directly impact the productivity of crops. The crop's phyllosphere (leaf lamina/chlorophyll) directly contributes to global photosynthesis and carbon fixation.

The microbial communities found on the phyllosphere are highly diverse. The microflora on the phyllosphere/phylloplane exhibit dynamic species composition and functional characteristics. However, there is a certain degree of specificity in terms of the microbial species that inhabit the crop's phyllosphere. The genetic makeup of the host plant influences the phyllosphere microflora. Atmospheric factors such as temperature, relative humidity, pH, moisture, radiation, and gaseous composition have an impact on the phyllosphere microbes, their communities, and physiological activities. Seasonal changes in phyllosphere microflora can be easily observed through analysis using selective media and/or 16S rRNA sequencing. The succession of microorganisms on foliage is worth noting.

Microbiology of the Agricultural Sky. K. R. Krishna (Ed)

Agronomic procedures such as foliar sprays of plant protection chemicals (fungicides, bactericides, insecticides), foliar nutrients (liquid NH_3, Urea, NH_4NO_3), herbicides, and aerial sprinkler irrigation have immediate effects on phyllosphere microbes, their density, and population. Systemic plant protection chemicals also affect phyllosphere exudations and microflora.

Phyllosphere microflora influences nutrient dynamics on the surface of foliage. Methano- and methylotrophs also inhabit the phyllosphere, potentially reducing methane (CH_4) in the troposphere, at least in the vicinity of crops. Beneficial microbes like biological-N fixers can be found on the phyllosphere, making significant contributions to the nitrogen cycle and related ecosystem functions.

Most airborne plant pathogens invade via phyllosphere, mainly through foliage. Phyllosphere, as such, is an abode to several plant pathogens of severe consequences to crops. Epidemics and pandemics are induced by airborne fungi/bacteria/viruses that invade via phyllosphere (e.g., *Puccinia graminis tritici*; Powdery mildew fungi, Brown spots of *Pseudomonas*, *Xanthomonas* blotches, and tobacco mosaic virus (TMV)). It is common to trace plant pathogens among the microbial assemblages of leaf lamina.

Biological control fungi/bacteria are another set of beneficial microbes that find a niche in the phyllosphere. Biocontrol organisms can physically exclude disease-causing microbes from the leaf surface, for example, by clogging their entry into leaf tissue. There are microbes that produce antagonistic chemicals, bacteriostatic agents, antibiotics, and more. Biocontrol agents (BCAs) definitely impact the microflora of the phyllosphere. Bacteriophages are sprayed on foliage to alter/control microbial pathogens. The phyllosphere harbors exudates that stimulate the growth of microbes found on foliage.

Overall, it is important to note that every molecule of foliar nutrient, moisture from the troposphere, and deadly plant pathogens that cause epidemics have to interact and cross the phyllosphere barrier.

6.1 INTRODUCTION TO CROP'S CANOPY

Firstly, the crop's canopy is an area within the lower troposphere. The crop's canopy extends from the soil surface into the extreme conditions of the lower troposphere. The crop's canopy is constantly interacting with tropospheric factors, starting from the plumule stage and continuing throughout the crop's growth until final senescence. This statement also applies to forest stands, natural vegetation of shrubs, herbs, and annual grasses. The term "phyllosphere" refers to the total surface area of the above-ground parts (canopy) of the plants. Microbes that thrive on the phyllosphere are also in constant interaction with each other. Phyllosphere microbes respond to gaseous and particulate matter present in the lower troposphere, as well as other environmental factors such as relative humidity, temperature, diurnal changes, precipitation, wind, turbulence, insects, birds, and other fauna.

There are several good reasons to learn more about microbes on the canopy of natural vegetation, forest plantations, vast expanses of field crops, and pastures. Microbes on the surface of vegetation must adapt, first and foremost, to the unpredictable environmental conditions in the open, known as the 'agricultural sky.' Foliage serves as a habitat for phyllosphere microbes, providing them with anchorage and some nutrients. The foliage also creates a favorable environment. However, phyllosphere microbes are still susceptible to the extreme conditions of the troposphere. They are exposed to antagonistic microbes that may periodically settle on the surface. Microbial succession can replace well-established microbial communities with new ones that thrive and flourish. This leads to significant changes in microbial diversity within the phyllosphere.

The term "phyllosphere" refers to the surface of foliage, which is inhabited by microbes on both the upper and lower surfaces of leaves. According to computer modeling, the phyllosphere is a vast area that exceeds the land surface. Vorholt (2012) states that the global leaf area is 508,630,100 km^2, and when considering both the upper and lower leaf surfaces, it adds up to 1,017,260,200 km^2. The phyllosphere is approximately twice the size of the land surface. The atmosphere and its components have a significant influence on the microbial population, diversity, and activity in this extensive area. The phyllosphere serves as a large habitat where microbes carry out numerous ecosystem functions. Undoubtedly, phyllosphere microbes have an immediate impact on cultivated crops, regardless of their location on the Earth's soil surface or in open fields.

Bringel & Couee (2015) point out that the phyllosphere is among the most congenial habitats for microbes to thrive and multiply. Furthermore, microbes found in the phyllosphere are often well adapted to the extremes of the habitat, if any. In fact, the phyllosphere microbiota are in constant interaction, both with crops and the troposphere. Additionally, phyllosphere microbes are in continuum with the gaseous exchange between the atmosphere, phyllosphere, and internal organs of plants. Microbes on foliage are indeed crucial components as they potentially affect levels of various gases such as CO_2, O_2, VOCs, etc. Every molecule of nutrients in soluble form or as a gaseous compound from the troposphere must transit via the phyllosphere before entering the internal organs of the plant. The phyllosphere is often well infested with diverse microbes. Physically, the phyllosphere is a region approximately 0.2 mm thick over the leaf's cuticular surface. It harbors nutrients, a variety of biochemicals, and whatever has been extraneously supplied to crops such as foliar fertilizers, pesticides, herbicides, pollutants, etc. Microbial pathogens that try to invade the palisade layer must pass through the phyllosphere. Similarly, phyllosphere microbes capable of fixing atmospheric nitrogen have to negotiate the extremes of tropospheric conditions, such as low nutrient concentrations, feeble amounts of organic matter, uncongenial temperature, and relative humidity. The phyllosphere is important when we intend to supply amendments via foliage, as well as its microbial component. Microbes can detoxify pesticides and make them less effective on insect pests. Phyllosphere microbes could overcome residual effects of pesticides and may antagonize pathogens spreading through foliage. The biological control of foliar plant pathogens obviously involves interaction between phyllosphere conditions, the pathogen, and the biocontrol agent (BCA). There are biocontrol microbes that thwart the penetration of the leaf surface by the pathogen (Andrews, 1992).

Knief et al. (2012) believe that most previous studies on phyllosphere microbes focus on tracing and identifying plant pathogens, mainly bacteria and fungi. Archaea are scarce in the phyllosphere. We need to investigate the reasons for this phenomenon. There are numerous non-pathogenic phyllosphere microbes whose dynamics are still unknown. They may be important for plant health, particularly in terms of nutrition. Some functions attributed to non-pathogenic phyllosphere microbiota include nitrogen fixation (Miyamoto et al., 2004), methane (CH_4) and methanol utilization (i.e., methylotrophy) (Delmotte et al., 2009), degradation of organic pollutants and pesticides (Ning et al., 2010), and suppression of pathogens through physical exclusion or via diffusive toxins or antibiosis.

Incidentally, there is a growing market for agricultural microbials. This growth is mainly due to the adoption of integrated control of diseases and pests. Most farmers prefer to apply microbial formulations via foliar mode, as suspensions in water. Therefore, atmospheric conditions and the ability of microbes to survive and be effective on the leaf surface are crucial. Reports by Research & Markets (2020) suggest that the demand for microbial BCAs is set to increase up to 11.6 billion US dollars by 2025. Microbial formulations with a bacterial component are common, followed by those with fungi and viruses. Currently, microbial BCAs are mostly being used on vegetables, fruit crops, and a few field crops.

Phyllosphere microbial assemblages harbor a wide diversity of yeasts and yeast-like organisms (Ruinen, 1963; de Menna, 1959). Extracellular lipid secretion has been observed in most of the samples collected from foliage.

6.1.1 PHYLLOSPHERE: DEFINITIONS AND EXPLANATIONS

The term "Phyllosphere" was first coined by F.T. Last while investigating fungi inhabiting leaf surfaces. Specifically, while understanding the etiology and life cycle of the fungal pathogen *Sporobolomyces* sp. affecting wheat and barley (Alklett et al., 2014; Last, 1955). In the case of surfaces of flowers and flower parts, it has been stated that microflora was identified on flowers as early as the 1800s (Boutroux, 1884; Hansen, 1890). Yet, there have not been many detailed investigations on microbes inhabiting flowers. Flowers of several commercial crops have also not been investigated for microbial assemblages (Alklett et al., 2014). Reviews that depict the progress of microbiology of flowers are scarce, if any. Microbes residing on stems, branches, and twigs have also been studied weakly.

A few representative definitions and/or explanations provided for the term 'phyllosphere' are considered here.

The phyllosphere is a habitat where microbes can survive and thrive. It is a vast area found globally, consisting of the above-ground parts of vegetation and crops (Chaudhary et al., 2017). Microbial interactions within the phyllosphere have an impact on the health of natural plant communities and the productivity of agricultural crops.

Another definition of 'phyllosphere' refers to the surfaces of leaves or the total above-ground surfaces of plants. It serves as a habitat for microorganisms, and the microbial species that inhabit the phyllosphere are known as 'epiphytes.' The aerial parts of extensive crop fields provide these microhabitats. Phyllosphere microbes are typically derived from the soil surface, atmospheric dust, and air spora.

The phyllosphere is an important ecosystem. It hosts a large and diverse microbial population. Phyllosphere microbiota play a critical role in protecting plants from diseases (Rastogi et al., 2013).

The phyllosphere is a region that supports a vibrant microbial community. Microbiota exhibit diverse types of physiological processes at the interface between plants and the troposphere (Parasuraman et al., 2019).

The phyllosphere is a habitat for bacteria, yeasts, filamentous fungi, algae, and protozoa. Microbes in the phyllosphere exhibit commensalism, mutualism, or antagonism within this habitat. Among the protists, bacteria are the predominant group, with a population size ranging from 10^2 to 10^{12} g^{-1} of leaf tissue.

The phyllosphere comprises the aerial portions of crops, including the leaves, stems, branches, flowers, and fruits. It is constantly exposed to the ambient atmosphere and supports a relatively large microbial community on its surface. The phyllosphere is considered a heterogeneous and dynamic environment for thriving microbes. Phyllosphere microbes are found on the above-ground portion of plants, mainly on stems, branches, leaves, flowers, and fruits. The microbial assemblage includes bacteria, fungi, archaea, protists, algae, etc. These microbes can either be advantageous or deleterious to the host plant (Gupta & Patil, 2021).

The phyllosphere refers to the above-ground surface of a plant. It serves as a habitat for microbes of diverse phylogenetic origins. The phyllosphere is further divided into the surface of foliage, known as the phylloplane/phyllosphere, and the surface of stems, branches, and twigs, known as the 'caulosphere.' The surface of flowers is referred to as the 'anthosphere,' while the surface of fruits is called the 'carposphere' (Shelake et al., 2019; Wikipedia, 2022; Droby & Visniewski, 2018). The phyllosphere plays a significant role in agricultural farming, despite being relatively nutritionally weak and physically extreme.

Thapa et al. (2017) explain the phyllosphere of cereal crops as the aerial parts, specifically the above-ground organs of plants. This includes the leaves, stem, branches, buds, flowers, bracts, calyx, sepals, fruits/pods, and panicles with seeds, as they develop. These organs provide a habitat for a diverse range of microbial inhabitants, including bacteria, actinomycetes, fungi, yeasts, viruses, archaea, protozoa, microscopic nematodes, and insects. The ability of the phyllosphere to selectively support microbes varies depending on the genotypes of the crop species.

There are microbiologists who believe that the 'phyllosphere' is a relatively unknown habitat where microbes thrive. The phyllosphere is a region that harbors a community of microbes that live in a symbiotic relationship with plant parts such as stems, branches, leaves, buds, and flowers. Phyllosphere microbes can survive both on the phylloplane (leaf lamina) and internally in the leaf tissue. The phyllosphere is an open system, where microbial species interact with each other as well as with the host and ambient atmospheric conditions. Several abiotic factors can affect phyllosphere microbes and their ecosystem functions (Carvahlo & Castillo, 2018).

6.2 PHYLLOSPHERE MICROBES

The phyllosphere, by definition, includes various regions on the surface of a crop's canopy. It encompasses the stem, branches, foliage (phyllosphere), flowers (anthosphere), pollen, fruits (carposphere), and seeds (spermopshere). However, this chapter does not intend to discuss these individual spheres in detail or exhaustively. Each one may require a separate chapter or volume. The focus of this chapter is on the phyllosphere and its microbial components. The term 'phyllosphere' is commonly used to refer to the area surrounding each individual leaf (foliage). However, depending on the context, 'phyllosphere' is also used to encompass the total surface area over the crop's canopy, including all above-ground parts of the plant.

Lindow & Brandl (2003) express that given the large expanses of vegetation/crops, their foliage, and the proportionate phyllosphere area, we have invested relatively less research time on it. Additionally, the phyllosphere is rich in both nutrients and microbes. The simple fact is that terrestrial leaf area is too large at $6.4 \times 10^8 km^2$ (Lindow & Brandl, 2003; Bulgerelli et al., 2013; Rastogi, 2012), at least two times the soil surface. The bacterial population on the phyllosphere region on the globe is about 10^{26} CFU (colony forming units) (Morris & Kinkel, 2002). The geo-biochemical processes attributable to global phyllosphere microbes are numerous. The impact of phyllosphere microbes on global crop production zones can be immensely important.

The sources of microbes (i.e., inoculum) for the phyllosphere are many. The 'Agrarian sky' or atmosphere to which it is always exposed is the major source. Drifting and settling microbes could land on the leaf lamina and other parts. Microbial diversity and load in the troposphere are important factors that affect phyllosphere microbial assemblages. Turbulence in the soil surface due to mild wind, farm vehicle movement, and farm operation disturbs soil/crops. They may hurl microbes onto the phyllosphere of crops. Whatever the mode of migration/spread of microbes, it is believed that in most farms, soil nearer the crops is the major source of microbes. In fact, microbial species traced in the phyllosphere often tally with those found in surface soil. Animals that brush past crops or birds that fly above the crop's canopy may add to phyllosphere microbes. Foliar sprays for irrigation, fertilizer supply, or even pesticide/fungicide distribution may also add to phyllosphere microbes. Low altitude wind in the troposphere is a supplier of microbial species on the leaf surfaces. Intercontinental wind (including jet-streams) also adds microbes to the phyllosphere of crops. It depends on geographical location. Precipitation like drizzle, rain, snow, frost, etc., may also add new microbial load on the phyllosphere. Next, it is the water bodies that may experience turbulent wind which adds to microbes in the

The phyllosphere, in fact, periodically receives microbes from the agrarian sky. Several microbial species may find a niche and establish themselves on the phylloplane/phyllosphere. At the same time, wind, turbulence, precipitation, and anthropogenic activity may disperse phyllosphere microbes to nearby and distant locations. The phyllosphere microbial population is diverse in terms of phylogeny and physiological manifestations. Bacteria, fungi, yeasts, viruses, protozoa, and archaea are often encountered in the phyllosphere microbial communities. Fungi, in particular, appear to be transient. Bacteria, on the other hand, have the highest population density. They are relatively small in size and can occur in small colonies, groups of cells, or even as single cells. Microbes are abundant on the phyllosphere during the growing season of crops, possibly due to ambient atmospheric conditions and the release of nutrients on the phylloplane (Andrews & Harris, 2000; Mercier & Lindow, 2000). Thompson et al. (1993, 1995) emphasize the need to study short-term and long-term changes in phyllosphere microbial assemblages, especially during the growing season of field crops. There are instances where phyllosphere bacterial assemblages depend on leaf structure, size, metabolites released through exudation, and those secreted by other microbes. The biochemical features exhibited in the phyllosphere/phylloplane region are indeed important (Yadav et al., 2005; Chaudhry et al., 2021).

Lindow & Brandl (2003) argue that we have gained significant knowledge about phyllosphere bacteria due to their dominance, diversity, and activity. The estimated bacterial population ranges from 10^6 to 10^7 cells/cm^2 (up to 10^8 cells/g of leaf) (Andrews & Harris, 2000; Beattie & Lindow, 1995; Hirano & Upper, 1989, 2000; Jacques & Morris, 1995; Jacques et al., 1993). Among the various nutrients, carbon/energy sources are particularly important. Mercier & Lindow (2000) discovered sugars in the leaf exudates, amounting to 0.2 to 10 g per leaf. This is believed to be sufficient to support a phyllosphere bacterial population in the range of 10^8 cells cm^{-2} area.

According to Vorholt (2012), the phyllosphere is a vast region, especially when considering the entire natural vegetation, forest plantations, and field crop production zones. Microbes that are found in the phyllosphere are often present in the troposphere, and *vice versa*, tropospheric microbes can commonly be found on the phyllosphere. Seasonal variations are to be expected.

The microbial load and diversity of the phyllosphere depend on several factors. Vorholt (2012) discovered that bacteria, specifically belonging to the classes *Alphaproteobacteria* and *Gammaproteobacteria,* were dominant on the phyllosphere. Phyla such as Bacteroides and Actinobacteria were frequently detected as well. These bacterial phyla/classes are also commonly found in tropospheric air samples, suggesting that they may originate from the same region as the phyllosphere microbes.

Phyllosphere microbes adopt certain mechanisms to ensure their survival on foliage. Vorholt (2012) lists the following mechanisms: (i) formation of bacterial cell aggregates, (ii) elaboration of chemicals that allow them to adhere and stay firmly on the leaf surface, and (iii) modified biochemical reactions that allow them to utilize nutrient sources available on the leaf and exudates (e.g., methanol, sugars, amino acids).

As stated earlier, the Phyllosphere encompasses not just the foliage and leaf surface (phylloplane), but also the area extending to 0.02 mm surrounding the lower and upper surface (phyllosphere). By definition, we also consider the surface area on stems, branches, twigs, flowers, fruits, and seeds as part of the phyllosphere. Microbes thrive on the surface of these plant parts. The nutritional status, environmental conditions, and diverse inoculum of microbes determine the total population supported by each organ on its surface.

The habitat above the leaf is often called the phyllosphere. The surface region of the leaf lamina is called the phylloplane, and the micro-habitat surrounding branches and twigs is called the 'Caulosphere' (Plate 6.1). The micro-habitat on the surface of fruits is called the 'Carposphere.' The thin area surrounding the seed is called the 'Spermosphere.' The habitat on the surface of flowers is known as the 'Anthosphere.' Anthosphere includes various parts of flowers such as the calyx, bracts, sepals, pistil, and pollen (Shelake et al., 2019; Plate 6.2). We find genetically diverse microbes in all the above habitats.

'Caulosphere' refers to the micro-habitat surrounding the stems and branches of trees, shrubs, and a large number of field crop species. The stems of trees are often woody, waxy, and have thick lignin deposits. They may not support a large population of microbes. However, much depends on the flow of nutrients and the availability of nutrients for microbes to flourish on the surface. Investigations have shown that bacteria, fungi, and yeasts are commonly found on the stems and barks (Plates 6.1-6.3). Bacteria such as *Pseudomonas, Flavobacterium,* and *Proteobacteria* have been observed in the caulosphere of trees. Yeast and fungal species encountered include

Saccharomyces, Candida, and *Lachancea.* Microbes in the caulosphere are constantly exposed to atmospheric factors such as light, temperature, UV radiation, moisture, etc. Mushroom spores are easily detected on barks.

Anthosphere is a transitory region that begins from the opening of the flower until maturity and senescence. The floral parts such as calyx, sepals, petals, stamens, style, ovary, and stigmas allow colonization by microbes (Plate 6.1). They function as habitats for microbes for a brief period. The floral surface region is supposed to be colonized by basidiomycetous yeasts, while pollen supports predominantly ascomycetous yeast. The nectar in floral parts supports diverse fungi. Usually, nectar-utilizing fungi are dominant among the microbial species found in the floral parts.

The pollen is a rich source of nourishment for microbes. Depending on the pollen physiology, surface characteristics, viability, nutrients, and moisture, microbes of diverse abilities thrive on it. The bacterial population is said to range from 10^6 to 10^9 g^{-1} pollen. Therefore, massive pollen drifts can transport a large population of microbes from one location to another. Epiphytotic bacterial species can occur singly or in colonies, or even clusters of colonies. This depends on the nature of the flower, its nutritional status, and environmental parameters (Shelake et al., 2019). While pollen may harbor commensals, symbionts, and disease-causing organisms, they could also serve as a vector operating in the agrarian sky.

Maniranjan et al. (2018) state that we have devoted more of our research efforts to understanding the microbiology of the rhizosphere and phyllosphere (meaning leaf lamina) than other parts of the plant. In the present context, it is worth noting that pollen and its surface also harbor diverse microbial flora. The source of microbes on the pollen could be from the troposphere or other parts of the flower, such as sepals, corolla, and calyx. Microflora from the soil surface that becomes airborne may deposit on the pollen surface. Raindrops, dew, mist, and precipitation may also contribute to the microflora on the pollen. Insects that repeatedly visit the flowers often add and redistribute microflora among flowers and their pollen. Furthermore, they point out that pollen, which is nutritive and has special modifications on its surface, may literally be an excellent abode for microbes.

Pollen from crop species like rye, rapeseed, crocus, and birch exhibited large diversity of microbes and succession patterns. Bacteria isolated from pollen surface belonged to *Proteobacteria*, *Actinobacteria*, *Acidobacteria*, and *Firmicutes* which were predominant. It seems that the type and characteristics of pollen have an effect on the microbial load. There are also differences in microbial diversity between pollen dispersed by wind, turbulence, and those carried by insects such as honeybees. Grain cereals, millets, and grasses may have distinct microflora on their pollen. For example, bacterial assemblages on *Setariaviridis* and *S. pumila* were different. Host genotypes also support distinct microflora (Rodrigues et al., 2018). The atmosphere, known as the agrarian sky, serves as a repository of diverse microbes from which host crops/genotypes select. Of course, the geographic location also affects the microbial flora in the atmosphere itself.

Carposphere is the region on the surface of fruits/pods produced by various plant species. It is a transitory zone that lasts from the initiation of fruit until maturity. The Carposphere is exposed to the ambient troposphere, and the microbes in the Carposphere constantly interact with troposphere conditions. Microbes and particulate matter (PM) that deposit from the troposphere may have an impact on the fruit itself. The nutritional composition of the fruit surface is dynamic and depends on its maturity. The surface can be rich in sugars, amino acids, and other easily available nutrients, resulting in high microbial diversity. Predominantly, bacteria, fungi, and yeasts are found in this habitat. Microbes in the Carposphere are exposed to atmospheric factors such as light, moisture, relative humidity (RH), UV radiation, and surface nutrients.

Microbes settle on fruit surfaces from the agrarian sky. The microbial deposits occur from the time fruits are formed in the outfield until maturity. For fruits left in the open after harvest, deposits of dust/microbes continue for a certain period until proper storage. The fruit surface harbors microbes of diverse origin. These epiphytes could be sourced from the soil surface. Also, the agrarian sky, including the troposphere, contributes microbes through still air, wind, breeze, PM, dusts, precipitation, and snow. Birds and aerial farm machines flying above the crop's canopy, such as drones, also introduce microbes. The epiphytic microbes may affect the quality of fruits. Fruit surface microbes may induce diseases that spoil the fruits during growth and maturity. This can occur on the tree or after harvest. There could be many biocontrol microbes harbored on fruit surfaces (carposphere). There may also be host species/genotype preferences for fruit surface microflora (Leff & Fierer, 2013). In such cases, we must screen for microbes on fruit surfaces and select ones that allow better preservation. Additionally, we should consider quality traits such as size, spots, specs, color of the rind, odor, etc.

Let us consider an example where the microflora on the surface of fruits has been investigated. According to Leff & Fierer (2013), most studies on fruit surfaces have aimed at identifying pathogens and their destructive nature on fruits, such as spots, cankers, and soft rots. The spread of human pathogens like *E. coli, Salmonella, Vibrio cholerae,* etc., has also been a major reason for studying fruit surface microbes. Therefore, our knowledge about the diversity of microbes related to fruit quality has been relatively limited. Overall, the fruit surface is home to a wide range of bacterial species. However, the microbial assemblages observed on fruits appear to be distinct based on the crop species. Often, a few bacterial species are predominant, while several others are found in small numbers. Additionally, some bacterial species may be restricted and not found at all in certain types of fruits.

Further, Leff & Fierer (2013) examined surface bacteria of a series of fruits like apples, grapes, peaches, strawberries, and mushrooms (fruiting body of fungus). They found *Actinobacteria, Bacteroidetes, Firmicutes,* and *Proteobacteria* to be predominant. In addition to the host species, farming procedures and conditions in the agrarian sky (troposphere) also affect the surface microbial assemblages. For instance, microbial assemblages on apples produced using organic farming methods differ from those produced using conventional pesticide application methods (Ottesen et al., 2009; Leff & Fierer, 2013; Brown, 2014). Certain vegetable species such as potatoes, sprouts, spinach, lettuce, tomatoes, peppers, and strawberries exhibit similar microbial assemblages (Ottesen et al., 2013). Surface soil serves as a major repository of microbes, which can contribute to the microbial presence on fruits depending on wind and turbulence.

The spermosphere refers to the region of small thickness (< 0.2 mm) surrounding the seed coat surface (i.e., rind). It harbors a diverse microbial assemblage, many of which can negatively impact seed quality if allowed to flourish.

Practically, seed dispersal occurs naturally or is aided by humans during the process of farming. Simultaneously, seed dispersal distributes microbes found on the seed coat/spermosphere. Atmospheric wind aids seed dispersal. Therefore, it also spreads microbes found in the spermosphere to different locations (*see* Nelson, 2017).

Airborne insects found in the atmosphere (agricultural sky) also support diverse types of microbes on their body surface. For example, honeybees carry pollen and microbes, while aerial vectors of disease-causing organisms, birds, and other ground fauna spread diverse species of microbes. These aspects are not being considered in detail in this book and may require exclusive treatises. Aerial vehicles like agricultural drones, blimps, and airplanes may also pick up microbes present in the vegetation on their surface and disperse them. This aspect needs detailed investigation.

Microbial assemblages are common on almost all crop species, as well as on plant species found in natural vegetation stretches (Beattie & Lindow, 1999). The bacteria present on the phyllosphere, without a doubt, differ in species diversity and total population. About two decades ago, Yang et al. (2000) stated that the phyllosphere microflora is more complex than we assume it to be. One of the reasons attributed to this complexity is the nutrient composition of the phyllosphere. For example, carbohydrates, organic acids, amino acids, and secondary metabolites may affect the microflora on the phyllosphere. Several other factors, such as plant species, cultivar, leaf age, leaf position, physiological status, foliar sprays conducted with different chemicals, and foliar irrigation, if any, may also affect the phyllosphere microflora (Andrews et al., 1980; Kinkel et al., 1996; O'Brien & Lindow, 1989; Wilson & Lindow, 1994). The tropospheric conditions experienced on the phyllosphere can be challenging for microbes to thrive. The troposphere can be considered 'extreme' in terms of relative humidity (RH), temperature, and nutrient availability. For example, leaf temperature at night could be 1–5°C, while in the mid-afternoon, it could reach 45–50°C. Yang et al. (2000) further state that tropospheric conditions and the availability of microbial inoculum of different species are crucial. Atmospheric microbes may alter the species diversity periodically, based on the bioaerosol and its composition. Some of these inferences are applicable to crops as well. For example, highly complex microbial assemblages and dynamic changes in microflora are to be expected on crop species like wheat, maize, rice, brassicas, etc.

In addition to the above inferences, Yang et al. (2000) and several others state that the techniques adopted may also be important. The culture-based methods may sometimes underestimate the number of microbial species detected (Hirano & Upper, 2000; Legard et al., 1994). The more recent 16S rRNA sequence-based identity allows us to rapidly check the microbial species. It also helps us in detecting the dynamic changes of microbial assemblages. The dynamics of microbial species in response to several different atmospheric factors could be deciphered accurately using 16S rRNA methods. Let us consider an example: Yang et al. (2000) detected over 17 microbial species on Citrus foliage using molecular methods. Few of the predominant microbial species found on the citrus phylloplane were *Aureobasidium pullulans*, *Lewiainfectoria*, *Morchella esculenta*, *Acinetobacter* sp., *Desulfurominaschoroethenica*, *Spingomonasadhaesiva*, *Clostridium bifermentans*, Uncultured delta proteobacterium, *Marinobacterhydrocarbonoclasticus*, Unidentified Cytophagales, Uncultured delta proteobacterium, Uncultured bacterium and *Enterobacter agglomerans*. We

PLATE 6.1 An example for anthosphere and caulosphere.

Note: Microbes on the surfaces of flowers and stems are responsible for several ecosystem functions. Arrows indicate the anthosphere and caulosphere.

Source: Roopashree B. N., & Sharath Kowligi, Eindhoven, Noord Brabant, Netherlands.

PLATE 6.2 An example of anthosphere of tulips.

Note: Arrows indicate petals and pollen that could harbor diverse microbes.

Source: Roopashree, B. N., & Sharath Kowligi, Eindhoven, Noord Brabant, Netherlands.

PLATE 6.3 The stem flow from the caulosphere introduces microbes and nutrients from plant surfaces into water bodies and soil.

Note: The above phenomena contribute to the soil microflora as water that drips carries surface microflora from foliage, stems, branches, twigs, and flowers. It also carries dissolved nutrients and organic-rich particulate matter. In fact, in certain geographic locations, N dynamics and microbial assemblages can be significantly influenced by stem flow. The agricultural sky serves as the source for both microbes and nutrients.

Source: Roopashree, B. N., & Sharath Kowligi, Eindhoven, Noord Brabant, Netherlands.

know very little, if at all, about the ecosystemic functions of these microbes either individually or collectively.

6.2.1 METHODS USED TO STUDY PHYLLOSPHERE MICROBES

Microbiologists have often employed multiple tests to determine the true identity of microbes found on various surfaces of crops. Culture-based methods, which rely on selective medium availability, are commonly used when there are indications of different phyla or genera of microbes. Preliminary microscopic observations are also highly valuable as they help narrow down our focus during the search for proper identification and classification of microbes. Cultural methods also assist in estimating the approximate population density of phyllosphere microbes per unit area. However, culture-based methods are time-consuming and not recommended when quick answers are required. In recent years, molecular markers such as 16S rRNA, 18S rRNA, and metagenomic methods have been widely utilized. These methods are routinely employed to assess microbial load, analyze periodic interactions, identify peak populations, and observe migratory trends, if any. Several authors have provided lists of microbial species found on the phyllosphere of different crop species (Alklett et al., 2014; Rastogi et al., 2013; Drahos, 1991). For instance, Bringel & Couee (2015) have compiled a list of plant host species and parts that can be investigated using molecular methods like 16S rRNA and 18S rRNA. Molecular markers have been used to assess the phyllosphere of various crops such as wheat, rice, brassicas, sugarcane, soybean, spinach, tomato, and cotton. Different plant organs, including apical leaves, mid leaves, stem, branches, flowers, buds, and sepals, have been examined to study microbial diversity and population using 16S rRNA procedures. Molecular procedures have also been utilized to assess environmental effects on phyllosphere microbes. For example, the effect of foliar sprays of nutrients, BNF organisms, herbicides, pesticides, etc., as well as the influence of temperature, light, CO_2 concentration, and radiation, has been assessed using 16S rRNA. This is done to identify changes in the microbial population. Overall, molecular approaches involve the use of high-speed, rapid throughput detection of microbes, making them extremely efficient. The sample size can be small, and even atmospheric samples with low population and gaseous matter can be analyzed using molecular methods. In cases where the sample size is too small or scarce, sequence amplification techniques are adopted (*see* Rastogi et al., 2013;

Knief, 2014; Knief et al., 2012; Muller & Ruppel, 2014; Muller et al., 2016). Scanning electron microscopic observations are useful when investigating the surface characteristics of plant organs and microbes in detail.

6.2.2 HOST PLANT GENETIC NATURE AND PHYLLOSPHERE MICROBES

Host plant genetics and its physiological manifestations have immense influence on the microbial assemblages possible in the phyllosphere. Even crop species grown as intercrops may show clear preferences for microbes on the leaf surface. The nature of the leaf surface, nutrients, exudates if any, and tropospheric microbial load determine, to a certain extent, the microbes found in the phyllosphere. Furthermore, the plant age and physiological status also affect the phyllosphere microbes.

Agricultural crops differ in their ability to support microbial assemblages, particularly in terms of total population and diversity (Inacio et al., 2002). The variation in microbial assemblages can even be observed among different genotypes of the same crop (Whipps et al., 2008). Batool et al. (2016) conducted a study to examine the phylloplane microbial assemblage and its impact on wheat crop growth. They investigated the differential effects of phylloplane microbes on three different wheat varieties throughout the season. Generally, the bacterial load on the phyllosphere increases as the growth season progresses. Clear differences were observed in the bacterial load among the wheat varieties. The wheat cultivar 'Sehar' supported the highest bacterial load of 1.1×10^9 CFU g^{-1} fresh leaf. This variety also exhibited higher quantities of auxins during the rapid growth phase of the crop. In general, bacteria with biological nitrogen-fixation ability and phosphate solubilization were observed. Some bacteria excreted HCN on the phyllosphere. Despite the complexities of phylloplane bacteria and their physiological manifestations, wheat varieties that produced more foliage showed a rapid growth rate and better productivity. They also harbored a higher population of phylloplane microbes, making them appear healthier. Productive wheat varieties had active phylloplane microbial assemblages. Prominent phylloplane microflora included *Bacillus, Microbacterium, Acinetobacter, Proteus,* and *Psychrobacter.* This study supports the idea that wheat varieties differ in their preferences for phylloplane microbes.

Rice genotypes investigated by Thapa et al. (2017) showed differences in the phyllosphere microbial population. The diversity of microbes also differed based on genotypes. These variations were attributed to leaf pigmentation, nutritional status of foliage, exudations, etc. Thapa et al. (2017) further reported that among the rice genotypes evaluated, they found that the nutritional status of the phyllosphere and the chemical nature of exudates affected the microbial diversity and total population. Hence, the efficacy of certain foliar inoculants of microbes may depend on the nutritional status of the phyllosphere. In fact, the population of phyllosphere microbes could be attributed to leaf-Fe, leaf-Mn, chlorophyll-b, and organic carbon traced on the lamina/internal tissue. Quantitative PCR analysis of the phyllosphere of rice showed that bacterial isolates were predominant. Cyanobacteria and archaea were also abundant, but seemed to show specificity to rice genotypes, perhaps dictated by the nutritional status of leaves. Cultural methods of enumeration, in general, showed that rice foliage supported 10^6 to 10^8 CFU per cm^{-2} in the phyllosphere. This is a relatively higher number attributable to the better nutritional status and exudations found on the phylloplane. It seems that about 60 bacterial isolates were identified as those producing auxins (Indole acetic acid) and siderophores.

Leaf characteristics of the host crop may affect the phyllosphere microbial assemblage. For example, *Arabidopsis* mutants with no wax or modified wax structure on the lamina were examined. The study showed that bacterial communities on waxy lamina differed from those detected on normal (non-waxy) leaves (Reisberg et al., 2013).

It is important to note that crop genotypes experience similar tropospheric conditions and are exposed to similar agronomic treatments and inputs. Similar microbial flora and load are prevalent in the agricultural sky. However, microbial assemblages are specifically preferred to suit a crop genotype. The genotype still influences the microbial assemblages in the phylloplane/phyllosphere. If we can decipher the microbial species preferred by the crop genotype, perhaps we can selectively apply them to crops. The intention is to produce healthy crops by taking advantage of genotype preferences.

6.2.3 PHYLLOSPHERE MICROBES OF FOREST TREES

Tree species, regardless of their geographic location and prevailing tropospheric conditions, are known to harbor diverse microbial assemblages from various phyla. Studies have shown that even a small area of leaf lamina can contain several hundred microbial species (Redford et al., 2010; Valverde et al., 2017). For instance, Valverde et al. (2017) reported the presence of different bacterial species on chestnut leaves, including a rich diversity of isolates from phyla such as *Firmicutes, Proteobacteria,* and *Actinobacteria.* These findings were obtained using non-culture based molecular methods. Many of the detected microbes were free-living biological nitrogen-fixers (Rico et al., 2014; Furkranz et al., 2008), while others exhibited antagonistic activity against leaf pathogens or had commensal relationships. *Dietzia* and *Lonsdalea* were frequently found on the leaf surface. It is evident that microbial assemblages on leaves have the potential to either enhance plant growth or contribute to the development of diseases. Understanding the microbial ecology of the phyllosphere is crucial for tree growth. Detailed information on leaf exudates is needed, as the secretions of phyllosphere microbes may be beneficial for tree crops. In conclusion, it is important to investigate microbial diversity, isolate and culture phyllosphere microbes, and assess their potential utility.

As mentioned earlier, chestnuts are susceptible to various microbes. However, our understanding of the microflora, their succession patterns, and their physiological effects is still limited. The microflora present on the surface of chestnut leaves, known as the phyllosphere microflora, may also produce unique metabolic products that can enhance the growth of seedlings and trees. It has been reported that phyllosphere microbes found on chestnut leaves are capable of producing phytohormones (Valverde et al., 2017). Additionally, beneficial microbes that can counteract plant pathogens have been identified in numerous tree and fruit plantations (Fokkema & Van Der Meulen, 1976; Fokkema, 1993).

The phyllosphere is an important component of vegetation, as it is involved in various plant physiological processes such as photosynthesis, foliar exudations, and gaseous exchange. Its significance is further emphasized by the fact that it provides a large surface area for diverse microbial communities. According to Lajoie et al. (2019), we have gained extensive knowledge about microbial interactions on the phyllosphere, particularly on the leaf lamina (known as the phylloplane). However, there is a lack of information regarding how the plant host influences the functions and interactions of microbial communities on the phyllosphere. It is crucial to understand whether there is a specific taxonomic affinity between crop species and the microbial assemblages that preferentially inhabit their leaf lamina. Metagenomic shotgun analysis has revealed that the functional diversity of microbial assemblages is influenced by the tree species present in the neotropical forests of Panama. In most cases, there is a selection of microbes based on their functional roles by the tree species. For instance, carbohydrate and amino acid metabolism, energy acquisition, and membrane transport were identified as predominant functions that determine the microbial selection to thrive on foliage. These preferences exhibited by tropical trees are not random, but rather specific. Furthermore, there is also a selection of microbial species based on the secondary metabolites produced. Sometimes, cell to cell adhesion process was key to harboring phyllosphere microbes. Elaboration of antibiotics could be a major focus while dealing with crops/genotypes susceptible to foliar diseases. Lajoie et al. (2019); and Bulgerelli (2013) further state that, tree species found in tropical Panama has microbial species like *Actinobacteria*, *Bacteroidetes*, *Firmicutes*,and *Proteobacteria.* There were unique tree and microbe species pairings based on functionality, not just taxonomic identities (*see* Lajoie, 2019; Kim et al., 2013). Such tree-microbe interactions may influence a few ecosystem services. Tree species-driven specific associations with phyllosphere microbes have been found on temperate trees (Laforest-Lapointe et al., 2016). Overall, Lajoie et al. (2019) state that adaptive mechanisms match the tropical tree species and the microbial assemblage on the phyllosphere. This may be crucial to tree growth. At the same time, we should realize that the atmosphere, i.e., agricultural sky, should serve as a valuable source of an assortment of microbial communities. This way, assemblages relevant to tree species could be generated.

However, Li et al. (2019) found that phyllosphere microbial assemblages of five different *Picea* species, given similar treatments, differed based on geographical location. The taxonomic composition of microbial community differed, while their functional aspects did not differ significantly. It seems that taxonomical composition did not have an impact on microbial processes recognized on the phyllosphere.

Barruso et al. (2021) have pointed out that the use of fertilizer-N as foliar sprays has increased in recent years. This is predominant in agricultural zones that support field crops such as cereals, legumes, and oilseeds, among others. However, a similar situation is observed even with natural vegetation, broad-leaved forests, and pine forest stands. Deposits of N via particulate matter (PM), smog, and dust have also increased. There is a greater tendency for foliage (phyllosphere) to encounter incessant deposition of N-containing PM. In their field trials using perennial tree species like Sessile Oak (*Quercus petraea*), high N deposits were received from the troposphere, which affected phyllosphere fungal assemblages. These effects were direct on microflora. Sometimes, N deposits modified the chemical nature of the phyllosphere, thus affecting the microbial assemblages on foliage.

6.3 ATMOSPHERIC FACTORS AND PHYLLOSPHERE MICROFLORA

The atmosphere, particularly the lower troposphere, has an immense influence on phyllosphere microbes. Vorholt (2012) clearly states that the structure of the microbial community on leaf surfaces is not randomly selected. The phyllosphere environment is characterized by a thick cuticle, a hirsute nature, and exudations from internal parts of the leaf. Additionally, abiotic factors such as relative humidity, temperature, radiation, and gaseous exchange phenomena affect the microbial phyla, classes, genera, and species that thrive on the leaves.

Finkel et al. (2012) have discovered that the phyllosphere bacterial community is influenced by geographical location. For instance, microbes from arid soils are frequently found on the leaves of plants grown in the Sonoran Desert. Soil effects may dominate regardless of the crop or genotype. Microbes on the soil surface in the vicinity of the crop's canopy are often traced on the phyllosphere.

The phyllosphere supports heterotrophic microbes that utilize carbonaceous compounds found on the leaf surface for energy. Additionally, the phyllosphere may harbor various species of photoautotrophic microbes such as green algae, blue green algae (BGA), and photoautotrophic bacteria. The light that reaches the phyllosphere can have different effects on the microbes present on the phylloplane. The impact of radiation from the sky on phyllosphere microbes may depend on its intensity and wavelength. Some bacteria and fungi produce different compounds in response to light. Light also influences several physiological processes of phyllosphere microbial communities, in addition to photosynthesis.

UV radiation can be harmful to microbes residing on the phyllosphere. However, certain microbes with specific adaptations like thick cell walls, ornamentations, pigmentations, and accumulations can tolerate UV radiation. These microbes are able to function normally in the phyllosphere despite the presence of UV radiation.

The rhizosphere is a region surrounding roots that exist in soil. Light, as a factor, is not perceived at all in the rhizosphere. An exception is perhaps aerial roots found on certain tree species. In contrast, phyllosphere microbes and the plant's foliage are constantly exposed to light during the day. Light, as an abiotic factor, operating from the agrarian sky, affects microbes and crops simultaneously. It also affects microbial interactions on the phylloplane (Carvalho & Castillo, 2018). There are phyllosphere microbes that utilize light as an energy source. They possess bacterial rhodopsin. Reports show that microbial rhodopsin collects light at a slightly different wavelength, so it does not interfere with the light spectrum absorbed by their host plant. Plants regularly utilize light for photosynthesis (carbon fixation). The plant leaf's photoreceptors operate at a narrow band of wavelength of radiation while photosynthesizing. In fact, light affects a series of metabolic processes related to photosynthesis.

Plant-microbe-light interactions and outcomes are complex (Carvalho & Folta, 2014; Alsanius, 2017). Several aspects of plant phyllosphere microbe interactions may be influenced by light. The exudations caused by plant metabolic processes in the phyllosphere can impact microbial behavior on the phylloplane. It is evident that bacteria, fungi, actinomycetes, and archaea may all experience harmful effects when exposed to UV radiation. Microbes that can efficiently repair DNA and rejuvenate DNA functions are able to survive, while others may perish. Phyllosphere that is exposed to ample sunlight may also receive UV radiation, potentially causing some phyllosphere microflora to suffer due to intense sunlight.

The direct effect of light on phyllosphere microbes requires careful consideration. Light directly impacts fungi, influencing processes such as hyphal elongation, conidiation, the formation of appressoria on leaf tissue, and invasion. This is particularly important when dealing with phyllosphere microbes that are fungal pathogens, such as leaf spot or rust fungi. Extensive research has been conducted on light-induced responses in fungal species like *Aspergillus, Candida, Neurospora, Fusarium, Phycomyces,* and others (Carvalho & Castillo, 2018). Light radiation plays a role in processes like conidia formation (Brandhoff et al., 2017), sclerotia formation, and fruiting body development. Light radiation also initiates certain metabolic processes, especially secondary metabolic pathways (Schumacher et al., 2014; Cohrs et al., 2016; Wang et al., 2010). In some cases, excessive light exposure can suppress the growth and sporulation of fungi. For example, light suppresses powdery mildew fungus on the phylloplane, inhibiting spore germination and reducing hyphal production (Suthaparan et al., 2012, 2014). Light also affects the circadian rhythms of fungi, such as *Neurospora crassa*. The duration of light exposure, as well as alternating day/night cycles and their lengths, have an impact on fungal physiology and its survival on the phyllosphere (Hivea et al., 2015).

The phyllosphere of greenhouse crops is exposed to both abiotic and biotic factors from the atmosphere above their canopies. One of these factors is artificial light, which has undergone changes to LED lamps. Therefore, Gharaie (2017) conducted a study on the effects of light (LED lamps) on the phyllosphere of ornamental crops grown in a greenhouse. They also examined the phyllosphere for biocontrol reactions of microbial assemblages against the pathogen *Botrytis cinerea.* Both culture-dependent methods and molecular techniques, such as 16S rRNA analysis, were used to identify the microbes on the phyllosphere. The upper leaves of the crops had a higher number of fungal species. The light treatments had a significant effect on the colonization of the phyllosphere by *Pseudomonads,* while the effect on actinomycete like *Streptomyces greasioviridis* was moderate.

Temperature is another abiotic factor that affects the microbial assemblages in the phyllosphere. The temperature of the leaf surface can have a definite effect on microbial diversity and population. The atmospheric temperature also has specific effects on growth rate, generation time, and metabolic products excreted on the leaf surface. In general, microbial interactions among the phyllosphere microbes can be regulated by the ambient temperature. Agrarian regions globally show variations in ambient temperature, which are based on geographic location, topography, season, diurnal patterns, clouds, and several other factors. All of these factors can directly affect the phyllosphere conditions and microbial load. The agroclimate has immediate effects on leaf initiation, phyllosphere conditions, and its nutrient status, such as exudates, temperature, moisture, etc. These same factors also affect the phyllosphere microflora. The phyllosphere microbial flora of crops grown in temperate or tropical regions may show natural adaptations to temperature regimes. Phyllosphere microbes in arid, hot regions with desiccating conditions should possess resistance to such extremes.

Phyllosphere microbes require proper anchorage and adherence to the phylloplane, which is the surface of the leaf. Microbial assemblages should be able to withstand breeze, strong wind, turbulence, and precipitation. Sometimes, even convection currents can affect microbes that reside on the crop's canopy. Once dislodged, microbes may find a different niche and be influenced by environmental factors. Say, on the soil surface or water bodies (*see* Plates 6.1 and 6.3). Adhesion properties of cell wall, slime secretion by microbes, wax by the leaf cuticle may allow microbes to attach properly to the leaf surface. Cell wall proper of phyllosphere microbial could affect proper anchorage on leaf surface.

6.4 FOLIAR SPRAY OF PLANT PROTECTION CHEMICALS AND ATMOSPHERIC POLLUTANTS INDUCE CHANGES IN PHYLLOSPHERE MICROFLORA

Foliar sprays of plant protection chemicals can impact the phyllosphere microbiome. Generally, these chemicals may decrease the overall microbial load on the foliage, including the population of pathogenic microbes. However, they can also affect non-target microbes and other organisms. Throughout a growing season, crops treated with plant protection chemicals may exhibit various unusual changes in phyllosphere microbial diversity, including a reduction in microbial load. Farmers often employ integrated control methods to minimize diseases and pests. Therefore, the combination of toxic chemicals, biological control agents (BCAs), and other agronomic procedures can collectively influence the composition of phyllosphere microbial communities.

For example, Cernava et al. (2019) have studied the phyllosphere microbiome of tea (*Camellia sinensis*). Firstly, they treated the plants with a chemical (sulfur pesticide formulation) and BCA (*Piriformospora indica*). This was done to understand the effect of foliar sprays of plant protection chemical/biological agent mixtures on phyllosphere microbial assemblages. They used 16S rRNA-based molecular methods to detect species diversity and quantify the microbial load on leaves. They were able to detect about 11 taxa of microbes belonging to *Proteobacteria* and *Actinobacteria* on plants sprayed with pesticide sulfur. On plants sprayed with BCA, they reported *Actinobacteria, Proteobacteria, Planctomycetes,* and *Verrucomicrobia.* They inferred that tea foliage supports a unique microflora, but it can be significantly affected when aerial sprays of chemicals or BCAs are applied. The impact of microbial load and diversity on tea plant growth needs to be further investigated. Similar studies have been reported previously on wheat crop's phyllosphere organisms using a combination of biocontrol microbe and sulfur-based pesticide (De Curtis et al., 2012). Fungicide sprays have also been found to affect several species of yeasts thriving on the phyllosphere (Buck & Burpee, 2092).

Farmers frequently adopt systemic fungicides. Systemic fungicides can be applied to foliage or injected into the soil. These chemicals are highly effective against bacterial and fungal plant pathogens. They are preferred over dusting on foliage. There is a lot to understand about the use of systemic chemicals and their impact on phyllosphere microbiota, particularly in terms of diversity and load. It is possible that by using systemic chemicals on the phyllosphere, we may actually suppress or completely remove several useful microbial species. For instance, Metalyxyl or Imidacloprid are potent systemic fungicides. Moulas et al. (2013) found a mild impact of these two systemic fungicides on the bacterial and fungal flora on the leaves. Imidacloprid, however, showed a stronger effect on the leaf-inhabiting fungal community. Molecular methods revealed that possible pathogens like ascomycetes

Erysiphae and *Cladosporium,* as well as a few basidiomycetes, survived on the leaf lamina. Interestingly, several species of yeast were stimulated in the phyllosphere.

Similarly, the application of Metalyxyl onto sweet pepper stimulated certain members of Enterobacteriaceae. There are examples involving pesticide application on other crops, such as vegetables. Zhang et al. (2009a, b) found that the application of cypermethrin on pepper and abamectin on broccoli did affect pathogens. However, they also had a detrimental effect on non-target microbial species residing on the phyllosphere. Systemic plant protection chemicals may affect epiphytic microbes found not only on leaves, but also on stems, branches, flowers, pollen, and even seeds, particularly if they persist for a longer period in the plant. Sometimes, it could actually be an interaction between the tropospheric microbial load/density and the persistence of the systemic effect of fungicide or pesticide. Periodic deposits of microbes on the phylloplane require attention. In a study by Wachowska (2009), the fungicide strobilurin strongly suppressed the epiphytotic growth of *Sporobolomyces roseus, Candida tropicalis,* and *Pichia anomala.* The residual effects of the fungicide continued.

Atmospheric pollution is increasing in high-input intensive farming zones. Additionally, it affects large expanses of crops. Primarily, aerially sprayed plant protection chemicals may drift and deposit on natural vegetation and/or non-target crops. The residual chemicals, which are buoyant and persist in the air, may eventually settle on the soil or crop canopy. This has an impact on phyllosphere microbiota, as well as detrimental effects on bio-aerosol microbes. Agronomic procedures such as crop residue burning release particulate matter laden with microbes. The combination of industrial smog, moisture, and particulate matter from crop residue burning is toxic to the crop's phyllosphere, particularly once the smog settles. Industrial emissions, including sulfur compounds, carbonaceous particles, and liquid toxicants, can have a highly suppressive effect on atmospheric microbes and phyllosphere assemblages.

Brengel & Couee (2015) state that epiphytic microbes on leaves could be of great utility in degrading/detoxifying toxic pollutants and pesticides. Atmospheric depositions also contain toxic compounds that reach the phyllosphere of crops. Such remediation using phyllosphere microbes can aim at reducing biochemicals unfavorable to crops. For example, phyllosphere microbes may reduce polycyclic aromatic compounds such as naphthalene, phenolics, nicotine, chloromethane, and isoprene (Bringel & Couee, 2015; Yutthammo et al., 2010; Sandhu, 2008). Several species of *Arthrobacter* found on the phyllosphere are able to degrade and reduce residual effects of fungicides like malathion, as well as herbicides such as glyphosate (Scheublin et al., 2013). Perhaps, the phyllosphere microbial community could be modified to our advantage. We should explore microbe-aided methods to degrade the toxic compounds found on the phyllosphere.

Let us consider a case study. Phenol and phenol-derived compounds are commonly observed pollutants in agricultural areas. Firstly, phenol can have detrimental effects on the biotic fraction of the troposphere in general. Phenol deposition on the phyllosphere can impact the microbial community present on the leaves. Phenol may suppress microbial activity. Organic solvents such as phenol, benzene, and toluene can affect bacterial cells by partitioning into the cell membrane, leading to the disruption of membrane integrity and function (Benndorf et al., 2001; TamLe et al., 2006; Saborit, 2009). Phenolic deposits can disrupt the structure and total population of the microbial community. However, there are also bacterial species found on the phyllosphere that have the ability to degrade phenol and its derivatives. Sandhu (2008) has isolated a few phyllosphere bacterial species that possess the ability to degrade atmospheric phenol. These microbes produce enzymes known as 'phenol hydroxylase.' *Pseudomonas fluorescens* CF 600 could detect phenol on the leaf surface as well as that present in the atmosphere (Shingler et al., 1989). Further, trails with C^{14} labeled phenol showed that phenol degrading strains of *P. flourascens* released 8 times more of CO_2 when compared with non-phenol degrading mutants sprayed on the leaf surface. This trait found in bacteria has a direct application in regulating the concentration of pollutants on the phyllosphere (Futamata et al., 2001). Any of the enzymes secreted by phyllosphere microbes that reduce pesticides, herbicides, or other toxic pollutants commonly encountered in the troposphere have applications in farming. Sandhu (2008) has further stated that the microbial community on the leaf surface generally harbors phenol-degrading species. About 10 isolates capable of degrading phenols were evaluated for efficacy. The phenol-degrading isolates were derived from genera like *Alcaligenes, Acinetobacteria,* and *Rhodococcus.* Incidentally, phenol hydroxylase was present in all of the gram-negative bacterial populations of the phyllosphere microbial community.

Tropospheric pollution has been experienced by farmers and gardeners for several decades. A study conducted in the mid-1980s by Khanna (1986) aimed to understand fluctuations in the population of fungi, actinomycetes, and bacteria. The objective was to observe the response of microbes to a polluted agrarian sky. The fungal population appears to increase in the presence of air pollution, while the bacterial and actinomycete populations decrease in the phyllosphere due to polluted air. However, there are a few microbial species such as *Curvularialunata, Acremonium* sp., and *Pestolata* sp. that can survive and thrive on the phyllosphere/phylloplane in a pollution-free agrarian sky.

Furthermore, there are reports supporting the notion that phyllosphere microbes are well adapted to agricultural conditions. Many of the species in the phyllosphere community may possess tolerance to agronomic practices such as pesticide or herbicide applications. It is possible that phyllosphere microbial communities have co-evolved with crops and cropping systems. Through natural selection, more isolates tolerant to pesticides/herbicides were able to thrive on the leaf surface. For instance, in the case of grapevines, Perazzolli et al. (2014) demonstrated that phyllosphere microbial communities exhibit resilience and adaptation, allowing them to withstand a certain level of pesticide/herbicide sprays. They may also tolerate the presence of biocontrol microbes used in integrated foliar disease management strategies alongside pesticides. In fact, in this particular case, the application of *Lysobacter* sp. as a biocontrol microbe in combination with Cu compounds did not impact the normal phyllosphere flora, as they were able to tolerate the presence of both the chemical and the biocontrol microbes.

Pesticides applied using foliar sprays can reach the phyllosphere. Additionally, pollutants from previously polluted or contaminated agricultural environments can also reach the phyllosphere as natural deposits. These pollutants may persist for a long time or degrade over time. The persistence and residual activity of pesticides on non-target microbes in the phyllosphere is a matter of concern. The phyllosphere microbiome often contains microbes that are susceptible to pesticides, biocontrol microbes/agents, and other toxic deposits received from the agricultural environment. However, there are crops that support both susceptible and tolerant microbes. Pesticides that are routinely applied may degrade on the phylloplane itself after a few days. For example, Kucharska et al. (2020) found that wheat crops grown in the Polish plains often support a variety of yeast species on their phyllosphere. Only certain combinations of pesticides suppressed the phyllosphere microbes, not all. A commercial mixture of Propiconazole, flusilazole, and carbendazim has been found to decrease the population of epiphytotic yeasts. However, the yeast population and diversity were restored within 10 days. There are other pesticide combinations that have been observed to stimulate the phyllosphere yeast population. Epoxiconazole, a commercial mixture of pesticides, has been found to stimulate yeasts on the phylloplane. Populations of *Aureobasidiumpollulans* and *Rhodotoruralaglutins* have actually been found to degrade pesticide formulations sprayed on leaves. Therefore, it is important to select the right fungicide and their mixtures if we intend to preserve yeasts while simultaneously reducing other pathogenic microbes that affect wheat leaves. Yeasts may also help in reducing the toxic effects of pesticides on other fungi found on the phyllosphere.

Practical farming, no doubt, is prone to applying large quantities of pesticides/herbicides. Therefore, we may need to apply the principles used with herbicide-tolerant crop genotypes to phyllosphere microbes as well. Genes for pesticide tolerance should be selected or cloned preferentially, especially if the microbes are useful as BCAs or as plant growth promoting microbes. The need of the hour is a herbicide or pesticide tolerant microbial assemblage.

Now let us consider some interesting phenomena related to crop pests versus phyllosphere microbes. Phyllosphere microbes engage in complex interactions with other biotic components. The outcome can have no impact on the crop, it can be advantageous, or even detrimental. There are phyllosphere microbes (e.g., *Pseudomonas* sp.) that can alter the host preference of chewing insects. Aerial sprays of pesticides may affect phyllosphere microbes and the plant's defense against insects (Humphrey et al., 2014; Karlsson et al., 2014; Ottesen et al., 2009; Glenn et al., 2015).

Bringle & Couee (2015) explain an interesting phenomenon. They state that certain Colorado beetles release microbes of the genera *Stenotrophomonas, Pseudomonas,* and *Enterobacter,* which predispose plants to susceptibility. These microbes actually modify the plant's defense responses. It appears that there are numerous instances where insects benefit from phyllosphere microbes. Phyllosphere microbes alter the plant's resistance and enable pests to establish themselves. These epiphytes impact the release of the plant's defense and volatile signal chemicals, making it easier for insect pests (Mason et al., 2014). In such cases, the outcome is determined by the complex interaction of phyllosphere chemicals, microbes, pests, and the atmosphere.

6.5 PHYLLOSPHERE MICROBES AND METHANE (CH_4) UTILIZATION

Microbes on the phyllosphere of several crops possess the ability to metabolize C1 compounds such as CH_4, methanol, formaldehyde, chloromethane, and more. This is a valuable trait, especially considering that CH_4 is one of the major greenhouse gases emitted. CH_4 emitted at the phyllosphere could be utilized by methylotrophs. According to Shivkumar et al. (2020), the phyllosphere supports a high population density of methylotrophic bacteria. The phyllospheres of crops are generally rich in Methylotrophic bacteria. Some of the dominant bacterial species identified on crops' phyllosphere include genera such as *Methylobacterium, Methylophilus, Methylibium, Hyphomicrobium, Methylocella, Methylocapsa,* and *Methylocystis.*

Methanotrophs and methylotrophs have been found on the phyllospheres of tree foliage such as pine, lilac, maple, oak, spruce, and apple (Doronina et al., 2004). *Methylobacterium extorquens* has been reported on the phyllosphere of Medicago (*Medicago truncatula*) leaves (Sy et al., 2005; Gourian et al., 2006), on strawberry (Belova et al., 2011), and on *Arabidopsis thaliana* (Verginer et al., 2010). Pigmented methylotrophic bacteria have been found on the leaves of various commercial crops such as sugarcane, pigeon pea, brassica oilseeds, potato, and radish (Meena et al., 2012). *Methylobacterium oryzae* on rice (Kwak et al., 2014) and *Methylobacterium radiotolerance* on groundnuts (*Arachis hypogea*) (Krishna Moorthy et al., 2018) are other examples. Several species of yeasts with methylotrophic ability have also been identified on crops (Kawaguchi et al., 2014). Further research is needed on methanotrophs due to their potential to reduce CH_4 concentration above the crop canopy, including the lower troposphere. It is important to focus on the proliferation of methanotrophs/methylotrophs on the phyllosphere and their dispersion into the troposphere. Factors such as wind, breeze, turbulence, and farm machinery movement in the fields may facilitate the migration of methylobacteria from one plant to another.

6.6 PHYLLOSPHERE MICROBIAL SUCCESSION

Microbial succession and its consequences on a microbiome, such as the phyllosphere, are worth examining. Perhaps, it is essential. This knowledge is applicable in several instances, such as developing suitable foliar sprays of nutrients, pesticides, fungicides, etc. Additionally, Marois & Coleman (1995) believe that knowledge about microbial succession is useful when designing biocontrol procedures and selecting biocontrol microbial species.

Microbial colonization of a crop's canopy, perhaps, follows similar principles and patterns that we already know for other biomes, like soil (Marois & Coleman, 1995). Although not entirely, microbial succession on the phyllosphere, in fact, follows patterns similar to others where nutrient inputs occur periodically. For example, inputs of organic particles, lignin compounds, mineral dust, fertilizer-based nutrients, etc. Nutrient utilization strategies, tolerance to uncongenial temperatures, low relative humidity (RH), desiccation, inundation with water, resistance to suppressive agents, and countering antagonists, etc., may actually determine the microbial succession that ensues. Microbial succession that induces the proliferation of biocontrol microbes is useful.

Microbial succession on the phylloplane is influenced by several abiotic and biotic factors. Firstly, we need to monitor the microbial species that colonize the leaf surface. The source of microbial assemblages is crucial. Often, in a cropped field, the soil is a major source of the phyllosphere microbial population. The microbiota keeps changing in response to ambient conditions. Phyllosphere nutrients, moisture, pH, and temperature change based on seasonality. Several other factors may also have a direct impact on microbial succession in the phyllosphere. For instance, Copeland et al. (2015) studied soybean and canola leaf surface microflora for 98 days. They investigated the microflora – its diversity and total population every week. The season did affect the microbial succession. However, the initial population depended on the soil surface microflora adjoining the crop's canopy.

There are reports dealing with microbial diversity and its dynamic changes on the leaf surface. To a certain extent, we know the causes for such dynamic changes in microbial assemblages on foliage. Factors that affect the microbial dynamics are, of course, many. Most of them are related to tropospheric conditions and the plant host itself. Kinkel (1997) points out that we need to accurately decipher aspects like the immigration, emigration, multiplication, senescence, and death of microbial components (species). We need methods that accurately quantify the microbial emigration/immigration on the phyllosphere. A good knowledge about the functional peculiarities of microbes that fluctuate in terms of population density is necessary. The agrarian sky, no doubt, has an influence on microbial assemblages on the phyllosphere.

Abiotic factors such as strong wind, breeze, turbulence, precipitation, and physical disturbance to crops due to agronomic practices (e.g., interculture, pruning, application of defoliators), including the movement of farm vehicles, could dislodge microbes on the phyllosphere. This can lead to impairment in microbial succession on leaves. Overall, the agrarian environment may impose stringent control on microbial succession within the crop's canopy.

6.7 PHYLLOSPHERE MICROBES AND BIOLOGICAL NITROGEN FIXATION (BNF)

The phyllosphere is a region that constantly interacts with tropospheric conditions. Phyllosphere microbes can utilize the various exudates available on leaf lamina. There may be many compounds in the phyllosphere that are specifically beneficial to biological nitrogen-fixing organisms. The total area of phyllosphere in global farming regions is quite large, which suggests that the population of phyllosphere microbes is expected to be substantial. These microbes play important roles in the ecosystem functions of crop canopies. It is well-known that nitrogen-fixing microbes occur on the leaf lamina and phyllosphere of crop plants, particularly in cereal grain crops, legumes, and other species. The presence of nitrogen-fixing microbes in the phyllosphere is highly relevant to global agriculture. However, we have not yet fully utilized the advantages of phyllosphere nitrogen-fixing microbes. We have not extensively manipulated these microbes on the phyllosphere, unlike what we have done with microbes that fix atmospheric nitrogen in the soil (e.g., *Bradyrhizobia, Azotobacter, Azospirillum,* BGA, etc.). It is possible that multi-dimensional interactions among phyllosphere nitrogen-fixing microbes, commensal microbes, pathogens, and nutrient transformers are vital. The goal should be to sustain a high microbial load and a higher ratio of nitrogen-fixing microbes on the phylloplane/phyllosphere.

Foliar sprays of microbes are not uncommon. The adoption of sprinklers, foliar fertilizer supply, aerial pesticide sprayers, and herbicide sprays may have massive effects on resident BNF microbes on phylloplanes of crops. Field crops with a short duration of 110 to 130 days may need to be quickly colonized by diverse microbes, including BNF organisms. Microbes are mainly received from natural deposition, but if not, they are artificially sprayed. Regardless of the source, microbes should establish, colonize, and function within a short period. This is necessary for the perceptible advantage from foliar BNF organisms. Numerous physicochemical and biogeochemical factors affect phyllosphere microbes, so we should consider several interacting factors when evaluating the phyllosphere microbes capable of biological N-fixation.

The foliar surface of several agricultural crops harbors N-fixing microbes and other species in a sizable number. For example, rice has been found to contain such microbes (Barraquio et al., 1997; Elbeltagy et al., 2001; Knauth et al., 2005). It has been suggested that the combination of high humidity, tropical temperatures, and shorter generation time supports the rapid multiplication of phyllosphere microbes. As a result, there is a higher density of microbes, including N-fixing microbes.

Free-living N-fixing bacteria, known as BNF bacteria, were cultured and evaluated for their ability to fix nitrogen on wheat leaves. These bacteria were obtained from the phyllosphere of crops such as rice, maize, wheat, and jute. The objective was to partially replace the use of external nitrogen fertilizers. It is difficult to quantify the exact amount of nitrogen derived from these BNF bacteria. However, they were found to promote the growth of crops. Factors such as establishment, niche identification, nutrient uptake from leaf exudates, and the surrounding environment on the wheat phyllosphere should be considered. Previous studies by Bessems (1973) and Bhurat & Sen (1968) have highlighted the importance of the host crop species and BNF microbes, especially when inoculating the phyllosphere of cereals with BNF organisms. This becomes even more crucial if the goal is to maximize nitrogen benefits (Ruinen, 1956, 1965; Pati & Chandra, 1981; Bessems, 1973).

Perennial forest stands found in temperate regions also support organisms in their phyllosphere that are capable of biological nitrogen fixation (BNF). For example, several decades ago, Jones (1970) discovered BNF organisms on the phyllosphere of Douglas Fir plantations. These phyllosphere BNF microbes could contribute to a portion of the annual nitrogen requirement. Forest tree species provide an excellent niche for various microbes to thrive on their foliage, including diazotrophic bacteria and blue-green algae (BGA). Microbiologists have examined this phenomenon using various methods, including the use of 16S rRNA gene sequences to identify the total biodiversity of microbes. Molecular analysis has also been employed to identify the occurrence of the *nif H* gene, which is associated with BNF microbes. Additionally, ^{15}N-based methods have been used to assess nitrogen accumulation resulting from BNF on the leaf surface. For instance, Furkranz et al. (2008) studied the microflora on the foliage of forest species such as *Carludovica drudei, Grias cauliflora,* and *Costus laevis*. They found cyanobacteria (BGA) at a lower population density. Among the leaf colonizers on these tropical tree species, many belonged to the α-Proteobacteria group. Some tree species exhibited a dominance of Azotobacteria on their phyllosphere. Vitousek et al. (2002) believe that phyllosphere nitrogen fixation by bacteria and BGA is crucial for sustaining several ecologically essential functions that occur in the phyllosphere. It may also be relevant to the maintenance of nitrogen pools in the forest region itself. Others, such as Pons et al. (2007), suggest that phyllosphere nitrogen fixation contributes to the maintenance of leaf nitrogen, biomass nitrogen, and even soil nitrogen in forest ecosystems. The symbiotic bacteria on the leaf surface actually supply nitrogen to bacteria, fungi, and archaea that are not capable of nitrogen fixation (Bentley, 1987; Lindow & Brandl, 2003; Lambais et al., 2006).

Forest trees encountered in higher latitudes often host epiphytes like mosses. These mosses grow on branches and hang freely in the sky. Interestingly, these mosses also harbor free-living N-fixers. Boreal shrubs and trees provide support for these mosses. The association between tree species and mosses seems to be influenced by the evolutionary history and spread of the host tree. Holland-Mortiz et al. (2021) discovered diverse bacterial species on the mosses, with a certain degree of specificity imposed by the tree species and mosses on the microflora. Several trees support N-fixing microbes, as proven by ^{15}N analysis. N-fixation trends are affected by light availability, temperature, and humidity in the phyllosphere and moss tufts. In addition to bacteria, cyanobacteria (blue-green algae) were also detected on the mosses. In tundra and boreal ecosystems, severe coldness, oxygen tensions, humidity, and moisture affect N-fixation, potentially on the leaf lamina, by bacterial and algal species. Adaptation to freezing conditions is crucial. The rates of atmospheric N fixed by phyllosphere microbes may vary based on both forest tree species/genotypes and the environment.

The diverse phyla of microbes found on the phylloplane/phyllosphere of tree crops, such as fruit trees, nuts, and timber, may possess the ability to fix atmospheric nitrogen. We need to maximize the potential of biological nitrogen fixation (BNF) in natural vegetation tracts and tree plantations. It is crucial to selectively introduce BNF microbes to the tree foliage. To achieve this, the troposphere should host a significant population of BNF microbes. Particulate deposits containing BNF microbes that reach the tree foliage play a vital role. They greatly contribute to the conversion of atmospheric nitrogen. Regular tree plantations, such as chestnuts, hazelnuts, citrus, apples, and mangoes, can all contribute to soil nitrogen if we enhance the population of BNF microbes on their leaves. Leaf exudates serve as carbon and energy sources, while the atmosphere provides nitrogen gas (N2). Therefore, we must improve the utilization of this fraction by utilizing BNF microbes.

6.8 PHYLLOSPHERE MICROBES AND CROP DISEASES

Microbial pathogens that invade host crops via foliage encounter the first line of defense at the phyllosphere. They must overcome phyllosphere microbial barriers and any chemical environment present. Studies conducted by the Biology group at the University of California, Berkeley suggest that we can construct microbiomes on the phyllosphere by spraying a specific mixture of a few microbial species. In their study, they tested a natural microbiome as well as one artificially developed, using 12 beneficial microbial species. They found that when microbial sprays were applied at a lower population count, it provided good protection against *Pseudomonas,* a pathogen. However, the application of fertilizers somehow disrupted the microbial assemblage, altering the nutrient balances in the phylloplane and resulting in a susceptible phyllosphere. For instance, tomato crops sprayed with fertilizers were more susceptible to leaf diseases compared to those not given fluid fertilizers via the phylloplane (Sanders, 2018). Furthermore, the researchers state that fertilizers induce changes in the stem microflora of tomatoes. Therefore, replenishing the natural microbiome that typically provides protection against foliar diseases is a recommended approach, which should be carefully planned and implemented. We are aware of phyllosphere microbiomes that physically avoid the establishment of pathogens, as well as those that release specific toxins and bacteriostatic agents (growth suppressants). Additionally, there are microbiomes that exude antibiotics to destroy pathogens. Options that alter microbial assemblages and result in enhanced resistance against foliar pathogens should be actively considered.

Crop diseases, particularly those infecting or thriving on the leaf surface, could alter the microflora that normally colonizes the phyllosphere of the host. For example, Suda et al. (2009) examined powdery mildew-infected and uninfected control leaves from Japanese cucumber plants. They evaluated the microbial diversity and population using culture media-based and 16S rRNA analysis methods. It appears that leaves infected with powdery mildew were predisposed to accommodate a higher density of microbial population. The functional diversity of microbes was also greater in fungal disease-affected leaves.

There are several crop diseases that have a direct impact on the phyllosphere microbiota. Fungal pathogens coexist with non-target commensals in the phyllosphere zone. Therefore, when fungicide is applied, it affects the disease-causing target pathogen. Additionally, fungicide may also affect other fungal species present on the leaf surface. Sometimes, leaves may host multiple fungal pathogens. Fungicide sprayed may therefore have different effects on each foliar pathogen.

Let's consider an example. Wheat is affected by a few Puccinia rust species, Septoria blotch, *Blumeria*-caused powdery mildew, and *Pyrenophora* leaf spots, among others. Yellow rust (*Puccinia striiformis*) is one of the serious fungal diseases. Knorr et al. (2019) evaluated at least three different fungicide formulations on *Puccinia striiformis* and other fungi residing on the phyllosphere. Broad-spectrum fungicides affect several species of fungi, both targeted and non-targeted. Knorr et al. (2019) also found that fungicide-treated crops showed a different microbial community compared to those without fungicide application. The targeted yellow rust fungus *P. striiformis* showed a significant decrease in population, while a few other fungi like *P. tritici-repentis* did not show a reduction in phyllosphere population, despite fungicidal treatment. Fungal population in the phyllosphere was estimated using visual scores, culturing methods/microscopic counts, as well as 16S rRNA-based molecular methods. Overall, the population and diversity of microbial assemblages on the leaf surface seem important. However, for fungicides to be effective on targeted fungal pathogens, timing, potency of the active ingredient, and dosage that reaches the phyllosphere are important (Knorr et al., 2019).

Similar findings have been reported regarding fungal flora and pathogens found on the phyllosphere of grape (Perazzolli et al., 2014; Pinto et al., 2014) and mango (de Jager et al., 2001). It is worth noting that fungicide sprays targeted at rust fungi, on average, offer an advantage of 0.5 to 1.5 t ha^{-1} grain ha^{-1} (Jorgensen, 2014).

6.8.1 PHYLLOSPHERE MICROFLORA AS BIOLOGICAL CONTROL AGENTS

Phyllosphere microbes must first adapt to environmental conditions, including nutrient availability in the phyllosphere. If they are biological control agents (BCAs), they must employ efficient mechanisms to counter pathogens. Biocontrol can be achieved through direct suppression of the pathogen or mediated through the host plant. Understanding microbial biocontrol strategies is crucial for effective pathogen control. In their overview, Legein et al. (2020) list mechanisms such as the production of antibiotic chemicals, hydrolytic enzymes, competition for nutrients, siderophores, modulation of hormones, and direct physical exclusion. These traits, including the production of antibiotics and hydrolytic enzymes, can be used as criteria for selecting suitable biocontrol microbial species (Mota et al., 2017). However, it is important to note that congenial atmospheric conditions are a fundamental requirement for any of these mechanisms to be effective.

There is a consistent effort to search for biocontrol microbes, particularly, from among phyllosphere microbial assemblages. The intention is to regulate or totally control leaf diseases or those pathogens that enter plant-host via phyllosphere. Also, pathogens that may multiply and perpetuate on the leaf lamina. There are indeed innumerable reports on this topic in the literature. A few often-coated biocontrol microbes are *Bacillus species*, *Pseudomonas species*, *Pantoea*, and *Trichoderma* (You et al., 2015; Trablesi& Mhamdi, 2013; Cha et al., 2016; Longa et al., 2008; Blacutt et al., 2016; Wilson & Lindow, 1993). Some of these have been developed into commercial formulations. They are suitable to spray on phyllosphere. Disease suppression on a host using BCAs is still not as simple. This is because the nature of the crop, disease incidence, and interaction with atmospheric factors affect the outcome. Let's consider an example. This is about Tobacco wildfire disease caused by *Pseudomonas syringae cv. tabaci*. Qin et al. (2019) report that tobacco wildfire disease is initiated when the pathogen (bacteria) deposits on leaves from the atmosphere. The severity of the disease actually depends on the rate of proliferation and total population of pathogenic bacteria on the phyllosphere. Furthermore, they state that the bacteria they utilized from the foliage assemblages offered excellent control of the tobacco wildfire disease. Here, it is the phyllosphere microbial assemblages that influence plant health. Qin et al. (2019) sprayed microbial BCAs every week during leaf development. This is because the microbial community on the leaf could be dynamic. Its complexity was investigated using molecular methods based on 16S rRNA sequence. Therefore, BCA has to be replenished to levels where it dominates the microbial assemblage. The idea presented here is that the highly complex and dynamic normal microflora, plus biocontrol microbes found on leaf lamina, may avoid infection and establishment of the bacterial pathogen, namely *P. syringae cv. tabaci*. They suspected that *Pantoea* and *Sphingomonas* found on the phyllosphere/phylloplane could be suppressing the pathogen. Incidentally, we should realize that the agricultural sky is the source of most, if not all, of the microbes detected on the tobacco. Troposphere microbes are indeed crucial during the natural suppression of diseases. This is also true while conducting biocontrol measures.

Thapa et al. (2017) have reported that a few genotypes of rice (*Oryza sativa*) supported phyllosphere microbes (community) that possessed antagonistic activity on known aerial pathogens like *Drechslera oryzae*, *Rhizoctonia solani* and *Magnaporthe oryzae*. Accentuating or inoculating antagonistic microbes is obviously beneficial as they help reduce crop disease incidence. It is important to note that when the phyllosphere supports a higher population of antagonists, there is a higher likelihood of their dispersion to other fields through wind and atmospheric turbulence. Antagonists will find new regions to spread and flourish, potentially reducing pathogens in proportion. Specific microbial species and their genotypes can be identified, multiplied, and evaluated against known aerial pathogens of rice varieties.

Antagonistic interactions among the phyllosphere microbial population are a natural phenomenon. This has been observed by many microbiologists and even utilized advantageously by practical farmers. Firstly, leaf-derived secondary metabolites secreted onto the phyllosphere zone may affect the microflora on the leaf surface. In addition, certain bacterial/fungal species colonizing the phyllosphere may produce antagonistic chemicals such as antibiotics, toxins, bacteriostatic agents, etc.

Karamanoli & Menkissoglu-Spiroudi (2005) examined several crop species for antagonistic chemicals originating from leaf metabolic processes. They also examined the bacterial pool of the phyllosphere. Significant differences in ice nucleating properties and antagonism of phyllosphere microflora were observed. Generally, the population of epiphytotic microbes ranged from 10^2 to 10^5 CFU g^{-1} leaf fresh weight. However, it decreased if antagonistic bacterial species were active and in greater number.

For example, Wei et al. (2016) explain that in the case of a bacterial BCA like *Bacillus subtilis,* its rapid spread, persistence, and effective control of pathogens are important. Sometimes, a sufficiently high population of biocontrol bacteria is applied to the foliage (phyllosphere). Yet, its survival rate may be low. Therefore, intermittent replenishment of the biocontrol microbe *B. subtilis* is recommended. Perhaps, 3–4 sprays in a crop cycle.

The crop species grown in the vicinity, i.e., in neighboring fields, react to troposphere microbes differently. Conversely, troposphere microbes may elicit a differential response from crop genotypes. For example, Kinkel (1997) states that mycobiota traced on crops are influenced by several factors related to the host's genetic constitution, its growth pattern, leaf exudates, etc. To a certain extent, the host determines the phyllosphere fungal community. Such a fungal assemblage may have benefits in terms of biological control of certain crop diseases. There is an interesting example related to the control of 'Black rot of Tea' by the aeromycoflora and phyllosphere microflora. Thoudam & Dutta (2012) have reported the occurrence of a few specific fungi above the tea gardens. The agricultural sky was sampled using a 'Hirst Trap.' They found *Aspergillus flavus*, *A. niger*, *Curvulria*sp., *Penicillium* sp., and *Trichoderma atroviridae*. Upon *in vitro* testing, these fungi showed clear antagonistic activity against the 'Black rot fungus of *Camelia sinensis.'* Next, they tried to culture a few phyllosphere fungi and evaluated their antagonistic effects on Black rot of tea. They found fungi like *A. flavus*, *A. niger*, *Penicillium* sp., *T.*

artoviridae. Foliar sprays of phyllosphere fungi reduced the effects of black rot. The antagonistic principle and their physiological effects on the black rot fungi need to be deciphered. This may lead to better control of black rot.

It is possible that in nature, several microbial species found in the phyllosphere exhibit antagonistic reactions to major pathogens. For example, rice blast caused by *Pyricularia oryzae* is a severe disease. However, Harsonawati et al. (2017) found that native actinomycetes, mostly of the genus *Streptomyces* and a few others, were suppressive on *P. oryzae*. They were able to isolate about 75 strains of *Streptomyces,* out of which approximately 18 isolates showed a strong antagonistic reaction to the blast fungus. Monitoring leaf blast and conducting prophylaxis with microbial sprays may be useful. It may also be beneficial to periodically monitor the presence of *Streptomyces* in the canopy and just above in the troposphere of rice crops, and then decide on control measures.

Bacterial blight of rice is a severe detriment to grain yield. It requires bactericidal sprays of high concentrations of active ingredient and large volumes as well. Biological control has been suggested to avoid excessive use of chemical pesticides. Since it is a leaf blight, the phyllosphere region was the first choice to search for microbes antagonistic to *Xanthomonas oryzae*. Let us consider an example. A few actinomycete species with the ability for antibiosis against the bacterial blight pathogen *Xanthomonas oryzae* have been tested. They belonged to *Streptomyces* sp., *S. luteogriesius, Actinomadura* sp., and *Kitasatospora azatica* (Ilsan et al., 2016). They were isolated from healthy leaves of rice plants at different stages of growth. The idea here is to spray microbial BCAs containing antagonistic actinomycetes to protect crops attacked by *X. oryzae*. It might be useful to apply prophylactic sprays on leaves so that the foliage harbors a large population of BCA at any time.

Bacteriophages are useful as Biological Control Agents (BCAs) when aiming to reduce the bacterial population in a microbiome. Bacteriophages that target a wide range of bacterial species have been identified and evaluated in laboratory conditions, as well as in natural phyllosphere surfaces and other locations. Bacteriophages are known to attack bacterial species with diverse functional abilities, including plant pathogenic bacteria. However, it is important to note that non-target bacteria may also be affected by bacteriophages, resulting in a reduction of the population of biocontrol bacterial species and others. Phyllosphere bacteriophages that attack BCAs like *Bacillus thuringiensis* or others can be considered, although spraying bacteriophages in this case may have disadvantages. The lytic activity of bacteriophages should be carefully considered, depending on the purpose. Phyllosphere bacteriophages have multiple facets and can be exploited to our advantage, as pointed out by Koskela & Taylor (2018). The activity of bacteriophages may be influenced by environmental conditions, and '*Phyllo-viruses*' could be utilized as drivers of the microbial community structure. They could be used for bacteriophage-mediated engineering of the phyllosphere microbiome. The survival and infectivity of bacteriophages used as BCAs should be taken into account, as several atmospheric factors can affect their survival and dispersal in cropped fields.

As mentioned earlier, the microbial load and diversity play a crucial role in preventing foliar diseases. The phyllosphere microbiome also serves as a habitat for beneficial microorganisms, such as bacteriophages. In fact, according to Alanin et al. (2022), the phyllosphere of wheat can be heavily infested with bacteriophages. Throughout the growth season, the wheat canopy can host these bacteriophages. The population and diversity of phyllosphere bacteria are important factors for the thriving of bacteriophages as beneficial microorganisms. Alanin et al. (2022) further state that through molecular analysis using 16S rRNA, they were able to identify over 850 virus operational taxonomic units in bacteriophages. The leaf lamina often exhibited 20×10^6 viral particles per leaf. Therefore, wheat phyllosphere naturally supports a large population of diverse bacteriophages, which can become active when they encounter a host bacterial cell. Bacteriophages routinely reduce the population of various species of bacterial hosts. Overall, it has been suggested that 'Phylloviruses' in general could have significant practical value in plant disease control.

Chestnut tree production has been popular in the European plains for several centuries, providing nuts and wood. However, chestnut trees are susceptible to various fungal diseases that primarily affect the foliage. One example is Chestnut blight, caused by the ascomycete *Cryphonectria parasitica,* which leads to cankers and dieback. Another is the oomycete *Phytophthora cinnamomi*, responsible for ink disease (Vettraino et al., 2005). This fungus blocks xylem vessels after entering through the leaves. In the natural environment, the chestnut phyllosphere harbors a diverse range of microbial species. Leaf populations have been observed to range from 10^6 to 10^7 cells cm^{-2} (Vacher et al., 2016; Vorholt, 2012). The microbial community typically includes bacteria, filamentous fungi, yeast, algae, and protozoa (Valverde et al., 2017). Efforts to identify antagonistic microbes from the leaf surface have been ongoing for many decades. The aim is to discover biocontrol microbes that can suppress foliar diseases. The phyllosphere provides a potential area for antagonistic interactions that can impede pathogen growth (Meyer & Leveau, 2012; Lindow & Leveau, 2002; Elad et al., 1999). Similarly, Barrera et al. (2020) have reviewed the usefulness of phyllosphere microbes in the biological control of foliar diseases affecting horticultural crops. The potential of phyllosphere microbes to suppress pathogens and maintain them below threshold levels should be further explored. For instance, the phyllosphere may have specific effects on fruit trees that warrant investigation. Kucheryava et al. (1999) identified several strains of *P. fluorescens* that can control *Venturia equalis,* the causal agent of apple scab disease. In other words, specific biological control agents can be developed for specific pathogens that attack crops.

Populus species are common in Northern hemisphere. The genus *Populus* has 25–30 species. They are mostly deciduous flowering plants of family Salicaceae. *Populus* trees are grown for timber, fuel, and as avenue trees. The male and female trees differ in several traits. In the present context, Liu et al. (2021) investigated the phyllosphere of these dioecious trees. They discovered microbial species that were unique to male and female *Populus cathayana.* Molecular sequencing methods, such as 16S rRNA, were utilized. For instance, bacterial species *Gemmata* and fungal species *Pringsheima* spp. were specifically associated with the phyllosphere of female *Populus* trees. Similarly, bacterial species *Chitinophaga* and fungal species *Phaeococcomyces* spp. were exclusively detected on the phyllosphere of male trees. Furthermore, the phyllospheres of male trees exhibited a predominance of bacterial genera like *Spirosoma* and *Amnibacterium*. Additionally, fungal species such as *Venturia, Suillus,* and *Elmerina* spp. were present. *Proteobacteria* and *Planctomycetes* were abundant in the phyllosphere of female *Populus* trees, while fungi like Ascomycetes and Basidiomycetes were frequently found in the phyllosphere of female trees. The exact cause of such differences in the phyllosphere of male and female trees needs to be deciphered.

6.9 BIO-STIMULANTS AND PHYLLOSPHERE MICROBIOME

Practical farming involves the use of foliar sprays containing a variety of plant protection chemicals and stimulants. There have been several reports on plant growth stimulants derived from extracts of various vegetation parts such as leaves, branches, twigs, flowers, and seeds. Protein hydrolysates from a wide range of sources have been evaluated as sprays on foliage (Colla et al., 2017a, b; Wang et al., 2019; Wagi & Ahmed, 2017). The foliar spray of these bio-stimulants is known to have an impact on crop growth. It primarily affects the microbial assemblages found on the phyllosphere and phylloplane. For instance, Luziatelli et al. (2019) evaluated the effects of vegetation-derived bio-stimulants on microbiome diversity and total population. Bio-stimulant sprays on leaves induce rapid growth of microbes like *Pantoea*, *Pseudomonas*, *Acinetobacter*, and *Bacillus genus* (Dutkiewicz, 2016). These phyllosphere microbes possessed biocontrol properties on plant pathogens (e.g., on *Fusarium*). A few others that were accentuated on leaf surface had plant growth stimulation properties. The purpose here is to improve the phyllosphere zone to the plant's advantage. We should also understand the effect of atmospheric microbes that deposit onto the leaf surface periodically. They too may have certain beneficial effects. This is in addition to the benefits derived from spraying bio-stimulants. Here, we have only considered the bio-stimulants sprayed onto foliage. We also restrict our focus to microbial assemblages on foliage. There are numerous reports on plant growth substances that are applied to the seed surface and those that operate in the rhizosphere. However, these are not within the scope of this volume.

KEYWORDS

- **atmospheric factors**
- **carposphere**
- **caulosphere**
- **crop diseases**
- **crop's canopy**
- **exudates**
- **foliar sprays**
- **leaf lamina**
- **microbial assemblages**
- **microbial diversity**
- **microbial succession**
- **microbiome**
- **nutrient dynamics**
- **phylloplane**
- **phyllosphere**
- **stem flow**

REFERENCES

Alanin, S., Djurhuus, A. M., Kot, W., Gobbi, A., & Hansen, L. H., (2022). Bacteriophages roam the wheat phyllosphere. *Viruses, 14*, 244 https://doi.org/10.3390/v14020244/.

Alklett, K., Hart, M., & Shade, A., (2014). The microbial ecology of flowers: An emerging frontier in phyllosphere research. *Botany, 92*, 253–266. dx.doi.org/10.1139/cjb-2013-0166 /.

Alsanius, B. W., Bergstrand, K. J., Hartmann, R., Gharaie, S., & Wohanka, W. M., (2017). Ornamental flowers in new light: Artificial lighting shapes the microbial phyllosphere community structure of greenhouse grown sunflowers (*Helianthus annuus* L.). *Scientia Horticultura, 216*, 234–247. doi: 10.1016/j. scienta.2017.01.022/.

Andrews, J. H., & Harris, R. F., (2000). The ecology and biogeography of microorganisms on plant surfaces. *Annual Review of Phytopathology, 38*, 145–180.

Andrews, J. H., (1992). Biological control in the phyllosphere. *Annual Review of Phytopathology, 30*, 603–635.

Andrews, J. H., Kenerley, C. M., & Nordheim, E. V., (1980). Positional variation in phylloplane populations within an apple tree canopy. *Microbial Ecology, 6*, 71–84.

Barraquio, W. L., Revilla, L., & Ladha, J. K., (1997). Isolation of endophytic diazotrophic bacteria from wetland rice. *Plant and Soil, 194*, 15–24.

Barrera, S., Sarango-Flores, S., & Montenegro-Gomez, (2020). The phyllosphere microbiome and its potential application in horticultural crops: A review. *RevistaColombiana De CienciasHorticolas, 13*(3). https://doi.org/10.17584/rcch.2019vl3i3.8405/.

Barruso, L., Bani, A., Pioli, S., Ventura, M., Panzacchi, P., Antonielli, L., Giammarchi, F., et al., (2021). Do aerial nitrogen depositions affect fungal and bacterial communities of oak leaves? *Frontiers in Microbiology, 12*, 633535. doi: 10.3389/fmicb.2021.633535/.

Batool, F., Rehman, Y., & Shahida, H. S., (2016). Phylloplane associated plant bacteria of commercially superior wheat varieties exhibit superior plant growth promoting abilities. *Frontiers in Life Science, 9*, 313–322. doi: 10.1080/21553769.2016.1256842/ https://doi.org/10.1080/21553769.2016.1256842/.

Beattie, G. A., & Lindow, S. E., (1995). The secret life of foliar bacterial pathogens on leaves. *AnnualReview of Phytopathology, 33*, 145–172.

Beattie, G. A., & Lindow, S. E., (1999). Bacterial colonization of leaves: A spectrum of strategies *Phytopathology, 89*, 353–359.

Belova, S. E., Baani, M., Suzina, N. E., Bodelier, P. L. E., Liesack, W., & Dedysh, S. N., (2011). Acetate utilization as a survival strategy of peat-inhabiting *Methylocystis* spp. *Environmental Microbiology Report, 3*(1), 36–46. https://doi.org/10.1111/j.1758-2229.2010.00180.x/.

Benndorf, D., Loffhagen, N., & Babel, W., (2001). Protein synthesis patterns in *Acinetobacter calcoaceticus* induced by phenol and catechol show specificities of responses to chemo stress. *FEMS Microbiology Letters, 200*, 247–252.

Bentley, B. L., (1987). Nitrogen fixation by epiphylls in a tropical rainforest. *Annals of Missouri Botanical Gardens, 74*, 234–241.

Bessems, E. P. M., (1973). Nitrogen fixation in the phyllosphere of graminae. *Agricultural Research Reports, 786.*

Bhurat, M. C., & Sen, A., (1968). Nitrogen fixing bacteria on the phyllosphere of wheat and pea. *Indian Journal of Agricultural Research, 38*, 319–325.

Blacutt, A. A., Mitchell, T. R., Bacon, C. W., & Gold, S. E., (2016). *Bacillus mojavensis*RRC101 lipopeptides provoke physiological and metabolic changes during antagonism against

Fusarium verticillioides. *Mol. Plant Microbe Interaction, 29*, 713–723. doi: 10.1094/MPMI-05–16–0093-R/.
Boutroux, L., (1884). Sur la conservation des ferments alcooliques dans la nature. *Annales Des Sciences Naturales: IV* (Vol. 17, pp. 145–209). Botanique, Paris, France.
Brandhoff, B., Simon, A., Dornieden, A., & Schumacher, J., (2017). Regulation conidiation in *Botrytis cinerea* involves the light-responsive transcriptional regulators BcLTF3 and BcREG1. *Current Genetics, 63*, 931–949. doi: 10.1007/s00294-017-0692-9/.
Bringel, F., & Couee, J., (2015). Pivotal roles of phyllosphere microorganisms at the interface between plant functioning and atmospheric trace gas dynamics. *Frontiers in Microbiology, 6*, 486. https://doi.org/10.3389/fmicb.2015.00486/.
Brown, E., (2014). The impact of systemic and copper pesticide applications on the phyllosphere microflora of tomatoes. *Journal of Science of Food and Agriculture, 95*, 1116–1125. doi: 10.1002/jsfa.7010/.
Buck, J. W., & Burpee, L. L., (2002). The effects of fungicides on the phylloplane yeasts populations of creeping bent grass. *Canadian Journal of Microbiology, 48*(6), 522–529.
Bulgarelli, D., Schlaeppi, K., Spaepen, S., Themaat, E., & Lefert, P., (2013). Structure and functions of the bacterial microbiota of plants. *Annual Review of Plant Biology, 64*, 807–838.
Carvalho, S. D., & Castillo, J. A., (2018). Influence of light on plant–phyllosphere interaction. *Frontiers in Plant Science. 9*, 1482. doi: 10.3389/fpls.2018.01482/.
Carvalho, S. D., & Folta, K. M., (2014). Environmentally modified organisms–expanding genetic potential with light. *Critical Review of Plant Science, 33*, 486–508. doi:.1080/07352689.2014.929929/.
Cernava, T., Chen, X., Krug, L., Yang, M., & Berg, G., (2019). The tea leaf microbiome shows specific responses to chemical pesticides and biocontrol applications. *Science of the Total Environment, 667*, 33–40.
Cha, J. Y., Han, S., Hong, H. J., Cho, H., Kim, D., Kwon, Y., Kwon, S. K., et al., (2016). Microbial and biochemical basis of a *Fusarium* wilt-suppressive soil. *International Society of Microbial Ecology Journal, 10*(1), 119–129. https://doi.org/10.1038/ismej.2015.95/.
Chaudhary, D., Kumar, R., Sihag, K., Rashmi, K., & Anjukumari, K., (2017). Phyllospheric microflora and its impact on plant growth: A review. *Agricultural Reviews, 38*, 51–59.
Chaudhry, V., Punge, P., Sengupta, P., Doehlemann, G., Parker, J. E., & Kemen, E., (2021). Shaping the leaf microbiota-microbe interactions. *Journal of Experimental Botany, 72*, 36–56. doi: 10.1093/jxb/eraa417/.
Cohrs, K. C., Simon, A., Viaud, M., & Schumacher, J., (2016). Light governs differentiation in the grey mould fungus *Botrytis cinerea* via the putative transcription factor BcLTF2. *Environ. Microbiol., 18*, 4068–4086. doi: 10.1111/1462-2920.13431/.
Colla, G., Cardarelli, M., Bonini, P., & Rouphael, Y., (2017b). Foliar applications of protein hydrolysate, plant and seaweed extracts increase yield but differentially modulate fruit quality of greenhouse tomato. *Hortscience, 52*, 1214–1220. doi: 10.21273/HORTSCI12200-17 /.
Colla, G., Hoagland, L., Ruzzi, M., Cardarelli, M., Bonini, P., Canaguier, R., et al., (2017a). Biostimulant action of protein hydrolysates: Unraveling their effects on plant physiology and microbiome. *Front. Plant Sci., 8*, 2202. doi: 10.3389/fpls.2017.02202/.
Copeland, J. K., Yuan, L., & Levghifard, M., (2015). Seasonal community succession of the phyllosphere microbiome. *Molecular Plant-Microbe Interactions, 28*, 274–285. doi: 10.1094/MPMI-10-14-0331-FI./.

De Curtis, F., De Cicco, V., & Lima, G., (2012). Efficacy of biocontrol yeasts combined with calcium silicate or Sulphur for controlling durum wheat powdery mildew and increasing grain yield components. *Field Crops Research, 134*(12), 36–46.
De Jager, E. S., Wehner, F. C., & Korsten, L., (2001). Microbial ecology of the mango phylloplane. *Microbial Ecology, 42*(2), 201–207.
De Menna, M. E., (1959). Yeasts from the leaves of pasture plants. *New Zealand J. Agric. Res.*, 2, 394–405.
Delmotte, N., Knief, C., & Chaffron, S., (2009). Community proteogenomics reveals insights into the physiology of phyllosphere bacteria. *Proceedings of the National Academy of Sciences of the United States of America* (Vol. 106, No. 38, pp. 16428–16433).
Doronina, N. V., Ivanova, E. G., Suzina, N. E., & Trotsenko, Y. A., (2004). Methanotrophs and methylobacteria are found in woody plant tissues within the winter period. *Microbiology, 73*(6), 702–709.
Drahos, D. J., (1991). Methods for the detection, identification, and enumeration of microbes. In: Andrews, J. H., (ed.), *Microbial Ecology of Leaves; Springer Contemporary Series* (pp. 135–157).
Droby, S., & Visniewski, M., (2018). *The Fruit Microbiome: The New Frontier for Post-Harvest Biocontrol and Post-Harvest Biology*. https://doi.org/10.1016/j.postharvbio.2018.03.004/.
Dutkiewicz, J., Mackiewicz, B., Lemieszek, M. K., Golec, M., & Milanowski, J., (2016). *Pantoea agglomerans*: A mysterious bacterium of evil and good. Part 1. Beneficial effects. *Annals of Agricultural and Environmental Medicine, 23*, 206–222. doi: 10.5604/12321966.1203879/.
Elad, Y., Bélanger, R. R., & Kohl, J., (1999). Biological control of diseases in the phyllosphere. In: Albajes, R., Lodovica, G. M., Van, L. J. C., & Elad, Y., (eds.), *Integrated Pest and Disease Management in Greenhouse Crops; Developments in Plant Pathology* (Vol. 14). Springer, Dordrecht. https://doi.org/10.1007/0-306-47585-5_24/.
Elbeltagy, A., Nishioka, K., Sato, T., Suzuki, H., Ye, B., Hamada, T., et al., (2001). Endophytic colonization and in plant nitrogen fixation by a *Herbaspirillum sp.* isolated from wild rice species. *Applied and Environmental Microbiology, 67*, 5285–5293.
Finkel, O. M., Burch, A. Y., Elad, T., Huse, S. M., Lindow, S. E., & Post, A. F., (2012). Distance-decay relationships partially determine diversity patterns of phyllosphere bacteria on tamrix trees across the Sonoran Desert. *Applied and Environmental Microbiology, 78*, 6187–6193. doi: 10.1128/aem.00888-12/.
Fokkema, N. J., & Van, D. M. F., (1976). Antagonism of yeast like phyllosphere fungi against *Septoria nodorum* on wheat leaves. *Netherlands Journal of Plant Pathology, 82*, 13–16.
Fokkema, N. J., (1993). Opportunities and problems of control of foliar pathogens with micro-organisms. *Pest Management Science, 77*, 411–416. https://doi.org/10.1002/ps.2780370416/.
Furkranz, M., Wolfgang, W., Andreas, R., Guy, A., Frank, R., & Angela, S., (2008). Nitrogen fixation by phyllosphere bacteria associated with higher plants and their colonizing epiphytes of a tropical lowland rainforest of Costa Rica. *International Society for Microbial Ecology (ISME), 2*, 561–570.
Futamata, H., Harayama, S., & Watanabe, K., (2001). Group-specific monitoring of phenol hydroxylase genes for a functional assessment of phenol-stimulated trichloroethylene bioremediation. *Applied and Environmental Microbiology, 67*, 4671–4677.

Gharaie, S., (2017). *Light-Phyllosphere Interactions in Greenhouse Grown Ornamental* (p. 75). Department of Biosystems. Swedish University of Agricultural Sciences, Alnarp, Sweden.

Glenn, D. M., Bassett, C., & Dowd, S. E., (2015). Effect of pest management system on 'Empire' apple leaf phyllosphere populations. *Scientia Horticultura, 183*, 58–65. doi: 10.1016/j.scienta.2014.12.009/.

Gourion, B., Francez-Charlot, A., & Vorholt, J. A., (2008). Phyllosphere is involved in the general stress response of *Methylobacteriumextorquens*AM1. *Journal of Bacteriology, 190*, 1027–1035. https://doi.org/10.1128/JB.01483-07/.

Gupta, R., & Patil, R., (2021). Phyllospheric microbes: Diversity, functions, interaction an applications. In: Yadav, A. N., et al., (eds.), *Current Trends in Microbial Biotechnology for Sustainable Agriculture, Environmental and Microbial Biotechnology*. https://doi.org/10.1007/978-981-15-6949-4_13/.

Hansen, E. C., (1890). Nouvelles recherches sur la circulation du *Saccharomyces apiculatus* dans la nature. *Annals of Science Naturales IV. Series Botany, 11*, 185–192.

Harsonowati, W., Astuti, R. I., & Wahyudi, A. T., (2017). Leaf blast disease reduction by rice phyllosphere actinomycetes producing bioactive compounds. *Journal of General Pathology, 83*, 98–108.

Hevia, M. A., Canessa, P., Müller-Esparza, H., & Larrondo, L. F., (2015). A circadian oscillator in the fungus *Botrytis cinerea* regulates virulence when infecting *Arabidopsis thaliana. Proceedings of National Academics. Science, U.S.A., 112*, 8744–8749. doi: 10.1073/pnas.1508432112/.

Hirano, S. S., & Upper, C. D., (1989). Dual variation in population size and ice nucleation activity of *Pseudomonas syringae* on snap bean leaflets. *Applied and Environmental Microbiology, 55*, 623–630.

Hirano, S. S., & Upper, C. D., (2000). Bacteria in the leaf ecosystem with emphasis on *Pseudomonas syringae*—A pathogen, ice nucleus, and epiphyte. *Microbiology and Molecular Biology, Review, 64*, 624–653.

Holland-Moritz, H., Julia, E. M., Lewis, L. R., Miller, S. M., Mack, M. C., Ponciano, J. M., McDaniel, S. F., & Fierer, N., (2021). The bacterial communities of Alaskan mosses and their contributions to N_2-fixation. *Microbiome, 9*(1), 53. doi: 10.1186/s40168-021-01001-4.

Humphrey, P. T., Nguyen, T. T., Villalobos, M. M., & Whiteman, N. K., (2014). Diversity and abundance of phyllosphere bacteria are linked to insect herbivory. *Molecular Ecology, 23*, 1497–1515. doi: 10.1111/mec.12657/.

Ilsan, N. A., Nawangsih, A. A., & Wahyndi, A. T., (2016). Rice phyllosphere actinomycetes as biocontrol agent of bacterial leaf blight disease on rice. *Asian Journal of Plant Pathology, 10*, 1–8.

Inacio, J., Pereira, P., De Carvalho, M., Fonseca, A., Amaral-Collaco, M. T., & Spencer-Martins, I., (2002). Estimation and diversity of phylloplane mycobiota on selected plants in a Mediterranean type ecosystem in Portugal. *Microbial Ecology, 44*, 344–353.

Jacques, M. A., & Morris, C. E., (1995). A review of issues related to the quantification of bacteria from the phyllosphere. *FEMS Microbiology and Ecology, 18*, 1–14.

Jacques, M. A., Kinkel, L. L., & Morris, C. E., (1993). Population sizes, immigration, and growth of epiphytic bacteria on leaves of different ages and positions of field-grown endive (*Cichorium endiva* var. *latifolia*). *Applied and Environmental Microbiology, 61*, 899–906.

Jones, J. B., Vallad, G. E., Iriarte, F. B., Obradović, A., Wernsing, M. H., Jackson, L. E., Balogh, B., et al., (2012). Considerations for using bacteriophages for plant disease control. *Bacteriophage,2*(4), 208–214. https://doi.org/10.4161/bact.23857/.

Jones, K., (1970). Nitrogen fixation in the phyllosphere of the Douglas fir, *Pseudotsugadouglasii*. *Annals of Botany, 34*, 239–144. https://doi.org/10.1093/oxfordjournals.aob.a084358/.

Jørgensen, L. N., Hovmoller, M. S., Hansen, J. G., Lassen, P., Clark, B., & Bayles, R., (2014). IPM Strategies and their dilemmas including an introduction to www.eurowheat.org. *Journal of Integrated Agriculture, 13*(2), 265–281.

Karamanoli, K., & Menkissoglu-Spiroudi, U., (2005). Bacterial colonization of the phyllosphere of nineteen plant species and antimicrobial activity of their leaf secondary metabolites against leaf associated bacteria. *Chemoecology, 15*, 59–67.

Karlsson, I., Friberg, H., Steinberg, C., & Persson, P., (2014). Fungicide effects on fungal community composition in the wheat phyllosphere. *PLoS One, 9*, e111786. doi: 10.1371/journal.pone.0111786/.

Kawaguchi, K., Yurimoto, H., Oku, M., & Sakai, Y., (2011). Yeast methylotrophy and autophagy in a methanol-oscillating environment on growing *Arabidopsis thaliana* leaves. *PLoS One, 6*(9), e25257. https://doi.org/10.1371/journal.pone.0025257/.

Khanna, K. K., (1986). Phyllosphere microflora of certain plants in relation to air pollution. *Environmental Pollution, 42*, 191–200.

Kim, M., Singh, D., Lai-hoe, A., Chun, J., & Adams, J. M., (2013). Distinctive phyllosphere bacterial communities in tropical trees. *Microbial Ecology, 63*, 674–681.

Kinkel, L. L., (1997). Microbial population dynamics on leaves. *Annual Review of Phytopathology,35*, 327–347. https://doi.org/10.1146/annurev.phyto.35.1.32/.

Kinkel, L. L., Wilson, M., & Lindow, S. E., (1996). Utility of microcosm studies for predicting phylloplane bacterium population sizes in the field. *Applied and Environmental Microbiology, 62*, 3413–3423. doi: https://doi.org/10.1128/aem.62.9.3413-3423.1996/.

Knauth, S., Hurek, T., Brar, D., & Reinhold-Hurek, B., (2005). Influence of different *Oryza* cultivars on expression of *nif H* gene pools in roots of rice. *Environmental Microbiology, 7*, 1725–1733.

Knief, C., (2014). Analysis of plant microbe interactions in the era of next generation sequencing technologies. *Frontiers in Plant Science, 216*. doi: 10.3389/fpls.2014.00216/.

Knief, C., Delmotte, N., & Chaffron, S., (2012). "Metaproteogenomic analysis of microbial communities in the phyllosphere and rhizosphere of rice." *International Society of Microbial Ecology Journal, 6*, 1378–1390.

Knorr, K., Jørgensen, L. N., & Nicolaisen, M., (2019). Fungicides have complex effects on the wheat phyllosphere mycobiome. *PLoS One, 14*(3), e0213176./.

Koskela, B., & Taylor, T. B., (2018). Multifaceted impacts of bacteriophages in the plant microbiome. *Annual Review of Phytopathology, 56*(1), 361–380. doi: 10.1146/annurev-phyto-080417-045858/.

Krishnamoorthy, R., Kwon, S. W., Kumutha, K., Senthilkumar, M., Ahmed, S., Sa, T., & Anandham, R., (2018). Diversity of culturable methylotrophic bacteria in different genotypes of groundnut and their potential for plant growth promotion. *Biotechnology, 8*, 275.

Kucharska, K., Wachowska, U., & Czaplicki, S., (2020). Wheat phyllosphere yeasts degrade propiconazole. *BMC Microbiology, 20*, 242. https://doi.org/10.1186/s12866-020-01885-6/.

Kucheryava, N., Fiss, M., Auling, G., & Kroppenstedt, R. M., (1999). Isolation and characterization of epiphytic bacteria from the phyllosphere of apple, antagonistic *in vitro* to *Venturia inaequalis* the causal agent of apple scab. *System and Applied Microbiology, 22*, 472–478.

Kwak, M. J., Jeong, H., Madhaiyan, M., Lee, Y., Sa, T. M., Oh, T. K., & Kim, J. F., (2014). Genome information of *Methylobacterium oryzae,* a plant-probiotic methylotroph methylotroph in the phyllosphere. *PLoS One, 9*(9), e106704/.

Laforest-Lapointe, I., Messier, C., & Kembel, S. W., (2016). Host species identity, site and time drive temperate tree phyllosphere bacterial community structure. *Microbiome, 4*, 27–32.

Lajoie, G., Maglione, R., & Kembel, S. W., (2019). *Drivers of Phyllosphere Microbial Functional Diversity in a Neotropical Forest.* doi: https://doi.org/10.1101/851485/.

Lambais, M. R., Crowley, D. E., Cury, D. E., & Rodrigues, R. R., (2006). Bacterial diversity in tree canopies of the Atlantic Forest. *Science, 312*, 1917.

Last, F. T., (1955). Seasonal incidence of *Sporobolomyces* on cereal leaves. *Transactions of British Mycological Society, 38*(3), 221–239. doi: 10.1016/S0007-1536(55)80069-1./.

Leff, J. W., & Fierer, N., (2013). Bacterial communities associated with the surfaces of fresh fruits and vegetables. *PLoS One*. https://doi.org/10.1371/journal.pone.00593 10/.

Legard, D. E., McQuilken, M. P., Whipps, J. M., Fenlon, J. S., Fermor, T. R., Thompson, I. P., Bailey, M. J., & Lynch, J. M., (1994). *Agriculture Ecosystem and Environment, 50*, 87–101.

Legein, M., Smets, W., Vandenheuvel, D., Eilers, T., Muyshondt, B., Prinsen, E., Samson, R., & Lebeer, S. L., (2020). Modes of action of microbial biocontrol in the phyllosphere. *Frontiers in Microbiology, 11*, 1619, pp. 1–18. doi: 10.3389/fmicb.2020.01619/.

Li, Y., Wu, X., Wang, W., Wang, M., Zhao, C., Chen, T., Liu, G., et al., (2019). Microbial taxonomical composition in spruce phyllosphere, but not community functional structure, varies by geographical location. *Peer Journal,* 7, e7376. doi 10.7717/peerj.7376/.

Lindow, S. E., & Brandl, M. T., (2003). Microbiology of the phyllosphere. *Applied Environment Microbiology, 69*, 1875–1883.

Lindow, S. E., & Leveau, J. H., (2002). Phyllosphere microbiology. *Current Opinions in Biotechnology, 13*, 238–243.

Liu, L., Lu, L., Li, L., Meng, Z., Dong, T., Peng, C., & Xu, X., (2021). Divergence of phyllosphere microbial communities between females and males of the dioecious *Populus cathayana. Molecular Plant-Microbes Interactions, 34*, 351–361. https://doi.org/10.1094/MPMI-07-20-0178-R/.

Longa, C. M. O., Pertot, I., & Tosi, S., (2008). Ecophysiological requirements and survival of a *Trichoderma atroviride* isolate with biocontrol potential. *Journal of Basic Microbiology, 48*(4), 269–277. https://doi.org/10.1002/jobm.20070 0396.

Luziatelli, F., Ficca, A. G., Colla, G., Baldassarre, Š. E., & Ruzzi, M., (2019). Foliar application of vegetal-derived bioactive compounds stimulates the growth of beneficial bacteria and enhances microbiome biodiversity in lettuce. *Frontiers in Plant Science, 10*, 60. doi: 10.3389/fpls.2019.00060/.

Maniranjan, B. A., Ratering, S., Rusch, V., Schwiertz, A., Plum, G., Cardinale, M., & Schnell, S., (2018). Bacterial microbiota associated with flower pollen is influenced by pollination type and shows a high degree of diversity and species specificity. *Environmental Microbiology, 18*, 5161–5174. https://doi.org/10.1111/1462-2920.13524/.

Marois, I. J., & Coleman, L. L., (1995). Ecological succession and biological control in the phyllosphere. *Canadian Journal of Botany, 73*, 11–17. https://doi.org/10.1139/b95-228/.

Mason, C. J., Couture, J. J., & Raffa, K. F., (2014). Plant-associated bacteria degrade defense chemicals and reduce their adverse effects on an insect defoliator. *Oecologia, 175*, 901–910. doi: 10.1007/s00442-014-2950-6.

Meena, K. K., Kumar, M., Kalyuzhnaya, M. G., Yandigeri, M. S., Singh, D. P., & Saxena, A. K., (2012). Epiphytic, pink-pigmented methylotrophic bacteria enhance germination and seedling growth of wheat (*Triticum aestivum*) by producing phytohormone. *Antonie VanLeeuwenhoek, 101*, 777–786.

Mercier, J., & Lindow, S. E., (2000). Role of leaf surface sugars in colonization of plants by bacterial epiphytes. *Applied and Environmental Microbiology, 66*, 369–374.

Meyer, K. M., & Leveau, J. H. J., (2012). Microbiology of the phyllosphere: A playground for testing ecological concepts. *Oecologia, 168*, 621–629.

Miyamoto, T., Kawahara, M., & Minamisawa, K., (2004). Novel endophytic nitrogen-fixing clostridia from the grass miscanthus sinensis as revealed by terminal restriction fragment length polymorphism analysis. *Applied and Environmental Microbiology, 70*, 6580–6586.

Morella, N. M., Gomez, A. L., Wang, G., & Leung, M. S., (2018). The impact of bacteriophages on phyllosphere bacterial abundance and composition. *Molecular Ecology Wiley Online Library*https://doi.org/10.1111/mec.14542/.

Morris, C. E., & Kinkel, L. L., (2002). Fifty years of phyllosphere microbiology: Significant contributions to research in related fields. In: Lindow, S. E., Hecht-Poinar, E. I., & Elliott, V., (eds.), *Phyllosphere Microbiology* (pp. 365–375). American Psychopathological Society Press, St. Paul, Minnesota, USA.

Mota, M. S., Cesar, B. G. C. B., Souza, I. T., & Moura, A. B., (2017). Bacterial selection for biological control of plant disease: Criterion determination and validation. *Brazilian Journal of Microbiology, 48*, 62–70.

Moulas, C., Petsoulas, C., Rousidou, K., Perrruchon, C., Karas, P., & Karpouzas, D. G., (2013). *Effects of Systemic Pesticides Imidacloprid and Metalaxyl on the Phyllosphere of Pepper Plants* (p. 8). Hindawi Publishing Corporation Article ID 969750. http://dx.doi.org/10.1155/2013/969750/.

Müller, T., & Ruppel, S., (2014). Progress in cultivation-independent phyllosphere microbiology. *FEMS Microbial Ecology, 87*, 2–17. doi: 10.1111/1574-6941.12198/.

Müller, T., Behrendt, U., & Ruppel, S., (2016). Fluorescent pseudomonads in the phyllosphere of wheat: Potential antagonists against fungal phytopathogens. *Current Microbiology, 72*, 383–389.

Nelson, E. B., (2017). The seed microbiome: Origins, interactions, and impacts. *Plant and Soil, 422*(1, 2), 7–34. 10.1007/s11104-017-3289-7/.

Ning, J., Bai, Z., & Gang, G., (2010). "Functional assembly of bacterial communities with activity for the biodegradation of an organo phosphorus pesticide in the rape phyllosphere." *FEMS Microbiology Letters, 306*(2), 135–143.

O'Brien, R. D., & Lindow, S. E., (1989). Effect of plant species and environmental conditions on epiphytic population sizes of *Pseudomonas syringae* and other bacteria. *Phytopathology, 79*, 619–627.

Ottesen, A. R., González, P. A., White, J. R., Pettengill, J. B., Li, C., & Allard, S., (2013). Baseline survey of the anatomical microbial ecology of an important food plant: *Solanum lycopersicum* (tomato). *BMC Microbiology, 13*, 114. doi: 10.1186/1471-2180-13-114/.

Ottesen, A. R., White, J. R., Skaltsas, D. N., Newell, M. J., & Walsh, C. S., (2009). Impact of organic and conventional management on the phyllosphere microbial ecology of an apple crop. *J. Food Prot., 72*, 2321–2325.

Parasuraman, P., Pattnaik, S., & Busi, S., (2019). Phyllosphere microbiome: Functional importance in sustainable agriculture. In: *Microbial Biotechnology in Agro-environmental Sustainability* (pp. 135–138). Elsevier Publishers, Holland. https://doi.org/10.1016/B978-0-444-64191-5.00010-9 /.

Pati, B. R., & Chandra, A. K., (1981). Effect of spraying phyllosphere bacterial isolates on wheat leaves. *Plant and Soil, 61*, 419–427.

Perazzolli, M., Antonielli, L., Storari, M., Puopolo, G., Pancher, M., Giovannini, O., Pindo, M., & Pertot, I., (2014). Resilience of the natural phyllosphere microbiota of the grapevine to chemical and biological pesticides. *Applied and Environmental Microbiology, 80*, 3585–3596. doi: 10.1128/AEM.00415-14/.

Pinto, C., Pinho, D., Sousa, S., Pinheiro, M., Egas, C., & Gomes, A. C., (2014). Unravelling the diversity of grapevine microbiome. *PLoS One, 9*(1). https://doi.org/10.1371/journal.pone.0085622/.

Pons, T. L., Perreijn, K., Van, Kessel, C., & Werger, M. J. A., (2007). Symbiotic nitrogen fixation in a tropical rainforest: ^{15}N natural abundance measurements supported by experimental isotopic enrichment. *New Phytologist, 173*, 154–167.

Qin, C., Tao, J., Liu, T., Liu, Y., Xiao, N., Li, T., Gu, Y., et al., (2019). Responses of phyllosphere microbiota and plant health to application of two different biocontrol agents. *Open AMB Express, 9*, 1–13. Springer Nature. https://doi.org/10.1186/s13568-019-0765-x.

Rastogi, G., Coaker, G. L., & Leveau, J. H., (2013). New insights into the structure and function of phyllosphere microbiota through high-throughput molecular approaches. *FEMS Microbiology Letters, 348*, 1–10.

Rastogi, G., Sbodio, A., Tech, J. J., Suslow, T. V., Coaker, G. L., & Leveau, J. H., (2012). Leaf microbiota in an agroecosystem: Spatio-temporal variation in bacterial community composition on field-grown lettuce. *International Society of Microbiology Ecology Journal. 6*, 1812–1822.

Redford, A. J., Bowers, R. M., Knight, R., Linhart, Y., & Fierer, N., (2010). The ecology of the phyllosphere: Geographic and phylogenetic variability in the distribution of bacteria on tree leaves. *Environmental Microbiology, 12*, 2885–2893. doi: 10.1111/j.1462-2920.2010.02258/.

Reisberg, E. E., Hildebrandt, U., Riederer, M., & Hentschel, U., (2013). Distinct phyllosphere bacterial communities on *Arabidopsis wax* mutant leaves. *PLoS One, 8*, e78613. doi: 10.1371/journal.pone.0078613/.

Research and Market, (2020). *Agricultural Microbials Market by Type, Function, Crop Type, Mode of Application, Formulation, and Region – Forecast to 2025* (pp. 1–6). Globe News Wire. https://www.globenewswire.com/news-release/2020/05/20/2036230/0/en/Agricultural-Microbials-Market-by-Type-Function-Crop-Type-Mode-of-Application-Formulation-and-Region-Forecast-to-2025.html/ (accessed on 19 January 2024).

Rico, L., Ogaya, R., & Terradas, J., (2014). Community structures of N_2-fixing bacteria associated with the phyllosphere of a Holm oak forest and their response to drought. *Plant Biology, 16*, 586–593.

Rodrigues, C. E., Mitter, B., Antonielli, L., Trognitz, F., Company, S., & Sessitch, A., (2018). Roots and panicles of the C4 model grasses *Setariaviridis* and *S. pumila* host distinct bacterial assemblages with core taxa conserved cross host genotypes and sampling sites. *Frontiers in Microbiology*. 10.3389/fmicb.2018.02708.

Ruinen, J., (1956). Occurrence of *Beijerinckia* species on the phyllosphere. *Nature London, 177*, 220–221.

Ruinen, J., (1963). The phyllosphere II. Yeasts from the phyllosphere of tropical foliage. *Antonie van Leuwenhoek, 29,* 425–438.

Ruinen, J., (1965). Nitrogen fixation in the phyllosphere. *Plant and Soil, 22*, 375–394.

Saborit, J. M. D., (2009). Effects of air pollution on citrus trees. *Tree and Forestry Sciences and Biotechnology Series.* Global Science Books (pp. 1–13). http://www.globalsciencebooks.info/Online/GSBOnline/images/0906/TFSB_3(SI1)/TFSB_3(SI1)92-104o.pdf (accessed on 19 January 2024).

Sanders, R., (2018). *Fertilizer Destroys Plant Microbiome's Ability to Protect Against Disease* (pp. 1–7). University of California, Berkeley, California, USA, Science News. https://vcresearch.berkeley.edu/news/fertilizer-destroys-plant-microbiomes-ability-protect-against-disease (accessed on 19 January 2024).

Sandhu, A., (2008). *Phosphorus-Mediated Bacterial Degradation of Airborne Phenol* (Vol. 15878, p. 89). Iowa State University, Ames, Iowa, USA Ph.D. Dissertation. https://lib.dr.iastate.edu/rtd/15878/(accessed on 19 January 2024).

Scheublin, T. R., Johan, H., & Leveau, J. H., (2013). Isolation of *Arthrobacter* species from the phyllosphere and demonstration of their epiphytic fitness. *Microbiology Open, 2*, 205–213. doi: 10.1002/mbo3.59/.

Schumacher, J., Simon, A., Cohrs, K. C., Viaud, M., & Tudzynski, P., (2014). The transcription factor BcLTF1 regulates virulence and light responses in necrotrophic plant pathogen *Botrytis cinerea. PLoS Genetics, 10*, e1004040. 10.1371/journal.pgen.1004040/.

Shelake, R. M., Pramanik, D., & Kim, J. Y., (2019). Exploration of plant-microbe interactions for sustainable agriculture in CRSPR era. *Microorganisms, 7*, 269. https://doi.org/10.3390/microorganisms7080269/.

Shingler, V., Franklin, F. C. H., Tsuda, M., Holroyd, D., & Bagdasarian, M., (1989). Molecular analysis of a plasmid-encoded phenol hydroxylase from *Pseudomonas* CF600. *Journal of General Microbiology, 135*, 1083–1092.

Shivakumar, N., Satishkumar, R., Selvakumar, G., Shyamkumar, R., & Arjunkumar, K., (2020). Phyllosphere microbiomes: Diversity, ecological significance, and biotechnological applications. In: *Plant Microbiomes for Sustainable Agriculture, Sustainable Development and Biodiversity* (pp. 113–172). Springer Nature Switzerland AG. https://doi.org/10.1007/978-3-030-38453-1_5/.

Suda, W., Nagasaki, A., & Shishido, M., (2009). Powdery mildew infection changes bacterial community composition in the phyllosphere. *Microbes and Environments, 24*, 217–223. doi: 10.1264/jsme2.ME09114/.

Suthaparan, A., Stensvand, A., Solhaug, K. A., Torre, S., Mortensen, L. M., Gadoury, D. M., et al., (2012). Suppression of powdery mildew (*Podosphaerapannosa*) in greenhouse roses by brief exposure to supplemental UV-B radiation. *Plant Dis., 96*, 1653–1660. doi: 10.1094/PDIS-01-12-0094-REease.

Suthaparan, A., Stensvand, A., Solhaug, K. A., Torre, S., Telfer, K. H., & Ruud, A. K., (2014). Suppression of cucumber powdery mildew by supplemental UV-bradiation in greenhouses can be augmented or reduced by background radiation quality. *Plant Disease, 98*, 1349–1357. doi, 10.1094/PDIS-03-13-0222-RE/.

Sy, A., Timmers, A. C., Knief, C., & Vorholt, J. A., (2005). Methylotrophic metabolism is advantageous for *Methylobacteriumextorquens* during colonization of *Medicago truncatula* under competitive conditions. *Applied and Environmental Microbiology, 71*, 7245–7252.

Tam Le, T., Eymann, C., Albrecht, D., Sietmann, R., Schauer, F., Hecker, M., & Antelmann, H., (2006). Differential gene expression in response to phenol and catechol reveals different metabolic activities for the degradation of aromatic compounds in *Bacillus subtilis*. *Environmental Microbiology, 8*, 1408–1427.

Thapa, S., Prasanna, R., Ranjan, K., Velmourougane, K., & Ramakrishna, B., (2017). Nutrients and host attributes modulate the abundance and functional traits of phyllosphere microbiome in rice. *Microbiological Research, 204*, 55–64.

Thompson, I. P., Bailey, M. J., Ellis, R. J., Lilley, A. K., McCormack, P. J., Purdy, K. J., & Rainey, P. B., (1995). Short-term community dynamics in the phyllosphere microbiology of field-grown sugar beet. *FEMS Microbiology and Ecology, 16*, 205–211.

Thompson, I. P., Bailey, M. J., Fenlon, J. S., Fermor, T. R., Lilley, A. K., Lynch, J. M., McCormack, P. J., et al., (1993). Quantitative and qualitative seasonal changes in the microbial community from the phyllosphere of sugar beet (*Beta vulgaris*). *Plant and Soil, 150*, 177–191.

Thoudam, R., & Dutta, B. K., (2012). Control of black rot of tea, *Camellia sinensis* (L.) Kuntze with mycoflora isolated from tea environment and phyllosphere. *Journal of Biological Control, 26*, 341–346.

Trabelsi, D., & Mhamdi, R., (2013). Microbial inoculants and their impact on soil microbial communities: A review. *Biomedical Research International*. https://doi.org/10.1155/2013/86324 0?.

Vacher, C., Hampe, A., & Porté, A. J., (2016). The phyllosphere: Microbial jungle at the plant climate interface. *Annual Review Ecology and Evolutionary Systematic, 47*, 1–24.

Valverde, A., González-Tirante, M., Medina-Sierra, M., Rivas, R., Santa-Regina, I., & Igual, J. M., (2017). Culturable bacterial diversity from the chestnut (*Castanea sativa* Mill.) phyllosphere and antagonism against the fungi causing the chestnut blight and ink diseases. *AIMS Microbiology, 3*(2), 293–314.

Verginer, M., Leitner, E., & Berg, G., (2010). Production of volatile metabolites by grape associated microorganisms. *Journal of Agricultural and Food Chemistry, 58*, 8344–8350.

Vettraino, A. M., Morel, O., & Perlerou, C., (2005). Occurrence and distribution of *phytophthora* species in European chestnut stands, and their association with ink disease and crown decline. *European Journal of Plant Pathology, 111*, 169–180.

Vitousek, P. M., Cassman, K., Cleveland, C., Crews, T., Field, C. B., Grimm, N. B., et al., (2002). Towards an ecological understanding of biological nitrogen fixation. *Biogeochemistry, 57*, 1–45.

Vorholt, J. A., (2012). Microbial life in the phyllosphere. *Nature Reviews: Microbiology, 10*, 828–840.

Wachowska, U., (2009). Activity of fungicides against epiphytic yeast-like fungi of winter wheat. *Polish Journal of Environmental Studies, 18*, 1171–1176.

Wagi, S., & Ahmed, A., (2017). Phyllospheric plant growth promoting bacteria. *Journal of Bacteriology and Mycology Open Access, 5*(1), 215–216.

Wang, D., Deng, X., Wang, B., Zhang, N., Zhu, C., & Jiao, Z., (2019). Effects of foliar application of amino acid liquid fertilizers, with or without *Bacillus amyloliquefaciens* SQR9, on cowpea yield and leaf microbiota. *PLoS One, 14*(9), e0222048. https://doi.org/10.1371/journal.pone.0222048/.

Wang, H., Jiang, Y. P., Yu, H. J., Xia, X. J., Shi, K., & Zhou, Y. H., (2010). Light quality affects incidence of powdery mildew, expression of defense-related and associated metabolism in cucumber plants. *European Journal of Plant Pathology, 127*, 25–135. doi, 10.1007/s10658-009-9577-1/.

Wei, F., Hu, X., & Xu, X., (2016). Dispersal of *Bacillus subtilis* and its effect on strawberry phyllosphere under open field and protection condition. *Nature Scientific Reports, 6*(22611), 1–9. doi: 10.1038/srep22611/.

Whipps, J. M., Hand, P., Pink, D., & Bending, G. D., (2008). Phyllosphere microbiology with special reference to diversity and plant genotype. *Journal of Applied Microbiology, 105*, 1744–1755.

Wikipedia, (2022). *Phyllosphere* (pp. 1–6). https://en.wikipedia.org/wiki/Phyllosphere/ (accessed on 19 January 2024).

Wilson, M., & Lindow, S. E., (1993). Interactions between the biological control agent *Pseudomonas fluorescens* strain A506 and Erwinia amy-lovora in pear flowers. *Phytopathology, 83*, 117–123.

Wilson, M., & Lindow, S. E., (1994). Coexistence among epiphytic bacterial populations mediated through nutritional resource partitioning. *Applied and Environmental Microbiology, 60*,4468–4477.

Wilson, M., & Lindow, S. W., (1993). Interaction between the biological control agent *Pseudomonas fluorascens* strain 506 and *Erwinia amylovora* in pear flowers. *Phytopathology, 83*, 117–123.

Yadav, R. K. P., Karamanoli, K., & Vokou, D., (2005). Bacterial colonization of the phyllosphere of Mediterranean perennial species as influenced by leaf structural and chemical features. *Microbial Ecology, 50*, 185–196.

Yang, C. H., Crowley, D. E., Borneman, J., & Keen, N. E., (2000). Microbial phyllosphere populations are more complex than previously realized. *Proceedings of National Academy of Science, USA, 98*, 3889–3894.

You, C., Zhang, C., Feng, C., Wang, J., & Kong, F., (2015). *Myroidesodoratimimus*, a biocontrol agent from the rhizosphere of tobacco with potential to control *Alternaria alternata. Biocontrol, 60*, 555–564. https://doi.org/10.1007/s1052 6-015-9654-7.

Yutthammo, C., Thongthammachat, N., Pinphanichakarn, P., & Luepromchai, E., (2010). Diversity and activity of PAH-degrading bacteria in the phyllosphere of ornamental plants. *Microbial. Ecology, 59*, 357–368. doi: 10.1007/s00248–009-9631-8/.

Zhang, B. Z., Bai, Z., & Hoefel, D., (2009a). The impacts of cypermethrin pesticide application on the non-target microbial community of the pepper plant phyllosphere. *Science of the Total Environment, 407*(6), 1915–1922.

Zhang, B., Tang, L., & Li, M., (2009b). Effect of abamectin insecticide on the microbial community in broccoli phyllosphere. *Environmental Science, 30*, 1292–1297.

CHAPTER 7

Aerial Microbes and Crop Diseases: Epidemics and Pandemics

ABSTRACT

The theme of this chapter is to demonstrate the detrimental impact of the agrarian environment on food grain production worldwide. The agrarian sky plays a role in the spread of diseases among natural vegetation and various crop species. It also contributes to the occurrence of epidemics and pandemics that devastate crops. If left unaddressed, the decline in crop productivity can be severe and recurring in every season. The agrarian sky facilitates the migration of plant pathogens and provides favorable conditions for their survival, propagation, and infection of crops. It serves as a habitat for microbial plant pathogens, particularly in regions with hot or unhealthy winds, which act as reservoirs for disease-causing agents. Wheat rust fungi, such as stem rust (*Puccinia graminis tritici*), yellow rust (*P. striiformis*), and leaf rust (*P. triticina*), are well-studied examples of airborne fungi. Some rust fungi can migrate over long distances and trigger epidemics, like sugarcane rust (*P. melancephala*). Coffee rust, caused by *Hemileia vastatrix,* is another airborne rust fungus that rapidly spreads and causes significant damage to coffee plantations. It is commonly recommended to plant rust-resistant genetic stocks and distribute them strategically in areas where urediospores travel through the agrarian sky. However, it is also important to implement measures that reduce the population of urediospores in the air above the crop canopy. Employing multiple effective methods to minimize the spread of diseases is a prudent approach.

There are highly airborne fungal pathogens, such as fusarium head blight (FHB) fungus (*Fusarium graminorum*). Fungal blasts, like Pyricularia

Microbiology of the Agricultural Sky. K. R. Krishna (Ed)

blast on staple cereals, are also mediated via airborne fungi. Fungal blights mediated by wind, water, and farm vehicles have affected major food crops. For example, Phytophthora blight of potatoes in the mid-1800s and Helminthosporium blight of rice in India in the 1930s-1940s, which was due to *Cochliobolus miyabeanus*. Factors operative in the agrarian sky, especially wind, precipitation, and lack of phyto-sanitation in fields, have induced such epidemics.

The agrarian sky is also equitable in supporting bacterial diseases. It has consistently aided epidemics/pandemics caused by bacterial plant pathogens. Troposphere factors like wind, jet-streams, storms, and precipitation mediate the migration of bacterial pathogens. Xanthomonas species are excellent examples of airborne/waterborne bacteria that induce diseases leading to epidemics. A few examples are Angular blotch of cotton caused by *X. campestris var. malvacearum* and Citrus canker by *X. campestris var. citri. Xanthomonas campestris var. campestris* causes cruciferous blotches. *Xanthomonas oryzae* has caused severe yield loss within Asian rice belts in 1960. *Pseudomonas* spp. is a common airborne microbial pathogen. It is transmitted via the wind, clouds, precipitation, foliar sprays, etc. There are also vector-mediated bacterial diseases that affect crops. For example, citrus greening (Huanglongbing disease) recently caused epidemics in Floridian citrus groves. It has earlier caused epidemics in Asia and has rapidly spread into American citrus groves. It is transmitted by an aerial Psyllid insect vector – *Diaphorina citri*.

The aerial environment in agrarian areas plays a significant role in the transmission of viral epidemics and pandemics to crops. In recent years, major cereal crops have experienced a decline in grain yield due to the presence of airborne insect vectors that disperse viruses. For instance, the wheat streak mosaic virus (WSMV) is transmitted by mites, while aphids, jassids, psyllids, and whiteflies also act as vectors for various viruses. Understanding their flight behavior, migratory patterns, and viral loads is crucial in assessing the risk they pose.

Several examples of airborne virus-vector combinations that contribute to the spread of crop epidemics include the maize yellow dwarf virus transmitted by aphids (*Ropalosiphum padi*), the Citrus tristeza virus transmitted by *Toxoptera citricida,* and the Banana bunchy top virus spread by *Pentalonia nigronervosa*. These viruses take advantage of the airspace above crops to propagate.

The agrarian sky constantly harbors a multitude of disease-causing agents. The presence and intensity of these microbial species in the air can vary. Assessing their virulence and their potential to reduce crop yield is essential in practical farming.

7.1 INTRODUCTION

Historically, 'Agrarian Sky' has the reputation of causing widespread crop devastation worldwide, regardless of geographic location, soil fertility status, crop species, or farming practices. This is due to the role of the 'Agrarian Sky' in harboring countless plant pathogens and facilitating their growth. The 'Agrarian Sky' has also played a role in the migration and spread of plant pathogens across different agroecosystems. It provides a favorable environment for these pathogens to rapidly infect crops. The detrimental impact of the 'Agrarian Sky' has been evident season after season, year after year, with long cycles of crop devastation caused by similar diseases recorded in historical records. Farmers have recognized the influence of the 'Agrarian Sky' and its biotic factors on crop health for millennia. Ancient Roman farmers, for example, identified winds in the 'Agrarian Sky' above their crops as 'warm or hot' or 'sickly,' indicating the presence of pathogens. Wheat fields, for instance, were attacked by rust when such disease-causing winds blew over the crops. Globally, there has not been a recent year without an epidemic or pandemic caused by airborne crop pathogens. We have witnessed epidemics affecting multiple crops in recent years, with many of them being mediated through the 'Agrarian Sky.'

The literature included in this chapter, inferences by crop disease experts, and combinations of related information presented actually intend to emphasize that the 'agrarian sky' poses a real threat to crops, particularly when it comes to crop diseases. It serves as an excellent medium for spreading microbes that can induce epidemics. The agrarian sky has the potential to create a massive pandemic, resulting in significant crop loss on a global scale. This situation indirectly prompts us to question why we are not manipulating the agrarian sky as much as we do with soil, crops, or even water to control diseases. Our ability to regulate the behavior of the agrarian sky immediately above the crop's canopy is crucial. Unfortunately, at present, we have limited knowledge about the various parameters and methods that need to be adopted.

What is an Epidemic or Pandemic Disease on Crop? Plant disease epidemics involve changes in disease intensity on a host population over time and space. Acquiring comprehensive information on this process is necessary. It helps us to understand the factors that cause epidemics (El Jorroudi et al., 2018). Furthermore, a thorough knowledge and detailed data about cropping systems practiced and risk factors may help farmers overcome massive losses due to epidemics. A good understanding of the pathogen's life cycle, intensity of inoculum, its spread via aerospace above the canopy, weather patterns, and the crop's stages most vulnerable to the pathogen are required. Such information aids in modeling and decision-making regarding prophylaxis, such as fungicidal spray schedules.

Jones (2008) attempts to explain that the domestication of plants, their rapid movement across different geographic locations, and their interaction with viruses *in situ* or in distant locations may have served as precursors for epidemics. New encounters between microbes and crops are always possible due to the aerial migration of disease propagules through aeolian processes. Additionally, climate change effects and atmospheric processes can induce or alter the interaction between plants and microbes. These factors have the potential to trigger epidemics or pandemics.

We have gained a significant understanding of various aspects of epidemics and pandemics caused by microbial pathogens in general. However, we require more information on strategies that can assist us in regulating airborne plant pathogens. We have not given the same level of attention to modifying the factors in the agrarian sky as we have to controlling and altering the soil/water phases of agricultural fields. Airborne pathogens can have devastating effects, potentially ruining the agrarian economies of multiple nations in one full swoop. Despite this, we have dedicated less effort to controlling aerial pathogens, particularly their migratory routes, infectivity, and tolerance to the harsh conditions of the troposphere, among other factors. There have indeed been several epidemics reported in the literature. Typically, the list begins with the Potato epidemic caused by *Phytophthora infestans* (de Bary), which was introduced into Ireland from Mexico in the 1840s (Campos & Ortiz, 2020; 2 Blades Foundation, 2021). In this chapter, we discuss the role of the agrarian sky in disseminating causal agents such as bacteria, fungi, and viruses, as well as the imitation of epidemics and pandemics. It is worth noting that the severity of an epidemic/pandemic is often greatly influenced by troposphere conditions. Aspects of pathogens such as virulence, intensity, rapid spread, and loss of grains/fruits are also important.

Ristaino et al. (2021) have demonstrated that crop disease outbreaks are becoming more frequent. Virulent strains are expanding in terms of area, disease intensity, and even persistence over multiple years. Reports from crop disease surveillance agencies indicate the emergence of several epidemics/pandemics. These disease flare-ups are causing significant losses in crops and food grains (2 Blades Foundation, 2022).

It is crucial to recognize that the agrarian aerospace has a profound impact on the epidemiology of microbial plant pathogens. Our focus should be on investigating methods to regulate this domain in order to mitigate the devastations caused by epidemics/pandemics. Interestingly, among microbes, it is the airborne fungi that pose the greatest threat to agricultural crops.

Discussions in the sub-sections of Section 7.2 make it clear that airborne fungi, whose survival and spread are partly supported by the troposphere, can be highly detrimental. In some cases, they can lead to devastating epidemics/pandemics.

7.2 AGRARIAN SKY AND MICROBIAL PATHOGENS OF CROPS

In nature, there is a vast abundance of airborne microbial plant pathogens that require attention. Many of these pathogenic microbes reside in the agrarian sky for their entire life cycle or at least one or a few stages of it. The agrarian sky can harbor multiple microbial plant pathogens simultaneously, in various combinations and proportions. In practical agriculture, we have encountered several diseases caused by microbes that operate at mild levels, which may not require intensive study and prophylaxis. However, there are several others that are devastating. The dispersal of microbial pathogens onto crops is often facilitated through the agrarian sky. Some pathogens may thrive in the troposphere for a certain period, such as the fungi that cause epidemics/pandemics and disperse over long distances. Globally, major agricultural regions that produce food grains and forest belts suffer severe damages due to epidemics and pandemics. These intense disease situations erupt periodically. If left unattended and unchecked, some microbial plant pathogens revisit the crops frequently in a cyclical manner.

The primary objective of this chapter is to provide evidence that the 'agrarian sky' serves as a habitat for significant epidemic/pandemic-causing microorganisms. Only a few examples of utmost importance to global agricultural enterprises are discussed. However, the subsequent sections include discussions on those epidemics that have historically caused devastating impacts. The microbial pathogens included in these discussions encompass fungi (such as rusts, blights, mildews, smuts, and ergots), bacteria (blotches, blights, and blasts), and viruses transmitted by aerial vectors (such as streak viruses, leaf curls, and stunts). Examples of epidemics and pandemics on major crops like wheat, maize, rice, legumes, oilseeds, and vegetables have been utilized to demonstrate the urgent need for greater attention to be given to the agrarian sky. Particularly, in terms of disease management.

Several species of archaea have been reported to occur in the agrarian sky (troposphere) (see Chapters 2 and 3). However, there are no reports of archaea being pathogenic or inducing epidemics on crop plants. It seems that they are rarely found on the phyllosphere. Perhaps we need to conduct a detailed investigation to determine the role of archaea found in the air as either deleterious or beneficial biotic agents on crops (Taffner et al., 2018; Jung et al., 2020).

7.2.1 AGRARIAN SKY AND RUST FUNGI ATTACKING AGRICULTURAL CROPS

The agrarian sky has played a crucial role in the causation and perpetuation of wheat epidemics/pandemics. Rusts, spots, blotches, blights, and blasts have occurred on the wheat crop. Perhaps, humans have noticed them since the period when wheat was domesticated, several millennia ago. However, wheat rust pathogens have evolved to become more aggressive through rapid genetic drift and acclimatization to newer wheat genotypes and cropping systems. They have so far withstood the effects of climate change, repeatedly causing the disease. Estimates indicate that neglecting the agrarian sky, i.e., ignoring rust fungi and the lack of prophylactic sprays, could reduce global wheat yield by 15%. According to Figueroa (2018), major rust fungi such as black stem rust, stripe or yellow rust, and leaf rust have been devastating on the wheat crop. A few recent additions to the list of fungal pathogens on wheat are Septoria and Helminthosporium blotches, FHB, and Pyricularia wheat blast. A favorable agrarian sky for rust and other fungi affecting wheat, along with a set of versatile and aggressive strains of pathogens, and a susceptible wheat crop at the seedling stage, create the right conditions for rapid disease dissemination. This, of course, leads to massive and large-scale loss of foliage and grain yield.

Wheat is an important cereal as it provides calories to over 1.2 billion people on Earth. Its cultivation zones are found in all continents. Globally, wheat covers slightly more than 219 million hectares. Abiotic stresses such as drought, floods, high ambient temperature, and severely cold conditions are detrimental to optimum yield. Climate change also has its share of deleterious effects, although not in all regions. In this context, we are concerned with rust fungi. Rusts of wheat are among the most investigated fungal plant pathogens. Rust fungi have the potential to cause epidemics and pandemics season after season. They are distributed worldwide. Rusts are airborne fungi with ability to attack major cereal (Graminaceous) food crops such as wheat, rye, triticale, barley, millets, etc. (Lorrain et al., 2018; McIntosh, 1995, 2009). Agricultural sky is the major source of wheat rust inoculum. Therefore, regular monitoring of aerospace above crops for urediniospores of various species is essential (FAO, 2104). Further, we should realize that a 15% grain loss due to rust epidemics means 5.8 billion US$ in exchequer (at 2014 price levels) (Figeuroa et al., 2018). A global surveillance program is needed to cover all wheat cultivating areas (Park et al., 2011). Aerial surveys using aerostats and satellite-based information networks appear to be almost mandatory (Park et al., 2011; Krishna, 2020; Listman, 2018). This suggestion aligns with initiatives already undertaken by agencies in the USA (Kansas State University), Australia (Queensland), and New Zealand wheat consortia. The objective is to monitor the agrarian sky for the yellow streak virus, which is transmitted aerially by the Russian Aphid. (Listman, 2018; Krishna, 2019).

Wheat crops grown in agricultural zones are susceptible to attack by three major rust fungi. Wheat Black Stem Rust, caused by *Puccinia graminis tritici* (Pgt) (Schumann & Leonard, 2000), is one of them. Another is Stripe Rust, also known as yellow stripe rust, caused by *Puccinis striformis tritici* (Pst) (Chen, 2005; Chen et al., 2014). Leaf Rust, caused by *Puccinia triticina,* is also referred to as Brown Leaf Rust (*P. recondite tritici*) (*see* Terefe et al., 2009). These rust fungi have evolved into different races, each with varying reactions to different wheat genotypes. The severity of their epidemics/pandemics also varies. During their fungal life cycle, each of these races is airborne for a certain period, with the urediniospores stage being critical as it is airborne. Rust spores are capable of withstanding harsh tropospheric conditions and can travel long distances with the wind, clouds, and precipitation patterns.

A thorough understanding of microbial load and diversity, specifically focusing on rust fungi, bacteria, and viruses that have the potential to cause diseases, is necessary when aiming to stabilize wheat crop yield. Rust fungi are undoubtedly airborne, with spores (urediniospores) being transported across the globe through high-altitude winds, tornadoes, jet-streams, clouds, and storms, facilitated by the agrarian sky. According to Nagarajan & Singh (1990), the long-distance transport and survival of various types of rust fungi spores are among the major causes of crop devastation. Wheat rust fungi can disperse through the troposphere for short distances between fields, and short-distance migration can also occur when farm vehicles move across fields. Migration can also be facilitated by aerial sprays of fluids, mild breezes, and turbulence. In fact, a single leaf blade of a wheat plant can sometimes carry millions of urediniospores capable of being lifted into the sky and spreading to other regions (Zadoks & Schein, 1979; Burdon, 1987). The long-distance migration of urediniospores plays a critical role in the regional spread of wheat rust (Brown & Hovmoller, 2002; Mahafee & Stoll, 2016; Mendgen, 1997).

Let us consider a few examples of regional dispersal of rust fungi (propagules) via tropospheric wind. In North America, wheat rusts are known to travel from the Southwest of the Great Plains to the Northeast via low altitude wind. Light-weight spores travel for over 800 km each year (Roelf et al., 1992; Golan & Pringle, 2017). Spores are tolerant to aerospace conditions. Spores spread the disease rapidly in a few days. Ultimately, the survival and regeneration of spores in the location where they land is important. Regional transport of rust fungi also occurs within the Australian continent (Luig, 1985; McGrath & Pennypackers, 1991). Regional movement of wheat rust has also been observed and routes marked in other continents like Africa, South Asia (Brown & Hovmoller, 2002), and the Indian Subcontinent (Nagarajan & Singh, 1990). Intercontinental long-distance movement of wheat rusts is among the best-studied examples of microbes that disperse via the sky. The aerospace above crops and oceans is traversed by lightweight urediniospores during intercontinental transit. The spores also disperse across continents by first ascending to higher altitudes in the sky. They then travel over distances of 8,000 to 15,000 km via wind and jet streams (Acosta-Martinez et al., 2018). The long-distance dust plumes created by the Sub-Sahara transport rust fungi to the Caribbean and Southern plains. These are prime examples of agrarian sky-mediated epidemics (Brown & Hovmoller, 2002). Rust fungi are also known to migrate from the Gobi region in Mongolia and the Fareast to the Northwest Pacific in America (Kellog & Griffin, 2006). Periodically, stem rust fungi originating from East Africa (Kenya), such as the Ug99 strain, are known to travel to West Asia via Arabia (Brown & Hovmoller, 2002; Smith et al., 2012, 2018; Morris & Sands, 2017). There are also reports of rust fungi enduring the agrarian sky and spreading from the Indus Plains to the Eastern Gangetic wheat belt (Nagarajan & Joshi, 1985).

Black stem rust (*Pgt*) of wheat has a worldwide distribution. It is rapidly transmitted in the troposphere through urediniospores. This basidiomycetous fungus is commonly found in relatively warm and moisture-filled skies. The stem, foliage, panicles, glumes, and even awns display red urediniospores. The aerial parts of a crop serve as a major source of rust fungal inoculum. The troposphere is regularly replenished with rust spores, easily carried by mild winds, turbulence, or convection currents (Figeuroa et al., 2018). This airborne fungus causes a reduction in grain formation. Infection leads to smaller panicles and grains (Leonard & Szabo, 2005). Planting resistant wheat varieties, adopting cropping systems that avoid areas with high stem rust spore concentrations, and using fungicidal sprays have helped control *Pgt* to some extent. However, an epidemic caused by *Pgt* can still result in the loss of millions of tons of grains (Figeuroa et al., 2018). During recent past, airborne *Pgt* has exposed the vulnerability of several commonly cultivated wheat cultivars across different locations around the world. Virulent strains have emerged in the agricultural landscape. Some of these, such as 'Ug 99,' have spread to different continents. The spread of virulent strains (e.g., Ug 99, Digalu) in the agricultural sector has significantly impacted food security in terms of the demand for wheat grains (Singh et al., 2011, 2015; Olivera Firpo et al., 2015, 2017; *see* BGRI Archive site, 2022).

Around 2017, the virulent stem rust strain known as 'Ug99' had already spread to regions in Iran, Sudan, Kenya, Ethiopia, Uganda, and South Africa. Several high-risk regions included Egypt, Iran, Iraq, Afghanistan, and Pakistan. Regions in the Indian subcontinent and Central Asia were also at risk, although to a relatively low level of susceptibility. There are still regions that have not been invaded by Ug99. Therefore, it is crucial to carefully monitor the crops' canopy and the agrarian sky above them in the European plains, as neglecting the aerospace above wheat growing regions can lead to disasters. It is important to avoid an agrarian sky infested with urediniospores of a virulent strain. The sanitation of the aerospace above crops in terms of disease propagules is emphasized here. The agrarian sky harbors over 10^6 microbes per cm^3, as reported by Wommack in 2010. Hence, it is essential to monitor the ratio of rust spores or disease-causing propagules to the total microbial count in the troposphere.

Agrarian sky can induce epidemics of large proportions repeatedly, in short intervals. Disease outbreaks could also be spaced well depending on the occurrence of virulent strains and/or susceptible hosts. Breakdown of crops' resistance to attacks by airborne fungi is common. Let us consider an example. A black stem rust outbreak (epidemic) was identified in 1899 in the Australian continent. Similar outbreaks of *Pgt* on Australian wheat were recorded in 1947, 1973, and more recently in 2014 (Pardey et al., 2013).

Let us now consider Stripe Rust of wheat, also known as 'Yellow Rust.' It is caused by *Puccinia striiformis f. sp. tritici* (*Pst*). Stripe rust is a devastating pathogen that leads to a complete loss of grain yield. In 2016, the global wheat grain loss due to *Pst* was estimated at US $2.7 billion. In Australia alone, the grain loss was estimated at 1.27 billion A$ in 2014 (Beddow et al., 2015; Murray & Brennan, 2009).

Favorable lower troposphere conditions in agrarian skies facilitate the massive accumulation of *Pst* spores. Wind, breeze, turbulence, dust storms, and jet streams further aid in its rapid dispersal. It is said that 88% of wheat genotypes sown anywhere in the world are susceptible to virulent forms of *P. striiformis*. The airspace above wheat crops serves as a significant reservoir of *Pst* races. In fact, genetically diverse populations of *Pst* are distributed throughout the wheat production zones worldwide. Natural sexual recombination events induce genetic diversity. Several races and virulent strains of *Pst* have been identified in North America, Europe, and Australia. Aggressive races of *Pst* with tolerance to uncongenial temperature, desiccation, and radiation has been noticed. Airborne urediniospores of *Pst* do migrate to shorter/longer distances via troposphere. Polymorphs of stripe rust fungus has been traced in several countries like China, Central Asia, Indo-Gangetic regions, etc.

No doubt, yellow stripe rust fungus is versatile. Genetic drifts generating new races have aided the pathogen in retaining a high degree of virulence. During the past decade, a few virulent strains have swept across wheat cultivating zones in the European plains. It seems that some of the rust fungal strains originated in the sub-Himalayan wheat belt in South Asia. Intercontinental migration via the agrarian sky, utilizing low and high-altitude wind and weather systems, has caused the spread of virulent strains of yellow stripe rust (Hovmoller et al., 2008; Hubbard et al., 2015; Plate 7.1).

Wellings (2011), in his review, states that during the past centuries, both the host and pathogen have experienced natural selection. They have undergone genetic modifications to adapt to each other. The airborne spores travel long distances. *Berberis* spp. is a collateral host that supports the perfect stage of the fungus (Jin et al., 2010). Furthermore, Wellings (2011) points out that genetic drifts and evolution of pathotypes during the past five decades have induced disease foci. This has profound consequences on global cereal production trends. The stripe rust fungus has often multiplied rapidly, inducing epidemics. However, the build-up of rust fungus depends on conditions prevalent in the atmosphere and the actual spread of spores/propagules in the crop's canopy. Regional yellow rust epidemics are common in wheat belts in the North American Plains (Chen et al., 2005, 2014).

Rust fungus, which is airborne through its asexual spores, has the potential to cause global pandemics on wheat. A collective survey of global wheat afflicted by yellow rust has indicated that airborne fungus can result in grain loss ranging from 0.1% to 5% through localized epidemics. Severe cases with 25% grain loss via rapid outbreaks have also been reported. *Puccinia striformis var tritici* is currently known to have spread rapidly and caused pandemics in the North American Great Plains (Brown et al., 2001; Chen et al., 2002; Chen, 2005; Plate 7.1). *Puccinia striiformis* has also spread to Argentinean wheat-producing regions (German et al., 2007), Asian Wheat belts (such as Indo-Gangetic and Terai in Nepal) (Singh et al., 2005), Northeast China (Chen et al., 2009; Wan et al., 2004; Wellings, 2011), and Australian Wheat growing regions (Wellings & McIntosh, 1990; Wellings, 2003, 2007a, b; Cromey, 1989; Beresford, 1982). Wheat production zones in Africa (Ethiopia, Kenya, South Africa) (Boshoff et al., 2002), Iran, and the Arabian Peninsula (Yemen), as well as Southern Italy (Vallega & Zatelli, 1979), have also been invaded by *Pst*. It also occurs in Western Europe (England) (Bayles et al., 1989). Knowledge about the survival trends of airborne spores is essential. Deciphering the aerial transit routes of *Pst* via low wind and high-altitude jet streams helps us understand the cereal growing regions that could become vulnerable to epidemics/pandemics. Aerial surveys of *Pst* propagules, sanitation of agrarian sky, and growing non-hosts or resistant versions of wheat are crucial if we intend to avoid yellow rust epidemics.

Wheat Yellow rust epidemics occurred in South Africa in the 1990s. Later, it was traced in Australia (Wellings et al., 2003, 2007a, b). A few strains of wheat yellow rust fungus found in the eastern USA showed virulence and aggressiveness beyond the known levels (Chen, 2005; Milus et al., 2006, 2008; Hovmoller & Justesen, 2007a, b). Hovmoller et al. (2008) provide a list of countries in different continents that the aggressive strains of wheat yellow rust virus (*P. striformis var tritici*) have invaded via aerospace. This includes North America, Europe, and even far-off locations in Australia.

The wheat yellow rust fungus is capable of overcoming host plant resistance genes. For example, a *Pst* strain with 15 virulence loci was traced in North America and Australia. Similarly, *Pst* strains that can overcome resistance genes of wheat genotypes have spread into the European wheat belt. Tolerant *Pst* strains that overcome host resistance genes have also been traced in Western and Central Asia (Hovmoller et al., 2008). The rust fungal strains, it seems, are capable of producing an excessively large number of spores, equivalent to twice the usual spore count. Perhaps, this trait helps them infect new wheat genotypes. The airborne nature of rust spores, their tolerance to harsh environments, and their ability to overcome host resistance (i.e., aggressiveness) together enable them to induce epidemics. Particularly, when conditions allow them to travel long distances in the troposphere.

Leaf rust of wheat is caused by *Puccinia triticina* (*Pt*) (Anikster et al., 1997; Bolton, 2008). The pathogen *Pt* withstands environmental vagaries. Congenial temperatures and moisture in the agrarian sky aggravate its attack on the wheat crop. Spores are airborne via wind, turbulence, dust storms, and tornadoes. Leaf rust disease spreads through wind and splashing precipitation. Optimal moisture and temperature ranging from 59–68°F, with wet weather, induce disease epidemics in 7–10 days. In the USA, leaf rust fungus overwinters on wheat in Southern states. It erupts in spring to infect the crop in North and Midwest states of the USA. Dispersal through troposphere wind is crucial to initiate new epidemics. Undoubtedly, the agrarian sky plays a vital role in the dispersal of leaf rust of wheat.

Leaf rust fungus exhibits high genetic diversity based on geographic location and predominant wheat cultivars in the region. Leaf rust causes a reduction in panicle size, grain number, and weight, leading to a loss of grain yield. In the USA, a grain yield loss equivalent to 350 million US dollars has been attributed to leaf rust each year. In Australia, the loss is about 12 million Australian dollars (Huerto-Espino et al., 2011, 2013; Murray & Brennan, 2009) (at price levels a decade and a half ago). The leaf rust fungus produces five different types of spores. Leaf rust displays constant genetic drift, and new recombinants appear periodically, causing severe disease outbreaks. Asexual spores (urediniospores) are contagious, and aerial dispersal causes disasters in wheat. Furthermore, the adaptability of the fungus to weather adds to the problems. Alternate hosts for leaf rust include *Thalictrum, Anchusa, Clematis,* and *Isopyrum* (Figeuroa et al., 2018). Alternate hosts can be a source for the aerial spread of the fungus and initiate new infections.

Soybean rust, caused by *Pakapspora pachyrhiza* and *P. meibomiae*, is distributed worldwide. *P. pachyrhiza* was originally a significant defoliator and yield reducer in Asia (Sconyers et al., 2006). It appeared in Africa in the 1990s. Intercontinental weather patterns, wind, jet streams, and hurricanes seemingly facilitated the movement of its urediniospores into the USA in 2004 (Bromfield, 1984; Sinclair & Hartman, 1999; Rupe and Sconyers, 2008). Rust spores are pale yellow-brown-hyaline, thick-walled, echinulate, and light in weight. They can remain airborne for extended periods and are known to survive warm temperatures in the sky but not freezing conditions. However, they can rapidly transit through the troposphere at different altitudes. The soybean rust fungus also produces teliospores (sexual spores) that aid in overwintering and survival in the field. Rust pustules on foliage

PLATE 7.1 Yellow rust of wheat caused by *Puccinia striiformis var tritici* (*Pst*).

Note: Wheat leaves display heavily infected yellow rust pustules. Urediniospores are loosely attached to the leaf lamina and detach easily with even the slightest wind or turbulence. Rust spores can travel longer distances if they reach higher altitudes in the troposphere.

Source: Dr. Kianian Shaharyar, Cereal Rust Laboratory, USDA, Minnesota, USA; https://www.ars.usda.gov/midwest-area/stpaul/cereal-disease-lab/docs/cereal-rusts/wheat-stripe-rust/.

spread quickly under humid conditions. This disease serves as an excellent example of an airborne fungal disease with the potential to cause epidemics/ pandemics. Rust disease has been attributed to pod yield losses of 30–80% (Sconyers et al., 2006). The soybean rust fungus has numerous alternative hosts in legume sub-families. Planting non-hosts in the dispersal pathway, using resistant genotypes, and implementing cultural practices are recommended as effective control measures. Additionally, exploring techniques to reduce rust spore density in the airspace should be considered.

Climate change and frequent episodes of precipitation events seem to exacerbate epidemics and pandemics. For instance, soybean rust fungus spreads rapidly due to cloud movement, hurricanes, storms, and regular seasonal rains. The airborne rust spores (urediniospores), being light, easily float into the troposphere and can travel longer distances (Cardens-McGill, 2021).

Peanut rust is caused by an airborne fungus, *Puccinia arachidis*. Its urediniospores are light, with a thick wall and ornamentation, allowing them to travel long distances and withstand harsh troposphere conditions. Initially confined to North America and China, *P. arachidis* spread to African peanut producing zones, such as the 'Groundnut Basin of Senegal,' in the 1970s. Currently, *P. arachidis* can be found in almost all peanut cultivating areas across different continents. The airborne urediniospores are prone to long-distance transit in the troposphere (Savary et al., 1988; Subramaniam et al., 1985). There are several *P. arachidis* tolerant or even resistant cultivars available in global germplasm collections. These cultivars should be planted in areas where the agrarian sky is infested with *P. arachidis* urediniospores. Routes of dissemination of urediniospores need to be mapped and marked, and only resistant cultivars of peanuts should be planted along these routes.

Sugarcane rust is caused by *Puccinia melamcephala*. It is an airborne fungus. Its urediniospores are traceable in the lower troposphere. Sugarcane rust was confined to locations in Africa and Asia until the late 1970s. Its spread to the Caribbean, Brazil, and other sugarcane growing regions was observed in 1978. The intercontinental (transatlantic) journey above seas seems to coincide with the transit of Saharan dust plumes. Purdy (1985) mentions that weather-related events and the tolerance of urediospores to harsh troposphere conditions are important. Wind velocity determines the rate at which urediospores reach the Caribbean from Western Sahara. Usually, urediospores may travel 2,000 to 3,000 km in 3–4 days. Initially, the rust fungus that traveled from Northwest Africa did not encounter resistant sugarcane genotypes. Hence, the pandemic was severe. However, during the past decades, sugarcane tolerant to rust fungi has been increasingly utilized.

Rust of coffee (*Coffea spp.*) is worldwide in distribution. The coffee rust is caused by the fungus *Hemileia vastatrix*. This fungus has caused epidemics and pandemics in Asia, Africa, and Meso-America. Aerial spores drift along with the wind and alight in distant locations, inducing fresh foci of infection. Coffee rust (*H. vastatrix*) can be devastating. It almost decimated coffee plantations around Lake Victoria and other parts of Kenya. It spread to Brazil and then on to Ecuador, Mexico, and Nicaragua in the 1970s. This fungus is capable of being transferred via tropospheric wind and other weather phenomena. The transport of infected seedlings has also dispersed the rust fungus (Schieber & Zentmeyer, 1984).

Most rust fungal species are airborne via urediniospores, teliospores, and hyphal segments. The spore stage is a crucial part of their life cycle. This is because they migrate in the extreme conditions that the troposphere presents. Fungal spores are the contagions that survive, float, drift, and travel in the agrarian sky. They reinfect upon descending on a crop's canopy. They exhibit innumerable natural adaptations during this process of aerial migration. They may become dormant to avoid ambient troposphere conditions. Spore size, its striations, ornamentations, spore wall thickness, and its ability to withstand harsh radiation and desiccation are important traits. A few other traits like lightness and rapid germination upon reaching a susceptible host need attention.

Kakishima et al. (1995, 2017) have reported details of an airborne fungus called *Coleaosporium plumeria*. This fungus causes rust on Plumeria tree species, which are common ornamental trees. The immediate spread of the fungus occurs through airborne spores. Fungal migration is facilitated by weather patterns such as El Nina and La Nina, clouds, wind, jet-streams, and typhoons. Kakishima et al. (2017) provided a detailed description of its intercontinental spread. The fungus, *C. plumeria,* has been traced in South Asia, Asia Pacific, Australasia, and a few other regions of the world. In recent years, anthropogenic factors such as human travel have also contributed to its intercontinental journey and spread. Studies based on samples and herbaria suggest that *Coloeosporium plumeria* occurs in the sky throughout the year, overcoming environmental challenges. This rust fungus was originally found in Central America, but Kakishima et al. (2017) mention that it has become virulent. It spread rapidly to other continents through aerial routes. Rust fungi are known to produce aerial spores that can withstand harsh environments and travel long distances via intercontinental wind patterns and dust storms. Preserved herbaria since the early 1900s suggest that it first occurred in the West Indies (Santo Domingo) in 1852. It was localized in the tropical Caribbean and Central America (Arthur, 1916, 1934; Sydow & Sydow, 1915, 1924). The prominent dispersal of this rust fungus occurred in the 1980s (Kakishima et al., 2017). Aerial sampling has shown that its severity has spread to even remote regions from India to Australia, New Zealand, and the South Pacific in the 1980s–1990s. Urediniospores are responsible for the intercontinental aerial spread. In the 1990s, it was traced in West Africa, India, and other subtropical regions of Asia (Holcomb & Aime, 2010; Yan et al., 2016).

Aerospace may not support all types of fungal pathogens. Long-distance dispersal, in particular, is a prerequisite for fungal pathogens that cause pandemics. Let's mention a few characteristics of spores of fungal pathogens (Golan & Pringle, 2017; 2 Blades Foundation, 2020). Leaf Rust of Coffee is caused by *Hemileia vastatrix*. Its urediospores are small, measuring 29.7–34.5 × 5–8 μm, and have a globose shape (Isard et al., 2005). Soybean Leaf Rust, induced by *P. pachyrhizi*, produces urediospores of size 18–34 × 15–24 μm. The spores are globose and have ornamentations (Carvalho et al., 2011).

Wheat rust fungi are known to travel very long distances and induce epidemics/pandemics in different continents. Please provide information on spore size, shape, weight, and ornamentations, if available. Wheat Stem Rust, caused by *Puccinia graminis tritici* (*Pgt*), produces urediospores of size 28–30 × 17.5 μm. These spores are globose (Anikster et al., 2005). Sugarcane Rust fungus (*P. melanoocephala*) travels intercontinental distances through high-altitude winds and jet streams. It produces urediospores of size 28–33 × 18–23 μm, which are light in weight and obovoid (Anikster et al., 2005). Wheat Yellow Rust is a devastating disease that experiences frequent epidemics and pandemics. It is caused by *P. striformis*. The airborne urediospores are ellipsoid and relatively small, measuring 14–36 × 13–21 μm. They possess a tough spore wall (Anikster et al., 2005). Clearly, rust fungi capable of causing epidemics/pandemics have various adaptations in their spores. Incidentally, it is the spores in aerospace that we need to quantify and study their virulence.

7.2.2 BLIGHT AND BLAST CAUSING FUNGAL PATHOGENS: EXAMPLES

Potato crops are exposed to attacks by microbial pathogens, including fungi, bacteria, and viruses. Many of these pathogens have caused severe outbreaks in the past, leading to epidemics. Some have even caused global pandemics. The airspace above fields that support potatoes can serve as a significant reservoir for these pathogens. This is why we refer to the 'agrarian sky' as a 'sickness spreading entity.'

Fungal epidemics on potatoes have been known for several decades (Campos & Ortiz, 2020). Historically, *Phytophthora infestans* caused a late blight epidemic in potatoes and other food crops, which is among the well-studied fungal outbreaks. During the 1800s, it devastated crops in North America and the European Plains. *P. infestans* has a wide host range, so the initial inoculum persists on a few different crop species. *Phytophthora* species are dispersed through several mechanisms (Ristaino & Gumpertz, 2018). These include: (i) soil-mediated dispersal; (ii) irrigation water and sprinkler-mediated spore dispersal in the air; (iii) dispersal of sporangia via stems, leaves, fruits, dead, and decaying residues of previous and present crops; (iv) aerial dispersal through sporulating foci present on foliage and fruits; and (v) human and farm implement movement in fields. The agrarian sky aids in the dispersal of sporangia from infected leaves, stems, tubers found in storage, culled piles, residue heaps, and surface soil. Spores germinate quickly within hours if a moist climate prevails. Sporangia disseminate through wind and rain events (Rupp & Jacobsen, 2017). We are primarily concerned with the aerial spread of sporangia in the fields. Sporangia are lightweight and can tolerate tropospheric conditions just above the crop's canopy. Spores quickly initiate disease in new areas. It is essential to maintain farm sanitation and cull affected plants/portions to prevent aerial spread of the fungus. Among the various species of the genus *Phytophthora,* aerial dispersal is particularly important with *P. infestans* causing late blight on potato and tomato (Zwankhuizen et al., 1998). The presence of sporulating lesions on leaves and their lift into the air through turbulence or wind is significant with *P. palmivora* affecting Cocoa (Gregory et al., 1984). Fungal spores found on above-ground plant parts are important with *P. parasitica* causing blight on Rhododendron (Benson & Campbell, 1985). Fields with the sporulating stage of the fungus, combined with breeze, wind, turbulence, and susceptible genotypes of the host crop, can lead to epidemics. A moist agrarian sky contributes to the virulence of the blight fungus.

Now, let us consider an epidemic (blight) causing fungus that affects global maize productivity. Corn blight fungus is distributed worldwide. It is caused by *Bipolaris maydis* (*Cochliobolus heterostrophus*), also identified and referred to as *Helminthosporium*. The genus *Bipolaris* has over 100 species. The fungal pathogen is an ascomycete. The aerial mycelium is gray to olivaceous black. It produces asexual conidia and sexual ascospores (Agrios, 2005). However, it is the airborne and highly contagious asexual conidia that are of concern (Chakraborty et al., 2020).

Bipolaris maydis has caused a severe epidemic in the Southern Plains of the USA. The Southern corn blight initially emerged as a severe infection in Texas and quickly spread to the Northern states of Minnesota, Dakota, and Iowa. It appears that male sterile lines of maize used in breeding are susceptible to three different races: Race T, Race O, and Race C. Each race causes slightly different but specific symptoms. Race C is pathogenic only on male sterile lines. Most maize genotypes with normal cytoplasm are susceptible to Race O, although the levels of susceptibility may vary (Bhandhari et al., 2017; Mubeen et al., 2017).

Bipolaris can also infect sorghum and teosinte. The epidemic is a result of farming practices that are favorable to the fungal pathogen. The fungal pathogen affects the stem, sheath, and ears of the plants. Therefore, leaving infected crop residue on the surface soil of a field is risky. Even mild wind and turbulence can carry conidia into the air, facilitating easy dissemination. The weather conditions above the canopy need to be conducive, with a warm troposphere (65°F to 90°F), wind, and moisture. Maize blight reached epidemic levels in 1968 and 1970, as well as several times in different regions of the world. In 1970, it caused a 15% loss of grains in the USA (Josephson et al., 1971; Gregory et al., 1978). In favorable warm and humid conditions in the tropics, grain loss can reach up to 70%. There are suggested control measures, such as planting cultivars tolerant to the blight fungus, adopting cropping systems that avoid repeated maize cultivation, incorporating or burning crop residue that may harbor fungal propagules, and avoiding seasons that are conducive to the rapid spread and development of the fungus.

Pyricularia sp. causes wheat blast (Danelli et al., 2019). The fungus produces asexual spores, known as conidia. Conidiospores are easily detached from the leaf surface due to mild wind or turbulence. These buoyant and lightweight conidia become airborne and can travel long distances to initiate new infections. Factors such as precipitation, relative humidity (RH), and ambient temperature can affect the survival and germination of conidia. Traps coated with Vaseline or paste can be used to capture conidia in the air. It is important to pay attention to the effects of climate change on the quantity of conidia produced by the fungus and their distribution in the troposphere.

Bipolaris oryzae causes brown spot on rice crops grown in tropical regions. The vertical profile of conidial distribution in the troposphere appears to be significant. Spores at lower altitudes tend to settle quickly on the crop canopy, resulting in shorter airborne distances.

The evaluation of pathogen load in the agricultural sky is necessary. Such data is crucial for forecasting epidemics caused by blast fungi like *Pyricularia* or blights caused by *Bipolaris* (Ortega et al., 2020). Understanding the presence of virulent genes in the fungal pathogen is also valuable.

7.2.3 FUSARIUM SPECIES INDUCED EPIDEMICS AND PANDEMICS: EXAMPLES

Wheat head blight is caused by a complex of fusarial species such as *Fusarium graminearum, F. culmorum, F. cerealis,* and *F. poae* (Schmale & Bergstrom, 2010). The pathogen complex predominantly infects shoots and spikes. Its host range includes wheat, barley, oats, and a few other small grain cereals. Additionally, this fungal pathogen releases a mycotoxin called deoxynivalenol in measurable quantities (ppm) in seed lots.

Fusarium Head Blight (FHB) is mainly caused by *Fusarium graminearum* (anamorph) (*Gibberella zeae* – teleomorph). It is one of the pathogenic fungi that can easily disperse in the air above crop fields. This is due to the production of lightweight aerial spores by the fungus during both asexual and sexual cycles. The asexual sporangia (macroconidia) can easily travel short distances to infect new wheat spikes. The fungus causes discoloration and the presence of chaffy grains.

The sexual spores are produced by *G. zeae* in the perithecia (i.e., asci). Ascospores are very light (Schmale & Bergstrom, 2010). They are released forcibly and in large numbers into the air. The FHB fungus is mostly homothallic. Heterothallic strains are rare on spikes. These spores travel short distances if they drop to the soil surface. Windy and turbulent conditions allow the levitation of asci into higher altitudes. They can reach not only long distances of a few hundred kilometers but may even transit intercontinental distances (Schmale & Bergstrom, 2010; Teli et al., 2016). The troposphere altitude reached by these ascospores is important. It seems that both short and long-distance dispersal of the pathogen in the airspace is significant in the dissemination of the disease. Additionally, we should note that once an area is infested with FHB fungi, the lower troposphere may harbor sickle-shaped conidia and ascospores in measurable numbers for a long period of time. This means that inoculum is present in the air at any time. It truly represents a *'sick agrarian sky'* in terms of FHB disease.

The Fusarium head blight (FHB) was first reported in 1900 on North American wheat. Severe epidemics occurred in 1915 and 1920. By 1919, the epidemic had reached parts of the Canadian Plains. FHB, with reduced severity, occurred in the 1950s, 1960s, and 1970s in the Great Plains. A major outbreak of FHB occurred in the 1980s, affecting wheat-producing regions in Canada (Saskatchewan and Manitoba) as well as from North Dakota to Kansas in the USA.

Host crop resistance has been suggested as a measure to control FHB (Ghimire et al., 2020). Biological control of the FHB pathogen has also been explored and suggested (Powell & Vujanovic, 2021). Additionally, predicting FHB incidence using computer modeling methods might be helpful (Shah et al., 2013). Pre and post-anthesis weather is particularly important for forecasting epidemics. The effects of climate change, El-Nino-Southern Oscillation, and related phenomena also impact FHB epidemics (Shah et al., 2013; Kriss et al., 2012).

Methods that reduce FHB inoculum density in the air above seedlings are worth trying. Perhaps, spraying chemicals that reduce FHB fungal spores or using laser and radiation to reduce aerial spore density are ideas worth exploring. It is crucial to focus these methods correctly above the canopy and at accurate distances. Destroying spore viability in the air is also important.

Fusarium oxysporum var cubense race 4 is a fungal pathogen on bananas. It causes Panama disease. Aerial conidia are highly contagious. They migrate rapidly from one field to other. In case of Cavendish bananas, race 4 of *F. odoratissimum* has spread rapidly in Asia. Then, it migrated to Africa and Latin American tropics (Cardenas-McGill, 2021). Epidemics occurred in Sri Lanka in 1868, in the 1920s, 1950s, and 1970s in South America. Prior to planting, it is important to consider the fungal ability to produce a large number of sporangia and asci, weather conditions, and the susceptibility of crop genotypes. It is worth attempting to monitor the agrarian sky for fusarial spores.

7.2.4 MILDEWS ON CROPS

Downy mildews on crops are caused by fungi that belong to the class Oomycetes. Most of them are classified under the family Peronosporaceae. Downy mildews can be devastating to food crops, especially cereals, as they form yellow spots on foliage. In severe cases, defoliation can be intense, eventually leading to plant death. The spores of the mildew fungus choke the xylem vessels. The fungus overwinters as oospores or hyphal segments.

Oospores germinate when the ambient temperature is between 50-75 degrees Fahrenheit and the relative humidity is 85%. The sporangia, which are light in weight, are dispersed by wind or splashing rainwater. Fungal spores may also be dispersed by farm vehicles. They germinate on a leaf surface. Zoospores, also light in weight, can be found in large numbers in the lower troposphere. The entire lifecycle of the fungus takes less than a week. The fungus can infect and form dense downy growth on the lower surface of all plants in a field, occurring within a short span of time. Undoubtedly, in the context of agrarian sky and its microbiological components, the population/density of spores in the troposphere air is important.

Powdery mildews are airborne fungi that spread rapidly through wind and rainwater. Hurricanes, continental winds, and jet streams also carry powdery mildew spores from one region to another. There are several species of powdery mildew fungi that attack different food and commercial crops. Powdery mildew of peas is caused by a complex of fungal species, namely *Erysiphepoligoni, E. pisi,* and *E. cichoracearum. Erysiphe trifolii* and *E. baeumleri* also cause this disease on peas (Fondevilla & Rubiales, 2012). Secondary infection of the host can be devastating and is caused by conidia. Conidia are produced in large numbers and are disseminated by tropospheric wind. Conidia germinate on leaf lamina and cause infection. Once again, in this case, agrarian sky mediates the fungal dispersal and induces disaster.

There are races of the powdery mildew fungus *Erysiphe* sp., which is the causal agent of powdery mildew in peas. However, the fungus also attacks crops such as cabbage, lucerne, pigeon peas, beans, coriander, and others. In the USA, it has been estimated that the loss of garden pea pods due to *Erysiphe* is around 20–23% each year. This affects pod filling and leads to a depreciation in pod weight.

Powdery mildews of grapevines are distributed worldwide. They pose a significant problem in grape-growing regions of the USA, such as California (Smith et al., 2022), the Ozarks region in Arkansas, Midwest states like Ohio, and even the Northeast (Ellis, 2022). The disease is widespread across the entire European plains, from Southern France to the Russian Steppes. In India, it is observed in all grape-growing regions. In Australia, the fungus is endemic to Western Australia and can rapidly spread to other regions (Taylor & Gordon, 2022). The disease is caused by an ascomycete called *Erisyphenecator* (*Uncinulanecator*). It primarily affects the green succulent tissue of grapevines, resulting in small white or ash-gray spots on the leaf lamina. These spots enlarge and eventually cover the entire leaf surface. Infected flowers may wither and drop without setting fruits, and severe cases of infection can lead to berry drops. The fungus is highly airborne. The ascospores released from cleistothecium (fruiting body) travel through the air and settle on the green tissue of grape vines in the vicinity. They germinate and release millions of asexual spores (conidia) in 7–8 days. Conidia are light and reinfect grape vines. Conidia disperse even over longer distances. It depends on the altitude they reach in the troposphere. Troposphere conditions affect the germination of conidia. A temperature of 68°F to 77°F and relative humidity of 85% are conducive for the spread and establishment of powdery mildew fungus. Undoubtedly, the agrarian sky poses a threat to grapes if it harbors a large number of powdery mildew spores. A high density of spores just above the vines could be dangerous. The migratory nature of the fungus in the air requires attention from grape vine experts, particularly to spray fungicides and reduce the inoculum.

7.2.5 SMUT AND ERGOTS ON CROPS

Smuts of cereals are usually internally or externally seed-borne diseases. For example, loose smut of wheat (*Ustilago nuda*), cob smut of maize (*U. maydis*), smut of barley (*U. tritici*), and smut of *Setaria italica* (*U. crameri*). However, wind, frequent or intermittent rains during the flowering and grain maturity period can aid in the dispersal of smuts and bunts in cereals. Reports from Australian wheat producing states suggest that previously infected rachis, grain, and residues can release spores. These spores float in the air and get blown by wind or rain to distant places within a field. Black, lightweight, powdery spores that drift from pre-infected spikes infect and establish on healthy spikelets (Thomas, 2021). Troposphere weather conditions play a vital role in the establishment of smuts on new panicles. Smut fungi show specificity towards host crop species and their genotypes. In the USA alone, over 140 different smut-causing fungi have been identified on cereals and grasses.

In addition, the agrarian sky can serve as a reservoir for disease propagules of smuts and ergots in millets and other small grain cereals. For instance, Sorghum ergot, caused by *Claviceps africana,* is an endemic disease in the millet-growing regions of Southern Africa. It is transmitted through airborne secondary conidiation. The honey dew panicles provide the conidial inoculum that is lifted into the atmosphere and carried by the wind. The rate of spread of ergot conidia in the agrarian sky appears to be similar to that observed for other airborne fungi causing epidemics, such as *Puccinia graminis tritici* (Frederickson et al., 1993). Clearly, ergots can be contagious and pose a significant risk. Undoubtedly, the agrarian sky contaminated with ergot spores is unhealthy and hazardous, capable of initiating an epidemic. Sorghum fields affected by ergot typically yield less than 1,000 kg of grains per hectare, whereas ergot-controlled fields can yield 3,000 kg per hectare (De Milliano et al., 1992). Windborne secondary conidia can be trapped and regularly monitored during flowering and panicle maturation as a preventive measure.

Pearl millet (*Pennisetum glaucum*) is susceptible to ergot caused by *Claviceps fusiformis*. It is primarily disseminated through seeds mixed with ergot propagules. Insects crawling on pre-infected panicles also disperse ergot spores. Rain splashes on panicles and wind that carries loose (drifting) spores rapidly disseminate the ergot fungus into the air. In severe cases, the air above the canopy of a crop may harbor a high density of ergot spores.

7.3 BACTERIAL DISEASES, EPIDEMICS, AND PANDEMICS OF CROPS

Maize is a staple cereal crop in many regions of the world. However, it is attacked by several airborne bacterial pathogens. A few of them have caused epidemics periodically in different geographic locations. Pseudomonads are common bacterial infections on maize, affecting foliage. For example, *Pseudomonasavenae* subsp. *avenae* causes bacterial blight, while Chocolate spot is caused by *P syringae* subsp. *coronafaciens.* Bacterial stripe is caused by *P. andropogonis.* Epidemics due to *Xanthomonas holicolus* have been reported. There are also Enterobacter and Erwinia sp. that bring about rots of maize tissue and cobs. Corn is also infected by *Spriosplasmakunkelii.* Bacterial plant pathogens are commonly dispersed through splashing rainwater, aerial irrigation sprinklers, and wind. In fact, agrarian sky mediates the dissemination of several bacterial pathogens of crops.

Bacterial infections on wheat are as important as those caused by other microbial agents, such as fungi or viruses. Duvellier et al. (1997) have listed several bacterial diseases that attack wheat crops at various stages. Many of them are airborne and epiphytotic on foliage. Therefore, mild wind, turbulence, or movement of farm vehicles aid in their dissemination to the troposphere. Otherwise, they fall from the crop's canopy to rest on the soil surface. Dissemination may also occur via irrigation sprinklers.

The most frequently encountered bacterial pathogens on wheat are *X. translucenspv. cerealis, X.t.translucens,* and *X.t secalis* (Hagborg, 1974). Pseudomonads like *Pseudomonas syringaepv.* syringae cause leaf blight and are distributed worldwide on wheat. *P. syringaepv. japonica* causes black node (Dye et al., 1980). Other bacterial infections that wheat crops suffer from include *Bacillus megaterium,* which causes Tan streak (Hosford, 1982); *Clavibactertritici,* which causes 'Tundu disease'; and *Erwinia raphontici,* which causes 'Pink seed' (McMullen et al., 1984; Forster & Bradbury, 1990). Pink seed is a mild disease and is considered opportunistic on wheat due to its broad host range. *Pseudomonas cichorii* causes stem rot, and *P. fuscovaginae* causes sheath rot.

As mentioned earlier, many of the bacterial pathogens are epiphytotic, making them easily released into aerosol via mild turbulence.

Rice crops are affected by a few bacterial pathogens with the potential to cause outbreaks, leading to epidemics of large proportions. Rice blight is caused by Xanthomonas oryzae, while Leaf Streak is caused by *X. oryzicola.* Other bacterial pathogens that attack rice crops include *Erwinia chrysanthemi,* which causes foot rot, *Burkholderia glumae,* which causes grain rot, and *Pseudomonas fuscovaginae,* which causes sheath rot.

Bacterial epidemics and pandemics on grain crops, vegetables, oilseeds, and other economically valuable commercial crops such as cotton or sugarcane are widely reported. We have accumulated extensive knowledge about bacterial blights, spots, angular spots, blotches, rots, and wilts caused by bacterial pathogens. Among the genera, *Xanthomonas, Pseudomonas,* and *Erwinia* are the most frequently encountered gram-negative bacterial pathogens. Bacterial pathogens are disseminated by various factors in the agricultural environment, including irrigation water, seeds, and transportation of products. Soil-borne or residue-borne bacterial agents can be easily dispersed by low-level winds that skim the soil surface and the troposphere just above it. Bacterial cells transit along with soil particles, organic dust, and other floating items. Turbulence on the soil surface easily disturbs bacterial cells, causing them to levitate into the airspace above the crop's canopy. Bacterial plant pathogens can reach epidemic proportions when they are rapidly dispersed under warm and moist troposphere conditions, especially during severe precipitation events with splashing droplets. Sprinkler irrigation and foliar sprays of chemicals can also carry a certain load of bacterial pathogens. As mentioned earlier, the troposphere can facilitate the rapid dispersal of several bacterial pathogens in crops. Warm, humid conditions with convective currents, drifting wind, clods, mists, fog, and even snow (e.g., *Pseudomonas aeruginosa, P. syringae*) may carry pathogens to different distances in cropped areas. Insect movement above the crop's canopy can also induce epidemics by distributing highly infective strains of bacteria.

Species of *Xanthomonas* that affect several crops should be considered. In a few cases, they have caused severe epidemics/pandemics. Periodic reports of such destructive diseases have been observed in all agrarian regions. The genus *Xanthomonas* is distributed worldwide. *Xanthomonas campestris pv. malvacearum* causes angular leaf spots on cotton crops. It was first reported in the Southern States of the USA in 1891. If left unattended, it can be devastating. It has caused epidemics in many cotton cultivating regions. The bacterial pathogen is easily disseminated through rainfall, wind turbulence, and farm vehicle movement. It requires high relative humidity and warm conditions (65°F–90°F) to develop into a severe disease (Kamareit et al., 2021). Crop loss is massive because the causal agent is not only widespread but also affects foliage, photosynthetic efficiency, and the main product, i.e., bolls. Undoubtedly, the presence of *Xanthomonas* in the agrarian sky above cotton fields can be deleterious.

Bacterial blight and Bacterial leaf streak are among the devastating diseases affecting rice cultivation in Southeast Asia (Saha et al., 2015). According to Saha et al. (2015), several virulent strains of *X.o. oryzae* have periodically emerged in rice-growing regions of Asia. Bacterial leaf blight (BLB) causes periodic epidemics in rice and is distributed worldwide. BLB has a significant impact on rice production in Asia, America (including the Caribbean), Australia, Africa, and other regions, resulting in 20–30% grain loss in certain rice-growing areas (Saha et al., 2015). Importantly, the spread of *X.o. oryzae* is facilitated by severe winds, turbulence caused by monsoon winds, and heavy rainfall. Host resistance has been suggested as a method to mitigate BLB.

X. oryzae var oryzae caused BLB became conspicuous with the introduction of high yielding cultivars of rice. Bacterial blight was first recognized in Japan in 1884. However, its presence and epidemic proportions were noticed in the 1960s and 1970s in Southeast Asia. *X. oryzae var. oryzae* has been attributed to grain yield losses ranging from 9.5% to 22% (Kumar et al., 2017). Rice blight is known for its rapid spread, particularly in warm/hot, humid, and windy conditions. Epidemics of *Xanthomonas campestris pv. oryzae* became more frequent after the 1960s, coinciding with the release of the stout-stemmed, high-tillering dwarf rice cultivar IR8. Although IR8 responded well to fertilizer-N and had better yields, it was susceptible to blight. Rice-producing regions in South and Southeast Asia experienced significant grain losses ranging from 20% to 50%. The combination of humid weather, windy conditions, splashing raindrops, and farm vehicles contributed to the rapid spread of the bacterial pathogen. Low altitude winds on the soil surface can lift soil particles, dust, and crop residues carrying blight-causing bacteria.

Xanthomonas campestris pv. glycine is a pathotype often traced in the troposphere above vast soybean expanses, in North America and other regions. A warm/hot, windy, and moist environment is conducive for bacteria to survive in the troposphere. It also promotes the virulence of the bacteria and facilitates the occurrence of epidemics. Soybean leaf pustules can reduce photosynthetic efficiency and lead to pod loss. Bacterial blight of lima beans is caused by *X. campestris pv. phaseoli*. The spread of the disease is facilitated by rainwater and wind.

Xanthomonas campestris pv. vesicatoria is a common occurrence on solanaceous vegetables like capsicum, tomato, brinjal, etc. Studies by McInnes et al. (1988) has shown that aerosol above the vegetable fields are carriers of pathogenic bacteria. The distribution of bacterial pathogens in the troposphere profile appears to be significant. Typically, the bacterial population is higher in the mid-afternoon compared to early morning or evening/night. The pathogens are transported by high-speed winds above the crop canopy. Sprinkler irrigation or rainfall events appear to wash down the airborne microbes. As a result, the population detected by air traps is lower after rain or sprinkler usage.

Xanthomonas campestris and *Pseudomonas syringe* survive epi-phytotically for extended periods, on foliage of vegetable crops. Agronomic procedures such as clipping can disturb bacteria on the canopy (foliage), releasing them into the agrarian sky. These bacteria are highly prone to being disseminated over longer distances through wind or turbulence. The agrarian sky undoubtedly serves as a repository for *X. campestris* populations. It is important to note that there are other means of Xanthomonas blight dissemination. It appears that the application of Mancozeb + copper sprays can reduce bacterial counts in the air and on the canopy. Additionally, it can also reduce populations of *Pseudomonas syringae*, the causal agent of bacterial brown spot. This reduction in bacterial populations is in addition to the control of Xanthomonas blight (McInnes et al., 1988).

On tomato, *X.c. pv. tomato* causes bacterial leaf spot. It is transmitted aerially, through sprinkler water/mist, rainfall events, and crop residue. McInnes et al. (1988) conducted intensive sampling of tomato fields and analyzed *X. campestris* counts on the canopy. They concluded that aerosolization of pathogenic bacteria was important, aiding the rapid spread of the bacterial pathogen above vegetable fields. Similar conclusions were drawn for *Pseudomonas syringae var. syringae,* which affected succulent tissues of tomatoes and sweet peppers.

Bacterial black rot of cabbage is caused by *X. campestris pv. campestris*. This bacterial pathogen also affects several other Brassicaceous crops. Weeds, particularly cruciferous ones, harbor *X.c. var campestris* in fields. Kuan et al. (1986) examined leaf samples with lesions and without lesions that could harbor the pathogenic bacteria. Assays have shown that, on average, leaf samples from Brassicaceae weeds contained 6.7×10^7 bacterial cells g^{-1} fresh weight. Heavy rainfall and low night temperatures reduced *X.c. campestris* population on weeds and the crop's canopy. However, air samples from the troposphere just above the crop's canopy showed 0.84 cells m^{-3} air. It is suggested that aerosol-based dispersal of *X.c. campestris* within the fields and to longer distances could be important.

Xanthomonas campestris pv. vascularum is a pathogen affecting sugarcane fields. Humid conditions in the crop's canopy allows for rapid build-up of pathogen's population. Bacterial infection results in low cane yield. There are indeed innumerable crops that are attacked by *X. campestris* and its pathotypes. In all above cases, agrarian sky holds the key for disease severity, at least, to certain extent.

Citrus canker has repeatedly flared into epidemics on different continents. It is caused by *X. campestris pv. citri*. The troposphere plays a role in the survival, dissemination, and build-up of inoculum that transits to neighboring fields. It also causes epidemics if conditions congenial to the pathogen prevail. There is no doubt that the bacterial genus *Xanthomonas* is deleterious to a wide variety of crop species. In fact, the genus *Xanthomonas* is versatile in rapidly securing a niche in agrarian regions. It utilizes transmission agents in the atmosphere such as wind, rain, storms, irrigation water, and crop residues. Farm operations conducted with low sanitation may induce epidemics. *Xanthomonas campestris var citri* can survive in aerospace for a length of time during transmission. Dry desiccating conditions with scanty moisture are detrimental to the pathogen. Overall, the agrarian sky and its factors have an impact on the survival, dissemination, disease causation, and intensity of bacterial pathogens such as *X. campestris*.

The above discussion makes it absolutely clear that 'Agrarian Sky' harbors bacterial plant pathogens like *Xanthomonas,* which poses a significant danger. It can serve as a source of crop sickness, epidemics, or even pandemics during a crop season. The risk of bacterial diseases extends beyond the soil profile or irrigation water, as it also lurks from above the crop, i.e., the sky.

7.4 VIRAL DISEASES OF CROPS MEDIATED BY AERIAL VECTORS

During recent decades, the wheat crop has suffered due to epidemics/pandemics caused by viruses. This is in addition to the rusts that have always been severe detriments to crop yield. Once again, the 'agrarian sky' that supports vectors like aphids, mites, and whiteflies is a repository and the cause for the spread of viral diseases. These insect vectors and the viruses they carry remind us just how devastating the agrarian sky could be on food grain-generating regions of the world, at any time of the year.

WSMV was first observed in Nebraska, in the Great Plains of the USA, on wheat, in the 1920s. It was initially referred to as the 'yellow virus' (Singh et al., 2018). WSMV can be a devastating virus transmitted through the air. WSMV is transmitted aerially by a mite called the wheat curl mite (*Aceria tosichella*) (Skoracka et al., 2018). Mites carrying the virus move from wheat crops to volunteers and other grass species, typically during the summer. The viral inoculum is then transmitted by the mite when fresh wheat is sown. The virus also affects several other cereal and grass species such as maize (*Zea mays*), barley (*Hordeum vulgare*), millet (*Panicum* sp.), foxtail millet (*Setaria italica*), and *Echinocloa* sp. (Chalupnıkova et al., 2017; Drab et al., 2014; French & Stenger, 2003; Stenger & French, 2009). WSMV is a single-stranded RNA virus belonging to the genus Tritimovirus (Potyviridae) (Singh et al., 2018). WSMV is distributed worldwide and has been found in wheat-growing regions of the Northern Great Plains, Brazilian Cerrados, Pampas of Argentina, European Plains, Turkey, Iran, and Australia (Hadi et al., 2011; Navia, 2013; Singh et al., 2018). The severity of this virus disease can vary depending on the wheat genotype cultivated, its tolerance to mites, and the environment (Agrarian sky).

The vector and WSMV are affected by weather conditions. The vector (mite) could become active and spread the virus from September to December. Dry conditions in February have also been linked to new outbreaks and increased disease severity (Webb, 2004).

When it comes to virus diseases (epidemics) in wheat cultivated in the Great Plains, Burrows et al. (2016) have clearly emphasized the importance of identifying 'Green Bridge' crops and other plant species. Avoiding 'Green Bridge' crops helps break the vector cycle and prevent the perpetuation of the virus. This is because 'green bridge' cereals serve as the main reservoir of viruses that affect wheat crops in the main field.

7.4.1 AERIAL INSECT VECTORS OF CEREAL VIRUSES

It is insufficient to solely investigate the microbial load and species diversity of pathogens present in the troposphere. The movement of microbial populations across agricultural skies, facilitated by vectors, is of utmost importance. Vectors that transmit cereal diseases are particularly significant in terms of global productivity. Cereal aphids, as highly effective dispersal agents of crop pathogens, impact various cereal species and others through their vectoring ability. These aerial vectors enable the worldwide spread of viral pathogens across cereal crops. Aphids themselves are pests that feed on crops by extracting plant sap. They can rapidly move from one field to another through short-distance flights, and during high-speed winds, they can be carried over long distances. They can even be transported to different continents through wind, jet streams, and weather patterns. The disease-causing potential of the "agricultural sky" is dependent on the vectors that carry the microbes throughout the crop canopy. Therefore, it may be necessary to develop a scale for measuring and quantifying the severity of the "disease-causing potential of the agricultural sky," taking into account all relevant parameters.

Parry (2013) states that there are at least 5,000 species of aphids, which can affect cereals, grasses, and related plant species. However, only a few of them are dreaded for their ability to consistently and rapidly transmit viruses on major cereals such as wheat, oats, barley, and other species. Blackman & Eastop (2000) list several aphid species whose ability to transmit viruses affects cereal crops. Most of these aphids can be found all over the globe in the troposphere and close to the crop's canopy. They include the Rose-Grain aphid (*Metopolophium dirhodum*), Grain Aphid (*Sitobion avenae*), Oat aphid (*Rhopalosiphum padi*), Maize Aphid (*Rhopalosiphum maidis*), Russian wheat aphid (*Diuraphis noxia*), Indian Grain Aphid (*Sitobion miscanthi*), Cereal aphid (*Sitobion fragariae*), and the Fescue grass aphid (*Metopolophium festucae*). In addition to aphids, there are jassids, psyllids, plant hoppers, whiteflies, beetles, bugs, and bees that are capable of spreading viruses that cause cereal diseases.

Mostly, the extent of insect vector migration is the basis for the dispersal of microbial pathogens. Aphid migration itself is explained as the ability of the insect vector (loaded with microbes) to move into areas that are not typically determined by their food habits (Johnson, 1969; Hulle, 1994). There are simulations and actual data available regarding flight behavior, specifically the flight patterns of aphids and other vectors that hover over cereal fields (Johnson, 1969; Parry, 2013; *see* Chapter 3–Section 3.3.1). These flight patterns and the distance covered by the vectors determine the speed of dispersal and the extent of damage caused by the microbial pathogen.

Virus-induced epidemics and pandemics can be devastating, resulting in reductions in food grain yield almost every year. The extent of crop loss due to epidemics can be attributed to various factors, including climate change, monocropping patterns, and the generation of genetic variants in nature. Jones (2021) states that in the past year alone, viral epidemics reduced grain productivity equivalent to 30 billion US dollars. Major losses occurred due to repeated virus epidemics in specific farming regions. Additionally, most viral epidemics were facilitated by aerial vectors, mainly insects. Therefore, continuous monitoring of insect vectors in the airspace above crop canopies is necessary.

There are several different lists of virus epidemics on major food crops, particularly on cereal food grains, potatoes, legumes, oilseeds, and plantations. Based on these lists, we can conclude that the agrarian environment facilitates the dissemination of viruses. Viral dissemination occurs through vectors. Jones (2021) has reviewed several viral diseases that greatly impact global grain productivity. Among the vectors, insects such as whiteflies, aphids, plant hoppers, thrips, honeybees, butterflies, and mites play a significant role in virus transmission. In other words, an agrarian environment with a sky above the crop's canopy consistently populated by these vectors is a sure recipe for inducing epidemics or pandemics caused by viruses.

When discussing virus-induced epidemics, Jones (2021) defines an epidemic as a disease of high intensity occurring in a specific area, especially where the causal agent is present for a considerable period. Resurgences are common after a period of low-intensity disease. An epidemic can be confined to a single field or a few fields in a small farming region. On the other hand, a pandemic is characterized by infections that originate in locations far from the initial focus, occurring due to the long transit of the pathogen.

A few examples of virus induced epidemics of major food crops as quoted by Jones (2021) are:

1. **Maize Yellow Dwarf Disease:** It is caused by Barley Yellow Dwarf Virus (BYDV) and Cereal Yellow Dwarf Disease (CYDV). The virus BYDV belongs to the Luteovirus genus, while CYDV belongs to the Polerovirus genus. These viruses affect the wheat crop (Triticum aestivum) in Europe, the Middle East, South Asia, the North American Plains, and Oceania. The airborne vector transmits this virus. The actual transmission of the virus occurs in the canopy and aerospace region. The main vectors, Aphids (*Ropalosiphum padi, R.*

maidis, Sitobion spp.), transmit the viruses. The disease can range from sporadic to widespread, depending on the intensity and extent of cropping.

2. **Wheat Streak Mosaic Virus (WSMV):** This virus has caused epidemics in Europe, Central and East Asia, Australasia, North, and South America. The leaf curl mite (Aceria tosichella) transmits the virus in the airspace above crops.

3. **Potato Necrosis Ringspot Disease:** It is caused by Potato Virus-Y, which is transmitted aerially by aphids (*Myzus persicae*). It causes epidemics in all continents, except Antarctica.

4. **Faba Bean Necrotic Yellow Disease:** The causal agent is Faba bean necrotic yellow virus, belonging to the genus Nanovirus. It is transmitted aerially by *Aphis crassivora, A. fabae,* and *Acyrthosiphon pisum*. The virus affects bean crops in Europe, the Middle East, North Africa, and South Asia.

5. **Tomato Brown Rugose Fruit Disease:** The causal agent belongs to the genus Tobamovirus. Tomato brown rugose fruit virus is transmitted by contact, not by sucking and injecting sap. Bees that visit during pollination spread the virus. It is known to cause epidemics in Europe, the Middle East, USA, and South America.

Next, let us consider a few examples of viruses that are prone to cause pandemics, on food crops. Again, Jones (2021) quotes the following:

1. **Maize Lethal Necrosis Virus:** It has caused pandemics on maize cultivated in East Africa, Southeast Asia, South America, and the European Plains. This disease is actually caused by a complex of Maize Mottle Virus (genus: Mechalomovirus) and Potyviruses. Beetles, thrips, and mites transmit the viruses through the air. Thrips, such as *Frankliniella williamsi* and the Eriophyid mite *Aceria tosichella*, are vectors.

2. **Rice Tungro Disease:** It is caused by a complex of viruses, specifically a mix of Tungrovirus and Waikavirus. The virus is transmitted through the air by leafhoppers such as *Nephotetix virscens*. It causes pandemics in rice crops cultivated in South and Southeast Asia, resulting in severe loss of rice grains. The agrarian sky in the rice

belts appears to harbor the Tungro virus, which can induce disease at any time when the vector multiplies in number.

3. **Sweet Potato Virus Disease:** There is a pandemic in Sub-Saharan Africa, the Middle East, Southeast Asia, Central America, and South America. The causal agents are a complex of viruses belonging to the genera Crinivirus, Potyvirus, and Ipomovirus. The virus is transmitted via whiteflies that hover above crops. Aphids like *Myzus persicae* also transmit the virus. The pandemic results in massive yield losses in large patches.
4. **Banana Bunchy Top Disease:** It is known to cause pandemics in Sub-Saharan Africa, the Middle East, the Indian subcontinent, Southeast Asia, and Oceania. It is caused by a virus of the genus Babuvirus. Aerial transmission is accomplished by the aphid *Pentalonia nigronervosa*. It causes devastating losses to plantation growth, with fruit formation almost nonexistent.
5. **Citrus Tristeza Virus:** This belongs to the genus Closterovirus. It is transmitted by aphids, such as *Toxoptera citricida. Aphis gossypii, A. spiraecola,* and *T. aurantia* are other vectors. The virus affects citrus groves in North and South America, the Middle East, the Indian Subcontinent, Southeast Asia, and Oceania. Citrus plants are usually killed due to tristeza, and the yield potential is almost reduced to zero.

The above list is not exhaustive. It is mentioned here solely to support the view that epidemics and pandemics induced by viruses acting singly or in complex mixtures are mostly transmitted via the agrarian sky. The sanitation of aerospace above the crop's canopy is crucial, both in terms of swiftly moving (flying) insects as well as virus propagules that they carry. The agrarian sky needs to be monitored regularly, and the aphids, whiteflies, mites, thrips, and beetles need to be kept under threshold.

There are indeed several concepts suggested by virologists to explain why, when, how, and where plant viruses reach epidemic proportions or become pandemic, migrating to distant locations. Intercontinental migration is considered a cause of pandemics. Our understanding of plant viruses with the potential to cause epidemics/pandemics has been reviewed by multiple authors since the 1900s. For instance, Thresh (1980, 2006) provides detailed explanations about the epidemiology and conditions required for viral pandemics. In nature, plant viruses can potentially cause diseases of varying severity, depending on the crop species, cropping systems, and environmental conditions. According to Klinkowski (1970), plant viruses can induce epidemics or pandemics when they migrate from their center(s) of origin and encounter new crops. Jones (2020) believes that such encounters are common in nature, including in agrarian regions, and may be major causes of viral virulence and rapid spread in those regions. It is important to recognize that the agrarian environment often serves as a medium for the spread of viruses with the potential to devastate crops. For example, out the 12–15 viral epidemics quoted by Jones (2020), most if not all, the mentioned viruses were dispersed by aerial vectors. Particularly, whiteflies (*Bemisia tabaci*) or aphids (*Aphis craccivora*) or plant hoppers and other species fly or drift freely above the crop's canopy and spread the viral inoculum. Furthermore, Jones (2021) suggests that the crop and the virus that cause epidemics might have co-evolved in the region of domestication. To substantiate this view, we need clear indications from DNA sequence analysis of both the crop and the virus. Jones quotes potato viruses – A, S, and Y, belonging to Potyviruses, that occur in the region of domestication, in the Andes mountains. New encounters of host crops and viruses could also induce pandemics.

Plant pathologists have tried to define the natural or man-made microbe-mediated epidemics and pandemics on crops grown in different continents. Here, we are focused on understanding the virus-induced epidemics/pandemics on food crops. Several factors affect the initiation and spread of epidemics on crops. Jones (2020), for example, explains that epidemics/pandemics may erupt in a location with an indigenous virus population that encounters introduced crops. The introduced crop somehow becomes susceptible, and the virus spreads rapidly across the region. This also happens when the virus becomes virulent and induces disease on vast stretches. Then, it is called a pandemic (Anderson et al., 2004; Fargette et al., 2006; Thresh, 2006; Jones, 2014; Hema et al., 2014).

There are several terms related to epidemics/pandemics that describe viral infective behavior. An 'emerging virus' is one that has caused epidemics in a new host. Usually, such a virus migrates from a previously known host and attacks a new crop species. Jones (2020) explains the emergence of a viral disease in a region as "the first appearance of a virus and its initial increase in incidence/geographic range." Then, there are "remerging viruses" that cause epidemics year after year or at intervals. This phenomenon is referred to as "re-emergence" or "remerging virus." It involves a virus that once caused serious disease problems but then declined in importance, only to suddenly increase in incidence and geographical distribution. Such viruses can cause significant damage to crops. The spill-over of viruses indigenous to an agrarian location into introduced crops is one of the important reasons for epidemics (Jones, 2020). 'Spillover' also refers to the dissemination of a plant virus from an infected host to a new host. The virus may not have encountered the new host previously. It is important to note that all of the variations in plant-insect vector and insect vector-virus relationships mentioned above are greatly influenced by the agrarian environment.

Plant viruses can spread to new areas when they come into contact with susceptible crop species/genotypes. A global pandemic occurs when these viruses are aided in their migration across continents by wind or aerial insect vectors, often through wind or jet streams. The key requirements for a pandemic are a highly virulent virus, a medium like wind or aerial insects that can rapidly disseminate the virus particles, and susceptible crops grown in small or large areas. The virus must also be able to withstand the environmental conditions during migration.

The main point to note is that epidemics and pandemics usually result in catastrophic crop losses (see Thresh, 2006). It is crucial to curtail the spread of the viral agent within its region, especially when it is at low infectivity or virulence levels. Jones (2020) and Hema et al. (2014) have documented several viral epidemics and pandemics observed in recent decades. Here are a few examples:

1. **Brown Streak Virus (Ipomovirus):** It is traced in East and Southern Africa and affects cassava cultivated on the African continent. It is spread via the vector, whitefly (*Bemisia tabaci*). Areas sown with susceptible cultivars get devastated if they support a large population of the vector (Thresh et al., 1994; Tomlinson et al., 2018; Legg et al., 2011).

2. **Cassava Mosaic Virus Complex (Begamovirus):** Cassava mosaic viruses are found in Sub-Saharan Africa, Madagascar, South India, and Southeast Asia. They affect cassava cultivation areas in West and East Africa, as well as Southern India. The viruses are transmitted by an insect vector, the whitefly (*Bemisia tabaci*) (Otim-Nape & Thresh, 1998; Legg et al., 2011; Legg & Fauquet, 2004; Patil & Fauquet, 2009). For over a century, cassava mosaic disease (CMD) has been affecting cassava cultivation in Africa, as explained by Legg et al. (2006, 2011). When outbreaks occur, the

current management strategies are deemed insufficient. The authors suggest that the use of CMD-resistant cassava genotypes can greatly regulate and reduce the occurrence of regional pandemics. Host plant resistance proves effective against the whitefly vector in the challenging agricultural environment. Another method is to monitor and control the spread of the *Bemisia tabaci* vector throughout the tropical regions of Africa.

3. **Groundnut Rosette Virus (Umbravirus):** The virus is traced in West Africa. It affects groundnuts grown in Sub-Saharan West Africa. The virus is transmitted by an Aphid (*Aphis crassivora*). An epidemic of the virus causes devastating loss of foliage and pods. Both the vector and the virus are airborne during the crop season.

4. **Maize Streak Virus (Mastervirus):** Maize streak mosaic virus is the causal agent found in Sub-Saharan and parts of Southern Africa. Several species of leafhoppers belonging to the genus *Cicadulina* are the aerial vectors. The agrarian sky that supports these vectors can induce devastation of maize crops.

5. **Rice Yellow Mottle Disease (Sobemovirus):** A virus has been traced in the rice growing regions of West Africa. It is spread by beetles and causes significant damage to foliage and grain in the region (*see* Jones, 2020).

6. **Rice Hoja Blanca Virus (Tenuivirus):** This virus is found in rice-growing locations within North and South America. The virus is vectored by the plant hopper (*Sogatodes orizicola*). It has the potential to devastate rice productivity in South America (see Jones, 2020). The plant hopper that hovers above the rice fields needs to be controlled in a timely manner.

Rice is predominantly grown in the Asian tropics and subtropics, and it is affected by several viruses. Not all infective agents are airborne, nor are they all transmitted by aerial vectors. A few infective viruses are seed-borne. However, rice, in general, is susceptible to the following viruses, many of which may not induce epidemics or pandemics but remain as mild infectious agents: ricing bunchy stunt virus, rice dwarf virus, rice black streak virus, rice grassy stunt virus, rice tenue virus, rice yellowing virus, and rice necrosis mosaic.

There are several vegetable species that are attacked by viruses. Many of these viral agents are actually picked up and rapidly spread in the aerospace by vectors. For example, in the case of tomatoes, Alfalfa Mosaic Virus and Beet Curly Top Virus are transmitted by plant hoppers, leaf hoppers, and aphids. Tomato Spotted Wilt Virus is transmitted by a few species of thrips (Damicone & Brandenberger, 2000). The dominant viral agent and aerial vector species may differ based on geographic location. Vectors and viruses may have multiple hosts, so they may drift to fields with susceptible crop species. In such cases, massive aerial sprays to control the aerial vector will be required.

Overall, the above examples of virus-caused epidemics, induced by introduced crops coming into contact with indigenous virus flora, can have a harsh impact on crops. The speed and severity of viral attacks depend on various factors in the agricultural environment. Factors such as wind, aerial vectors and their population density, their ability to transmit, and general weather conditions in the agricultural area should be considered. It is crucial to monitor the agricultural environment just above the crop canopy to detect vectors as early as possible. Prophylactic insecticide sprays are necessary.

Viroids are infectious to numerous crop species and can even cause epidemics. Some viroids are transmitted through the air and can withstand harsh conditions in the troposphere to some extent. Several viroids that can cause diseases in field crops, vegetables, and fruit crops are actually transmitted by aerial vectors (Hadidi et al., 2022; Di Serio et al., 2017). The aphid species *Myzus persicae,* for example, is a vector for potato spindle tuber viroid (PSTV) (De Bokx et al., 1981; Salazar et al., 1995; Querci et al., 1997; Syller et al., 1997). Similarly, the aphid *Macrospihum euphorbia* mediates the spread of PSTVd, apple chlorotic fruit spot viroid (ACFSVd), and tomato planta macho viroid (TPMVd) (Galindo et al., 1986). The moth species *Cydiapomonella* transmits ACFSVd (Leichtfried et al., 2020). Whiteflies, such as *Bemisia* sp., also transmit PSTVd. Bumblebees (*Bambus ignites*) spread tomato chlorotic dwarf viroid (TCDVd) (Candresse et al., 2010) and tomato apical stunt viroid (TASVd) (Matsuura et al., 1986). European honeybees and even pollen can act as aerial vectors for viroids. Viroids that affect peaches, such as peach latent mosaic viroid (PLMVd), are known to be transmitted along with pollen (Barba et al., 2007; Desvignes, 1986). Tropospheric wind, turbulence, and raindrops can carry viroids from one location to another in the troposphere.

7.5 EPIDEMICS AND PANDEMICS ON CROPS CAUSED BY AIRBORNE MICROBES: A FEW EXAMPLES

Major food grain-producing regions all over the world are becoming susceptible to epidemics and pandemics. Genetic drifts of pathogenic microbes, genetic breakdown in crop resistance, monocropping trends, congenial weather, and a high load of microbial pathogens in the aerospace have been a few reasons. Additionally, the effects of climate change are contributing to the buildup of pathogens. It is reported that wheat, rice, and corn have each experienced annual grain losses ranging from 20–30% in the past decade.

Global aerial monitoring systems are in place for only a few diseases such as wheat rusts, WSMV, late blight of potatoes, cassava mosaic virus, etc. Aerial spore density or insect vectors can be periodically monitored, if possible, through electronic and manual means. There are internet-based networks connected via low altitude balloons, aerostats, and drones to disseminate information regarding the progress of epidemics/pandemics in different continents (Cardenas-McGill, 2021; see Krishna, 2022). This is similar to the aerostat-based network that monitors international borders (US Department of Homeland Security-Science and Technology Directorate, 2018).

Knowledge about epidemiology, particularly the dispersal patterns of microbes that cause epidemics or pandemics, is of immense relevance (Lamondia & Aylor, 2001). Aylor (2001, 2003, 2004) has studied the epidemiology of airborne microbial pathogens and their impact on disease causation, as well as the grain loss caused by such epidemics/pandemics. They have evaluated three models that deal with the dispersal of airborne microbes, especially plant pathogens (Aylor & Shaw, 2000). These models include short-distance dispersal (lagrangian particle trajectory), intermediate-distance dispersal (approximately 1 km distance), and long-distance dispersal of over 50 km from the focus of infection or inoculum of the pathogen. In the case of long-range dispersal, factors such as easy levitation and biometrics of spores, including spore viability, germination, wall thickness, tolerance to desiccation, UV-radiation, and survival period in the troposphere, are important (Aylor, 1998). Even among long-distance airborne microbes, there are pathogens that travel for just 1-2 km in the air, such as Apple scab caused by *Venturia inequalis*. Asexual spores of *Phytophthora infestans* disperse up to about 50 km through air, splashing water, and movement of farm vehicles. There are also other pathogens that transit for distances of 700 to 800 km, for example, *Perenosporatabacina* that causes mold attack on Tobacco foliage. Intercontinental migration by airborne microbial pathogens, such as wheat rusts, sugarcane rusts, or streak virus via vectors, can induce periodic epidemics and pandemics. The rust spores are well adapted to harsh climates in the lower and upper troposphere, allowing them to travel distances of 2,000 to 15,000 km (Aylor, 2004). It is crucial to regularly monitor the microbial load and diversity of plant pathogens causing epidemics. Failure to do so can have devastating effects on farmers' livelihoods and result in the destruction of crops. Plant pathogenic microbes undergo rapid genetic mutations, which can generate virulent forms. Therefore, it is essential to regularly monitor the agrarian sky (e.g., Ug 99 strain of *Pgt*).

The fecundity, virulence, and survival in air and on the crop's surface are aspects that need consideration while predicting severity. The epidemic-causing fungi may produce a few trillion spores. This ensures that spores get disseminated despite harsh climatic factors. Harsh troposphere reduces spore population, but it is known that *Pgt* produces 25 trillion spores in one hectare of wheat crop. Downy mildew may release 225 million spores per hectare into the air. Bunt of wheat releases about 6 to 12 million spores per hectare. Smut of maize seems to release 1,25,000 trillion spores from the ears. Therefore, we have to assess spore population with the right perspective. Spore density in the air and the spread of epidemics need careful consideration. We may have to analyze the population of the pathogen in the entire profile of the troposphere from the soil surface to a certain height. Recurrence of epidemics in short intervals leads to the buildup of the pathogen in the aerospace. It may form a niche in an area if susceptible crops and genotypes are planted regularly. Monocultures provide a better opportunity for epidemic microbes to survive and flourish.

The agrarian sky can be notorious. It may seem to harbor no or low density of epidemic-causing pathogens for a couple of days or weeks. However, many of them migrate rapidly through wind and jet streams. This phenomenon swiftly introduces microbial pathogens into an area. As a result, if one estimates a high density of pathogens in the agrarian sky, they may just move away from the area rapidly in the troposphere due to the same climatic parameters. Therefore, we must understand the migratory behavior of these pathogens.

Globally, the top few microbial plant pathogens that are important agents causing epidemics or pandemics on food crops vary from year to year. The dominance of microbial pathogens that induce epidemics may be transitory, lasting for a certain period of time. The population, virulence, and migratory trends of epidemic-causing microbes fluctuate. The devastation caused by microbial plant pathogens changes with the season and location. We can attribute a portion of this manifestation to the aerospace above crops, known as the 'agrarian sky.' As discussed in Chapter 2, for a pandemic to become prominent, long-distance or intercontinental migrations are necessary. Atmospheric mechanisms such as dust plumes, jet streams, drifting vectors, etc., need to be considered. Some recent epidemics/pandemics of profound consequences to crops in different continents include (*see* 2 Blades Foundation, 2021; Krishna, 2022).

Farmers worldwide have experienced epidemics and pandemics caused by microbes. A few have been devastating to crops and farmers' livelihoods. For farmers, the recovery of yield levels and normal financial returns has taken extended periods. A couple of airborne plant pathogens have almost brought the entire human population and domesticated animals in the region to the brink.

In fact, the earliest reports of an epidemic pertain to *Phytophthora* late blight. It affected the carbohydrate source in Ireland and Britain in the 1840s. Perhaps the most famous among the crop epidemics/pandemics is the Late Blight of Potato caused by *Phytophthora infestans*. It restricted carbohydrate supply to regions in Northern Europe and the American continent. The pathogen spread via wind and raindrop splashes. Farms with low phytosanitation were severely affected.

The first recorded instances of crop pandemics that spread through the agrarian sky date back to the 1800s. Some examples include the Late blight of Potato in Ireland during the 1840s and 1850s, and the rust pandemic caused by *Hemeliavastetrix* in coffee plantations in Sri Lanka in 1868 (2 Blades Foundation, 2021). The Urediniospores of coffee rust are highly contagious and airborne. In the 1870s and 1880s, grapes in Southern France suffered from downy mildew pandemics caused by *Plasmoparaviticola.* The grape vines took a decade to recover. The causal agent is airborne and produces a large quantity of sporangia. These sporangia erupt through stomata and float in the air. As a result, sporangia spread through wind and rainwater splashing.

In the early 1900s, Brown rust of wheat wreaked havoc on wheat farmers throughout North America and Europe. *Puccinia graminis tritici* is the cause of Brown rust of wheat, also known as stem rust. The fungus spreads through the air during its uredial stage. The asexual spores can travel several hundred to thousands of kilometers in the air. The vast expanse of the sky has facilitated the rapid spread of pandemics. The urediniospores are able to tolerate the harsh environment of the troposphere and can survive extreme climates. When these spores land on a susceptible host, they induce rust disease. Initially, there were setbacks when high-yielding wheat cultivars were grown as they were susceptible to rust. However, since those early setbacks, consistent breeding for resistance has allowed us to accumulate wheat genotypes that are tolerant to brown rust. Improving genetic resistance to rust and planting non-host or resistant lines of wheat are effective methods for controlling brown rust. It is generally recommended to monitor the entire profile of the troposphere (altitude) for urediospores. This helps us to provide early warnings to farmers and regions about an impending epidemic. Throughout the past century, epidemics caused by wheat rust fungus occurred in 1916, the 1930s, the 1950s, and most recently in 1999 (2 Blades Foundation, 2021).

During the 1920s until the end of the 1950s, the airborne inoculum of *Fusarium* caused massive epidemics of Panama Disease on bananas. This disease is caused by *Fusarium oxysporum vas cubense* race 4. The epidemic was particularly severe in the tropical regions of Central and South America. The sickly agrarian sky played a part in rapidly distributing the sickle-shaped conidia, aiding the spread of the disease and causing the epidemic (2 Blades Foundation, 2021).

The Northeast plains in India experienced a major epidemic caused by *Cochliobolus miyabeanus* (*Helminthosporium oryzae* = *Bipolaris* sp.) during the 1930s and 1940s. This epidemic resulted in the brown spot disease of rice (IRRI, 2021; 2 Blades Foundation, 2021; Groth & Hollier, 2012). The native human population heavily relied on rice as a staple food and carbohydrate source, leading to famine and large-scale migration of farmers to safer regions for rice cultivation.

Epidemics of bacterial blight on rice were observed in the Asian tropics during the 1960s. The Southern corn blight epidemic became prominent in the Northern Plains of the USA during the 1970s, caused by *Xanthomonas oryzae var oryzae*.

An epidemic of coffee rust reappeared, but this time in Central and South America alone, in the 1970s. It was caused by virulent strains of *Hemelia-vastetrix* (Scheiber & Zentmeyer, 1984).

An epidemic on bananas was noticed again in the 1970s. Panama Disease of Banana caused by *Fusarium oxysporum var cubense* race 4 recurred in the 1990s. The epidemic covered large areas in Southeast Asia.

Citrus canker caused by *Xanthomonas campestris var citri* reached epidemic proportions from the 1990s to 2010. The citrus groves in Florida, USA were affected. It depreciates the quality of the fruits and also reduces photosynthetic efficiency.

Several crop species have experienced epidemics/pandemics in recent years (2010 to present). The airspace above the crops, wind, and water have served as dispersal agents. During the past decade (2010 to present), Citrus greening disease (Huanglongbing disease) has severely affected citrus groves in Florida. Citrus greening is caused by a bacterium called *Candidatus liberibacter*sp. It initially affected Asian citrus groves before spreading to Florida's citrus belt. The bacterial pathogen is transmitted by a Psyllid vector called *Diaphorinacitri,* which is an aerial vector. Other notable epidemics include Wheat stripe (Pst) in 2010, which affected the wheat belt in the Northern Great Plains. The coffee rust epidemic reappeared in South America in 2012. An epidemic of Maize Necrosis was reported in the USA in 2011. Wheat blast, traced in India and Bangladesh, is a recent epidemic (Singh et al., 2021). It affected wheat in the plains from 2016 to 2020. Wheat blast is caused by the pathotype triticum of the fungus *Magnaporthe oryzae (syn Pyricualria oryzae lineage triticum)* and is disseminated through wind and rain splash. The above list of epidemics/pandemics is not exhaustive. These examples serve to demonstrate that the agrarian sky can be deadly to crops if it harbors microbial pathogens with the potential to cause pandemics.

Overall, farmers should have a better understanding of the aerial behavior of pathogens that cause epidemics/pandemics. It is not enough to only know about diseases that are seedborne, soil borne, or waterborne. Being able to manipulate microbial plant pathogens, their population, genetic diversity, and functions in the air or even before they spread in the free troposphere would greatly benefit global farming.

Crops rely on the agrarian sky, which includes the aerospace above the crop canopy, just as much as they rely on soil, water, and their own genetic makeup and physiological abilities to adapt to the environment. The sky provides countless advantages to crops. However, crops are also susceptible to harmful biotic factors, such as aerial plant pathogens. It is evident from the above explanation that aerial microbial plant pathogens can have devastating effects on crops, leading to epidemics/pandemics and crop losses. Therefore, it is crucial not to overlook the agrarian sky (aerospace) above crops.

KEYWORDS

- **airborne microbes**
- **aphids**
- ***Bipolaris maydis***
- **blights**
- **blotches**
- **epidemics**
- **ergots**
- ***Fusarium oxysporum***
- **genetic drift**
- **insect vectors**
- **jassids**
- **mildews**
- **pandemics**
- ***Pseudomonas***
- **psyllids**
- ***Puccinia graminis***
- **rust fungi**
- **smut**
- **virulence**
- ***Xanthomonas oryzae***
- **yellow rust**

REFERENCES

2 Blades Foundation, (2021). *Major Crop Disease Outbreaks: Disease Epidemics Are an Ongoing Concern* (pp. 1–12). https://2blades.org/major-crop-disease-outbreaks/ (accessed on 19 January 2024).

Acosta-Martinez, V., Van, P. S., Moore-Kucera, J., Baddock, M. C., & Zobeck, T. M., (2015). Microbiology of wind eroded sediments: Current knowledge and future. *Aeolian Research, 18*, 99–113.

Agrios, G. N., (2005). Plant diseases caused by fungi. In: *Plant Pathology* (5thedn., pp. 467, 468). Elsevier Academic, Amsterdam, Netherlands.

Anderson, P. K., Cunningham, A. A., Patel, N. G., Morales, F. J., Epstein, P. R., & Daszak, P., (2004). Emerging infectious diseases of plants: Pathogen pollution, climate change and agrotechnology drivers. *Trends in Ecology and Evolution, 19*, 535–544.

Anikster, Y., Bushnell, W., Roelfs, A., Eilam, T., & Manisterski, J., (1997). *Puccinia recondita* causing leaf rust on cultivated wheats, wild wheats, and rye. *Canadian Journal of Botany, 75*, 2082–2096.

Anikster, Y., Eilam, T., Bushnell, W. R., & Kosman, E., (2005). Spore dimensions of *Puccinia* species of cereal hosts as determined by image analysis. *Mycologia, 92*, 474–484.

Arthur, J. C., (1916). Uredinales of Porto Rico based on collections Whetxel, H. H., & Olive, E. E., *Mycologia, 8*, 55–104.

Arthur, J. C., (1934). Manual of the rusts in United States and Canada reprinted with a supplement by G. B. Cummins, 1962 Hafner Publishing Company, New York.

Aylor, D. E., & Shaw, R. H., (2000). Integrating spore dispersal and models of turbulence. In: *Proceedings of 24th Conference on Agricultural and Forest Meteorology* (pp. 174, 175). American. Meteorological Society, Boston, MA., USA.

Aylor, D. E., (1990). The role of intermittent wind in the dispersal of fungal pathogens. *Annual Review of Phytopathology, 28*, 73–92.

Aylor, D. E., (1998). Biometeorological parameters governing long-distance spread of plant disease by atmospheric transport of inoculum. In: *Proceedings of 13th Conference of Biometeorology and Aerobiology* (pp. 311, 312). American Meteorological Society. Boston, MA., USA.

Aylor, D. E., (2001). Aerobiology of fungi in relation to capture and release by plants. In: Lindow, S. E., Hecht-Poinar, E. I., & Elliot, V. J., (eds.), *Phyllosphere Microbiology* (pp. 341–361). APS Press, St. Paul, MN. USA.

Aylor, D. E., (2003). Spread of plant disease on a continental scale: Role of aerial dispersal of pathogens. *Ecology, 84*, 1989–1997.

Aylor, D. E., (2004). *Analysis of Risk of Regional Spread of Plant Disease Airborne Inoculum* (pp. 1–12). Project report. Connecticut Agricultural Experimental Station, Storrs, Connecticut, USA.

Barba, M., Ragozzino, E., & Faggioli, F., (2007). Pollen transmission of peach latent mosaic viroid. *Journal of Plant Pathology, 89*, 287–289.

Bayles, R. A., Channell, M. H., & Stigwood, P., (1989). *Yellow Rust of Wheat* (11–17). United Kingdom Pathogen Virulence Survey, 1989 Annual Report.

Beddow, J. M., Pardey, P. G., Chai, Y., Hurley, T. M., Kriticos, D. J., Braun, H. J., Park, R. F., et al., (2015). Research investment implications of shifts in the global geography of wheat stripe rust. *Nature Plants, 1*, 15–32.

Benson, D. M., & Campbell, C. L., (1985). Spatial pattern of *Phytophthora* root rot and dieback of azalea in container-grown nursery stock. *Plant Disease, 69*, 1049–1054.

Beresford, R. M., (1982). Stripe rust (*Puccinia striiformis*), a new disease of wheat in New Zealand. *Cereal Rusts Bulletin, 10*, 35–45.

BGRI Archive Site, (2022). *About Borlaug Global Rust Initiative* (p. 118). https://globalrust.org/about-bgri/ (accessed on 19 January 2024).

Bhandhari, R. R., Aryal, L., Sharma, S., Acharya, M., Pokhrel, A., Apar, G. C., Kaphle, S., et al., (2017). Screening of maize genotypes against southern leaf blight (*Bipolaris maydis*) during Summer Season in Nepal. *World Journal of Agricultural Research,5*, 31–41. doi: 10.12691/wjar-5-1-5/.

Blackman, R., & Eastop, V., (2000). *Aphids on the World's Crops: An Identification and Information Guide* (2nd edn., p. 476). Chichester: Wiley.

Bolton, M. D., Kolmer, J. A., & Garvin, D. F., (2008). Wheat leaf rust caused by *Puccinia triticina*. *Molecular Plant Pathology, 9*, 563–575. https://doi.org/10.1111/j.1364-3703.2008.00487.x/.

Boshoff, W. H. P., Pretorius, Z., & Van, N. B. D., (2002). Establishment, distribution, and pathogenicity of *Puccinia striiformis* f sp. tritici in South Africa. *Plant Disease, 86*, 485–492.

Bromfield, K. R., (1984). Soybean rust. *Monograph* (Vol. 11, p. 65). The American Phytopathological Society, St. Paul, MN, USA.

Brown, J. K. M., &, & Hovmoller, M. S., (2002). Aerial dispersal of pathogens on the global and continental scales and its impact on plant disease. *Science, 297*, 537–541.

Brown, W. M., Hill, J. P., & Velasco, J. R., (2001). Barley yellow rust in North America. *Annual Review of Phytopathology, 39*, 367–384.

Burdon, J. J., (1987). *Diseases and Plant Population Biology* (p. 224). Cambridge University Press, Cambridge, England.

Burrows, M., Thomas, C., McRobrts, N., Bostock, R. M., Coop, L., & Stack, J., (2016). Coordination of diagnostic efforts in the great plains: Wheat virus survey and modeling of disease onset. *Plant Disease, 100*, 1037–1045.

Campos, H., & Ortiz, O., (2020). *The Potato Crop: Its Agricultural, Nutritional and Social Contribution to Humanoids* (pp. 351–430). Springer Open Book. https://doi.org/10.1007/978-3-030-28683-5/.

Candresse, T., Marais, A., Tassus, X., Suhard, P., Renaudin, I., Leguay, A., Poliakoff, F., & Blancard, D., (2010). First report of tomato chlorotic dwarf viroid in tomato in France. *Plant Disease, 9*, 633.

Cardenas-McGill, S., (2021). *Plant Pandemics Threaten Future Food Supply* (pp. 1–7). Futurity. https://www.futurity.org/plant-pandemics-crops-food-supply-2571952-2/ (accessed on 19 January 2024).

Carvalho, C. R., Fernandes, R. C., Carvalho, G. M. A., Barreto, R. W., & Evans, H. C., (2011). Crypto-sexuality and the genetic diversity paradox in coffee rust, *Hemileia vastatrix* (pp. 1–12). *PLoS One, 6*, e26387.

Chakraborty, P., Chakraborty, A., & Gupta-Bhattacharya, S., (2020). Dispersal of airborne pathogenic conidia of *Bipolaris oryzae* inciting brown spot disease of paddy in West Bengal, India. *Journal of Tropical Agriculture, 58*, 197–208.

Chalupnıkova, J., Kundu, J. K., Singh, K., Bartakova, P., & Beoni, E., (2017). Wheat streak mosaic virus: Incidence in field crops, potential reservoir within grass species and uptake in winter wheat cultivars. *Journal of Integrated Agriculture, 16*, 60345–60357.

Chen, W. Q., Wu, L. R., Liu, T. G., & Xu, S. C., (2009). Race dynamics, diversity, and virulence evolution in *Puccinia striiformis f sp. tritici* causal agent of wheat stripe rust in China from 2003 to 2007. *Plant Disease, 96*, 1093–1101.

Chen, W., Wellings, C., Chen, X., Kang, Z., & Liu, T., (2014). Wheat stripe (yellow) rust caused by *Puccinia striiformis f. sp. tritici*. *Molecular Plant Pathology, 15*, 433–446.

Chen, X. M., (2005). Epidemiology and control of stripe rust (*Puccinia striiformis f sp. tritici*) on wheat. *Canadian Journal of Plant Pathology, 27*, 314–337.

Chen, X. M., Moore, M., Milius, E. A., Long, D. L., Line, R. F., Marshall, D., & Jackson, L., (2002). Wheat stripe rust epidemics and races of *Puccinia striiformis f. sp. tritici* in the United States in 2000. *Plant Disease, 86*, 39–46.

Cromey, M. G., (1989). Occurrence and effects of stripe rust in wheat spikes in New Zealand. *New Zealand Journal of Crop and Horticultural Science, 17*, 155–158.

Damicone, J. P., & Brandenberger, L., (2000). *Common Diseases of Tomato* (pp. 1–8). Oklahoma Cooperative Extension Service, Stillwater, Oklahoma, USA.

Danelli, A. L. D., Fernandes, J. M. C., Maciel, J. L. N., Boaretto, C., & Forcelini, C. A., (2019). Monitoring *Pyricularia* sp. airborne inoculum in Passo Fundo, Rio Grande do Sul – Brazil. *Summa Phytopathologica, 45*, 361–367.

De Bokx, J. A., & Pirone, P. G., (1981). Transmission of potato spindle tuber viroid by aphids. *Netherlands Journal of Plant Pathology, 87*, 31–34.

De Milliano, W. A. J., (1992). Sorghum diseases in southern Africa. In: *Sorghum and Millet Diseases: A Seeond World Review* (Vol. 19, pp. 335, 336). International Crops Research Institute for the Semi-Arid Tropics, Hyderabad, India (Abstract).

Desvignes, J. C., (1986). Peach latent mosaic and its relation to peach mosaic and peach yellow mosaic virus diseases. *Acta Horticulture, 193*, 51–57.

Di Serio, F., Li, S. F., Pallas, V., Owens, R. A., Randles, J. W., Sano, T., Verhoeven, T. J., et al., (2017). Viroid taxonomy. In: Hadidi, A., Flores, R., Randles, J. W., & Palukaitis, P., (eds.), *Viroids and Satellites* (pp. 135–146). Academic Press: Oxford, UK.

Drab, T., Svobodov, A. E., Ripl, J., Jarosova, J., Rabenstein, F., Melcher, U., & Kundu, J. K., (2014). SYBR green I based RT-qPCR assays for the detection of RNA viruses of cereals and grasses. *Crop Pasture Science, 65*, 1323–1328.

Duveiller, E., Fucikovsky, L., & Rudolph, K., (1997). *The Bacterial Diseases of Wheat: Concepts and Methods of Disease Management* (pp. 1–86). Mexico, D.F.: CIMMYT. https://repository.cimmyt.org/bitstream/handle/10883/1227/64767.pdf?sequence=1&isAllowed=y (accessed on 19 January 2024).

Dye, D. W., Bradbury, J. F., Goto, M., Hayward, A. C., Lelliott, R. A., & Schroth, M. N., (1980). International standards for naming pathovars of phytopathogenic bacteria and a list of pathovar names and pathotype strains. *Review of Plant Pathology, 59*, 153–168.

El Jarroudi, M., Kouadio, L., Tychon, B., El Jooroudi, M., Junk, J., Bock, C., & Delfosse, (2018). Modeling the main fungal diseases of winter wheat: Constraints and possible solutions. *Advances in Plant Pathology*. doi: 10.5772/intechopen.75983/.

Ellis, M. A., (2022). *Powdery Mildew of Grapes* (pp. 1–3). Ohio State Agricultural Extension, Columbus Ohio, USA. https://ohioline.osu.edu/factsheet/plpath-fru-37/ (accessed on 19 January 2024).

FAO, (2014). *Strengthening Capacities and Promoting Collaboration to Prevent Wheat Rust Epidemics-Wheat Rust Disease-Global Program* (pp. 1–35). Food and Agricultural Organization of the United Nations, Rome, Italy. https://www.fao.org/documents/card/en/c/1a3efebf-ac3f-4cf0-a7c5- (accessed on 19 January 2024).

Fargette, D., Konate, G., Fauquet, C., Muller, E., Peterscmitt, M., & Thresh, J. M., (2006). Molecular ecology and emergence of tropical plant viruses. *Annual Review of Phytopathology, 44*, 235–260.

Figueroa, M., Hammond-Kossack, K. E., & Solomon, P. S., (2018). A review of wheat diseases. *Molecular Plant Pathology, 19*, 1–14. doi: 10.1111/mpp.12618/.

Fondevilla, S., & Rubiales, D., (2012). Powdery mildew control in pea: A review. *Agronomy for Sustainable Development, 32*, 401–409.

Forster, R. L., & Bradbury, J. F., (1990). Pink seed of wheat caused by *Erwinia rhapontici* in Idaho. *Plant Disease, 74*, 81.

Frederickson, D. E., Mantle, P. G., & De Milliano, W. A. J., (1993). Windborne spread of ergot disease (Claviceps Africana) in sorghum, lines in Zimbabwe. *Plant Pathology, 42*, 368–377.

French, R. C., & Stenger, D. C., (2003). Evolution of wheat streak mosaic virus dynamics of population growth within plants may explain limited variation. *Annual Review of Phytopathology, 41*(1), 199–214. doi: 10.1146/annurev.phyto.41.052002.095559/.

Galindo, J., Lopez, M., & Aguilar, T., (1986). Significance of *Myzuspersicae* in the spread of tomato planta macho viroid. *Fitopatologica Brasilia, 11*, 400–410.

German, S., Barcellos, A., Chaves, M., Kohli, M., Campos, P., & De Viedma, L., (2007). The situation of common wheat rusts in the Southern Cone of America and perspectives for control. *Australian Journal of Agricultural Science, 58*, 620–630.

Ghimire, B., Sapkota, S., Bahri, B. A., Martinez-Espinoza, A. D., Buck, J. W., & Mergoum, M., (2020). *Fusarium* head blight and rust diseases in soft red winter wheat in the Southeast United States: State of the art, challenges, and future perspective for breeding. *Frontiers in Plant Science, 11*, 1080. doi: 10.3389/fpls.2020.01080/.

Golan, J. J., & Pringle, A., (2017). Long-distance dispersal of fungi. *American Society of Microbiology Spectrum, 5*(4), FUNK-0047-2016. doi: 10.1128/microbiolspec. FUNK-0047-2016/.

Gregory, L. V., Ayers, J. V., & Nelson, R. R., (1978). **Predicting yield losses in corn from southern corn leaf blight**. *Phytopathology, 68*, 517–521.

Gregory, P. H., Griffin, M. J., & Maddison, M. R., (1984). Cocoa black pod: A re-interpretation. *Growers Bull., 35*, 5–22.

Groth, D., & Hollier, C., (2012). *Brown Spot of Rice* (pp. 1–5). Plant Pathology Department, Louisiana State University, LSU-Agriculture, Louisiana, USA. https://www.lsuagcenter.com/~/media/system/4/6/2/f/462fd54617eaff51c0fdeb50d5feccdd/pub3108brownspothighres.pdf/ (accessed on 19 January 2024).

Hadi, B. A. R., Langham, M. A. C., Osborne, L., & Tilmon, K. J., (2011). Wheat streak mosaic virus on wheat: Biology and management. *Journal of Integrated Pest Management, 2*, 1–5.

Hadidi, A., Sun, L., & Randles, J. W., (2022). Modes of viroid transmission. *Cells, 11*, 719 https://doi.org/10.3390/cells11040719/.

Hagborg, W. A. F., (1974). Notes on bacterial diseases of cereals and some other crop plants. *Canadian Plant Disease Survey, 54*, 129–151.

Hema, M., Sreenivasulu, P., Patil, B. L., Kumar, L., & Reddy, D. V. R., (2014). Tropical Food legumes: Virus diseases of economic importance and their control. *Advance in Virus Research, 90*, 431–455. https://doi.org/10.1016/B978-0-12-801246-8.00009-3Get rights and content.

Hirst, J. M., & Stedman, O. J., (1963). Dry liberation of fungal spores by raindrops. *Journal of General Microbiology, 33*, 335–344.

Holcomb, G. E., & Aime, M. C., (2010). First report of *Plumeria* spp. rust caused by *Coleopsporium plumeriae* in Louisiana and Malaysia and *Catheranthus roseus*, a new host of this rust. *Plant Disease, 94*, 272–723.

Hosford, Jr. R. M., (1982). White blotch incited in wheat by *Bacillus megaterium pv. cerealis*. *Phytopathology, 72*, 1453–1459.

Hovmøller, M. S., & Justesen, A. F., (2007a). Appearance and interpretation of atypical phenotypes of *Puccinia striiformis* f. sp. *tritici* in NW-Europe. *Australian Journal of Agricultural Research, 58*, 518–524.

Hovmøller, M. S., & Justesen, A. F., (2007b). Rates of evolution of a virulence phenotypes and DNA markers in a northwest European population of *Puccinia striiformis* f. sp. *tritici. Molecular Ecology, 16*, 4637–4647.

Hovmoller, M. S., Yahaouii, A. H., Milus, E. A., & Justessen, A. S., (2008). Rapid global spread of two aggressive strains of a wheat rust. *Molecular Ecology*. doi: 10.1111/j.1365-294X.2008.03886.x/.

Hubbard, A., Lewis, C. M., Yoshida, K., Ramirez-Gonzalez, R. H., De Vallavieille-Pope, C., Thomas, J., Kamoun, S., et al., (2015). Field pathogenomics reveals the emergence of a diverse wheat yellow rust population. *Genome Biology, 16*, 23.

Huerta-Espino, J., Singh, R. P., Bhavani, S., Njau, P., Autrique, E., Herrera-Foessel, S. A., et al., (2013). *Frequency of Ug99 Resistant Wheat Lines Derived from Segregating Populations Selected Under the Mexican and Mexico-Kenya Shuttle Breeding Schemes* (p. 74). In: BGRI (Borlaug Global Rust Initiative) 2013 Technical Workshop, New Delhi (India).

Huerta-Espino, J., Singh, R. P., Germán, S., McCallum, B. D., Park, R. F., Chen, W. Q., et al., (2011). Global status of wheat leaf rust caused by *Puccinia triticina. Euphytica, 179*, 143–160. doi: 10.1007/s10681-011-0361-x.

Hullé, M., Coquio, S., & Laperche, V., (1994). Patterns in flight phenology of a migrant cereal aphid species. *Journal of Applied Ecology, 31*, 49–58.

IRRI, (2021). *Rice Knowledge Bank.* International Rice Research Institute, Manila, Philippines http://www.knowledgebank.irri.org/training/fact-sheets/pest-management/diseases/item/brown-spot/ (accessed on 19 January 2024).

Isard, S. A., Gage, S. H., Comtois, P., & Russo, J. M., (2005). Principles of the atmospheric pathway for invasive species applied to soybean rust. *Bioscience, 55*, 851–861. http://dx.doi.org/10.1641/0006-3568(2005)055 [0851:POTAPF]2.0.CO;2. /.

Jin, Y., Szabo, L. J., & Carson, M., (2010). Century old mystery of *Puccinia striformis* life history solved with the identification of *Barberis* as alternate host. *Phytopathology, 100*, 432–435.

Johnson, C. G., (1969). *Migration and Dispersal of Insects by Flight* (p. 763). London: Methuen & Co Ltd.

Jones, R. A. C., (2008). Plant virus emergence and evolution: Origins, new encounter scenarios, factors driving emergence, effects of changing world conditions, and prospects for control. *Virus Research, 141*, 113–130. https://doi.org/10.1016/j.virusres.2008.07.028 /.

Jones, R. A. C., (2014). Plant virus ecology and epidemiology: Historical perspectives, recent progress, and future prospects. *Annals Applied Biology, 164*, 320–347.

Jones, R. A. C., (2020). Disease pandemics and major epidemics arising from new encounters between Indigenous viruses and introduced crops. *Viruses, 12*, 1388. doi: 10.3390/v12121388 www.mdpi.com/journal/viruses/ (accessed on 19 January 2024).

Jones, R. A. C., (2021). Global plant virus disease pandemics and epidemics. *Plants, 10*, 233 https://doi.org/10.3390/plants10020233/.

Josephson, L. M., Graves, C. R., Kinier, H. L., & Hilty, J. W., (1971). **Reduction in yield of corn from southern corn leaf blight.***Plant Disease Reporter, 58*, 219, 220.

Jung, J., Tafner, J., Siobkim, J., Berg, G., & Minryu, C., (2020). *Archaea, Tiny Helpers of Land Plants* (pp. 1–12). https://doi.org/10.1016/j.csbj.2020.09.005/.

Kakishima, M., Ji, J. X., Zhao, P., Wang, Q., Li, Y., & McKenzie, E. H. S., (2017). Geographic expansion of a rust fungus on *Plumeria* in Pacific and Asian countries. *New Zealand Journal of Botany, 55*(2), 178–186. doi: 10.1080/0028825X.2017.1281316/.

Kakishima, M., Kobayashi, T., & McKenzie, E. H. C., (1995). A warning against invasion of Japan by the rust fungus, *Coleopsporiumplumeriae*, on *Plumeria. Forest Pests, 44*, 144–147.

Kamareit, B., Rothrock, C., Nicholas, R., Price, P., Mehl, H., Dufault, D., Bart, R., et al., (2021). *Identification and Management of Bacterial Blight of Cotton* (pp. 1–7). https://www.cottoninc.com/cotton-production/ag-research/plant-pathology/management-bacterial-blight-cotton/ (accessed on 19 January 2024).

Kellogg, C. A., & Griffin, D. W., (2006). Aerobiology and the global transport of desert dust. *Trends in Ecology and Evolution, 21*, 638–644.

Klinkowski, M., (1970). Catastrophic plant diseases. *Annual. Review of Phytopathology, 8*, 37–60.

Krishna, K. R., (2019). *Unmanned Aerial Vehicle Systems in Crop Production: A Compendium* (p. 675). Apple Academic Press Inc. Waretown, New Jersey, USA.

Krishna, K. R., (2020). *Aerial Robotics in Agriculture: Parafoils, Blimps, Aerostats and Kites* (pp. 227–324). Apple Academic Press Inc. Palm Bay, Florida, USA.

Krishna, K. R., (2022). *The Agricultural Sky: A Concept to Revolutionize Farming* (p. 438). Apple Academic Press Inc. Palm Bay, Florida, USA.

Kriss, A. B., Paul, P. A., & Madden, L. V., (2012). Variability in *Fusarium* head blight epidemics in relation to global climate fluctuations as represented by the El-Nino-Southern oscillation and other atmospheric patterns. *Phytopathology, 102*, 55–64. doi: 10.1094/PHYTO-04-11-0125./.

Kuan, T. L., Minsavage, G. V., & Schad, N. W., (1986). Aerial dispersal of *Xanthomonas campestris* pv. *campestris* from naturally infected *Brassica campestris. Plant Disease, 70*, 409–413.

Kumar, S., Meshram, S., & Sinha, A., (2017). Bacterial diseases in rice and their eco-friendly management. *International Journal of Agricultural Sciences and Research, 7*, 31–42.

Lamondia, J. A., & Aylor, D. E., (2001). Epidemiology and management of a periodically introduced pathogen. *Biological Invasions, 3*, 273–282.

Large, E. C., (1940). *The Advance of the Fungi* (p. 488). Jonathan Cape, London.

Legg, J. P., Jeremiah, S. C., Obiero, Maruthi, M. N., Ndvetabula, I., Okao-Okubja, Bouwenmeester, H., et al., (2011). Comparing the regional epidemiology of the Cassava mosaic and cassava brown streak virus pandemics in Africa. *Virus Research, 159*, 161–170. doi: 10.1016/j.virusres.2011.04.018/.

Legg, J. P., Owoi, B., Sseruwagi, & Ndunguru, J., (2006). Cassava mosaic virus diseases in east and Central Africa: Epidemiology and management of a regional pandemic. *Advances in Virus Research, 67*, 355–418. https://doi.org/10.1016/S0065-3527(06)67010-3/.

Legg, J., & Fauquet, C., (2004). Cassava mosaic Gemini viruses in Africa. *Plant Molecular Biology, 56*, 585–599.

Leichtfried, T., Reisenzein, H., Steinkellner, S., & Gottsberger, R. A., (2020). Transmission studies of the newly described apple chlorotic fruit spot viroid using a combined RT-qPCR and droplet digital PCR approach. *Archives of Virology, 165*, 2665–2671.

Leonard, K. J., & Szabo, L. J., (2005). Stem rust of small grains and grasses caused by *Puccinia graminis. Molecular Plant Pathology, 6*, 99–111.

Listman, M., (2018). *Research Program on Wheat CGIAR* (pp. 1–3). Consultative Group on International Agricultural Research, FAO HQ, Rome, Italy.

Lorrain, C., Cristine, K., Dos, S. G., Germain, H., Hackers, A., & Duplessis, S., (2018). Advances in understanding obligate biotrophy in rust fungi. *New Phytologist, 222*, 1190–1206. doi: 10.1111/nph.15641/.

Luig, N. H., (1985). Epidemiology in Australia and New Zealand. In: Roelfs, A. P., & Bushnell, W. R., (eds.), *The Cereal Rusts Volume H* (pp. 301–328). Academic Press, Orlando.

Madden, L. V., (1992). Rainfall and the dispersal of fungal spores. *Advances of Plant Pathology, 8*, 39–79.

Mahaffee, W. F., & Stoll, R., (2016). The ebb and flow of airborne pathogens: Monitoring and use of disease management decisions. *Phytopathology Review, 106*, 420–431.

Matsuura, S., Matsushita, Y., Kozuka, R., Shimizu, S., & Tsuda, S., (2010). Transmission of tomato chlorotic dwarf viroid by bumblebees (*Bombus* ignites) in tomato plants. *European Journal of Plant Pathology, 126*, 111–115.

McGrath, M. T., & Pennypackers, S. P., (1991). Reduction in the rate and duration of grain growth in wheat due to stem rust and leaf rust. *Phytopathology, 81*, 778–787.

McInnes, T. B., Gitatis, R. D., McCarter, S. M., Jaworski, C. A., & Phatak, S. C., (1988). Airborne dispersal of bacteria in tomato and pepper transplant fields. *Plant Disease, 72*, 575–579.

McIntosh, R. A., (2009). History and status of the wheat rusts. In: *Proceedings of the Borlaug Global Rust Initiative* (pp. 11–23). Technical Workshop BGRI Cd Obregon, Mexico.

McIntosh, R. A., Wellings, C. R., & Park, R. F., (1995). *Wheat Rusts: An Atlas of Resistance Genes* (p. 200). CSIRO Publications, Victoria, Australia.

McMullen, M. P., Stack, R. W., Miller, J. D., Bromel, M. C., & Youngs, V. L., (1984). *Erwinia rhapontici*, a bacterium causing pink wheat kernels. *Proceedings of the North Dakota Academy of Science, 38*, 78.

Mendgen, K., (1997). The uridinales. In: Caroll, G. C., (ed.), *The Mycota. Plant Relations Part B* (pp. 79–94). Springer, Berlin, Germany.

Milus, E. A., Kristensen, K., & Hovmøller, M. S., (2008). Evidence for increased aggressiveness in a recent widespread strain of *Puccinia striiformis* f. sp. *tritici* causing stripe rust wheat. *Phytopathology, 99*, 89–94.

Milus, E. A., Seyran, E., & McNew, R., (2006). Aggressiveness of *Puccinia striiformis* f. sp. *tritici* isolates in the South-Central United States. *Plant Disease,* 90,847–852.

Morris, C., & Sands, D., (2017). Impacts of microbial aerosols on natural and agro-ecosystems: Immigration, invasions, and their consequences. In: Delort, A. M., & Amatom, P., (eds.), *Microbiology of Aerosols* (pp. 269–279). John Wiley and Sons, Inc., Hoboken, New Jersey, USA.

Mubeen, S., Rafique, M., Hussain, F., Muhamad, R., Hasaan, M., & Chaudary, J., (2017). Study of southern corn leaf blight (SCLB) on maize genotypes and its effect on yield. *Journal of Saudi Society of Agriculture, 10*, 210–217. https://doi.org/10.1016/j.jssas.2015.06.006/.

Murray, G. M., & Brennan, J. P., (2009). Estimating disease losses to the Australian wheat industry. Australasia. *Plant Pathology, 38*, 558–570.

Nagarajan, S., & Joshi, L. M., (1985). Epidemiology in the Indian subcontinent. In: Roelfs, A. P., & Bushnell, W. R., (eds.), *The Cereal Rusts: Diseases, Distribution, Epidemiology, and Control* (Vol. 2, pp. 371–402). Academic Press, Orlando, FL, USA.

Nagarajan, S., & Singh, D. V., (1990). Long distance dispersion of rust pathogens. *Annual Review of Phytopathology, 28*, 139–153.

Navia, D., De Mendonca, R. S., Skoracka, A., Szydło, W., Knihinicki, D., Hein, G. L., Da Silva, P. P. R., et al., (2013). Wheat curl mite, *Aceriatosichella,* and transmitted viruses: An expanding pest complex affecting cereal crops. *Experimental and Applied Acarology, 59*, 95–143.

Olivera, F. P., Newcomb, M., Flath, K., Sommerfeldt-Impe, N., Szabo, L., Carter, M., Luster, D., & Jin, Y., (2017). Characterization of *Puccinia graminis f. sp. tritici* isolates derived from an unusual wheat stem rust outbreak in Germany in 2013. *Plant Pathology, 66*, 1258–1266. doi: 10.1111/ppa.12674/.

Olivera, F. P., Newcomb, M., Szabo, L., Rouse, M. N., Johnson, J. L., Gale, S. W., Luster, D., et al., (2015). Phenotypic and genotypic characterization of race TKTTF of *Puccinia graminis f. sp. tritici* that caused a wheat stem rust epidemic in southern Ethiopia in 2013/14. *Phytopathology, 105*, 917–928.

Ortega, S. F., Ferrocino, I., Adams, I., Silvestri, S., Spadaro, D., Gullino, M. L., & Boonham, N., (2020). Monitoring and surveillance of aerial mycobiota rice paddy through DNA metabarcoding and qPCR. *Journal Fungi, 6*, 372–378.

Otim-Nape, G. W., & Thresh, J. M., (1998). The current pandemic of cassava mosaic virus disease in Uganda. In: Jones, G., (ed.), *The Epidemiology of Plant Diseases* (pp. 423–443). Kluwer: Dordrecht, The Netherlands.

Pardey, P. G., Beddow, J. M., Kriticos, D. J., & Hurley, T. M., (2013). Right sizing stem-rust research. *Science, 340*(6129), 147–148. doi: 10.2307/41942540/.

Park, R., Fetch, T., Hodson, D., Jin, Y., Nazari, K., Prashar, M., & Pretorius, Z., (2011). International surveillance of wheat rust pathogens: Progress and challenges. *Euphytica, 179*, 109–117.

Parry, H. R., (2013). Cereal aphid movement: General principles and simulation modeling. *Movement Ecology, 1,* 14. doi: 10.1186/2051-3933-1-14/.

Patil, B., & Fauquet, C., (2009). Cassava mosaic Gemini viruses: Actual knowledge and perspectives. *MolecularPlantPathology, 10*, 685–701.

Powell, A. J., & Vujanovic, V., (2021). Evolution of *Fusarium* head blight management in wheat: Scientific perspectives on biological control agents and crop genotypes protocooperation. *Applied Science, 11*, 8960. https://doi.org/10.3390/app11198960/.

Purdy, L. H., (1985). Sugarcane rusts. In: Roelfs, A. P., & Bushncll, W. R., (eds.), *The Cereal Rusts* (pp. 237–256). Academic Press, Orlando, Florida, USA.

Querci, M., Owens, R. A., Bartolini, I., Lazarte, V., & Salazar, L. F., (1997). Evidence for heterologous encapsidation of potato spindle tuber viroid in particles of potato leafroll virus. *Journal of General Virology, 18*, 1207–1211.

Ristaino, J. B., & Gumpertz, M. L., (2018). New frontiers in the study of dispersal and spatial analysis of epidemics caused by species in the genus *Phytophthora. Annual Review of Phytopathology, 38*, 541–576.

Ristaino, J. B., Anderson, P. K., Bebbard, D. P., Braumane, K. A., Cunniffef, N. K., Fedorof, N. V., Finegold, C., et al., (2021). The persistent threat of emerging plant disease pandemics to global food security. *Proceedings of National Academy of Sciences USA, 118*(23), e2022239118 https://doi.org/10.1073/pnas.2022239118/ pp. 1- 9.

Roelfs, A. P., Singh, R. P., & Saari, E. E., (1992). *Rust Diseases of Wheat: Concepts and Methods of Disease Management* (p. 89). International Centre for Maize and Wheat. El Baton, Mexico.

Rowell, J. B., & Romig, R. W., (1966). Detection of urediospores of wheat rusts in spring rains. *Phytopathology, 56*, 807–811.

Rupe, J., & Sconyers, L., (2008). Soybean rust. *The Plant Health Instructor* (pp. 1–17). doi: 10.1094/PHI-I-2008-0401-01/.

Rupp, J., & Jacobsen, B., (2017). *Bacterial and Fungal Diseases of Potato and Their Management* (pp. 1–10). Montana State Extension Service, Bozeman, Montana, USA.

Sache, I., (2000). Short distance dispersal of wheat rust spores by wind and rain. *Agronomie, 20*, 757–787.

Saha, S., Garg, R., Biswas, A., & Rai, A. B., (2015). Bacterial diseases of rice: An overview. *Journalof Pure and Applied Microbiology, 9*, 725, 726.

Salazar, L. F., Querci, M., Bartolini, I., & Lazarte, V., (1995). Aphid transmission of potato spindle tuber viroid assisted by potato leaf roll virus. *Fitopatologia, 30*, 56–58.

Savary, S., Bosc, I. P.,Noirot, M., & Zadoks, I. C., (1988). Peanut rust in West Africa: A new component in a multiple pathosystem. *Plant Disease, 72*, 1001–1009.

Schieber, E., & Zentmeyer, G. A., (1984). *Coffee Rust in the Western Hemisphere Plant Disease, 68*, 89–93.

Schmale, III. D. G., & Bergstrom, G. C., (2010). *Fusarium* head blight in wheat. *The Plant Health Instructor* (pp. 1–8). doi: 10.1094/PHI-I-2003-/.

Schumann, G. L., & Leonard, K. J., (2000). Stem rust of wheat (black rust). *American Phytopathology Society the Plant Health Instructor*. doi: 10.1094/PHI-I-2000-0721-01/.

Sconyers, L. A., Kemerait, R. C., Brock, J., Phillips, D. V., Jost, P. H., Sikora, E. J., Gutierrez-Estrada, E., et al., (2006). *Asian Soybean Rust Development in 2005: A Perspective from the South-eastern United States* (pp. 1–23). http://apsnet.org/edcenter/apsnetfeatures/Pages/SoybeanRustDev.aspx/ (accessed on 19 January 2024).

Shah, D. A., Molineros, J. E., Paul, P. A., Willyerd, K. T., Madden, L. V., & De Wolf, E. D., (2013). Predicting *Fusarium* head blight epidemics with weather-driven pre- and post-anthesis logistic regression models. *Phytopathology, 103*, 906–919. doi: 10.1094/PHYTO-11-12-0304-R.

Sinclair, J. B., & Hartman, G. L., (1999). Soybean rust. In: Hartman, G. L., Sinclair, J. B., & Rupe, J. C., (eds.), *Compendium of Soybean Diseases* (4th edn., pp. 25, 26). APS Press, St. Paul, MN, USA.

Singh, K., Wegulo, S. N., Skoracka, A., & Kumar, J., (2018). Wheat streak mosaic virus: A century old virus with rising importance worldwide. *Molecular Plant Pathology, 19*, 2193–2206.

Singh, P. K., Gayathri, N. C., Roy, C., Roy, K. K., He, X., Tembo, B., Xu, K., et al., (2021). Wheat blast: A disease spreading by Intercontinental jumps and its management strategies. *Frontiers in Plant Science, Plant Breeding Section*, pp. 1–23. https://doi.org/10.3389/fpls.2021.710707/.

Singh, R. P., Hodson, D. P., Huerta-Espino, J., Jin, Y., Bhavani, S., Njau, P., Herrera-Foessel, S., et al., (2011). The emergence of Ug99 races of the stem rust fungus is a threat to world wheat production. *Annual. Review of Phytopathology, 49*, 465–481.

Singh, R. P., Hodson, D. P., Jin, Y., Lagudah, E. S., Ayliffe, M. A., Bhavani, S., Rouse, M. N., et al., (2015). Emergence and spread of new races of wheat stem rust fungus: Continued threat to food security and prospects of genetic control. *Phytopathology, 105*, 872–884.

Singh, R. P., William, H. M., Huerta-Espino, J., & Rosewarne, G., (2005). Wheat rust in Asia: Meeting the challenges with old and recent technologies. In: *New Directions for a Diverse Planet, Proceedings of the 4th International Crop Science Congress* (pp. 1–13). Brisbane, Australia.

Skoracka, A., Rector, B. G., & Hein, G. L., (2018). The interface between wheat and the wheat curl mite, *Aceriatosichella*, the primary vector of globally important viral diseases. *Frontiers in Plant Science, 9*, 1098. doi: 10.3389/fpls.2018.01098/.

Smith, D. J., Jaffe, D. A., Birmele, M. N., Griffin, D. W., Schuerger, A. C., & Hee, J., (2012). Free tropospheric transport of microorganisms from Asia to North America. *Microbial Ecology, 64*, 973–985.

Smith, D. J., Ravichandar, J. D., Jain, S., Griffin, D. W., Yu, H., Tan, Q., Thissen, J., et al., (2018). Airborne bacteria in earth's lower stratosphere resemble taxa detected in the troposphere: Results from a new NASA aircraft bioaerosol collector (ABC). *Frontiers in Microbiology, 9*, Article 1752. https://doi.org/10.3389/fmicb.2018.01752/.

Smith, R. J., Bettiga, L. J., & Gubler, W. D., (2022). *Pest Management Guidelines: Grapes* (Pub. No.: 3448, p. 75). University of California, Davis, CA, USA, UC ANR Publication. **University of California, IPM.** https://www2.ipm.ucanr.edu/agriculture/grape/powdery-mildew/ (accessed on 19 January 2024).

Stenger, D. C., & French, R. C., (2009). Wheat streak mosaic virus genotypes introduced to Argentina are closely related to isolates from the American Pacific Northwest and Australia. *Archives of Virology, 154*, 331–336. doi 10.1007/s00705-008-0297-1/.

Subrahmanyam, P., Reddy, L. J., Gibbons, R. W., & McDonald, D. O., (1985). Peanut rust: A major threat to peanut production in the semiarid tropics. *Plant Disease, 69*, 813–819.

Sydow, P., & Sydow, H., (1915). *Monographia Uredinearum* (Vol. 3). Lipsiae: Fratres Borntaeger.

Sydow, P., & Sydow, H., (1924). *Monographia Uredinearum* (Vol. 4). Lipsiae: Fratres Borntaeger.

Syller, J., Marczewski, W., & Pawlowicz, J., (1997). Transmission by aphids of potato spindle tuber viroid encapsidated by potato leafroll luteovirus particles. *European Journal of Plant Pathology, 103,* 285–289.

Taffner, J., Erlacher, A., Bragina, A., Berg, C., Eichinger, C., & Berg, G., (2018). *What is the Role of Archaea in Plants?* New insights from vegetation of the Alpine Bogs. doi: https://doi.org/10.1128/mSphere.00122-18/.

Taylor, A., & Gordon, C., (2022). *Powdery Mildew of Grapevine in Western Australia* (pp. 1–35). Department of Primary Industries and Regional Development, Agriculture and Food, South Perth, Australia. https://www.agric.wa.gov.au/table-grapes/powdery-mildew-grapevines-western-australia/ (accessed on 19 January 2024).

Teli, B., Chattopadhyay, A., Meena, S. C., Gangwar, P., & Pandey, S. K., (2016). Present status of *fusarium* head blight of wheat and barley in India. In: Vaish, S. S., (ed.), *Diseases of Wheat and Their Management* (pp. 79–92). Astral International Publishers, New Delhi, India.

Terefe, T., Paul, I., Mebalo, J., Naicke, K., & Meyer, L., (2009). Occurrence and pathogenicity of *Puccinia triticina* on wheat in South Africa during 2007. *South African Journal of Plant and Soil, 26*(1), 51–54. doi: 10.1080/02571862.2009.10639933/.

Thomas, G. J., (2021). *Smut and Bunt Diseases of Cereals- Biology, Identification, and Management* (pp. 1–6). Department of Primary Industries- Agriculture and Food. Government of Western Australia, https://www.agric.wa.gov.au/barley/smut-and-bunt-diseases-cereal-biology-identification-and-management/ (accessed on 19 January 2024).

Thresh, J. M., (1980). The origins and epidemiology of some important plant virus diseases. *Applied. Biology, 5*, 1–65.

Thresh, J. M., (2006). Crop viruses and virus diseases: A global perspective. In: Cooper, J. I., Kuehne, T., & Polischuk, V. P., (eds.), *Virus Diseases and Crop Biosecurity* (pp. 9–32). Springer: Dordrecht, The Netherlands.

Thresh, J. M., Fargette, D., & Otim-Nape, G. W., (1994). Effects of African cassava mosaic Gemini virus on the yield of cassava. *Tropical. Science. London, 34,* 26.

Tomlinson, K. R., Bailey, A. M., Alicai, T., Seal, S., & Foster, G. D., (2018). Cassava brown streak disease: Historical timeline, current knowledge, and future prospects. *Molecular Plant Pathology, 19,* 1282–1294.

US Department of Homeland Security-Science and Technology Directorate (2018). *Tethered Aerostat System Application Note* (pp. 1–18). https://www.dhs.gov/sites/default/files/publications/TetheredAerostat_AppN_0913-508.pdf (accessed on 19 January 2024).

Vallega, V., & Zitelli, G., (1979). Epidemics of yellow rust in Italy. *Cereal Rusts Bulletin, 6,* 17–22.

Wan, A., Zhao, Z., Chen, X., He, Z., Jin, S., Jia, Q., Yao, G., et al., (2004). White stripe rust epidemic and virulence of *Puccinia striiformis f. sp. tritici* in China in 2002. *Plant Disease, 88,* 896–904.

Webb, C. A., (2004). *Epidemiological Factors Impacting the Development of Wheat Streak Mosaic Virus Outbreak.* Kansas State University, Manhattan, USA., Masters Dissertation. https://krex.k-state.edu/dspace/handle/2097/39127/ (accessed on 19 January 2024).

Wellings, C. R., & McIntosh, R. A., (1990). *Puccinia striiformis*f. sp. *tritici* in Australia: Pathogenic changes during the first 10 years. *Plant Pathology, 39,* 316–325.

Wellings, C. R., (2007a). *Puccinia striiformis* in Australia: A review of the incursion, evolution, and adaptation of stripe rust in the period 1979–2006. *Australian Journal of Agricultural Research, 58,* 567–575.

Wellings, C. R., (2007b). *Puccinia striiformis f. sp. tritici* in Australasia: Pathogenic changes during the first 10 year. *Plant Pathology, 39,* 316–325.

Wellings, C. R., (2011). Global status of stripe rust: A review of historical and current threats. *Euphytica, 179,* 129–141.

Wellings, C. R., Wright, D. G., Keiper, F., & Loughman, R., (2003). First detection of wheat stripe rust in Western Australia: Evidence for a foreign incursion. *Australasian Plant Pathology, 32,* 321–322.

Womack, A. M., Bohannan, B. J. M., & Green, J. L., (2010). Biodiversity and biogeography of the atmosphere. *Philosophical Transactions of Royal Society – B, 365,* 3645–3653.

Yan, J., Wu, P. S., Shi, Z. W., & Wu, Y., (2016). A new record of *Coleosporium* (Uredinales) in China. *Mycosystems, 25,* 327, 328.

Zadocks, J. C., & Schein, R., (1979). *Epidemiology and Plant Disease Management* (p. 427). Oxford University Press, New York, USA.

Zwankhuizen, M. J., Govers, F., & Zadoks, J. C., (1998). Development of potato late blight: Disease foci, disease gradients, and infection sources. *Phytopathology, 88,* 754–763.

CHAPTER 8

Farm Procedures and Aerial Microbes

ABSTRACT

Tropospheric conditions affect agricultural procedures. The crops, cropping systems, and series of farm procedures, from land development to final harvest, are influenced, to some extent, by the prevailing tropospheric conditions. Conversely, each farm procedure adopted by farmers has an impact on tropospheric conditions. In this chapter, we focus on how microbes in the troposphere are affected by various farm operations. Post-harvest operations such as seed processing, de-hulling, and winnowing also contribute to dust and microbe emissions in the troposphere. Farm activities classified as agro industry, which are based on farm products, also release particulate matter (PM) and microbes into the agrarian sky. Livestock production procedures also release particle matter laden with microbes into the troposphere.

Farm operations result in the emission of particulate matter (PM; sand grains, soil particles, silicate minerals) of various sizes (PM_{10}, PM_5, $PM_{2.5}$, and nanoparticles), soil organic dust, crop residue, and microbes into the troposphere.

Soil tillage is among the major farm procedures that disturb the soil and its microflora. It aids in the emission of microbes with diverse genetic nature and physiological abilities into the air. If the soil is a repository of plant pathogens, then tillage may spread them to other areas through small dust plumes, drifts, and levitation caused by it. Most of the agricultural operations in open dryland fields, orchards, and forest plantations do contaminate the troposphere with microbes. The levitation of particulate matter (PM) and microbes has an influence on the migratory trends of microbes released into the troposphere. Crop residue burning is a recent trend in many agrarian regions, affecting the microflora and chemical composition of the agrarian sky. Farm operations induce emission, levitation, and migration

Microbiology of the Agricultural Sky. K. R. Krishna (Ed)

of plant pathogens and beneficial microbes like biological nitrogen-fixers in the troposphere. Certain farm operations involve the direct application of biocontrol microbes on the canopy of crops, which may also alter the microflora in the immediate vicinity within the troposphere. In severe cases, farm operations may altogether alter the microflora in terms of microbial density, phylogenetic, and genetic diversity.

Almost every farm vehicle and its accessories harbor a large load of microbes that can reach the troposphere and transit to longer distances, both via the vehicles themselves and via wind and precipitation. Farm vehicles can be a habitat for innumerable microbial species, including plant pathogens. If not curtailed using strict sanitation, these plant pathogens have the potential to induce epidemics.

In recent years, agricultural drones have been rapidly adopted in all agrarian regions. Similar to ground vehicles, agricultural drones can serve as aerial vectors of plant disease propagules. Microbial plant pathogens found on the drone's body surface can easily disperse through the troposphere. In the case of 'sprayer drones,' the formulation storage tank, formulation itself, and sprayer nozzles can become contaminated. This leads to contamination of the airspace wherever drones fly over cropped fields. Microbes literally become airborne via the contaminated drone (aerial vehicle). Robotic farm vehicles, such as GPS-guided autonomous tractors used for soil tillage, planting, interculture, combine harvesters, and grain processors, can contribute to disease spread within a farm and outside the farming zone.

Livestock enclosures, feed lots, animal pens, animals, and farm worker activity release several microbial species into the air. Firstly, they may reach the air within the farm enclosures. Particulate matter (PM) and microbes, including commensals (heterotrophic) and pathogens of animals, farm workers, and crops, can escape into the atmosphere. The use of scrubbers and disinfection of enclosures usually reduces microbial contamination of the troposphere in the vicinity. The troposphere above farm animal and poultry enclosures can be a source of several respiratory disease-causing microbes (bacteria, fungi, viruses). If proper sanitation procedures are not adopted, feed lots may disseminate pasture crop diseases.

8.1 FARM PROCEDURES AND AERIAL MICROBES: AN INTRODUCTION

The troposphere is a region that has a major impact on the agricultural crop production tactics adopted by farmers (Krishna, 2022). The influence of the troposphere on crops and farm procedures is significant. *Vice versa,* crop production procedures also have an immense influence on troposphere conditions and various phenomena. In this chapter, we focus on the role that microbes play in an open field with crops, orchards, or fallows. The interactive phenomenon involving troposphere factors, crops, soil surface, moisture sources, and microbes is worth investigating. The troposphere has a significant influence on the microbes found in the surface soil and crops. The troposphere is a region where crops and microbes both have to negotiate. The evolution of farming practices versus the role of atmospheric microbes needs attention. Global farming emits particulate matter (PM), organic dust, and microbes with diverse metabolic abilities. The rates of emission, composition of microflora, and their dissemination rates may differ depending on troposphere conditions.

Farming can be a significant source of both organic and inorganic dust, as well as PM (particulate matter) and attached microbes. The emission of PM varies in rate and duration, ultimately reaching the agricultural sky. The most noticeable sources of PM include fields with row crops, orchards, farm vehicles and equipment used during outdoor field operations, crop burning, pesticide spraying, herbicide application, and combine harvesting. Additionally, wind blowing over crops in open fields contributes to PM dispersion (HSE, 2007).

Livestock production units, feeding lots, and animal pens also release PM that carries microbes. These microbes can be either commensals or pathogens that affect both livestock and farmers working in the animal sheds. For instance, microbes that cause respiratory diseases in animals or humans can be easily detected in the air within farmhouses. It is evident that every farm operation related to crop production and livestock maintenance in enclosed spaces has some level of association with microbial activity. Microbes in the troposphere, air inside greenhouses, and animal enclosures are of significant concern. Although their regulation is not well-established.

Industrial activities, transportation vehicles, and mining are considered important sources of pollution in the atmosphere. These activities often result in the release of pollutants and microbes that can drift into agricultural areas. In many cases, agricultural regions are located in close proximity to industrial zones, which means that agricultural pursuits are exposed to both pollutants and airborne microbes from industries. Erisman et al. (2007) remind us that agricultural activities themselves can also generate aerosols containing potentially pathogenic microbes for plants, farmers, and farm workers. Rural communities often experience pollution in the agricultural sky due to sand particles, organic dust, and various species of microbes. Therefore, in industrialized regions, agricultural crops are frequently exposed to pollutants and harmful microbes originating from both farms and industries.

According to the World Health Organization (1999, 2000), sand grains, mineral dust, organic dust from crops, wood flour, cotton, winnowing dust, and pollen can all carry significant amounts of microbial cells. These particles can remain suspended in the atmosphere or be transported to other locations, potentially reaching higher altitudes through wind and convection currents. Both particulate matter (PM) and microbes can have serious impacts on crops, orchards, livestock, and the health of farm workers. It is important to note that the presence and dissemination of microbes from farmland into the atmosphere depend on factors such as geographic location, topography, soil characteristics, climate, season, moisture levels in the troposphere, crops, and agronomic practices adopted (Jager, 2005).

8.2 AGRICULTURAL OPERATIONS, PARTICULATE MATTER (PM) EMISSION, AND MICROBIAL CONTAMINATION OF AGRARIAN SKY

There are several reports that substantiate that PM emissions from agricultural zones are a worldwide phenomenon (Peterfalvi et al., 2018). Basic factors to be considered while forecasting dust emissions and their intensity during crop production need attention. These factors include the origin of particles, physicochemical properties of soil, biotic aspects of PM, meteorological conditions, agricultural procedures, etc. Detailed reports exist that deal with PM emissions due to tillage of soils. Similarly, the effects of other farm procedures like discing, plowing, harrowing, and in-season procedures that affect PM emissions have been studied (Cassel et al., 2003; Ottl et al., 2005).

According to Arslan & Aybek (2013), during the cultivation of crops, dust and microbe emissions to the air occur due to operations such as conventional soil tillage, clod crushing, leveling, furrow or bed preparation, seeding, fertilizer inoculation, pesticide application, and harvesting. However, the tendency to adopt minimum tillage or no-tillage practices is supposed to reduce dust formation and levitation into the agrarian sky. They have further mentioned that long-term experimentation has provided insights into the extent of dust/microbes released into the air by each farm operation. Gaffney & Yu (2003) found that dust laden with microbes emitted by discing, ridging, planting seeds, interculture, and harvesting are not the same. The extent of microbial load and diversity noticed in the dust above crops like cotton (Arslan & Aybek, 2011), wheat, or almond orchards varies. Therefore, it is the combined effect of specific crops and field procedures that decide the microbial flora in the agrarian sky. Let us consider a specific location: the San Joaquin Valley of California. The PM_{10} emission estimates for land preparation (discing, leveling, furrow opening) were 13,000 tons per year. For harvesting, it was 13,300 tons per year. Soil tillage is supposedly a crucial step that decides the extent of fugitive dust and microbes emitted during crop production.

It has been observed that the size of aerosol particles is influenced by farm operations. This is in addition to the characteristics of the soil and the cropping systems that are implemented. The size of the particles appears to have an immediate impact on the types of bacterial species that are attached and their population (Lighthart, 1984, 1997). Leaf debris and larger particles undoubtedly support higher bacterial counts. Bacterial species such as *Pseudomonas aeruginosa* are commonly found in the larger particulate matter (PM). It should be noted that the size of emitted PM also varies depending on the terrain conditions in general (Lighthart & Schaffer, 1994; Schaffer & Lighthart, 1997). Farms tend to emit more microbes during daytime hours, particularly in the summer. The larger particles (>7 μ) contain a higher population of microbes compared to the smaller ones with a size of <1.1 μ.

8.2.1 LAND DEVELOPMENT

Land and its topography, land usage history, soil type, characteristics of its profile, depth of land cutting, and extent of soil turning affect dust emissions. The emission of microbes adhering to PM into the atmosphere is also affected by land preparation methods.

Land rich in organic matter and microbial activity will obviously release more microbes into the atmosphere. Land development involves a series of earth-moving operations, which require heavy vehicles (e.g., bulldozers, excavators, tractors) and attached equipment. The removal of natural vegetation from previously forested areas requires the use of tree-cutting equipment. Heavy vehicle-supported operations are needed for shrub removal and the removal of understory ground flora (plants/weeds). Such farm operations release massive quantities of microbe-laden dust into the troposphere.

During the development of farmland for crop production, particularly cereals, legumes, and other annuals, heavy disc plows are used for initial plowing. This process results in the formation of numerous clods and the release of soil particles, dust, and lightweight organic debris. These fractions become airborne during farm operations, and particulate matter ($PM_{2.5}$ to PM_{10}) may even reach higher altitudes, such as the upper troposphere. The upper troposphere experiences significant alterations in microbial load, flora, and genetic diversity when such procedures are carried out, especially in areas that were previously undisturbed. Aeration stimulates the activity of microbial flora, leading to the loss of carbon as gaseous CO_2. Consequently, these procedures can permanently modify tropospheric conditions and cause changes in the microbial population and diversity in the air.

Land leveling often requires the use of heavy tillage and levelers, which inevitably result in the emission of large amounts of dust and microbes into the atmosphere. Additionally, activities such as marking field layouts, constructing farm roads, channels, and bunds also contribute to the release of soil and dust into the atmosphere. The amount of soil microbes lost to the troposphere depends on the native microflora and the microbial load of sand or organic dust particles.

8.2.2 BURNING NATURAL VEGETATION, CROPS, STUBBLES, WEEDS, AND VOLUNTEERS IN THE FIELD.

Crop residue and stubble burning are widespread in areas that support field crops such as cereals. This practice offers a quick, cost-effective, and efficient way to clear stubbles, weeds, and volunteers from the previous season, while also enriching the soil with minerals. However, crop stubble burning releases excessive amounts of PM 10, 2.5, and even finer particles. It also introduces toxic compounds, CO_2, and microbes into the lower troposphere. These microbes can travel to other locations through air currents, winds, and jet streams. Some of these microbes may be pathogens that affect crops and livestock, and they can also pose health risks to farm workers and the general population. The combination of irritants in the emissions and microbes can lead to epidemics among animals and humans. According to Yadav (2019), large-scale crop residue and stubble burning, which helps farmers save on expenses, has become increasingly popular in regions that produce field crops, such as the Indo-Gangetic Plains. It has been observed that stubbles of crops like wheat, rice, sugarcane, and small grain millets are burned to clear the fields. However, this practice creates a hazardous situation where the "agrarian sky" becomes polluted with pollutants. Crop residue burning induces PM at levels higher than the threshold, as well as pathogens that affect crops. Farmers and their workers are also at risk of respiratory disorders. Smoke containing PM and microbes emitted during wildfires or created by farmers during crop production can be a source of harmful pathogens. These pathogens can affect livestock and farm workers (Sever, 2021). Microbial pathogens from previous crops can be detected in the atmosphere. The entire profile of the troposphere, from the soil surface and canopy to several hundred feet in altitude, can harbor harmful microbes. These microbes can become infused into the atmosphere and settle, potentially leading to the spread of epidemics and pandemics. Crop burning disrupts the cycling and continuum of microbial flora between the soil surface, troposphere layers, and the sky in agricultural areas. This can result in significant changes in the microbial flora and their activity in the troposphere. Creating discontinuity in troposphere microbes can be beneficial, particularly in controlling microbial plant pathogens that affect crops.

As mentioned earlier, crop residue burning is currently a practice implemented in fields immediately after crop harvest. Stubbles remaining after harvest are burned, particularly those that may hinder soil tillage operations. Large-scale burning is the most effective method to remove these stubbles. The purpose of stubble burning is to provide farmers with more time between two crops. However, this practice releases significant amounts of particulate matter (PM) and associated microbes that survive the burning process. These particles and microbes can be found in the air, clouds, and even at higher altitudes above the crops. Some estimates of $PM_{2.5}$ levels in the agricultural air of Northwest USA are available. Soil tillage contributes to 38% of the total $PM_{2.5}$, while approximately 35% is attributed to crop residue burning, including natural vegetation zones. Sulfate aerosols from industries and farming regions account for about 20% of the $PM_{2.5}$, while emissions from vehicles contribute to around 2%, and cooking with fossil fuels contributes to 1% (Torigoe et al., 2000; Jimenez, 2006). Interestingly, $PM_{2.5}$–PM_{10} often contain a wide variety of microbes. Additionally, burning a ton of stubble results in a loss of 5.5 kg of nitrogen, 2.3 kg of phosphorus, 25 kg of potassium, and 1 kg of sulfur, as well as organic carbon.

There are suggestions to reduce crop burning. Furthermore, a prominent suggestion is to adopt seeders that can function in fields with standing stubble. The use of zero-tillage seed drills is also recommended. Using a paddy straw chopper and incorporating stubble is a good idea. Sometimes, burning is used to remove ground flora and turf, if any. Burning vegetation releases large quantities of dust, carbonaceous particles, sulfur-containing organic debris, and microbes. This process is known to cause smog and leads to the greenhouse effect. Aerial microbes in the 'smog' could induce acid rain if mixed with clouds and raindrops. Microbes with the ability to transform various carbon compounds and sulfur-containing particles are easily traced in such 'smog.'

8.2.3 SOIL TILLAGE AFFECTS DYNAMICS OF MICROBES IN SOIL AND AGRARIAN SKY

In this section, we are considering PM emissions from agricultural soils and from standing crops due to various farming procedures. The assumption is that the emitted PM does contain microbes from these surfaces/sources. The extent of microbes released to the troposphere and their migration to different locations or return to the soil is dependent on the microbial load and diversity of the PM. There is clear data available about the microbial aspects of PM. On a generalized basis, higher PM emissions result in a higher loss of microbes into the air.

There are nations/regions that support vast expanses of agricultural crop production. For example, Peterfalvi (2018) states that Hungary is primarily an agrarian nation. About 56% of its area is plowed and cropped each year/season. Therefore, PM emission, chemical pollution, and microbial contamination of the air are derived primarily from farming procedures. During the period 2000 to 2015, PM10plus adhering microbes emitted each year by the Hungarian crop production zones were significant. It fluctuated between 5.8 and 6.7 Gg. It was attributable to soil tillage and in-season farm procedures. Next, manure application and soil loosening emitted PM_{10} that fluctuated between 3.2 and 4.3 Gg each year, from 2000 to 2015. The total PM_{10} emission from agricultural activities, during 2000 to 2015, in the Hungarian plains fluctuated between 9.6 and 10.1 Gg. Such high amounts of PM emission occur when soils are dry, sandy, and organic dust is also high. We should note that it is not just PM10 that spreads into the troposphere. A large population of microbes from the agroecosystems fills the troposphere. Microbes migrate to different distances. There could be several plant pathogens that get dispersed this way. Microbial contamination of the troposphere occurs naturally and without special effort by humans.

Reports from European farming zones, specifically in Belgium, suggest that PM emissions could be high at 7.5 kg $ha^{-1}yr^{-1}$ for total suspended particles (TSP). Approximately 2.0 to 3.1 kg $ha^{-1}yr$ of this is attributed to PM_{10} (Bogman et al., 2007). It is reported that agricultural activities account for 35% of TSP emissions and 24% of PM_{10} emissions. These aerosols contain a significant amount of microbes derived from chernozemic soils and crop canopies. Fine particles from the soil also contribute pollutant chemicals and microbes to the troposphere. Row cropping is known to contribute significantly to PM and microbe-laden aerosols. According to Madden et al. (2008), failure to adhere to air quality standards is a contributing factor to PM and microbial contamination of the air.

Let us consider another example. Peterfalvi et al. (2018) mention that soil tillage intensity depends on the crop species intended to be planted. Additionally, farm procedures during the crop season are dependent on the crop species. The total PM emission depends on the extent of farming enterprises in a region. The extent of each crop species grown also affects PM emissions. For instance, discing emits 78 mg m^{-3} PM if the crop is cotton, 313 mg m^{-3} if it is garbanzo, 380 mg m^{-3} if it is melons, 545 mg m^{-3} if it is tomatoes, and 1,375 mg m^{-3} if it is wheat (Petefalvi et al., 2018; Casell et al., 2003).

Primarily, it is suspected that PM_{10} emitted from agricultural activity may contain solid mineral particles containing Al, Fe, C, Mg, Si, sulfates, nitrates, ammonium ions, inorganic ions, elemental C particles, heavy metal particles, pesticide residues, radioactive material, and a variety of diverse microbes thriving on the PM. Soil tillage accentuates the emission of PM_{10} along with microbes. The PM_{10} along with microbes from tillage is rich in organic fraction if it is derived from soils rich in organic fraction. Another cause is crop burning. Crop burning primarily releases a large fraction of organic particles laden with microbes (Peterfalvi et al., 2018; Holmen et al., 2001a, b.).

Generally, soil tillage, crop harvesting (combining), and post-harvest processing, if any, conducted in the open field emit a large fraction of PM and attached microbes. Primary plowing emits 120 mg m^{-3}, harrowing emits 82 mg m^{-3}, discing emits 137 mg m^{-3}, and interculture emits 186 mg m^{-3} (Kasumba et al., 2011; Ottle et al., 2005; Wang et al., 2010). PM_{10} plus microbe emission patterns can also be characterized by dividing the total crop season into soil tillage, growing season, harvesting season, and post-harvest processing (Chen et al., 2015). It can be monitored throughout the year as well. In China, for example, Chen et al. (2017) found that PM_{10} emissions are higher during April to October, which coincides with land preparation, in-season, and harvest period. Diurnal changes in PM_{10} emissions have also been noted in Europe (Hoffmann & Funk, 2015).

Soil erosion due to air (wind) or water forces acting via the agrarian sky is synonymous with the loss of microbes from the field. It is proportionate to the microbial load of soil particles/organic dust carried away from the field location. Microbes lost from the field could have significant consequences for the conduct of several ecosystem functions. Several biochemical transformations of the organic fraction of soil could proceed at low rates or even cease to occur if certain microbial species are lost to the agrarian sky. There are also pathogens and symbionts that could be lost due to wind or water erosion. Tillage and the lack of soil conservation practices like contour bunding, mulching, clod formation, etc., induce the loss of soil plus microbes. Excessive repeated tillage may make the situation difficult for conserving soil microbes. Acosta-Martinez & O'Brien (2013), for example, state that in the USA, wind-eroded soils have been examined for several decades. Estimates for the amount of surface inorganic particles and organic dust detached from field soil are available. However, data about microbes critical for soil biogeochemical transformations are lacking.

Wind tunnel experiments have shown us that several bacterial species, fungi, and virus particles can be detected in the wind (Acosta-Martinez & O'Brien, 2013). Fields supporting onions, beets, potatoes, and other vegetables lose soil and microbes due to wind blowing over the fields. This situation is accentuated when tillage operations are intense. Intensely tilled soil surfaces are prone to wind erosion. In fact, pyrosequencing methods allow us to identify several microbial genera/species. *Proteobacteria* are often associated with coarse particles and are dispersed only over shorter distances due to tillage or wind. It seems that microbial species are crucial for the normal conduct of C and N transformations in surface soil, and these microbial species are also lost. Fine particles that are levitated due to tillage and/or wind often carry microbes like *Bacteroidetes*. A combination of deep or surface tillage plus wind is prone to adding microbes to the troposphere at low altitudes. Conversely, soils suffer a loss of microbes.

Whenever we observe PM emissions from soil or cropped fields, the underlying assumption is that sand or organic dust particles carry diverse physiological microbes. These microbes can have various consequences for crops, livestock, and humans. As mentioned earlier, soil tillage is one of the main causes of PM emissions in agricultural areas (Funk et al., 2008; Hinz, 2007; Schenker, 2004). Most soil types used for crop production are disturbed through soil tillage. There are very few soil types that do not require preparatory soil tillage before planting. Soil tillage causes the shearing of clods, smaller aggregates, particulate dust matter, and sand. This process releases both PM and microbes into the air. Munch et al. (2020) state that among different natural (e.g., wind, water) or man-made factors that disturb soil, soil tillage is perhaps the most potent. It leads to a significant loss of soil structure and aggregation. Tillage easily releases soil microbes into the atmosphere. They also mention that soil tillage of different intensities has varying abilities to release microbes into the air. Soil tillage is believed to produce six times more dust than wind blowing over the same field (Munch et al., 2020; Goossens et al., 2004).

There are several soil-related operations in farms that release PM and microbes into the air. The extent of emissions varies depending on soil characteristics such as texture, moisture content, and tillage intensity (Cassel et al., 2003; Chen et al., 2015, 2017; Ottl & Funk, 2007). Seasonal variations in PM and microbe emissions are possible, based on soil moisture variations that occur due to seasons. For example, soil tillage operations of similar intensity conducted on the same field/soil can result in 10 times higher PM emissions compared to those in spring. Farm operations like light tilling may generate about 2 kg ha^{-1} of PM, while planting, which causes less disturbance than tilling, liberates a mere 0.2 kg ha^{-1}. Deep tillage involving discing, turning, and furrow or channel creation operations emit over 10 kg ha^{-1} of PM (Cassel et al., 2003; Munch et al., 2020). Clod crushing and leveling using different implements also release PM and microbes into the air (Madden et al., 2008).

Microbes detected in soil, water bodies, and the agrarian sky are interrelated, to a certain extent. It is this continuum that gets altered to different levels when we adopt mild or intense tillage procedures. Zero-tillage may conserve the soil plus microbe complex *in situ*.

Arslan & Aybek (2013) suggest that dust emissions from soil tillage and other farm activities must be identified and reduced. We should not underestimate or neglect the consequences of the loss of dust and microbes into the aeolian region. Mostly, it is soil tillage and other operations that disturb, turn over, and cause silt/sand particles to levitate into the troposphere. Conservation agriculture is expected to reduce mineral and organic dust emissions. Conservation tillage often results in less disturbance and microbial emissions compared to conventional deep tillage. Conservation also reduces the chances of microbial dissemination via the aeolian process. Undoubtedly, domestic animals and human farm laborers need to avoid excessive dust and microbial density in the air. Soil tillage on drier soils emits more sand particles, organic dust, and attached microbes than moist arable soil with optimum soil texture.

Madden et al. (2008) and Arslan & Aybek (2013) state that traditional tillage practices are among the most disruptive on the soil profile. Therefore, they emit a large amount of dust laden with microbes into the air. Standard tillage practices in many farms worldwide are the dustiest agricultural operations (Schlatter et al., 2018). Interestingly, conservation tillage reduced

PM emissions to the troposphere by 35–52% compared to standard tillage. No-tillage farms are also valuable suggestions as they significantly reduce PM_{10} emissions.

Farm equipment plays a significant role in the emission of dust and microbes into the air. It also determines the level of exposure that farm workers have to airborne dust and soil/aerial microbes. Whether it is manual labor or the use of tractors and attachments like discs or tines, tillage exposes farmers to dust and aerial microbes. Even modern autonomous (robotic), GPS-guided heavy tractors that conduct strip tillage are not exempt from the loss of topsoil, particulate matter (PM), and microbes. In cases where the soil is rich in organic matter, organic dust that can increase the microbial load in the atmosphere is emitted rapidly. Similarly, when tilling soils that support cover crops or thick pastures, the organic fraction in the troposphere increases (Schenker, 2000; Nieuwenhuijsen et al., 1998). Often, the troposphere PM_{10} levels exceed the threshold of 10 mg m^{-3}. Farmers who use mechanical tillage are exposed to silica material, sand particles, and harmful microbes (Swanepoel et al., 2010). For instance, the concentration of respirable quartz particles can be reduced from 2 mg m^{-3} to 0.5 mg m^{-3} if a farmer stays in a closed cabin with steering during tillage operations. According to Kirkhorn & Garry (2000), using an enclosed cabin while conducting soil tillage helps avoid dust inhalation. A cabin with air filters on a tractor is effective in preventing the inhalation of PM_{10}–$PM_{2.5}$. It also prevents microbial contamination inside the cabins of tractors/combines.

8.2.4 ORGANIC MATTER APPLICATION, EMISSION OF PARTICULATE MATTER (PM) AND MICROBES

Organic matter inputs to the soil surface and its incorporation may release lightweight dust particles carrying microbes. These microbes can serve as cellulose digesters and may include pathogens for crops, animals, and farm workers' respiratory tracts. Regardless of the microbe species, they are initially lifted into the troposphere. These particles then migrate shorter distances through saltation or creeping, moving from one location to another. The introduction of organic matter and soil disturbance undoubtedly alters the microbial load and species diversity in both the soil profile and the air. Typically, several tons or cartloads of organic matter are spread on the soil surface.

The application of organic manure itself leads to the emission of lightweight dust particles into the atmosphere, ranging from 0.05 to 8.37 kg/ha depending on soil texture and moisture levels. The shear force generated by tillage operations on organic matter is actually significant. Particle size is another factor that determines the levitation of microbes into the troposphere. Aero-microbiologists and epidemiologists have often detected antibiotic-resistant microbes in dust derived from organic matter. There appears to be an increasing prevalence of antibiotic-resistant microbes in the air above livestock farming zones, animal houses, and crop production fields where animal-derived manure is consistently applied. Antibiotic-resistant bacteria have been found in the soil for more than a year, even after crop harvest (Friese et al., 2013; Hartman et al., 2017; Meyer et al., 2013). The dispersal of microbes originating from organic manure to human settlements is facilitated by wind. The application of contaminated organic matter is considered an important source of crop diseases. Plant pathogens often overwinter on dry organic matter. After the supply of organic matter to fields, wind acts as a sorting tool, separating particulate matter based on size and weight. Larger dust particles with a higher percentage of microbes may settle on the soil or crop canopy in the vicinity, while finer dust can travel longer distances and cause respiratory diseases (Blanco-Canqui et al., 2006; Nerger et al., 2017; Rühlmann et al., 2006).

8.2.5 DRYLANDS AND LOSS OF SOIL PARTICLES AND MICROBES

Agrarian soils in dry regions often consist of dust particles (PM) that can be released into the air through agricultural activities. Farm operations such as plowing, interculture, sprinkler irrigation, and the use of harvesters can disturb the soil and crops in the field. This results in the loss of soil particles (PM_{10}, $PM_{2.5}$) and microbes. Wind-induced erosion of soil is a process that occurs in the sky above drylands. There are three types of processes involved in wind-induced soil erosion and the loss of surface soil particles and microbes. Soil particles are lost through suspension, saltation, and creep. Suspension refers to the loss of soil particles through levitation caused by wind force. Levitated soil particles, along with microbes on their surface, may float for a while and then settle on the ground. Saltation is the main process through which surface soil particles and microbes are lost. The soil particles are thrown to a far-off distance by the wind. They reach a certain height and then fall to the ground after traveling through the air. Creep is a process in which larger soil particles, unable to be lifted into the air, move by crawling on the soil surface. They shift slowly from one location to another. This movement is induced by the wind that skims the surface of the fields (*see* USDA Natural Resources Conservation, 2007; Plate 8.1). Dryland soils are prone to experiencing the loss of soil through all three processes.

Wind-induced erosion of soil and dust is often encountered in fields with dry sandy soils, soils with a high sand and silt fraction, loose soils with low aggregation, and dusty soils with organic debris (Plate 8.1). The loss of microorganisms from the surface soil depends on the microbial load of each sand and dust particle. Therefore, whenever we experience topsoil loss, it indicates a potential loss of microflora from the soil to the air or other locations. In the context of agrarian sky, the wind erosion-induced loss of soil dust, silt, and organic matter is significant. The loss of attached microbes is an additional detrimental effect. The loss of microbes can be measured in terms of both total population and diversity.

An assumption here is that by reducing the loss of particles of size PM_{10} to $PM_{2.5}$, we can proportionately reduce the loss of minerals, organic fraction, and microbes. There are several farm procedures that should be adopted, some of which need to be modified to reduce soil erosion. For example, USDA experts dealing with dryland soil in Nevada suggest the following:

- Adopting chemi-fertigation and combining farm procedures related to pesticide and fungicide sprays can help reduce soil disturbance by minimizing farm vehicle movement associated with sprinkler irrigation.
- Tillage should be based on soil moisture conditions. Tillage must be timed to minimize the loss of PM_{10} and the attached microbes. Tractor movement in the field during tillage should be minimized by combining farm procedures. Soil loosening, weeding, interculture, and fertilizer granule placement can be combined, for example.
- Farm vehicle movement and their usage could be reduced to lessen soil erosion.
- Farm operation could be avoided during high-wind days.
- Seeding should be done when soil moisture conditions are optimum to avoid soil loss caused by disturbance during dry seeding.
- Reduced or no tillage conserves soil particles, dust, and microbes *in situ*.
- Soil roughening and leaving cloddy soil are techniques used to prevent excessive wind erosion. Other farming practices such as wind barriers, cover crops, mulches, cross wind ridges, planting rows, strip cropping (including perennial crops), etc., also help reduce soil erosion and the loss of surface microbes to the troposphere.

Wind erosion of farm soils and those in other regions may also lift PM_{10}. Particles could be embedded with microbial pathogens that affect crops, farm animals, and humans (Plate 8.1). Lessening PM_{10} also reduces the dispersal of pathogens and allergic constituents. Overall, farm procedures that reduce soil erosion and loss of particle matter will also reduce the loss of microflora. Furthermore, it reduces their migration to other locations. The loss of microbes to the air (troposphere) actually affects the stoichiometry of several physicochemical transformations relevant to nutrients in soil/air.

PLATE 8.1 Soil erosion from fields with dry sandy/silty soils supporting hay production, near Reno, in Nevada.

Source: United States Department Agriculture-Natural Resource Conservation Program, Reno, Nevada. https://www.nrcs.usda.gov/Internet/FSE_DOCUMENTS/nrcs144p2_037150.pdf; Photo in public domain.

8.3 AERIAL SPRAYS OF MICROBES

In the general course, microbes contaminating the irrigation water could form aerosols and survive. Then, microbes would spread in the troposphere based on drifts. Many of these microbial species could be plant pathogens. If the microbial species happens to be a crop pathogen, then it spreads the disease rapidly. Inactivation of pathogenic microbes is delayed because of congenial relative humidity (RH) and aerated conditions. A study in New Zealand aimed at deciphering the survival and dispersal pattern of microbes, particularly when they are seeded into water meant for the sprinkler irrigation system. They introduced two microbes, namely *Bacillus subtilis* and *Serratia entomophila,* into the sprayer system. Upon spraying, both got dispersed to an equal area from the sprayer. Closer to the sprayers, the microbial load in the air (aerosols) and soil was higher. It decreased as we moved away from the sprinklers. It was glaring to find that irrespective of the sprinkler types utilized, the encountered pathogens were found in the sampler/traps. This meant that farmers would easily inhale the aerosols studded with harmful microbes (Donnison et al., 2004). Basically, irrigation water used in aerial sprinklers has to be safe with regard to human pathogens, particularly coliform bacteria. It seems coliform bacteria are encountered most frequently in irrigation water (Akinde et al., 2016).

Whenever sprinklers are used, they can affect aerosol microbes, relative humidity (RH), temperature, radiation, and other factors that are relevant to the survival and spread of microbes. Leaf conditions may become favorable for pathogens that cause disease epidemics. It is said that disease severity can even increase if sprinklers are applied to a crop that is already affected by disease. Rust, leaf blight, and blotches may become more severe if the aerial conditions on the foliage are consistently humid and warm. Viral particles can also be efficiently transmitted if there is a high number of insects in the crop canopy (Café-Filho et al., 2018).

Foliar application of liquid dairy manure is a common practice among farmers. They utilize various methods such as traveling sprayer guns, center-pivot irrigators, or surface methods. There are clear advantages to using sprinkler methods, such as reducing runoff. However, trials and regular practice have shown that applying dairy manure via sprinkler can lead to contamination of the troposphere with aerosols. These aerosols may contain pathogens such as *E. coli, Campylobacter, Listeria, Giardia,* etc. (Borcardt & Burch, 2016).

Spray irrigation conducted using spray guns, traveling sprinklers, or center pivots undoubtedly results in the intermixing of aerial microbes with water droplets. These microbes can originate from various sources such as soil, drainage channels, etc. Katznelson & Teltch (1976) found that enteric bacteria were well dispersed in sprinkler water. The major coliform bacteria identified were *Salmonella, E. coli,* and *Erwinia* sp. Microbes were even detected up to 60 meters away from the sprinklers. On windy days, it is possible that they disperse even further.

We can attempt to introduce beneficial microbes, such as biological nitrogen-fixing (BNF) organisms, into sprinkler irrigation systems. This would ensure their uniform distribution onto the soil surface. Additionally, there are other beneficial microbes known as biocontrol agents (BCAs) that can be applied directly to the phyllosphere. Instead of implementing a separate schedule for sprinkler irrigation, these phyllosphere microbes could be sprayed using the existing sprinkler system.

However, it is important to consider the possibility of microbiologically induced corrosion of the sprinkler equipment. Microbes like microalgae, bacteria, and fungi can accelerate corrosion, leading to the deterioration of the sprinklers (O'Leary, 2017). These microbes can originate from the troposphere, soil surface, or be brought to the location by other farm vehicles. It is also worth noting that the sprinklers themselves may contribute to the settling of a variety of microbes found in the troposphere immediately above the crops.

8.3.1 AERIAL APPLICATION OF BIOFERTILIZER INOCULUM

Microbes are almost always encountered in the lower troposphere. These microbes exist as bioaerosols in the agrarian sky. Microbes of sizes 2.5 μ encountered in the particulate matter (PM) make up about 3 to 11% of the biomass, by weight, in the troposphere. It appears that approximately 25% of insoluble PM consists of microbes. Bacteria can be found at a population density of 10^4, while fungi can be found at a population density of 10^3 in the air (Boreson et al., 2004; Peccia & Henderson, 2006; Jones & Harison, 2004). The source of these aerosolized microbes is the soil surface or the canopy of crops. These microbes, with diverse physiological abilities, are recycled during agronomic procedures such as sprinkler irrigation or rainfall. They may naturally settle on soils when they are close to the soil surface. There are also farm procedures that involve microbial inoculation of the soil, where microbial inoculants are applied via aerial sprays. During such processes, we temporarily increase the tropospheric microbial load until a large fraction of the microbes settle on the soils or crop's foliage.

We observe a form of microbial recycling when sprinklers are utilized or when microbes naturally settle on the soil surface. The microbes on the soil surface are then lifted and dispersed into the atmosphere. Agronomic practices such as soil tillage, open air winnowing, and feed lot preparation in animal farms also contribute to the recycling of microbes. There is a suggestion to study the "airscape" in relation to microbes in general. Based on these airscape maps (Isard, 2001), we can apply microbial inoculants to crops. It is worth noting that "airscape" is a term similar to "landscape," but it pertains to the aerospace above the crop's canopy.

Microbes, especially biological nitrogen fixers, are applied as aerial sprays onto seedlings and the entire canopy. This is part of the in-season agronomic procedure. Firstly, it affects the microbial load in the air above the crops. Additionally, when the sprayed material settles on the soil surface, it contributes to the population of nitrogen-fixing microbes in the soil. The foliar sprays of nitrogen-fixing formulations typically contain asymbiotic nitrogen fixers such as *Azotobacter* sp. and *Azospirillum* sp. For instance, a foliar spray of a mixture of nitrogen-fixing species had positive effects on canola plants (Ahmadirad et al., 2016). The external application of nitrogen-fixing inoculum on the canopy/phyllosphere may impact the ratio of nitrogen-fixing microbes compared to other microbial species on the leaf surface. Moreover, when considering the "airscape" above a field, we are intentionally altering the distribution ratio of microbes.

8.3.2 AERIAL APPLICATION OF BIOCONTROL AGENTS (BACTERIA, FUNGI, BACTERIOPHAGES)

Reports indicate that up to 40% crop loss in some regions is attributable to plant pathogens/pests. Among control measures, biocontrol methods are being promoted in many locations practicing intensive crop production. Biocontrol using aerial sprays of microbes is preferred. This is because microbial agents are safer than chemical pesticides that leave toxic residues. Biological control seems to be the best alternative if we aim to achieve disease suppression. BCAs bring disease suppression without any environmental hazards. Microbial BCAs may also populate the airspace above crops. However, we can monitor them periodically to determine if the population density is appropriate. Accordingly, we can replenish BCAs (Nega, 2014).

BCAs are also microbes, but intentionally added to the troposphere near the crop's canopy, mostly using sprinklers. This measure is adopted when the plant pathogens are susceptible to biocontrol species. According to Shafi et al. (2017), there are several *Bacillus* species whose potential as BCAs of microbial plant pathogens of crops has yet to be fully utilized. There are numerous *Bacillus* species that are also classified as Plant Growth Promoters when applied to the crop's foliage.

BCAs, mainly bacteria (e.g., *Bacillus thuringiensis*), fungi (e.g., *Metarrhizium anisopleae*) and viruses (e.g., *Nuclear Polyhedrosis Virus*-NPV), reduce insect pest populations in crops. They are periodically sprayed onto the air and canopy. BCAs cause less harm to farm animals, general fauna, and humans. BCAs like bio-fungicides (*Trichoderma* sp.) and bio-pesticides (e.g., *B. sphaericus*) are gaining popularity. Commercial formulations of *B. thuringiensis* include Bactur, Dipel, Thuricide, Bactospiene, etc. (Usta, 2013; Frampton et al., 2012).

Bacteriophages have been considered effective biological control agents (BCAs) for crop diseases since the mid-1900s. Some advantages of using bacteriophages are their host specificity and ability to persist in the environment. Bacteriophages multiply when they encounter a host bacterial population. They are easy to handle and can be sprayed onto the crop's canopy using sprinklers (Jagdale & Kapudnis, 2012). However, it should be noted that the application of phages can alter the diversity of soil and tropospheric bacterial populations.

Bacteriophages were discovered in the early 20th century. Since then, efforts have been made to utilize them to our advantage in controlling bacterial diseases that affect crops, farm animals, and humans (Jones et al., 2012). The use of bacteriophage sprays to control bacterial plant pathogens is a promising idea (Rebekah et al., 2012). It could eventually become a valuable agronomic procedure. By introducing phage formulations into the troposphere via sprinklers, we can modify the microbial flora and density. Bacterial viruses (phages) are sprayed onto crops that may harbor bacterial inoculum or already suffer from severe disease. Xanthomonads are bacterial pathogens with the potential to cause significant losses in several crop species. Various species of *Xanthomonas* could be examined for their susceptibility to bacteriophages. A prophylactic spray of bacteriophages on soil and crops at the seedling stage can reduce the population of bacterial plant pathogens (Frampton et al., 2012; Stefani et al., 2021; Lang et al., 2007; Marco & Stall, 1983). Obtaining a high population density of bacteriophages on the leaf lamina is crucial for their effectiveness. This is especially true if we aim to control bacterial pathogens in the troposphere. The timing of aerial sprays of bacteriophages, which are highly specific to hosts (plant pathogens), is important. Undoubtedly, aerial sprays of bacteriophages expose them to UV radiation and the vagaries of the troposphere. Therefore, their survival and effectiveness may decrease over time, both in the air and on the leaf lamina. One possibility is to coat bacteriophage particles with agents that provide protection against harmful radiation. There are suggestions to spray bacteriophages along with non-pathogenic bacterial hosts, so that they can multiply and maintain a certain population density on leaves (Jones et al., 2012).

8.4 COMBINE HARVESTING, POST-HARVEST PROCESSING, AND MICROBES IN THE TROPOSPHERE

Harvesting stems, panicles, and grains, followed by grain threshing, de-husking, winnowing, and blowing operations generate large quantities of PM. Microbes resting on farm vehicles and equipment are also released into the atmosphere. Undoubtedly, the troposphere becomes polluted with organic dust of various sizes. Microbial density may temporarily change in the region. Microbes can even drift over a certain distance. For instance, farm operations such as harvesting, whether done manually or using combine harvesters, emit dust. Grain dust itself can be a severe irritant and allergenic to farmers. It may also carry plant pathogens that spread to locations farther in the region. Dust from panicles, grains, or husks from cereals usually harbors bacteria, fungi (spores, hyphal bits), and viruses. Additionally, microscopic and small insects and their parts are present. Pollen laden with microbes is also encountered (The Swedish National Board of Occupational Safety & Health, 1994; HSE, 1998, 2005, 2007; Spankie & Cherrie, 2012).

Harvesting is not just a strenuous activity for farmers. Field work requires long hours at a stretch in the open with crops, crop dust, pollutants, and harmful microbes, if any, in the air. This is true even with the use of combine harvesters, automatic drying devices, milling, carting, and transport. Each of these farm operations that finally yield grains could be deleterious to both crops and humans in the farming community. The final harvest from crops may release pathogens into the air along with particulate dust. This serves as a source (inoculum) for the next crop or those grown in the vicinity. The pollutants (PM_{10}–$PM_{2.5}$) affect farmers by causing allergies and respiratory diseases (Aybek et al., 2010). For example, Arslan & Aybek (2011) mention that farmers operating harvesters without cabins are exposed to 18–41 mg m^{-3} PM. This is 2.4 times higher than the daily permissible allowance. While farmers inside the cabins of combine harvesters are exposed to only 0.2 to 2.5 mg m^{-3} of dust, which is 25% of the permissible levels of dust emissions. There are other procedures that follow the combining process. Grain carting, for example, is known to release a significant amount of PM and microbes into the air, estimated at 1 to 40 mg m^{-3}, which is 4 times the permissible air quality standards. Grain drying, on the other hand, releases 6 times more dust particles into the troposphere than what is considered permissible, estimated at 4 to 57 mg m^{-3} of air. It is interesting to note that milling and packing processes may also emit dust (PM_{10}) and cause problems for both the next crop and farmers (Aybek et al., 2009; Arslan & Aybek, 2011; Arslan et al., 2010; Lee et al., 2006).

8.5 PLANTATION PROCEDURES, DUST, AND MICROBE EMISSIONS INTO AGRARIAN SKY

Orchard operations related to soil tillage, interculture, pruning, spraying, and harvesting can be highly specific. They depend on factors such as terrain, soil type and characteristics, and the crop being grown. However, there are certain commonalities in terms of the presence of aeolian dust, microbes, and changes in tropospheric conditions that these operations can cause. Arslan & Aybek (2013) note that medium-sized tractors are often used for interculturing, spraying, and applying fertilizers, crop residues, and organic mulches from various sources. These activities release particulate matter (PM) into the air. For instance, pesticide spraying can result in the release of droplets, dust particles, toxic pollutants, and various species of microbes into the airspace above the trees. Measurements have shown droplets smaller than 100 μm, as well as fine organic particles with microbes ranging from 100 to 200 μm in size. These are commonly detected after spraying (CIGR Handbook of Agricultural Engineering, 1999). Agricultural spraying releases particles and moisture that are prone to easily drift in the atmosphere. As a result, it can serve as a source of pollutants and harmful microbes. Larger droplets containing microbes, however, tend to settle closer to the ground due to gravity.

In an orchard, not all dust/microbe emissions affect crops and farmers. For example, the respirable fraction of pollutants, dust, microbes, and moisture is only 0.05 mg m^{-3} of the total emission of 13 mg m^{-3} (Popendorf et al., 1982). Even then, a useful suggestion for farm workers and tractor operators is to use good air filters (e.g., carbon bed air filters) in the vehicle's cabin. Alternatively, they should wear masks to avoid contagious microbes found in the atmosphere. This helps prevent respiratory problems such as bronchitis, pneumonia, allergies, and congestion. Furthermore, in a farm with mid-season crops, attention should also be given to dust and irritants emanating from foliage. A drenching spray can settle a large volume of dust/microbes onto the soil surface.

Orchard procedures may induce microbes to levitate and localize at different altitudes in the troposphere. For example, harvesting procedures may alter the PM content of the troposphere. Consequently, the microbial population and diversity are also affected. There are disease-causing microbial pathogens that may drift in the air. Inhalable dust from orchards (e.g., apple, almond, peaches, grapes) may affect the health of farmers. The dust created due to farm operations may contain respiratory disease-causing microbes. Lee et al. (2004, 2006), for example, report that PM in dust plumes above almond orchards reached 154 mg m^{-3}. Mechanical harvesting released inhalable dust equivalent to 52.7 mg m^{-3}. During manual harvest, farm workers are exposed more to PM and microbes. This is because they are closer to the point where dust/PM emanates. Of course, dust emission may differ based on the crop species harvested.

Grapevine experts at the University of California, Riverside aimed to understand the effect of procedures such as copper application and mowing on the bacterial population over the grapevine canopy. They also aimed to understand the effect of ice-nucleating bacteria in creating frost on the vines. For example, the occurrence of *Pseudomonas aeruginosa,* an ice-nucleating bacteria, was prolific. Plant protection procedures like copper application reduced ice-nucleating bacteria by 10-fold compared to the control group (not sprayed with Cu). As a consequence, the occurrence of frost also reduced when copper was sprayed. Grapevines were able to withstand temperatures as low as –3°C in the ambient atmosphere.

McGourty (2017) examined grapevines that were mowed and unmowed. However, the interculture procedure, such as mowing, did not show differences in the bacterial population. Petri plates with selective medium (agar) were placed in both mowed and unmowed plots. In fact, they found that the bacterial population increased by 2-fold in plots that were mowed. They attribute this to the disturbance of soil, ground flora, and the crop's canopy caused by mowing machines. This action disperses bacteria in all directions, including towards the grapevine foliage and canopy. Mowing, along with chemicals like Cu preparations, is needed to reduce ice nucleating bacteria. This will help reduce any potential frost damage.

8.6 LIVESTOCK PRODUCTION AND AERIAL MICROBES

There are several reports about outdoor aerial microbial flora encountered in locations dealing with intense or moderately intense crop production. Such data on microflora is also available for dryland farming zones, arid zones with scanty natural vegetation, and crops. Researchers have examined the aeolian dust for particle sizes ranging from $PM_{2.5}$ to PM_{10}. The aim was to decipher the microbial population and diversity. For example, Ruiz-Gill et al. (2020) report the occurrence of *Proteobacteria, Actinobacteria, Firmicutes,* and *Bacteroidetes*. These bacterial phyla are common. In the context of this chapter, we are interested in outdoor aerospace above crops, particularly the microflora released due to farming procedures and crop production trends. However, in this section, we are also concerned with indoor aerospace above 'farm animal enclosures.' Farm animal enclosures may release a significant amount of microbes into the atmosphere.

The airspace within farm enclosures, where animals such as bovine, swine, ovine, and poultry are housed, cannot be compared to the open sky above agricultural fields. However, these enclosures do release a significant amount of PM_{10} and PM_5-sized dust into the surrounding airspace. The airspace above the animals in these enclosures often contains dust particles generated from various activities involved in animal rearing. It is possible that the air within these enclosures may harbor microbial species due to the activities taking place. Numerous studies have reported the presence of microbial species in the airspace above farm animals housed in enclosures. Additionally, we can find microbes and particulate matter derived from the soil in the ambient troposphere above grazing areas. Inside the enclosures, most of the microbes present are either pathogenic or commensal. The air carries dust particles primarily originating from the materials found on the floor. External sources such as feed, water, and other ingredients contribute to the presence of microbes from outside the animal enclosures. Some of these microbes may escape the enclosures and reach the surrounding troposphere.

Animal husbandry procedures also affect the microbial flora. Cattle farms emit dust and microbes into the air. The emissions depend on various factors such as cattle behavior, number, feed lots, type of shed, etc. The quality of the air in the feed lots is important for several reasons. For example, it may harbor commensals or pathogens. Microbes from the cattle shed may later be emitted into the ambient air. Let's consider an example. Wilson et al. (2002) used dust track equipment to monitor the air above feedlot pens that housed 1,400 cattle. They collected dust particles of size PM 2.5 during the peak emission period (i.e., in the afternoon). They found several non-pathogenic gram-positive bacteria and fungi. They suspect that gram-negative bacteria could be present but were not culturable in the selective medium used. Fungi were relatively low in population in the cattle shed air. The total population and diversity, it seems, depended on the type of feedlot pen and cattle activity in the region.

Livestock-related emission of microbes may need further attention from epidemiologists concerned with both human and crop diseases (Cassey et al., 1980; Gibbs et al., 2010). Cattle grazing in open pastures may carry, disperse, and induce fresh infection foci. In fact, cattle feed lots may harbor microbial pathogens that affect crops, pastures, and other vegetation found in the vicinity. The troposphere that receives these pathogens may be a source for new infections on crops.

Examinations of swine houses have revealed the release of dust particles of various sizes. The particle sizes ranged from 3.3 μm to 4.7 μm. Approximately 31% of bacterial colony farming units observed were attributed to 4.7 μm sized aeolian dust. Around 15% of bacteria were found residing in particles of 3.3 μm (Curtiss et al., 1975; Zucker et al., 2000; Predicala & Urban, 2002; Yuan et al., 2010; Gibbs et al., 2010; Jerez et al., 2011). White et al. (2019) suggest that the bacterial and fungal flora found in the aerosols above swine farms have not been sufficiently characterized in many regions. The microbial diversity of sediments found in the vicinity of swine houses requires attention as they may carry both human and plant pathogens. In their study, approximately 28 different bacterial species were identified, which could potentially spread into open crop production zones. *Vice versa*, i.e., microbes found in the aerosol above cropped fields may also settle into farmhouses.

Aerospace within the broiler enclosures supports different bacterial species (Quarles & Gentry, 1970). The aerospace is undoubtedly contaminated by harmful microbes such as coliforms (*Escherichia coli*) and respiratory disease-causing bacterial complexes like *Mycoplasma gallisepticum,* etc. (Carlson & Whetham, 1968; Bradford et al., 2013). Litter samples and air examined in farm enclosures supporting broilers revealed the presence of *Paecilomycesvarioti, Trichoderma spp., Aureobasidium pullulans,* and *Hyalodendronlignicola* (Dennis & Gee, 1973). Dust from the ground surface, released due to bird activity and farming procedures, caused the microbes to become airborne and enter the aerospace.

The extent to which microbe-laden dust emissions from farmhouses are responsible for crop, human, and farm animal diseases is still to be deciphered in many regions. The airborne movement of microbes from one farm to another is said to spread epidemics. Effective control measures actually depend on our knowledge about dust particles, their size, and composition. The microbial load and diversity that escape from livestock houses into the troposphere need attention (Zhao et al., 2011, 2014). Yang (2018) mentions that, in a European setting, farmhouses could emit less dust and microbes if scrubbers are placed at the points of emission (Aarnink et al., 2009; Ogink et al., 2007; Radon & Danuser, 2002). Scrubbers reduced aerial contamination by farm livestock houses. Furthermore, dust emission from broiler houses could be reduced if rape seed oil is used (Aarnink et al., 2008). The bacteria emitted into the tropospheric air were reduced by 46–85% due to the use of scrubbers. The emissions of PM (particulate matter) were also reduced, with reductions ranging from 61% to 93% for PM_{10} and from 47% to 90% for $PM_{2.5}$. Additionally, the emission of ammonia (NH3) was reduced by about 70% to 100% when scrubbers were used in livestock production (Jerez et al., 2011). The implementation of scrubbers significantly reduces the emission of pathogenic microbes into the troposphere, resulting in lower numbers of coliform bacteria such as *Escherichia coli, Enterobacter* sp., and others. This is particularly observed in the aerosols above and around livestock houses where scrubbers are utilized. The presence of aerosols with little to no density of pathogenic microbes is always preferred. Clearly, scrubbers have the ability to modify the microbial load in the troposphere to more desirable levels.

Animal production systems are among the sources of PM and microbes that reach different altitudes in the troposphere (Jager, 2005; Bradford et al., 2013). Furthermore, PM found on the floor or in the air within enclosures may be similar to a great extent. PM in piggery houses, cattle sheds, or poultry enclosures is derived from soil particles, bedding materials like hay, fecal matter, litter, and feed such as chopped forage and feed concentrates, among others. It is evident that farm workers are affected by the polluted air in the animal enclosures, with respiratory disorders being the most common microbe-caused diseases.

In a farm with both crops and animal enclosures, soil particles usually dominate the troposphere. Within animal enclosures, we may encounter both aerial microbes from the open field and those related to animal rearing. Typically, animal enclosures generate PM of a size that can be easily inhaled, leading to respiratory diseases (Reed et al., 2006; Lee et al., 2006). For instance, in poultry enclosures, farm workers may be exposed to inhalable dust laden with harmful bacteria, ranging from 0.02 to 81 mg m^{-3}. To reduce the fine fraction of inhalable dust, methods such as pelleting feed or applying oil to reduce dust formation are employed. Coating the feed with soybean oil is a common practice (Guarino et al., 2007; EPA, 2009, 2012a, b).

The agricultural field supports diverse microbes in the troposphere. Most often, the source of microbes at low altitudes is the soil surface layers, crops, and water bodies. Sometimes drifts, wind, and rain may bring in the new microbial population from elsewhere. There are indeed several agronomic procedures that disturb soil. They induce dust plumes of PM and allow them to levitate, to higher altitude. Farmers, no doubt, are often exposed to PM of different sizes ranging from fine PM_{25} to $PM_{2.5}$–PM_{10}. These carry microbes that can have consequences for the health of farmers, crops, and potentially impact other biotic factors such as animals, birds, and insects. Grain storage facilities, animal feed storage facilities, animal houses with feed lots, and other farm structures can also harbor a wide variety of microbes in the air. These microbes could be harmful. For instance, the population of dust-borne microbes ranged from 129.2 to 1289.9 × 10^3 m^{-3} of air. Flour dust was found in high quantities in the air. Bacteria were the most prevalent in the air above grain storage areas, while fungi and actinomycetes were less common. *Erwinia herbicola* gram-negative rods were the most common aerial microbes in grain bins. In animal farms, such as hatcheries, the range of aerial microbes was from 225.5 to 595.4 × 10^3 m^{-3}. In broiler houses, it was around 7751.5 × 10^3 m^{-3} of air. *Staphylococcus* was commonly found in the air, followed by *Corynebacter* and *Streptococci* (Dutkiewicz, 2013).

8.7 AGRO-INDUSTRIES, PARTICULATE MATTER (PM), AND MICROBES

Agricultural produce-based industries, especially the processing, manufacturing, and transport systems, release PM, organic dust, and microbes. The aerosol above these agro-industries requires attention as it often harbors microbes. Grain milling, flour mills, hullers, feed processing, shellers, and packing industries related to agro products can release large quantities of dust and microbes into the air. Facilities such as storage enclosures (grain storage), vegetable storage, market yards, and forage lots may also release microbes. The aerosol above these locations may harbor microbes corresponding to the activity taking place. For instance, in Turkey, the troposphere aerosol above cotton ginners had a higher content of fine particles derived from cotton. Arslan & Aybek (2011) note that while there are thresholds and standards stipulated for fine particles in general, there are none specifically for agro industries. The potency of respirable dust from agro-industry may vary depending on the encountered microbial pathogen and its virulence.

In general, the threshold for the total nuisance limit is 15,000 particles of size PM_{10}. For the lower respiratory system, it is 5,000 particles of $PM_{2.5}$. The nuisance limit for respirable dust from grain processing industries is 15,000 particles of PM_{10} and 5,000 particles of $PM_{2.5}$. For raw cotton around the ginning industry, it is 1,000 particles of PM_{10}. For spinning mills, it is 200 particles of $PM_{2.5}$, and for weaving locations, it is 750 particles of $PM_{2.5}$ (OSHA, 2010; Arslan & Aybek, 2013; Funk et al., 2019). Thresholds for PM are specific based on location and industry in the vicinity.

8.8 AERIAL FARM VEHICLES AND MICROBES IN TROPOSPHERE

All farm vehicles plying on the ground or field carry a certain microbial load on the surface. They may carry a higher load during specific operations such as soil tillage, interculture, inoculating with BNF inoculum, fertilizers, pruning, combine harvesting, etc. The population density and diversity of microbes on the surface of farm vehicles, as well as those distributed in the air and soil due to turbulence, need proper evaluation. We should also expect disease-causing organisms on the body surface of farm vehicles. A tractor moving from a disease-afflicted cropped field to another has the potential to rapidly spread the disease.

Here, we discuss an interesting forecast regarding vector-mediated diseases caused by microbes. It is important to note that it is not just viruses or bacterial vectors such as aphids, whiteflies, jassids, psyllids, or beetles that should be considered. In recent years, there has been a revolution in the agricultural industry, particularly in the use of small planes known as 'Drones or Unmanned Aerial Vehicles-UAVs' (*see* Krishna, 2015, 2018, 2022). Agricultural drones have the capability to capture high-resolution images of crops, regardless of the time of year, season, or day/night. These images provide valuable spectral data that can be used to detect nutritional and/or moisture deficiencies, as well as early signs of crop diseases. Additionally, agricultural drones can be used to spray plant protection chemicals on disease-affected crops. However, it is important to note that these very same small aircraft, whether fixed-winged or copters, could also serve as a source of crucial microbial inoculum (pathogens). This microbial inoculum could later lead to devastating crop losses. As the usage of drones becomes more common, these aerial machines could potentially be responsible for crop disasters. Drones have the potential to spread epidemics or create pandemic conditions by carrying microbial pathogens from one area to another. Therefore, it is crucial to maintain good sanitary practices to ensure that these aerial vehicles remain free of any microbial contamination. One approach is to regularly employ sprayer drones to apply plant protection chemicals, fungicides, antibiotics, and insecticides, effectively disinfecting the agricultural environment. However, before implementing this method, it is essential to ensure that the drones and their surfaces are free from microbes.

Currently, sprayer drones equipped with pesticide/fungicide tanks and sprayer bars are being rapidly introduced in various farming regions. These drones distribute water, formulations, and any microbes present in them. Often, the microbes found on the body, wings, propellers, and other parts of the drones are dispersed into the troposphere. Small agricultural drones typically fly at altitudes ranging from 100 to 500 feet above the crop canopy. Consequently, microbes present on the drone's surface and in the tank's formulations/water can be carried to a small altitude within the troposphere. This can disturb the existing microbes in the air. Additionally, the air turbulence created by the drone's movement can further spread the microbes already present in the troposphere. Some of these microbes may also settle on the soil surface due to gravity.

There are agricultural drones that cover short distances. There are others that traverse a path to cover 50 to 500 km. Drones cover a large number of hectares of cropped land. Agricultural drones are also used to cover several kilometers of farmland. Their potential to distribute the microbial inoculum (plant pathogens) is proportional to the distance they can endure/cover during flight.

KEYWORDS

- **aerial microbes**
- **agents**
- **agro industry**
- **agronomy**
- **combine harvesting**
- **dry lands**
- **emissions**
- **farm procedures**
- **feed lots**
- **land preparation**
- **livestock production**
- **plant pathogens**
- **post-harvest**
- **respiratory diseases**
- **ridging**
- **seed processing**
- **soil erosion**
- **soil tillage**
- **winnowing**

REFERENCES

Aarnink, A. J. A., Zhao, Y., Van, H. T., & Hol, A., (2009). *Reduction of Emission of Fine Dust by a Uniqfill Multi-Stage Scrubber* (p. 19). Wageningen UR Livestock Research, Lelystad, The Netherlands.

Aarnink, A., Van, H. J., Van, H. T., Zhao, Y., & Ellen, H., (2008). Dust reduction in broiler houses by spraying rapeseed oil. In: *The Eighth International Symposium of Livestock Environment, Iguassu Falls, Brazil* (pp. 133–138).

Acosta-Martinez, V., & O'Brien, D., (2013). *Travel Patterns of Wind-Blown Soil Microbes*. United States Department of Agriculture Research Service. https://www.researchgate.net/publication/280072359_Travel_Patterns_of_Wind-Blown_Soil_Microbes/ (accessed on 19 January 2024).

Ahmadirad, S., Gholamhoseini, R. M., Ghalavand, A. G., Asgharzadeh, & Dolatabadian, A., (2016). *Canola Grown Under Different Nitrogen Regions*. https://doi.org/10.1016/j.rhisph.2016.08.006/.

Akinde, S. B., Sunday, A. A., Adeyemi, F. M., Fakayode, J. B., Oluwajide, O. O., Adebunmi, A. A., Oloke, J. K., & Adebooye, C. E., (2016). Microbes in irrigation water and fresh vegetables: Potential pathogenic bacteria assessment and implications for food safety. *Journal of ABSA International, 21*(2), 89–97. doi: 10.1177/1535676016652231/.

Animal Hygiene and Sustainable Livestock Production, (pp. 355–359). Vienna, Austria. https://www.isah-soc.org/userfiles/downloads/proceedings/2011_Proceeding1.pdf (accessed on 19 January 2024).

Arslan, S., & Aybek, A., (2011). PM10, PM2.5 and PM1 concentrations in cotton ginners. In: *13th International Congress and Agricultural Mechanization and Energy, Istanbul, Turkey, Tarim Makinalan Bilimi Dergisis (Journal of Agricultural Machinery Science), 7,* 113–119.

Arslan, S., & Aybek, A., (2013). *Particulate Matter Exposure in Agriculture* (pp. 1–33). Intech Open. http://dx.doi.org/10.5772/50084/.

Arslan, S., Aybek, A., & Ekerbiçer, H., (2010). Measurement of personal PM10, PM2.5 and PM1.0 exposures in tractor and combine operations and evaluation of health disturbances of operators. *Journal Agricultural Sciences, 16*, 104–115.

Aybek, A., Arslan, S., & Ekerbiçer, H., (2010). *Determination of Particulate Matter Concentrations and Health Risks in Agriculture and Agri-industry*. Final Report (In Turkish), Turkish Scientific and Technological Council Project No: 107 O 513.

Aybek, A., Arslan, S., & Genç, Ş., (2009). *The Effect of PM10 and PM2.5 Pollution on Feed Millworkers.* (IAEC Ref. 250, available on CD), International Agricultural Engineering Conference, Bangkok, Thailand.

Blanco-Canqui, H., Lal, R., Post, W. M., Izaurralde, R. C., & Shipitalo, J., (2006). Organic carbon influences on soil particle density and rheological properties. *Soil Science Society of America Journal, 70*, 1407–1414. https://doi.org/10.2136/sssaj2005.0355/.

Bogman, P., Cornelis, W., Rollé, H., & Gabriels, D., (2007). *Prediction of TSP and PM10 Emissions from Agricultural Operations in Flanders, Belgium.* Dust Conf 2007, Maastricht, Netherlands. https://www.researchgate.net/publication/254715564_Prediction_of_TSP_and_PM10_emissions_from_agricultural_operations_in_Flanders_Belgium (accessed on 19 January 2024).

Borcardt, M. A., & Burch, T. R., (2016). *Airborne Pathogens from Dairy Manure Aerial Irrigation and Human Health Risk – Soil Science Extension (wisc.edu)* (pp. 1–4). United States Department Agriculture, U.S. Dairy Forage Research Centre. https://www.ars.usda.gov/ARSUSERFILES/50901500/NEWSLETTERS/2016SPRING/PDF/36MANUREPATHOGENS.PDF/ (accessed on 19 January 2024).

Boreson, J., Dillner, A. M., & Peccia, J., (2004). Correlation bioaerosol load with PM 2.5 and PM 10 concentrations: A comparison between natural desert and urban fringe aerosols. *Atmospheric Environment, 38*, 6029–6041.

Bradford, S. A., Morales, V. L., Zhang, W., Harvey, R. W., Packman, A. I., Mohanram, A., & Welty, C., (2013). Transport and fate of microbial pathogens in agricultural settings. *Critical Reviews in Environmental Science and Technology, 43*, 775–893. doi: 10.1080/10643389.2012.710449/.

Café-Filho, A. C., Lopes, C. A., & Rossato, M., (2018). *Management of Plant Disease Epidemics with Irrigation Practices*. http://dx.doi.org/10.5772/intechopen.78253/.

Carlson, H. C., & Whentham, C. R., (1968). Coliform bacteria in chicken broiler house dust and their possible relationship to coli-septicemia. *Avian Diseases, 12*(2), 297–302.

Cassel, T., Trzepla-Nabaglo, K., & Flocchini, R., (2003). *PM10 Emission Factors for Harvest and Tillage of Row Crops.* International Emission Inventory Conference 'Emission

Inventories – Applying New Technologies.' San Diego. https://www3.epa.gov/ttn/chief/conference/ei12/poster/cassel.pdf/ (accessed on 19 January 2024).
Cassey, J. A., Curriero, F. C., Cosgrove, S. E., Nachman, K. E., & Schwartz, B. S., (1980). High-density livestock operations, crop field application of manure, and risk of community-associated methicillin-resistant staphylococcus aureus infection in Pennsylvania. *JAMA International Medicine, 173*(21), 1980–1990. doi, 10.1001/jamainternmed.2013.10408/.
Chen, W., Tong, D. Q., Zhang, S., Zhang, X., & Zhao, H., (2017). Local PM 10 and PM 2.5 emission inventories from agricultural tillage and harvest in North-Eastern China. *Journal of Environmental Science, 57*, 15–23.
Chen, W., Tong, D., Zhang, S., Dan, M., Zhang, X., & Zhao, H., (2015). Temporal variability of atmospheric particulate matter and chemical composition during a growing season at an agricultural site in northeastern China. *Journal of Environmental Science,* 38. 133–141.
CIGR Handbook of Agricultural Engineering, (1999). *Plant Production Engineering* (Vol. III, p. 660). Edited by CIGR—The International Commission of Agricultural Engineering. Editor: Bill A. Stout, Co-Editor: Bernard Cheze, Published by ASAE. https://www.scribd.com/document/314884929/Cigr-Handbook-Vol-3 (accessed on 19 January 2024).
Curtiss, S. E., Drummond, D. J., & Grunloh, D. J., (1975). Relative and qualitative aspects of aerial bacteria and dust in swine houses. *Journal of Animal Science, 41*, 1512–1520; doi: 10.2527/jas1975.4151512x. (March 24th, 2024)
Dennis, C., & Gee, J. M., (1973). Microflora of broiler house litter and dust. *Journal of General Microbiology, 78*, 101–107.
Donnison, A., Ross, C., Noonan, M., Fisher, G., & Waller, S., (2004). Bacterial survival and dispersal in spray irrigation aerosols. *New Zealand Journal of Agricultural Research, 47*, 575–585.
Dutkiewicz, J., (2013). Exposure to dust bacteria in agriculture 1. Environmental studies. *Archives of Environmental Health- An International Journal, 33*, 250–259.
EPA, (2009). *Air Quality Index: A Guide to Air Quality and Your Health.* EPA-456/F-09.
EPA, (2012a). *Particulate Matter.* http://epa.gov/ncer/science/pm/ (accessed on 19 January 2024).
EPA, (2012b). *Agriculture: Animal Feeding Operations.* http://epa.gov/ncer/science/pm/ (accessed on 19 January 2024).
Erisman, J. W., Bleeker, A., Hensen, A., & Vermeulen, A., (2007). Agricultural air quality in Europe and the future perspectives. *Atmospheric Environment, 42*, 3209–3217.
Frampton, R. A., Pitman, A. R., & Fineran, P. C., (2012). Advances in bacteriophage-mediated control of plant pathogens. *International Journal of Microbiology, 2012*, 326452.doi: 10.1155/2012/326452/.
Friese, A., Schulz, J., Laube, H., Von, C. S., Hartung, J., & Roesler, U., (2013). *Faecal Occurrence and Emissions of Livestock-Associated Methicillin-Resistant Staphylococcus Aureus (laMRSA) and ESbl/AmpC Producing E. coli from Animal Farms in Germany, 126*, 175–180.
Funk, R., Papke, N., & Hör, B., (2019). Wind tunnel tests to estimate PM10 and 706 PM2.5-emissions from complex substrates of open-cast strip mines in Germany. *Aeolian Research, 39*, 23–32. https://doi.org/10.1016/j.aeolia.2019.03.003/.
Funk, R., Reuter, H. I., Hoffmann, C., Engel, W., & Öttl, D., (2008). Effect of moisture on fine dust emission from tillage operations on agricultural soils. *Earth Surface Processes and Landforms Journal Geomorphology Research Group, 33*, 1851–1863.

Gaffney, P., & Yu, H., (2003). *Computing Agricultural PM10 Fugitive Dust Emissions Using Process Specific Emission Rates and GIS* (pp. 1–10). US EPA Annual Emission Inventory Conference, San Diego, California, USA. https://www3.epa.gov/ttnchie1/conference/ei12/fugdust/yu.pdf (accessed on 19 January 2024).

Gibbs, S. G., Green, C. F., Tarwater, P. M., Mota, L. C., Mena, K. D., & Scarpino, P. V., (2006). Isolation of antibiotic-resistant bacteria from the air plume downwind of a swine confined or concentrated animal feeding operation. *Environ. Health Perspectives, 114*(7), 1032−1037.

Goossens, D., (2004). *Wind Erosion and Tillage as a Dust Production Mechanism on North European Farmland* (pp. 15–40). Wind Erosion. Dust Dynamics. Observations and Simulations. Model. ESW Publication. Department. Environmental. Science. Erosion. Soil Water Conservation. Group, Wageningen. University. Wageningen.

Guarino, M., Jacobson, L. D., & Janni, K. A., (2007). Dust reduction from oil-based feed additives. *Applied Engineering in Agriculture, 23*(3), 329–332.

Hartmann, A., Amoureux, L., Locatelli, A., Depret, G., Jolivet, C., Gueneau, E., & Neuwirth, C., (2012). Occurrence of CTX-M producing *Escherichia coli* in soils, cattle, and farm environment in France (Burgundy Region). *Front. Microbial., 3.* https://doi.org/10.3389/fmicb.2012.00083.

Hinz, T., (2007). *Particulate Matter in and from Agriculture*. Bundesforschungsanstalt für Landwirtschaft (FAL). Get full web address.

Hoffmann, C., & Funk, R., (2015). Diurnal changes of PM10-emission from arable soils in NE-Germany. *Aeolian Research,* 17, 117–127.

Holmén, B. A., James, T. A., Ashbaugh, L. L., & Flocchini, R. G., (2001a). LIDAR assisted measurement of PM10 emissions from agricultural tilling in California's San Joaquin Valley – part I: Lidar. *Atmospheric Environment,* 35, 3251–3264.

Holmén, B. A., James, T. A., Ashbaugh, L. L., & Flocchini, R. G., (2001b). LIDAR assisted measurement of PM10 emissions from agricultural tilling in California's San Joaquin Valley – Part II: Emission factors. *Atmospheric Environment,* 35, 3265–3277.

HSE, (1998). *Grain Dust.* Guidance Note EH66. https://www.hse.gov.uk/pubns/eh66.htm (accessed on 19 January 2024).

HSE, (2005). *Respiratory Protective Equipment at Work: A Practical Guide*. Series Code: HSG53. https://www.hse.gov.uk/pubns/books/hsg53.htm (accessed on 19 January 2024).

HSE, (2007). *Controlling Grain Dust on Farms*. Agriculture Information Sheet No 3 (rev).

Isard, S. A., & Gage, S. H., (2001). *Flow of Life in the Atmosphere: An Airscape Approach to Understanding Invasive Organisms* (p. 304). Michigan State University Press, Detroit, MI, USA.

Jagdale, S., & Kapudnis, B., (2012). Bacteriophage liquid formulation: A potential green tool for the management of pomegranate bacterial blight. *Biological Control, 158*, 104597 https://doi.org/10.1016/j.biocontrol.2021.104597 /.

Jager, A. C., (2005). *Exposure of Poultry Farm Workers to Ammonia, Particulate Matter and Microorganisms in the Potchefstroom District, South Africa* (p. 78). MSc Dissertation, North- West University, South Africa. https://repository.nwu.ac.za/handle/10394/988/ (accessed on 19 January 2024).

Jerez, S. B., Zhang, Y., & Wang, X., (2011). Spatial and temporal distributions of dust and ammonia concentrations in a swine building. *Transactions of the ASABE, 54*(5), 1873–1891.

Jimenez, J. R., (2006). *Aerosol Characterization for Agricultural Field Burning Smoke* (p. 141). PhD Dissertation. Washington State University. https://rex.libraries.

wsu.edu/esploro/outputs/doctoral/Aerosol-characterization-for-agricultural-field-burning/99900877131201842/ (accessed on 19 January 2024).
Jones, A. M., & Harrison, R. M., (2004). The effects of meteorological factors on atmospheric bioaerosol concentrations—A review. *Science and Total Environment, 326*, 151–180.
Jones, J. B., Vallad, G. E., Iriarte, F. B., Obradovic, A., Wernsing, M. H., Jackson, L. E., Blogh, B., et al., (2012). Considerations for using bacteriophages for plant disease control. *Bacteriophage, 24,* 208–214.
Kasumba, J., Holmen, B. A., Hiscox, A., Wang, J., & Miller, D., (2011). Agricultural PM10 emissions from cotton field disking in Las Cruces, NM. *Atmospheric Environment,* 45, 1668–1674.
Katznelson, E., & Teltch, B., (1976). Dispersion of enteric bacteria by spray irrigation. *Journal of Water Pollution; Control Federation, 48*, 710–716. https://www.jstor.org/stable/i25038552 (accessed on 19 January 2024).
Kirkhorn, S. R., & Garry, V. F., (2000). Agricultural lung diseases. *Environmental Health Perspectives 108: Supplement, 4*, 705–712.
Krishna, K. R., (2015). *Agricultural Drones: A Peaceful Pursuit* (p. 486). Apple Academic Press Inc., New Jersey, USA.
Krishna, K. R., (2018). *Unmanned Arial Vehicles Systems in Crop Production* (p. 698). Apple Academic Press Inc. Palm Bay, Florida, USA.
Krishna, K. R., (2022). *The Agricultural Sky: A Concept to Revolutionize Farming* (p. 486). Apple Academic Press inc. Palm Bay, Florida, USA.
Lang, J. M., Gent, D. H., & Schwartz, H. F., (2007). Management of *Xanthomonas* leaf blight of onion with bacteriophages and a plant activator. *Plant Disease, 91*(7), 871–878.
Lee, K., Lawson, R. J., Olenchock, S. A., Vallyathan, V., Southard, R. J., Thorne, P. S., Saiki, C., & Schenker, M. B., (2004). Personal exposures to inorganic and organic dust in manual harvest of California citrus and table grapes. *Journal of Occupational and Environmental Hygiene, 1*, 505–514.
Lee, S. A., Adhikari, A., Grinshpun, S. A., McKay, R., Shukla, R., & Reponen, T., (2006). Personal exposure to airborne dust and microorganisms in agricultural environments. *Journal of Occupational and Environmental Hygiene, 3*, 118–130.
Lighthart, B., & Shaffer, B. T., (1994). Bacterial flux from chaparral into the atmosphere in mid-summer at a high desert location. *Atmospheric Environment, 28*, 1267–1274.
Lighthart, B., (1984). Microbial Aerosols: Estimated contribution of combine harvesting to an airshed. *Applied Environmental Microbiology, 47*, 430–432.
Lighthart, B., (1997). The ecology of bacteria in the alfresco atmosphere. *FEMS Microbial Ecology, 23*, 263–274.
Madden, N. M., Southard, R. J., & Mitchell, J. P., (2008). Conservation tillage reduces PM10 emissions in dairy forage rotations. *Atmospheric Environment, 42*, 3795–3808.
Marco, G. M., & Stall, R. E., (1983). Control of bacterial spot of pepper initiated by strains of *Xanthomonas campestris* pv. *vesicatoria* that differ in sensitivity to copper. *Plant Disease, 67*, 779–781. http://dx.doi.org/10.1094/PD-67-/.
McGaurty, G., (2017). *Frost, Cover Crops and Ice Nucleating Bacteria* (pp. 1–4). University of California, Riverside. https://www.lodigrowers.com/frost-cover-crops-ice-nucleating-bacteria/ (accessed on 19 January 2024).
Meyer, E., Gastmeier, P., Deja, M., & Schwab, F., (2013). Antibiotic consumption and resistance: Data from Europe and Germany. International. *Journal of Medical Microbiology: Special Issue Antibiotic Resistance, 303*, 388–395. https://doi.org/10.1016/j.ijmm.2013.04.004/.

Münch, S., Papke, N., Thiel, N., Nübel, U., Siller, P., Roesler, U., Biniasch, O., Funk, R., & Amon, T., (2020). Effects of farmyard manure application on dust emissions from arable soils. *Atmospheric Pollution Research, 11*, 1610–1624. doi: https://doi.org/10.1016/j.apr.2020.06.007./.

Nega, A., (2014). Review on concepts in biological control of plant pathogens. *Journal of Biology, Agriculture and Healthcare, 4*, 1–33.

Nerger, R., Funk, R., Cordsen, E., & Fohrer, N., (2017). Application of a modeling approach to designate soil and soil organic carbon loss to wind erosion on long-term monitoring sites (BDF) in Northern Germany. *Aeolian Research, 25*, 135–147.

Nieuwenhuijsen, M. J., Kruize, H., & Schenker, M. B., (1998). Exposure to dust and its particle size distribution in California agriculture. *American Industrial Hygiene Association Journal, 58*, 34–38.

O'Leary, T., (2017). *How to Identify Microbes that Contaminate my Fire Sprinkler System* (pp. 1–6). Hugeonot Laboratories, USA. https://www.linkedin.com/pulse/how-identify-microbes-contaminate-my-fire-sprinkler-system-o-leary (accessed on 19 January 2024).

Ogink, N. W. M., & Hahne, J., (2007). Removal of dust fractions by air scrubbers in livestock operations. In: *Proceedings of DustConf. 2007, International Conference in Maastricht, Netherlands* (pp. 1–12).

OSHA, (2010). *Occupational Safety and Health Standards* (p. 45). Available at: http://www.osha.gov/ (accessed on 19 January 2024).

Öttl, D., & Funk, R., (2007). PM emission factors for farming activities by means of dispersion modeling. *Landbauforsch. Völkenrode., 308*, 173–177.

Öttl, D., Funk, R., & Sturm, P., (2005). PM emission factors for farming activities. In: *Proceedings of the 14th Symposium Transport and Air Pollution* (pp. 1–36). Technical University Graz, Austria.

Peccia, J., & Hernandez, M., (2006). Incorporating polymerase chain reaction-based identification, population characterization, and quantification of microorganisms into aerosol science: A review. *Atmospheric Environment, 40*, 3941–3961.

Peterfalvi, N., Keller, B., & Magyar, M., (2018). PM10 emission from crop production and agricultural soils. *Agro chemistry and Soil Science, 67*, 143–159. doi: 10.1556/0088.2018.67.1.10/.

Popendorf, W. J., Pryor, A., & Wenk, H. R., (1982). Mineral dust in manual harvest operations. *Annals of American Conference Government Industrial Hygiene, 2*, 101–115.

Predicala, B. Z., & Urban, J. E., (2002). Assessment of bioaerosols in swine barns by filtration and impaction. *Current Microbiology, 44*, 136–140.

Quarles, C. L., & Gentry, R. F., (1970). Bacterial contamination in poultry houses and its relationship to egg hatchability. *Poultry Science, 49*(1), 60–66.

Radon, K., & Danuser, B., (2002). Air contaminants in different European farming environments. *Annals of Agricultural and Environmental Medicine, 9*, 41–48.

Rebekah, A. F., Andrew, R. P., & Peter, C. F., (2012). Advances in bacteriophage-mediated control of plant pathogens. *International Journal of Microbiology* (p. 11). Article ID 326452. Hindawi Publishing Corporation. doi: 10.1155/2012/326452/.

Reed, S., Quartararo, M., Kift, R., Davidson, M., & Mulley, R., (2006). *Respiratory Illness in Farmers Dust and Bioaerosols Exposures in Animal Handling Facilities* (p. 72). A report for the Rural Industries Research and Development Corporation, RIRDC Publication No 06/107.

Rühlmann, J., Körschens, M., & Graefe, J., (2006). A new approach to calculate the particle density of soils considering properties of the soil organic matter and the mineral matrix. *Geoderma, 130*, 272–283. https://doi.org/10.1016/j.geoderma.2005.01.024//.

Ruiz-Gill, T., Acuna, J. A., Fujiyoshi, S., Jun, N., D. T., Maruyama, F., & Jorquera, A., (2020). Airborne bacterial communities of outdoor environments and their associated influencing factors. *Environment International, 145*, 106156. https://doi.org/10.1016/j.envint.2020.106156//.

Schaffer, B. T., & Lighthart, B., (1997). Survey of culturable airborne bacteria at four diverse locations in Oregon: Urban, rural, forest, and coastal. *Microbial Ecology, 34*, 167–177.

Schenker, M. B., (2000). Exposures and health effects from inorganic agricultural dusts environmental health perspectives. *Occupational and Environmental Lung Diseases, 108*, 661–664.

Schenker, M. B., (2004). Personal exposures to inorganic and organic dust in manual harvest of California citrus and table grapes. *Journal of Occupational and Environmental Hygiene, 1*, 505–514. https://doi.org/10.1080/15459620490471616/.

Schlatter, D. C., Schillingerr, W. F., Bary, A. I., Sharrat, B., & Paulitz, T. C., (2018). Dust-associated microbiomes from dryland wheat fields differ with tillage practice and biosolids application *Atmospheric Environment, 185*, 29–40.

Sever, M., (2021). *Wildfire Smoke Seeds the Air with Potentially Dangerous Microbes* (pp. 1–7). Science News Explores. https://www.snexplores.org/article/wildfire-smoke-seeds-the-air-with-potentially-dangerous-microbes/(accessed on 19 January 2024).

Shafi, J., Tian, H., & Mingshan, J., (2017). Bacillus species as versatile weapons for plant pathogens: A review. *Biotechnology and Biotechnological Equipment, 31*(3), 446–459. doi: 10.1080/13102818.2017.1286950/.

Spankie, S., & Cherrie, J. W., (2012). Exposure to grain dust in Great Britain. *Annals of Occupational Hygiene* (Vol. 56, No. 1, pp. 25–36).

Stefani, E., Obradovi´c, A., Gaši´c, K., Altin, I., Nagy, I. K., & Kovács, T., (2021). Bacteriophage-mediated control of phytopathogenic xanthomonads: A promising green solution for the future. *Microorganisms, 9*, 1056. https://doi.org/10.3390/microorganisms9051056/.

Swanepoel, A. J., Rees, D., Renton, R., Swanepoel, C., Kromhout, H., & Gardiner, K., (2010). Quartz exposure in agriculture: Literature review and South African survey. *Annals of Occupational Hygiene, 54*(3), 281–292.

The Swedish National Board of Occupational Safety and Health. (1994). *Organic Dust in Agriculture* (Vol. 11, p. 21). General Recommendations of the Swedish National Board of Occupational Safety and Health on Organic Dust in Agriculture, AFS. https://www.google.co.in/books/edition/Organic_Dust_in_Agriculture/smtetwAACAAJ?hl=en/(accessed on 19 January 2024).

Torigoe, K., Hasegawa, S., Numata, O., Yazaki, S., Matsunaga, M., Boku, N., Hiura, M., & Ino, H., (2000). Influence of emission from rice straw burning on bronchial asthma in children. *Pediatrics, 42*, 143.

USDA Natural Resources Conservation, (2007). *Fugitive Dust: A Guide to the Control of Windblown Dust on Agricultural Lands in Nevada* (pp. 1–32). United States Department of Agriculture, Reno, Nevada.

Usta, C., (2013). *Microorganisms in Biological Pest Control: A Review* (pp. 287–317). Intech Open. http://dx.doi.org/10.5772/55786/.

Wang, J., Miller, D. R., Sammis, T. W., Hiscox, A. L., Yang, W. L., & Holmén, B. A., (2010). Local dust emission factors for agricultural tilling operations. *Soil Sci.,* 175(4), 194–200.

White, J. K., Nielsen, J. L., & Madsen, A. M., (2019). Microbial species and biodiversity in settling dust within and between pig farms. *Environmental Research, 171*, 558–567. https://doi.org/10.1016/j.envres.2019.01.008/.

Wilson, C. S., Morrow-Tesch, J., Straus, D. C., & Cooley, J. D., (2002). Airborne microbial flora in a cattle feedlot. *Applied and Environmental Microbiology, 68*, 3238–3242. Doi: 10.1128/AEM.68.7.3238–3242.2002/.

World Health Organization, (1999). *Hazard Prevention and Control in the Work Environment: Airborne Dust.* Geneva- WHO/SDE/OEH/99.14.

World Health Organization, (2000). *Air Quality Guidelines for Europe* (2nd edn., p. 91). WHO Regional Publications. European series.

Yadav, R. S., (2019). *Stubble Burning: A Problem for the Environment, Agriculture and Humans* (pp. 1–4). Down to Earth. https://www.downtoearth.org.in/blog/agriculture/stubble-burning-a-problem-for-the-environment-agriculture-and-humans-64912 (accessed on 19 January 2024).

Yang, Z., (2018). *Measuring Airborne Microorganisms and Dust from Livestock Houses* (p. 218). PhD Thesis, Wageningen University, Wageningen, the Netherlands. https://www.wur.nl/en/show/Measuring-Airborne-Microorganisms-and-Dust-from-Livestock-Houses.htm(accessed on 19 January 2024).

Yuan, W., Chai, T. J., & Miao, Z. M., (2010). ERIC-PCR identification of the spread of airborne *Escherichia coli* in pig houses. *Sci. Total Environ., 408*, 1446−1450.

Zhao, Y., Aarnink, A. J. A., De Jong, M. C. M., & Groot, K. P. W. G., (2011). Measuring airborne microorganisms and dust from livestock production systems. In: *XV ISAH Congress, 2011*.

Zhao, Y., Aarnink, A. J. A., De Jong, M. C. M., & Groot, K. P. W. G., (2014). Airborne microorganisms from livestock production systems and their relation to dust. *Critical Reviews in Environmental Science and Technology, 44*(10), 1071−1128. https://doi.org/10.1080/10643389.2012.746064/.

Zucker, B. A., Trojan, S., & Muller, W., (2000). Airborne gram-negative bacterial flora in animal houses. *Journal of Veterinary Medicine, Series B, 47*, 37–46. doi: 10.1046/j.1439-0450.2000.00308.x/.

Index

B

C

F

Q

R

S